MW01274910

Saïd Abbas, Mouffak Benchohra, John R. Graef, Johnny Henderson
**Implicit Fractional Differential and Integral Equations**

# De Gruyter Series in Nonlinear Analysis and Applications



## Volume 26

Saïd Abbas, Mouffak Benchohra, John R. Graef,
Johnny Henderson

# Implicit Fractional Differential and Integral Equations

Existence and Stability

DE GRUYTER

**Mathematics Subject Classification 2010**
Primary: 26A33, 34A08, 34A37, 34A60, 34B15, 34B37, 34D20, 34K05, 34K20, 34K32, 34K37, 35B35, 35R11, 39A12, 39A30, 45B05, 45D05, 45K05, 45M10

**Authors**
Prof. Dr. Saïd Abbas
Tahar Moulay University of Saïda
Laboratory of Mathematics
Geometry, Analysis, Control and Applications
P.O. Box 138, EN-Nasr
20 000 Saïda
Algeria

Prof. Dr. Mouffak Benchohra
Djillali Liabes University of Sidi Bel-Abbès
Laboratory of Mathematics
P.O. Box 89
22 000 Sidi-Bel-Abbes
Algeria

Prof. Dr. John R. Graef
University of Tennessee at Chattanooga
Department of Mathematics
415 EMCS Building
615 McCallie Ave
Chattanooga, TN 37 403
USA

Prof. Dr. Johnny Henderson
Baylor University
Department of Mathematics
97 328 One Bear Place
Waco, TX 76 798-7328
USA

ISBN 978-3-11-055313-0
e-ISBN (PDF) 978-3-11-055381-9
e-ISBN (EPUB) 978-3-11-055318-5
Set-ISBN 978-3-11-055382-6
ISSN 0941-813X

**Library of Congress Cataloging-in-Publication Data**
A CIP catalog record for this book has been applied for at the Library of Congress.

**Bibliographic information published by the Deutsche Nationalbibliothek**
The Deutsche Nationalbibliothek lists this publication in the Deutsche Nationalbibliografie; detailed bibliographic data are available on the Internet at http://dnb.dnb.de.


Typesetting: le-tex publishing services GmbH, Leipzig
Printing and binding: CPI books GmbH, Leck
♾ Printed on acid-free paper
Printed in Germany

www.degruyter.com

MIX
Papier aus verantwortungsvollen Quellen
FSC
www.fsc.org
FSC® C083411

Saïd Abbas dedicates this book to the memory of his father, Abdelkader Abbas. Mouffak Benchohra makes his dedication to the memory of his father, Yahia Benchohra. John R. Graef dedicates the book to his wife, Frances. Johnny Henderson dedicates the book to the memory of his advisor, Professor Lloyd K. Jackson.

# Preface

Fractional differential and integral equations generalize integrals and derivatives to noninteger orders. They were found to play a fundamental role in the modeling of a considerable number of phenomena in many areas, including electrical engineering, electronics, telecommunications, electromagnetism, electrochemistry, thermal engineering, mechanics, mechatronics, rheology, automatic control, robotics, signal processing, image processing, biology, biophysics, physics, and economics. The purpose of this book is to present detailed results about the existence and stability of various classes of fractional differential and integral equations and inclusions involving the Caputo fractional derivative and the Hadamard fractional integral. Some equations contain delays that may be finite or infinite. Others are subject to an impulsive effect. The tools used include classical fixed point theorems and new ones such as Darbo's and Mönch's fixed point theorems, as well as some Gronwall and Pachpatte lemmas. Each chapter concludes with a section devoted to notes and bibliographical remarks; all abstract results are illustrated with examples.

The content of this book complements the existing literature in fractional calculus. It is useful for researchers and graduate students for research, seminars, and advanced graduate courses in pure and applied mathematics, engineering, biology, and all other applied sciences.

We owe a great debt of gratitude to E. Alaidarous, W. A. Albarakati, S. Bouriah, M. A. Darwish, E. M. Hilal, J. E. Lazreg, G. M. N'Guérékata, J. J. Nieto, A. Petrusel, S. Sivasundaram, M. S. Souid, and J. J. Trujillo for their collaboration in research related to the problems considered in this book. We express our deepest thanks to the De Gruyter Mathematics Acquisitions Editor, Dr. Apostolos Damialis, and to the De Gruyter Project Editors, Konrad von Brück and Nadja Schedensack, for their combined attention and encouragement in getting this monograph to final publication.

Saïda, Algeria — S. Abbas
Sidi Bel Abbes, Algeria — M. Benchohra
Chattanooga, Tennessee, USA — J. R. Graef
Waco, Texas, USA — J. Henderson

https://doi.org/10.1515/9783110553819-201

# Contents

# Introduction

Fractional calculus is a generalization of differentiation and integration to the arbitrary (noninteger) order fundamental operator $D^{\alpha}_{a+}$, where $\alpha$, $a$, $\in \mathbb{R}$. The concept of fractional differential and integral equations has a long history. One may wonder what meaning may be ascribed to the derivative of a fractional order, that is, $\frac{d^n y}{dx^n}$, where $n$ is a fraction. In fact, the French mathematician l'Hôpital himself considered this very possibility in a correspondence with Leibniz. In 1695, in a letter to l'Hôpital, Leibniz raised the following question: *Can the meaning of derivatives with integer order be generalized to derivatives with noninteger orders?* l'Hôpital was somewhat curious about that question and replied with another question to Leibniz: "*What if the order is* $\frac{1}{2}$?" In a letter dated September 30, Leibniz replied: "$d^{\frac{1}{2}} x$ *would be equal to* $x\sqrt{dx : x}$. *This is an apparent paradox from which, one day, useful consequences will be drawn.*" Thus, September 30, 1695, marks the exact date of birth of the *fractional calculus*! Therefore, the fractional calculus has its origin in the works of Leibniz, l'Hôpital (1695), Bernoulli (1697), Euler (1730), and Lagrange (1772). Some years later, Laplace (1812), Fourier (1822), Abel (1823), Liouville (1832), Riemann (1847), Grünwald (1867), Letnikov (1868), Nekrasov (1888), Hadamard (1892), Heaviside (1892), Hardy (1915), Weyl (1917), Riesz (1922), P. Levy (1923), Davis (1924), Kober (1940), Zygmund (1945), Kuttner (1953), J. L. Lions (1959), Liverman (1964), and others developed the basic concept of fractional calculus.

Several approaches to fractional derivatives exist, for example, Riemann–Liouville (RL), Hadamard, Grunwald–Letnikov (GL), Weyl, and Caputo. The Caputo fractional derivative is well suited to the physical interpretation of initial conditions and boundary conditions. We refer readers, for example, to the books [35, 63, 78, 137, 181, 187, 200, 209, 210, 219, 239], the articles [46, 43, 47, 73, 72, 85, 91, 89, 94, 97, 101, 102, 103, 104, 180, 241], and references therein.

In 1783, Leonhard Euler made his first comments on fractional order derivatives. He worked on progressions of numbers and introduced for the first time the generalization of factorials to the *gamma* function. A little more than 50 years after the death of Leibniz, Lagrange, in 1772, indirectly contributed to the development of exponent laws for differential operators of integer order, which can be transferred to arbitrary order under certain conditions. In 1812, Laplace provided the first detailed definition of a fractional derivative. Laplace states that a fractional derivative can be defined for functions with representation by an integral; in modern notation it can be written as $\int y(t)t^{-x}dt$. A few years later, Lacroix worked on generalizing the integer order derivative of the function $y(t) = t^m$ to fractional order, where $m$ is some natural number. In modern notation, the integer order $n$th derivative derived by Lacroix can be given as

$$\frac{d^n y}{dt^n} = \frac{m!}{(m-n)!}t^{m-n} = \frac{\Gamma(m+1)}{\Gamma(m-n+1)}t^{m-n}, \quad m > n\,,$$

https://doi.org/10.1515/9783110553819-202

where $\Gamma$ is *Euler's gamma function* defined by

$$\Gamma(\varsigma) = \int_0^\infty t^{\varsigma-1} e^{-t} dt, \quad \varsigma > 0 .$$

Thus, replacing $n$ with $\frac{1}{2}$ and letting $m = 1$, one obtains the derivative of order $\frac{1}{2}$ of the function $y$:

$$\frac{d^{\frac{1}{2}} y}{dt^{\frac{1}{2}}} = \frac{\Gamma(2)}{\Gamma\left(\frac{3}{2}\right)} t^{\frac{1}{2}} = \frac{2}{\sqrt{\pi}} \sqrt{t} .$$

Euler's gamma function (or Euler's integral of the second kind) has the same importance in fractional order calculus, and it is basically given by the integral

$$\Gamma(z) = \int_0^\infty t^{z-1} e^{-t} dt .$$

The exponential provides the convergence of this integral at $\infty$. The convergence at zero obviously occurs for all complex $z$ from the right half of the complex plane ($\mathrm{Re}(z) > 0$).

This function is a generalization of a factorial in the following form:

$$\Gamma(n) = (n-1)! .$$

Other generalizations for values in the left half of the complex plane can be obtained in the following way. If we replace $e^{-t}$ by the well-known limit

$$e^{-t} = \lim_{n\to\infty} \left(1 - \frac{t}{n}\right)^n$$

and then use $n$-times integration by parts, we obtain the following limit definition of the gamma function:

$$\Gamma(z) = \lim_{n\to\infty} \frac{n! n^z}{z(z+1)\ldots(z+n)} .$$

Therefore, historically the first discussion of a derivative of fractional order appeared in a calculus written by Lacroix in 1819.

It was Liouville who engaged in the first major study of fractional calculus. Liouville's first definition of a derivative of arbitrary order $v$ involved an infinite series. Here, the series must be convergent for some $v$. Liouville's second definition succeeded in giving a fractional derivative of $x^{-a}$ whenever both $x$ and $a$ are positive. Based on the definite integral related to Euler's gamma integral, the integral formula can be calculated for $x^{-a}$. Note that in the integral

$$\int_0^\infty u^{a-1} e^{-xu} du ,$$

if we make change the variables $t = xu$, then

$$\int_0^\infty u^{a-1} e^{-xu} du = \int_0^\infty \left(\frac{t}{x}\right)^{a-1} e^{-t} \frac{1}{x} dt = \frac{1}{x^a} \int_0^\infty t^{a-1} e^{-t} dt .$$

Thus,

$$\int_0^\infty u^{a-1} e^{-xu} du = \frac{1}{x^a} \int_0^\infty t^{a-1} e^{-t} dt .$$

Through the gamma function we obtain the integral formula

$$x^{-a} = \frac{1}{\Gamma(a)} \int_0^\infty u^{a-1} e^{-xu} du .$$

Consequently, by assuming that $\frac{d^v}{dx^v} e^{ax} = a^v e^{ax}$, for any $v > 0$,

$$\frac{d^v}{dx^v} x^{-a} = \frac{\Gamma(a+v)}{\Gamma(a)} x^{-a-v} = (-1)^v \frac{\Gamma(a+v)}{\Gamma(a)} x^{-a-v} .$$

In 1884, Laurent published what is now recognized as the definitive paper on the foundations of fractional calculus. Using Cauchy's integral formula for complex valued analytical functions, and a simple change of notation to employ a positive $v$ rather than a negative $v$, will now yield Laurent's definition of integration of arbitrary order

$${}_{x_0}D_x^\alpha h(x) = \frac{1}{\Gamma(v)} \int_{x_0}^{x} (x-t)^{v-1} h(t) dt .$$

The Riemann–Liouville differential operator of fractional calculus of order $\alpha$ is defined as

$$(D_{a+}^\alpha f)(t) := \begin{cases} \dfrac{1}{\Gamma(n-\alpha)} \left(\dfrac{d}{dt}\right)^n \displaystyle\int_a^t (t-s)^{n-\alpha-1} f(s) ds, & \text{if } n-1 < \alpha < n , \\ \left(\dfrac{d}{dt}\right)^n f(t), & \text{if } \alpha = n , \end{cases}$$

where $\alpha, a, t \in \mathbb{R}$, $t > a$, $n = [\alpha] + 1$, $[\alpha]$ denotes the integer part of the real number $\alpha$, and $\Gamma$ is the gamma function.

The Grünwald–Letnikov differential operator of fractional calculus of order $\alpha$ is defined as

$$(D_{a+}^\alpha f)(t) := \lim_{h \to 0} h^{-\alpha} \sum_{j=0}^{[\frac{t-a}{h}]} (-1)^j \binom{\alpha}{j} f(t - jh) .$$

Binomial coefficients with alternating signs for positive values of $n$ are defined as

$$\binom{n}{j} = \frac{n(n-1)(n-2)\dots(n-j+1)}{j!} = \frac{n!}{j!(n-j)!} .$$

For binomial coefficient calculations, we can use the relation between Euler's gamma function and factorials given by

$$\binom{\alpha}{j} = \frac{\alpha!}{j!(\alpha-j)!} = \frac{\Gamma(\alpha)}{\Gamma(j+1)\Gamma(\alpha-j+1)}\,.$$

The Grünwald–Letnikov definition of differintegral starts from classical definitions of derivatives and integrals based on infinitesimal division and limit. The disadvantages of this approach are its technical difficulty in computations and in proofs and with the large restrictions on functions (see [262]).

The Caputo (1967) differential operator of fractional calculus of order $\alpha$ is defined as

$$({}^cD^{\alpha}_{a+}f)(t) := \begin{cases} \dfrac{1}{\Gamma(n-\alpha)}\displaystyle\int_a^t (t-s)^{n-\alpha-1}f^{(n)}(s)ds, & \text{if } n-1<\alpha<n\,, \\ \left(\dfrac{d}{dt}\right)^n f(t), & \text{if } \alpha = n\,, \end{cases}$$

where $\alpha$, $a$, $t \in \mathbb{R}$, $t > a$, and $n = [\alpha] + 1$. This operator was introduced in 1967 by the Italian mathematician Caputo.

This consideration is based on the fact that for a wide class of functions, the three best known definitions (GL, RL, and Caputo) are equivalent under some conditions (see ([160]). Unfortunately, fractional calculus still lacks a geometric interpretation of integration or differentiation of arbitrary order. We refer readers, for example, to books such as [23, 36, 35, 78, 161, 181, 187, 200, 209, 219, 239], the articles [46, 43, 47, 73, 72, 85, 89, 94, 97, 101, 102, 103, 180, 241], and the references therein.

In June 1974, Ross organized the "*First Conference on Fractional Calculus and Its Applications*" at the University of New Haven and edited its proceedings [227]. Subsequently, in 1974, Spanier published the first monograph devoted to *Fractional Calculus* [209]. Integrals and derivatives of noninteger order and fractional integrodifferential equations have found many applications in recent studies in theoretical physics, mechanics, and applied mathematics. There is a remarkably comprehensive encyclopedic-type monograph by Samko, Kilbas, and Marichev that was published in Russian in 1987 and in English in 1993 [239] (for more details see [197]). Works devoted largely to fractional differential and integral equations include the book by Miller and Ross (1993) [200], Podlubny (1999) [219], Kilbas et al. (2006) [181], Diethelm (2010) [137], Mainardi (2010) [198], Ortigueira (2011) [210], Abbas et al. (2012, 2015) [23, 36, 35], Baleanu et al. (2012) [78], Zhou (2014, 2016) [263, 264], Almeida et al. (2015) [59], Sabatier et al. (2015) [237], Povstenko (2015) [222, 221], Umarov (2015) [246], Cattani et al. (2016) [122], Goodrich and Peterson (2016) [144], and Uchaikin and Sibatov (2016) [243].

Since the second half of the twentieth century, the study of fractional differential and integral equations has made great strides (Oldham and Spanier 1974, Samko et al. 1993, Miller and Ross 1993, Kiryakova 1994, Gorenflo and Mainardi 1997, Podlubny 1999, Kilbas et al. 2006). Thanks to these advances, fractional differentiation has been applied in many areas: Electrical engineering (modeling of motors, modeling of transformers, skin effect), electronics, telecommunications (phase locking loops), electromagnetism (modeling of complex dielectric materials), electrochemistry (modeling of batteries, fuel cells, and ultracapacitors), thermal engineering (modeling and identification of thermal systems), mechanics, mechatronics (vibration insulation, suspension), rheology (behavior identification of materials, viscoelastic properties), automatic control (fractional order PID, robust control, system identification, observation and control of fractional systems), robotics (modeling, path tracking, path planning, obstacle avoidance), signal processing (filtering, restoration, reconstruction, analysis of fractal noise), image processing (fractal environment modeling, pattern recognition, edge detection), biology, biophysics (electrical conductance of biological systems, fractional modeling of neurons, muscle modeling, lung modeling), physics (analysis and modeling of diffusion phenomenon), and economics (analysis of stock exchange signals). In these applications, fractional differentiation is often used to model phenomena that exhibit nonstandard dynamical behaviors with long memory or with hereditary effects. We will now present a brief survey of applications of fractional calculus in science and engineering.

*The Tautochrone Problem.* This example was studied for the first time by Abel in the early nineteenth century. It was one of the basic problems where the framework of the fractional calculus was used, although it is not essentially necessary.

*Signal and Image Processing.* In the last decade, the use of fractional calculus in *signal processing* has increased tremendously. In signal processing, the fractional operators are used in the design of differentiators and integrators of fractional order, fractional order differentiator *FIR* (finite impulse response), *infinite impulse response* (IIR)-type digital fractional order differentiator, a new *IIR*-type digital fractional order differentiator (*DFOD*), and for modeling speech signals. The fractional calculus allows for edge detection, enhances the quality of images, and has interesting possibilities in various image enhancement applications such as image restoration, image denoising, and texture enhancement. It is used, in particular, in satellite image classification and astronomical image processing.

*Electromagnetic Theory.* The use of fractional calculus in electromagnetic theory has emerged in the last two decades. In 1998, Engheta [139] introduced the concept of fractional curl operators, and this concept was extended by Naqvi and Abbas [205]. Engheta's work gave birth to a new field of research in electromagnetics, namely, *fractional paradigms in electromagnetic theory*. Nowadays fractional calculus is widely used in electromagnetics to explore new results; for example, Faryad and Naqvi [141] have used fractional calculus for the analysis of a rectangular waveguide.

*Control Engineering.* In industrial environments, robots must execute their tasks quickly and precisely, minimizing production time, and the robustness of control systems is becoming imperative these days. This requires flexible robots working in large workspaces, which means that they are influenced by nonlinear and fractional order dynamic effects.

*Biological Population Models.* The problems of the diffusion of biological populations occur nonlinearly, and fractional order differential equations are appearing more and more frequently in various research areas.

*Reaction–Diffusion Equations.* Fractional equations can be used to describe some physical phenomena more accurately than classical integer order differential equations. Reaction–diffusion equations play an important role in dynamical systems in mathematics, physics, chemistry, bioinformatics, finance, and other research areas. There has been a wide variety of analytical and numerical methods proposed for fractional equations [196, 258], for example, finite difference methods [127], finite element methods, the Adomian decomposition method [226], and spectral techniques [194]. Interest in fractional reaction–diffusion equations has increased.

In recent years, there has been a significant development in the theory of fractional differential and integral equations. It was brought about by its applications in the modeling of many phenomena in various fields of science and engineering, such as acoustics, control theory, chaos and fractals, signal processing, porous media, electrochemistry, viscoelasticity, rheology, polymer physics, optics, economics, astrophysics, chaotic dynamics, statistical physics, thermodynamics, proteins, biosciences, and bioengineering. Fractional derivatives provide an excellent instrument for the description of memory and hereditary properties of various materials and processes. See, for example, [24, 25, 45, 55, 88, 98, 79, 80, 125, 159, 161, 197, 216, 238, 242, 248].

Fractional differential equations with nonlocal conditions have been discussed in [44, 50, 138, 152, 126, 206, 207] and the references therein. Nonlocal conditions were initiated by Byszewski [118] when he proved the existence and uniqueness of mild and classical solutions of nonlocal Cauchy problems (*C.P.* for short). As remarked by Byszewski [116, 117], nonlocal conditions can be more useful than the standard initial conditions to describe some physical phenomena.

Two measures are most important. The Kuratowski measure of noncompactness $\alpha(B)$ of a bounded set $B$ in a metric space is defined as the infimum of numbers $r > 0$ such that $B$ can be covered with a finite number of sets of diameter smaller than $r$. The Hausdorff measure of noncompactness $\chi(B)$ is defined as the infimum of numbers $r > 0$ such that $B$ can be covered with a finite number of balls of radii smaller than $r$. Several authors have studied measures of noncompactness in Banach spaces. See, for example, books such as [58, 81, 71], the articles [62, 83, 84, 93, 103, 105, 163, 202], and the references therein.

Recently, considerable attention has been paid to the existence of solutions of boundary value problems (BVPs) and boundary conditions for implicit fractional differential equations and integral equations with Caputo fractional derivatives. See, for example, [47, 51, 57, 56, 74, 94, 95, 97, 103, 164, 177, 188, 190, 191, 189, 241, 260] and the references therein.

Functional implicit differential and integral equations involving Caputo fractional derivatives were analyzed recently by many authors; see, for instance, [20, 17, 15, 14, 26, 34, 43, 90, 91, 92, 102, 104, 107, 108] and the references therein.

Ordinary and partial fractional differential and integral equations are one of the useful mathematical tools in both pure and applied analysis. There has been a significant development in ordinary and partial fractional integral equations in recent years; see the monographs of Abbas et al. [23, 36, 35], Appell et al. [67], Banaś and Mursaleen [82], Miller and Ross [200], Podlubny [219], and the papers by Abbas et al.[8, 7, 5], Banaś et al. [81, 83], and the references therein.

During the last 10 years, impulsive differential equations and impulsive differential inclusions with different conditions have been intensely studied by many mathematicians. The concept of differential equations with impulses were introduced by Milman and Myshkis in 1960 [201]. This subject was, thereafter, extensively investigated. Impulsive differential equations have become more important in recent years in some mathematical models of real-world phenomena, especially in biological or medical domains and in control theory; see, for example, the monograph of Graef et al. [148], Lakshmikantham et al. [186], Perestyuk et al. [215], and Samoilenko and Perestyuk [240]; several articles have also been published, for example, see [48, 57, 76, 90, 87, 86, 100, 105, 106, 157, 251, 252, 250, 254] and the references therein.

In the theory of functional differential and integral equations, there is a special kind of data dependency: Ulam, Hyers, Aoki, and Rassias [234]. The stability of functional equations was originally raised by Ulam in 1940 in a talk given at the University of Wisconsin. The problem posed by Ulam was the following: Under what conditions does there exist an additive mapping near an approximately additive mapping? (For more details see [244, 245]). The first answer to Ulam's question was given by Hyers in 1941 in the case of Banach spaces in [166]. Thereafter, this type of stability has been known as the Ulam–Hyers stability. The Hyers theorem was generalized by Aoki [65] for additive mappings and by Rassias [224] for linear mappings by considering an unbounded Cauchy difference. In 1978, Rassias [224] provided a remarkable generalization of the Ulam–Hyers stability of mappings by considering variables. A generalization of the Rassias theorem was obtained by Gavruta [142]. The concept of stability for a functional equation arises when we replace the functional equation by an inequality that acts as a perturbation of the equation. Thus, the stability question of functional equations is how the solutions of the inequality differ from those of the given functional equation. Considerable attention has been devoted to the study of the Ulam–Hyers and Ulam–Hyers–Rassias stability of all kinds of functional equations; one may consult the monographs [167, 172]. Bota-Boriceanu and Petrusel [112], Petru et al. [217], and Rus [234] discussed the Ulam–Hyers stability for operator equations and inclusions. Ulam stability for fractional differential equations with Caputo derivatives are proposed by Wang et al. [253]. More details from a historical point of view and recent developments of such stabilities are reported in the monographs [128, 167, 172, 174, 223, 225] and the papers [6, 24, 61, 92, 168, 170, 171, 173, 176, 182, 234, 252, 253, 254].

In this book we are interested in the existence and stability of solutions to initial and BVPs for functional differential and integral equations and inclusions that involve Caputo's fractional derivative and Hadamard's fractional integral. The book is arranged and organized as follows.

In Chapter 1, we introduce notations, definitions, and some preliminary notions. In *Section 1.1*, we give some concepts from the theory of Banach spaces; in *Section 1.2* we recall some basic definitions and facts on the theory of fractional calculus. *Section 1.3* recalls some properties of set-valued maps. In *Section 1.4*, we give some properties of the measure of noncompactness. *Section 1.5* presents definitions and examples concerning the phase space. *Section 1.6* is devoted to fixed point theory; here we give the main theorems that will be used in the following chapters. In *Section 1.7*, we give other auxiliary lemmas.

In Chapter 2, we will be concerned with the existence and stability of solutions for some classes of nonlinear implicit fractional differential equations (NIFDEs). In *Section 2.2*, we prove some results concerning the existence and stability of solutions for a system of NIFDEs, and *Section 2.3* is concerned with the existence and stability results for NIFDE with nonlocal conditions. In *Section 2.4*, we present other existence results for NIFDE in Banach spaces. *Section 2.5* is devoted to the existence and stability results for perturbed NIFDEs with finite delay. In *Section 2.6*, we establish a sufficient condition for the existence and stability of solutions of a system of neutral NIFDEs with finite delay.

In Chapter 3, we will be concerned with the existence and stability of solutions for some classes of impulsive NIFDEs. In *Section 3.2*, we establish some existence and stability results for impulsive NIFDE with finite delay. *Section 3.3* is devoted to the existence and stability results for impulsive NIFDEs with finite delay in Banach space. In *Section 3.4*, we prove existence and stability results for perturbed impulsive NIFDEs with finite delay. The last section is devoted to proving other existence and stability results for neutral impulsive NIFDEs with finite delay.

In Chapter 4, we prove sufficient conditions for the existence and stability of solutions for some classes of BVPs for NIFDEs. In *Section 4.2*, we establish some existence and stability results for BVP for NIFDEs with $0 < \alpha \leq 1$. In *Section 4.3*, we prove results for BVPs for NIFDEs with $1 < \alpha \leq 2$. In *Section 4.4*, we give some stability results for BVPs for NIFDEs. *Section 4.5* is devoted to other stability results for BVPs for NIFDEs in Banach spaces. In the last section, we prove the existence of $L^1$-solutions of BVPs for NIFDEs with local and nonlocal conditions.

In Chapter 5, we prove some existence and stability results for some classes of BVPs for impulsive NIFDEs. In *Section 5.2*, we give some existence and stability results for impulsive NIFDEs. *Section 5.3* is devoted to other existence results for impulsive NIFDEs in Banach spaces.

In Chapter 6, we shall give results about the integrable solutions for implicit fractional differential equations. In *Section 6.2*, we give some existence results for integrable solutions of NIFDEs. *Section 6.3* is devoted to $L^1$-solutions of NIFDEs with nonlocal conditions. In *Section 6.4*, we give some existence results for integrable solutions for NIFDEs with infinite delay. *Section 6.4* is devoted to other existence results for integrable solutions for NIFDEs.

In Chapter 7, we shall prove some existence results for some classes of partial Hadamard fractional integral equations and inclusions. In *Section 7.2*, we give some existence results for a class of functional partial Hadamard fractional integral equations. *Section 7.3* is devoted to existence results for Fredholm-type Hadamard fractional integral equations. In *Section 7.4*, we use the upper and lower solutions method for partial Hadamard fractional integral equations and inclusions.

In Chapter 8, we shall present results on the stability of solutions for partial Hadamard fractional integral equations and inclusions. In *Section 8.2*, we give some Ulam stability results for partial Hadamard fractional integral equations. *Section 8.3* is devoted to some global stability results for Volterra-type partial Hadamard fractional integral equations. In *Section 8.4*, we prove some Ulam stability results for Hadamard fractional integral equations in Frèchet spaces. In *Section 8.5* we present some Ulam stability results for Hadamard partial fractional integral inclusions via Picard operators.

In Chapter 9, we present results on the stability of solutions for Hadamard–Stieltjes fractional integral equations. In *Section 9.2*, we prove results on the stability of solutions for Hadamard–Stieltjes fractional integral equations. *Section 9.3* is devoted to global stability results for Volterra-type fractional Hadamard–Stieltjes partial integral equations. In *Section 9.4*, we prove some Ulam stability results for a class of Volterra-type nonlinear multidelay Hadamard–Stieltjes fractional integral equations.

In Chapter 10, we prove some results on the Ulam stability for random Hadamard fractional integral equations. In *Section 10.2*, we present results on the stability of solutions for partial Hadamard fractional integral equations with random effects. *Section 10.3* is devoted to global stability results for Volterra–Hadamard random partial fractional integral equations. In *Section 10.4*, we prove the existence and Ulam stability for multidelay Hadamard fractional integral equations in Frèchet spaces with random effects.

*Keywords and phrases:* Differential and integral equations, implicit differential equation, fractional order, left-sided mixed Riemann–Liouville integral, Riemann–Liouville and Caputo fractional order derivatives, Hadamard fractional integral, solution, upper and lower solutions, boundary value problem, initial value problem, nonlocal conditions, contraction, existence, uniqueness, Banach space, ARéchet space, phase space, impulse, finite delay, infinite delay, fixed point, attractivity, Ulam–Hyers–Rassias stability.

# 1 Preliminary Background

In this chapter, we introduce notations, definitions, and preliminary facts that will be used in the remainder of the book. Some notations and definitions from fractional calculus, definitions and properties of measures of noncompactness, and fixed point theorems are presented.

## 1.1 Notations and Definitions

Let $C(J, \mathbb{R})$ be the Banach space of all continuous functions from $J := [0, T];\ T > 0$ to $\mathbb{R}$ with the usual norm

$$\|y\| = \sup_{t \in J} |y(t)| ,$$

and let $L^1(J, \mathbb{R})$ denote the Banach space of functions $: J \to \mathbb{R}$ that are Lebesgue integrable with the norm

$$\|y\|_{L_1} = \int_0^T |y(t)| dt .$$

**Definition 1.1** ([131]). A map $f: J \times \mathbb{R} \times \mathbb{R} \longrightarrow \mathbb{R}$ is said to be $L^1$-Carathéodory if
**(i)** the map $t \longmapsto f(t, x, y)$ is measurable for each $(x, y) \in \mathbb{R} \times \mathbb{R}$,
**(ii)** the map $(x, y) \longmapsto f(t, x, y)$ is continuous for almost all $t \in J$,
**(iii)** for each $q > 0$ there exists $\varphi_q \in L^1(J, \mathbb{R})$ such that

$$|f(t, x, y)| \le \varphi_q(t)$$

for all $|x| \le q,\ |y| \le q$ and for a.e. $t \in J$.

The map $f$ is said to be of Carathéodory if it satisfies just (i) and (ii).

**Definition 1.2.** An operator $T: E \longrightarrow E$ is called compact if the image of each bounded set $B \subset E$ is relatively compact, i.e., $\overline{T(B)}$ is compact. $T$ is called a completely continuous operator if it is continuous and compact.

**Theorem 1.3** (Kolmogorov compactness criterion [133]).
*Let $\Omega \subseteq L^p(J, \mathbb{R})$ and $1 \le p \le \infty$. If*
**(i)** *$\Omega$ is bounded in $L^p(J, \mathbb{R})$ and*
**(ii)** *$u_h \longrightarrow u$ as $h \longrightarrow 0$ uniformly with respect to $u \in \Omega$,*
*then $\Omega$ is relatively compact in $L^p(J, \mathbb{R})$, where*

$$u_h(t) = \frac{1}{h} \int_t^{t+h} u(s) ds .$$

https://doi.org/10.1515/9783110553819-001

## 1.2 Fractional Calculus

**Definition 1.4** ([35, 181, 219]). The Riemann–Liouville fractional (arbitrary) order integral of the function $h \in L^1([a,b],\mathbb{R}_+)$ of order $\alpha \in \mathbb{R}_+$ is defined by

$$I_a^\alpha h(t) = \frac{1}{\Gamma(\alpha)}\int_a^t (t-s)^{\alpha-1}h(s)ds\,,$$

where $\Gamma(.)$ is the Euler gamma function. If $a = 0$, we write $I^\alpha h(t) = h(t) * \varphi_\alpha(t)$, where $\varphi_\alpha(t) = \frac{t^{\alpha-1}}{\Gamma(\alpha)}$ for $t > 0$, $\varphi_\alpha(t) = r$ for $t \le 0$, and $\varphi_\alpha \to \delta(t)$ as $\alpha \to 0$, where $\delta$ is the delta function.

**Definition 1.5** ([35, 181, 219]). The Riemann–Liouville fractional derivative of order $\alpha > 0$ of function $h \in L^1([a,b],\mathbb{R}_+)$ is given by

$$(D_{a+}^\alpha h)(t) = \frac{1}{\Gamma(n-\alpha)}\left(\frac{d}{dt}\right)^n \int_a^t (t-s)^{n-\alpha-1}h(s)ds\,.$$

Here $n = [\alpha]+1$ and $[\alpha]$ denotes the integer part of $\alpha$. If $\alpha \in (0,T]$, then

$$(D_{a+}^\alpha h)(t) = \frac{d}{dt}I_{a+}^{1-\alpha}h(t) = \frac{1}{\Gamma(1-\alpha)}\frac{d}{ds}\int_a^t (t-s)^{-\alpha}h(s)ds\,.$$

**Definition 1.6** ([35, 181]). The Caputo fractional derivative of order $\alpha > 0$ of a function $h \in L^1([a,b],\mathbb{R}_+)$ is given by

$$({}^cD_{a+}^\alpha h)(t) = \frac{1}{\Gamma(n-\alpha)}\int_a^t (t-s)^{n-\alpha-1}h^{(n)}(s)ds\,,$$

where $n = [\alpha]+1$. If $\alpha \in (0,1]$, then

$$({}^cD_{a+}^\alpha h)(t) = I_{a+}^{1-\alpha}\frac{d}{dt}h(t) = \int_a^t \frac{(t-s)^{-\alpha}}{\Gamma(1-\alpha)}\frac{d}{ds}h(s)ds\,.$$

The following properties are some of the main ones of fractional derivatives and integrals.

**Lemma 1.7** ([200]). *Let $\alpha > 0$ and $n = [\alpha]+1$. Then*

$$I^\alpha({}^cD^\alpha f(t)) = f(t) - \sum_{k=0}^{n-1}\frac{f^k(0)}{k!}t_k\,.$$

**Lemma 1.8** ([181]). *Let $\alpha > 0$; then the differential equation*

$${}^cD^\alpha h(t) = 0$$

*has the solution*

$$h(t) = c_0 + c_1t + c_2t^2 + \cdots + c_{n-1}t^{n-1}, \quad c_i \in \mathbb{R},\ i = 0,1,2,\ldots,n-1,\ n = [\alpha]+1\,.$$

**Lemma 1.9** ([181]). *Let $\alpha > 0$; then*

$$I^{\alpha c}D^{\alpha}h(t) = h(t) + c_0 + c_1 t + c_2 t^2 + \cdots + c_{n-1}t^{n-1} ,$$

*for arbitrary $c_i \in \mathbb{R}$, $i = 0, 1, 2, \dots, n-1$, $n = [\alpha] + 1$.*

**Proposition 1.10** ([181]). *Let $\alpha, \beta > 0$. Then we have*
**(1)** *$I^{\alpha}: L^1(J, \mathbb{R}) \to L^1(J, \mathbb{R})$, and if $f \in L^1(J, \mathbb{R})$, then*

$$I^{\alpha}I^{\beta}f(t) = I^{\beta}I^{\alpha}f(t) = I^{\alpha+\beta}f(t) .$$

**(2)** *If $f \in L^p(J, \mathbb{R})$, $1 \le p \le +\infty$, then $\|I^{\alpha}f\|_{L^p} \le \frac{T^{\alpha}}{\Gamma(\alpha+1)}\|f\|_{L^p}$.*
**(3)** *The fractional integration operator $I^{\alpha}$ is linear.*
**(4)** *The fractional order integral operator $I^{\alpha}$ maps $L^1(J, \mathbb{R})$ to itself continuously.*
**(5)** *If $\alpha = n \in \mathbb{N}$, then $I_0^{\alpha}$ is the n-fold integration.*
**(6)** *The Caputo and Riemann–Liouville fractional derivatives are linear.*
**(7)** *The Caputo fractional derivative of a constant is equal to zero.*

Now we recall some definitions and properties of Hadamard fractional integration and differentiation. We refer to [153, 181] for a more detailed analysis.

**Definition 1.11** ([153, 181]). The Hadamard fractional integral of order $q > 0$ for a function $g \in L^1([1, a], \mathbb{R})$ is defined as

$$({}^H I_1^r g)(x) = \frac{1}{\Gamma(q)} \int_1^x \left(\ln \frac{x}{s}\right)^{q-1} \frac{g(s)}{s} ds$$

provided the integral exists.

Analogous to the Riemann–Liouville fractional calculus, the Hadamard fractional derivative is defined in terms of the Hadamard fractional integral in the following way. Set

$$\delta = x\frac{d}{dx}, \quad q > 0, \quad n = [q] + 1 ,$$

where $[q]$ is the integer part of $q$, and

$$AC_\delta^n := \{u\colon [1, a] \to \mathbb{R}\colon \delta^{n-1}[u(x)] \in AC[1, a]\} .$$

**Definition 1.12** ([153, 181]). The Hadamard fractional derivative of order $q$ applied to the function $w \in AC_\delta^n$ is defined as

$$({}^H D_1^q w)(x) = \delta^n ({}^H I_1^{n-q} w)(x) .$$

It has been proved (e.g., Kilbas [[178], Theorem 4.8]) that in the space $L^1([1, a], \mathbb{R})$, the Hadamard fractional derivative is the left-inverse operator to the Hadamard fractional integral, i.e.,

$$({}^H D_1^q)({}^H I_1^q w)(x) = w(x) .$$

Analogously to the Caputo partial fractional integral and derivative [36, 35], we can define the Hadamard partial fractional integral and derivative. Also, the Hadamard partial fractional derivative is defined in terms of the Hadamard partial fractional integral.

**Definition 1.13.** Let $r_1, r_2 \geq 0$, $\sigma = (1,1)$, and $r = (r_1, r_2)$. For $w \in L^1(J, \mathbb{R})$, define the Hadamard partial fractional integral of order $r$ by the expression

$$({}^H I_\sigma^r w)(x,y) = \frac{1}{\Gamma(r_1)\Gamma(r_2)} \int_1^x \int_1^y \left(\ln \frac{x}{s}\right)^{r_1-1} \left(\ln \frac{y}{t}\right)^{r_2-1} \frac{w(s,t)}{st} dt ds \,.$$

By $1 - r$ we mean $(1 - r_1, 1 - r_2) \in (0,1] \times (0,1]$. Denote by $D_{xy}^2 := \frac{\partial^2}{\partial x \partial y}$ the mixed second-order partial derivative.

**Definition 1.14.** Let $r = (r_1, r_2) \in (0,1] \times (0,1]$ and $u \in L^1(J)$. Define the Hadamard fractional order derivative of order $r$ of $u$ by the expression

$${}^H D_\sigma^r u(x,y) = D_{xy}^2 [xy D_{xy}^2 ({}^H I_\sigma^{1-r} u)](x,y) \,.$$

**Definition 1.15.** Let $\alpha \in (0, \infty)$ and $u \in L^1(J)$. The partial Hadamard integral of order $\alpha$ of $u(x,y)$ with respect to $x$ is defined by

$${}^H I_{1,x}^\alpha u(x,y) = \frac{1}{\Gamma(\alpha)} \int_1^x \left(\ln \frac{x}{s}\right)^{\alpha-1} \frac{u(s,y)}{s} ds \quad \text{for almost all } x \in [1,a] \text{ and all } y \in [1,b].$$

Analogously, we define the integral

$${}^H I_{1,y}^\alpha u(x,y) = \frac{1}{\Gamma(\alpha)} \int_1^y \left(\ln \frac{y}{s}\right)^{\alpha-1} \frac{u(x,s)}{s} ds \quad \text{for all } x \in [1,a] \text{ and almost all } y \in [1,b].$$

**Definition 1.16.** Let $\alpha \in (0,1]$ and $u \in L^1(J)$. The Hadamard fractional derivative of order $\alpha$ of $u(x,y)$ with respect to $x$ is defined by

$${}^H D_{1,x}^\alpha u(x,y) = \frac{\partial}{\partial x} \left[ x \frac{\partial}{\partial x} ({}^H I_{1,x}^{1-\alpha} u) \right](x,y) \quad \text{for almost all } x \in [1,a] \text{ and all } y \in [1,b] \,.$$

Analogously, we define the derivative of order $\alpha$ of $u(x,y)$ with respect to $y$ by

$${}^H D_{1,y}^\alpha u(x,y) = \frac{\partial}{\partial y} \left[ y \frac{\partial}{\partial y} ({}^H I_{1,y}^{1-\alpha} u) \right](x,y) \quad \text{for all } x \in [1,a] \text{ and almost all } y \in [1,b] \,.$$

## 1.3 Multivalued Analysis

Let $(X, \|\cdot\|)$ be a Banach space and $K$ be a subset of $X$. We use the notation

$$\begin{aligned}
\mathcal{P}(X) &= \{K \subset X : K \neq \emptyset\}, \\
\mathcal{P}_{cl}(X) &= \{K \subset \mathcal{P}(X) : K \text{ is closed}\}, \\
\mathcal{P}_{b}(X) &= \{K \subset \mathcal{P}(X) : K \text{ is bounded}\}, \\
\mathcal{P}_{cv}(X) &= \{K \subset \mathcal{P}(X) : K \text{ is convex}\}, \\
\mathcal{P}_{cp}(X) &= \{K \subset \mathcal{P}(X) : K \text{ is compact}\}, \\
\mathcal{P}_{cv,cp}(X) &= \mathcal{P}_{cv}(X) \cap \mathcal{P}_{cp}(X).
\end{aligned}$$

Let $A, B \in \mathcal{P}(X)$. Consider $H_d : \mathcal{P}(X) \times \mathcal{P}(X) \longrightarrow \mathbb{R}_+ \cup \{\infty\}$ the Hausdorff distance between $A$ and $B$ given by

$$H_d(A, B) = \max\{\sup_{a \in A} d(a, B), \sup_{b \in B} d(A, b)\},$$

where $d(A, b) = \inf_{a \in A} d(a, b)$ and $d(a, B) = \inf_{b \in B} d(a, b)$. As usual, $d(x, \emptyset) = +\infty$.

Then $(\mathcal{P}_{b,cl}(X), H_d)$ is a metric space and $(\mathcal{P}_{cl}(X), H_d)$ is a generalized (complete) metric space [184].

**Definition 1.17.** A multivalued operator $N : X \to \mathcal{P}_{cl}(X)$ is called:
*(a)* $\gamma$-Lipschitz if there exists $\gamma > 0$ such that

$$H_d(N(x), N(y)) \leq \gamma d(x, y), \quad \text{for all } x, y \in X;$$

*(b)* a contraction if it is $\gamma$-Lipschitz with $\gamma < 1$.

**Definition 1.18.** A multivalued map $F : J \to \mathcal{P}_{cl}(X)$ is said to be measurable if, for each $y \in X$, the function

$$t \longmapsto d(y, F(t)) = \inf\{d(x, z) : z \in F(t)\}$$

is measurable.

**Definition 1.19.** The selection set of a multivalued map $G : J \to \mathcal{P}(X)$ is defined by

$$S_G = \{u \in L^1(J) : u(t) \in G(t), \quad \text{a.e. } t \in J\}.$$

For each $u \in C$, the set $S_{F \circ u}$ known as the set of selectors from $F$ is defined by

$$S_{F \circ u} = \{v \in L^1(J) : v(t) \in F(t, u(t)), \quad \text{a.e. } t \in J\}.$$

**Definition 1.20.** Let $X$ and $Y$ be metric spaces. A set-valued map $F$ from $X$ to $Y$ is characterized by its graph $Gr(F)$, the subset of the product space $X \times Y$ defined by

$$Gr(F) := \{(x, y) \in X \times Y : y \in F(x)\}.$$

**Definition 1.21.** Let $(X, \|\cdot\|)$ be a Banach space. A multivalued map $F\colon X \to \mathcal{P}(X)$ is convex (closed) if $F(X)$ is convex (closed) for all $x \in X$.

The map $F$ is bounded on bounded sets if $F(\mathcal{B}) = \cup_{x\in\mathcal{B}} F(x)$ is bounded in $X$ for all $\mathcal{B} \in \mathcal{P}_b(X)$, i.e., $\sup_{x\in\mathcal{B}}\{\sup\{|y| : y \in F(x)\}\} < \infty$.

**Definition 1.22.** A multivalued map $F$ is called upper semicontinuous (u.s.c.) on $X$ if for each $x_0 \in X$ the set $F(x_0)$ is a nonempty, closed subset of $X$ and for each open set $U$ of $X$ containing $F(x_0)$ there exists an open neighborhood $V$ of $x_0$ such that $F(V) \subset U$. A set-valued map $F$ is said to be u.s.c. if it is so at every point $x_0 \in X$. $F$ is said to be completely continuous if $F(\mathcal{B})$ is relatively compact for every $\mathcal{B} \in \mathcal{P}_b(X)$.

If the multivalued map $F$ is completely continuous with nonempty compact values, then $F$ is u.s.c. if and only if $F$ has closed graph (i.e., $x_n \to x_*$, $y_n \to y_*$, $y_n \in G(x_n)$ imply $y_* \in F(x_*)$).
The map $F$ has a fixed point if there exists $x \in X$ such that $x \in Gx$. The set of fixed points of the multivalued operator $G$ will be denoted by $FixG$.

**Definition 1.23.** A measurable multivalued function $F\colon J \to \mathcal{P}_{b,cl}(X)$ is said to be integrably bounded if there exists a function $g \in L^1(\mathbb{R}_+)$ such that $|f| \le g(t)$ for almost all $t \in J$ for all $f \in F(t)$.

**Lemma 1.24** ([165]). *Let $G$ be a completely continuous multivalued map with nonempty compact values. Then $G$ is u.s.c. if and only if $G$ has a closed graph (i.e., $u_n \to u$, $w_n \to w$, $w_n \in G(u_n)$ imply $w \in G(u)$).*

**Lemma 1.25** ([192]). *Let $X$ be a Banach space. Let $F\colon J \times X \longrightarrow \mathcal{P}_{cp,cv}(X)$ be an $L^1$-Carathéodory multivalued map, and let $\Lambda$ be a linear continuous mapping from $L^1(J, X)$ to $C(J, X)$. Then the operator*

$$\Lambda \circ S_{F\circ u}\colon C(J, X) \longrightarrow \mathcal{P}_{cp,cv}(C(J, X)),$$
$$w \longmapsto (\Lambda \circ S_{F\circ u})(w) := (\Lambda S_{F\circ u})(w)$$

*is a closed graph operator in $C(J, X) \times C(J, X)$.*

**Proposition 1.26** ([165]). *Let $F\colon X \to Y$ be an u.s.c. map with closed values. Then $Gr(F)$ is closed.*

**Definition 1.27.** A multivalued map $F\colon J \times \mathbb{R} \times \mathbb{R} \to \mathcal{P}(\mathbb{R})$ is said to be $L^1$-Carathéodory if

**(i)** $t \to F(t, x, y)$ is measurable for each $x, y \in \mathbb{R}$;
**(ii)** $x \to F(t, x, y)$ is u.s.c. for almost all $t \in J$;
**(iii)** for each $q > 0$ there exists $\varphi_q \in L^1(J, \mathbb{R}_+)$ such that

$$\|F(t, x, y)\|_{\mathcal{P}} = \sup\{|f| : f \in F(t, x, y)\} \le \varphi_q(t)$$
$$\text{for all } |x| \le q,\ |y| \le q \text{ and for a.e. } t \in J\,.$$

The multivalued map $F$ is said to be Carathéodory if it satisfies (i) and (ii).

**Lemma 1.28** ([145]). *Let $X$ be a separable metric space. Then every measurable multivalued map $F\colon X \to \mathcal{P}_{cl}(X)$ has a measurable selection.*

For more details on multivalued maps and the proof of the known results cited in this section, we refer interested reader to the books of Aubin and Cellina [68], Deimling [134], Gorniewicz [145], and Hu and Papageorgiou [165].

## 1.4 Measure of Noncompactness

We will define the Kuratowski (1896–1980) and Hausdorff (1868–1942) measures of noncompactness ($\mathcal{MNC}$ for short) and give their basic properties. Let us recall some fundamental facts of the notion of measure of noncompactness in a Banach space.

Let $(X, d)$ be a complete metric space and $\mathcal{P}_{bd}(X)$ be the family of all bounded subsets of $X$. Analogously denote by $\mathcal{P}_{rcp}(X)$ the family of all relatively compact and nonempty subsets of $X$. Recall that $B \subset X$ is said to be bounded if $B$ is contained in some ball. If $B \subset \mathcal{P}_{bd}(X)$ is not relatively compact, (precompact) then there exists an $\epsilon > 0$ such that $B$ cannot be covered by a finite number of $\epsilon$-balls, and it is then also impossible to cover $B$ by finitely many sets of diameter $< \epsilon$. Recall that the diameter of $B$ is given by

$$\operatorname{diam}(B) := \begin{cases} \sup\limits_{(x,y)\in B^2} d(x, y), & \text{if } B \neq \phi\,, \\ 0, & \text{if } B = \phi\,. \end{cases}$$

**Definition 1.29** ([183]). Let $(X, d)$ be a complete metric space and $\mathcal{P}_{bd}(X)$ be the family of bounded subsets of $X$. For every $B \in \mathcal{P}_{bd}(X)$, we define the Kuratowski measure of noncompactness $\alpha(B)$ of the set $B$ as the infimum of the numbers $d$ such that $B$ admits a finite covering by sets of diameter smaller than $d$.

**Remark 1.30.** It is clear that $0 \le \alpha(B) \le \operatorname{diam}(B) < +\infty$ for each nonempty bounded subset $B$ of $X$ and that $\operatorname{diam}(B) = 0$ if and only if $B$ is an empty set or consists of exactly one point.

**Definition 1.31** ([81]). Let $X$ be a Banach space and $\mathcal{P}_{bd}(X)$ be the family of bounded subsets of $X$. For every $B \in \mathcal{P}_{bd}(X)$, the Kuratowski measure of noncompactness is the map $\alpha\colon \mathcal{P}_{bd}(X) \to [0, +\infty]$ defined by

$$\alpha(B) = \inf\{r > 0\colon\ B \subseteq \cup_{i=1}^{n} B_i \text{ and } \operatorname{diam}(B_i) < r\}\,.$$

The Kuratowski measure of noncompactness satisfies the following properties:

**Proposition 1.32** ([81, 83, 183]). *Let $X$ be a Banach space. Then for all bounded subsets $A, B$ of $X$ the following assertions hold:*

1. *$\alpha(B) = 0$ implies $\overline{B}$ is compact ($B$ is relatively compact), where $B$ denotes the closure of $B$.*

2. $\alpha(\phi) = 0$.
3. $\alpha(B) = \alpha(\overline{B}) = \alpha(\operatorname{conv} B)$, *where* $\operatorname{conv} B$ *is the convex hull of* $B$.
4. *monotonicity:* $A \subset B$ *implies* $\alpha(A) \leq \alpha(B)$.
5. *algebraic semi-additivity:* $\alpha(A+B) \leq \alpha(A)+\alpha(B)$, *where* $A+B = \{x+y\colon x \in A; y \in B\}$.
6. *semihomogeneity:* $\alpha(\lambda B) = |\lambda|\alpha(B)$, $\quad \lambda \in \mathbb{R}$, *where* $\lambda(B) = \{\lambda x\colon x \in B\}$.
7. *semi-additivity:* $\alpha(A \cup B) = \max\{\alpha(A), \alpha(B)\}$.
8. *semi-additivity:* $\alpha(A \cap B) = \min\{\alpha(A), \alpha(B)\}$.
9. *invariance under translations:* $\alpha(B + x_0) = \alpha(B)$ *for any* $x_0 \in X$.

**Lemma 1.33** ([151]). *If* $V \subset C(J, E)$ *is a bounded and equicontinuous set, then*
(i) *the function* $t \to \alpha(V(t))$ *is continuous on* $J$ *and*

$$\alpha_c(V) = \sup_{0 \leq t \leq T} \alpha(V(t))\,;$$

(ii) $\alpha\left(\int_0^T x(s)ds\colon x \in V\right) \leq \int_0^T \alpha(V(s))ds$,

*where*

$$V(s) = \{x(s)\colon x \in V\}, \quad s \in J\,.$$

The following definition of measure of noncompactness appeared in Banaś and Goebel [81].

**Definition 1.34.** A function $\mu\colon \mathcal{P}_{bd}(X) \longrightarrow [0, \infty)$ will be called a measure of noncompactness if it satisfies the following conditions:
1. $\operatorname{Ker} \mu(A) = \{A \in \mathcal{P}_{bd}(X)\colon \mu(A) = 0\}$ is nonempty and $\operatorname{Ker} \mu(A) \subset \mathcal{P}_{rcp}(X)$.
2. $A \subset B$ implies $\mu(A) \leq \mu(B)$.
3. $\mu(\overline{A}) = \mu(A)$.
4. $\mu(\operatorname{conv} A) = \mu(A)$.
5. $\mu(\lambda A + (1-\lambda)B) \leq \lambda\mu(A) + (1-\lambda)\mu(B)$ for $\lambda \in [0, 1]$.
6. If $(A_n)_{n \geq 1}$ is a sequence of closed sets in $\mathcal{P}_{bd}(X)$ such that

$$X_{n+1} \subset A_n \ (n = 1, 2, \ldots)$$

and

$$\lim_{n \to +\infty} \mu(A_n) = 0\,,$$

then the intersection set $A_\infty = \bigcap_{n=1}^{\infty} A_n$ is nonempty.

**Remark 1.35.** The family $\operatorname{Ker} \mu$ described in 1 is said to be the kernel of the measure of noncompactness $\mu$. Observe that the intersection set $A_\infty$ in 6 is a member of the family $\operatorname{Ker} \mu$. Since $\mu(A_\infty) \leq \mu(A_n)$ for any $n$, we infer that $\mu(A_\infty) = 0$. This yields that $\mu(A_\infty) \in \operatorname{Ker} \mu$. This simple observation will be essential in our further investigations.

Moreover, we introduce the notion of a measure of noncompactness in $L^1(J)$. We let $\mathcal{P}_{bd}(J)$ be the family of all bounded subsets of $L^1(J)$. Analogously, denote by $\mathcal{P}_{rcp}(J)$

the family of all relatively compact and nonempty subsets of $L^1(J)$. In particular, the measure of noncompactness in $L^1(J)$ is defined as follows. Let $X$ be a fixed nonempty and bounded subset of $L^1(J)$. For $x \in X$, set

$$\mu(X) = \lim_{\delta \to 0} \left\{ \sup \left\{ \sup \left( \int_0^T |x(t+h) - x(t)| dt \right), \ |h| \leq \delta \right\}, \quad x \in X \right\}. \tag{1.1}$$

It can be easily shown that $\mu$ is a measure of noncompactness in $L^1(J)$ [81]. For more details on the measure of noncompactness and the proof of the known results cited in this section, we refer the reader to Akhmerov et al. [58] and Banaś et al. [81, 83].

## 1.5 Phase Spaces

In this section, we assume that the state space $(\mathcal{B}, \|\cdot\|_{\mathcal{B}})$ is a seminormed linear space of functions mapping $(-\infty, 0]$ to $\mathbb{R}$ and satisfying the following fundamental axioms introduced by Hale and Kato in [154].

**$(A_1)$** If $y: (-\infty, b] \to \mathbb{R}$ and $y_0 \in \mathcal{B}$, then for every $t \in J$ the following conditions hold:
- (i) $y_t \in \mathcal{B}$.
- (ii) $\|y_t\|_{\mathcal{B}} \leq K(t) \int_0^t |y(s)| ds + M(t)\|y_0\|_{\mathcal{B}}$.
- (iii) $|y(t)| \leq H\|y_t\|_{\mathcal{B}}$, where $H \geq 0$ is a constant, $K: J \to [0, \infty)$ is continuous, $M: [0, \infty) \to [0, \infty)$ is locally bounded, and $H, K, M$ are independent of $y(\cdot)$.

**$(A_2)$** For the function $y(\cdot)$ in **$(A_1)$**, $y_t$ is a $\mathcal{B}$-valued continuous function on $J$.

**$(A_3)$** The space $\mathcal{B}$ is complete.

Use the notation $K_b = \sup\{K(t) : t \in J\}$ and $M_b = \sup\{M(t) : t \in J\}$.

**Remark 1.36.** 1. $(A_1)$(ii) is equivalent to $|\phi(0)| \leq H\|\phi\|_{\mathcal{B}}$ for every $\phi \in \mathcal{B}$.
2. Since $\|\cdot\|_{\mathcal{B}}$ is a seminorm, two elements $\phi, \psi \in \mathcal{B}$ can satisfy $\|\phi - \psi\|_{\mathcal{B}} = 0$ without necessarily $\phi(\theta) = \psi(\theta)$ for all $\theta \leq 0$.
3. From the equivalence in the first remark, we can see that for all $\phi, \psi \in \mathcal{B}$ such that $\|\phi - \psi\|_{\mathcal{B}} = 0$. We necessarily have that $\phi(0) = \psi(0)$.

We now present some examples of phase spaces. For other details see, for instance, the book by Hino et al. [162].

### 1.5.1 Examples of Phase Spaces

**Example 1.37.** *Let us define the following spaces:*
*$BC$ the space of bounded and continuous functions defined from $(-\infty, 0] \to E$;*
*$BUC$ the space of bounded and uniformly continuous functions defined from $(-\infty, 0] \to E$;*

$C^{\infty} := \{\phi \in BC\colon \lim_{\theta\to-\infty}\phi(\theta)$ *exist in* $E\}$;
$C^{0} := \{\phi \in BC\colon \lim_{\theta\to-\infty}\phi(\theta) = 0\}$, *endowed with the uniform norm*

$$\|\phi\| = \sup\{|\phi(\theta)|\colon \theta \le 0\}\,.$$

*We have that the spaces* $BUC$, $C^{\infty}$ *and* $C^{0}$ *satisfy conditions (A1)–(A3). However,* $BC$ *satisfies (A1) and (A3), but (A2) is not satisfied.*

**Example 1.38.** *Let* $g$ *be a positive continuous function on* $(-\infty, 0]$. *We define:*
$C_g := \{\phi \in C((-\infty,0]), E)\colon \frac{\phi(\theta)}{g(\theta)}$ *is bounded on* $(-\infty, 0]\}$,
$C_g^0 := \{\phi \in C_g\colon \lim_{\theta\to-\infty}\frac{\phi(\theta)}{g(\theta)} = 0\}$ *endowed with the uniform norm*

$$\|\phi\| = \sup\left\{\frac{|\phi(\theta)|}{g(\theta)}\colon \theta \le 0\right\}\,.$$

*Then we have that the spaces* $C_g$ *and* $C_g^0$ *satisfy condition (A3). We consider the following condition on the function* $g$:
$(g_1)$ *For all* $a > 0$, $\sup_{0\le t\le a}\sup\{\frac{\phi(t+\theta)}{g(\theta)}\colon -\infty < \theta \le -t\}$.
*Then* $C_g$ *and* $C_g^0$ *satisfy conditions (A1) and (A2) if* $(g_1)$ *holds.*

**Example 1.39.** *The space* $C_\gamma$ *for any real positive constant* $\gamma$ *is defined by*

$$C_\gamma := \{\phi \in C((-\infty,0]), E)\colon \lim_{\theta\to-\infty} e^{\gamma\theta}\phi(\theta) \text{ exist in } E$$

*endowed with the norm*

$$\|\phi\| = \sup\{e^{\gamma\theta}|\phi(\theta)|\colon \theta \le 0\}\,.$$

*Then in the space* $C_\gamma$ *axioms (A1)–(A3) are satisfied.*

## 1.6 Some Fixed Point Theorems

In this section, we give the main fixed point theorems that will be used in subsequent chapters.

**Definition 1.40** ([60]). Let $(M, d)$ be a metric space. The map $T\colon M \longrightarrow M$ is said to be Lipschitzian if there exists a constant $k > 0$ (called a Lipschitz constant) such that

$$d(T(x), T(y)) \le kd(x, y) \quad \text{for all } x, y \in M\,.$$

A Lipschitzian mapping with a Lipschitz constant $k < 1$ is called a contraction.

**Theorem 1.41** (Banach's fixed point theorem [149]). *Let* $C$ *be a nonempty closed subset of a Banach space* $X$. *Then any contraction mapping* $T$ *of* $C$ *to itself has a unique fixed point.*

**Theorem 1.42** (Schauder fixed point theorem [149]). *Let* $E$ *be a Banach space,* $Q$ *a convex subset of* $E$, *and* $T\colon Q \longrightarrow Q$ *a compact and continuous map. Then* $T$ *has at least one fixed point in* $Q$.

**Theorem 1.43** (Burton and Kirk fixed point theorem [115]). *Let $X$ be a Banach space and $A, B\colon X \to X$ two operators satisfying*
**(i)** *$A$ is a contraction,*
**(ii)** *$B$ is completely continuous.*
*Then either*
- *The operator equation $y = A(y) + B(y)$ admits a solution or*
- *the set $\Omega = \{u \in X\colon u = \lambda A(\frac{u}{\lambda}) + \lambda B(u)\}$ is unbounded for $\lambda \in [0, 1]$.*

In the next definition, we will consider a special class of continuous and bounded operators.

**Definition 1.44.** Let $T\colon M \subset E \longrightarrow E$ be a bounded operator from a Banach space $E$ to itself. The operator $T$ is called a $k$-set contraction if there is a number $k \geq 0$ such that

$$\mu(T(A)) \leq k\mu(A)$$

for all bounded sets $A$ in $M$. The bounded operator $T$ is called condensing if $\mu(T(A)) < \mu(A)$ for all bounded sets $A$ in $M$ with $\mu(M) > 0$.

Obviously, every $k$-set contraction for $0 \leq k < 1$ is condensing. Every compact map $T$ is a $k$-set contraction with $k = 0$.

**Theorem 1.45** (Darbo fixed point theorem [81]). *Let $M$ be a nonempty, bounded, convex, and closed subset of a Banach space $E$ and $T\colon M \longrightarrow M$ a continuous operator satisfying $\mu(TA) \leq k\mu(A)$ for any nonempty subset $A$ of $M$ and for some constant $k \in [0, 1)$. Then $T$ has at least one fixed point in $M$.*

**Theorem 1.46** (Mönch's fixed point theorem [49, 202]). *Let $D$ be a bounded, closed, and convex subset of a Banach space such that $0 \in D$, $\alpha$ the Kuratowski measure of noncompactness, and $N$ a continuous mapping of $D$ to itself. If the implication $[V = \overline{\operatorname{conv}} N(V) \quad \text{or} \quad V = N(V) \cup \{0\}]$ implies $\alpha(V) = 0$ holds for every subset $V$ of $D$, then $N$ has a fixed point.*

For more details, see [49, 64, 145, 149, 183, 257].

**Theorem 1.47** (Nonlinear alternative to Leray–Schauder type [149]). *Let $X$ be a Banach space and $C$ a nonempty convex subset of $X$. Let $U$ be a nonempty open subset of $C$, with $0 \in U$ and $T\colon \overline{U} \to C$ a continuous and compact operator.*
*Then, either*
*(a) $T$ has fixed points or*
*(b) there exist $u \in \partial U$ and $\lambda \in (0, 1)$ with $u = \lambda T(u)$.*

**Theorem 1.48** (Martelli's fixed point theorem [199]). *Let $X$ be a Banach space and $N\colon X \to \mathcal{P}_{cl,cv}(X)$ an u.s.c. and condensing map. If the set $\Omega := \{u \in X\colon \lambda u \in N(u)$ for some $\lambda > 1\}$ is bounded, then $N$ has a fixed point.*

**Theorem 1.49** ([70]). *Let $(X, \|\cdot\|_n)$ be a Fréchet space, and let $A, B: X \to X$ be two operators such that*
*(a) $A$ is a compact operator;*
*(b) $B$ is a contraction operator with respect to a family of seminorms $\{\|\cdot\|_n\}$;*
*(c) the set $\{x \in X: x = \lambda A(x) + \lambda B(\frac{x}{\lambda}), \quad \lambda \in (0,1)\}$ is bounded.*
*Then the operator equation $A(u) + B(u) = u$ has a solution in $X$.*

Next, we state two multivalued fixed point theorems.

**Lemma 1.50** (Bohnenblust–Karlin 1950 [111]). *Let $X$ be a Banach space and $K \in \mathcal{P}_{cl,cv}(X)$, and suppose that the operator $G: K \to \mathcal{P}_{cl,cv}(K)$ is u.s.c. and the set $G(K)$ is relatively compact in $X$. Then $G$ has a fixed point in $K$.*

**Lemma 1.51** (Covitz–Nadler [130]). *Let $(X, d)$ be a complete metric space. If $N: X \to \mathcal{P}_{cl}(X)$ is a contraction, then $FixN \neq \phi$.*

## 1.7 Auxiliary Lemmas

We state the following generalization of Gronwall's lemma for a singular kernel.

**Lemma 1.52** ([256]). *Let $v: [0, T] \to [0, +\infty)$ be a real function and $w(\cdot)$ a nonnegative, locally integrable function on $[0, T]$. Assume that there exist constants $a > 0$ and $0 < \alpha < 1$ such that*

$$v(t) \leq w(t) + a \int_0^t (t-s)^{-\alpha} v(s) ds .$$

*Then there exists a constant $K = K(\alpha)$ such that*

$$v(t) \leq w(t) + Ka \int_0^t (t-s)^{-\alpha} w(s) ds \quad \text{for every } t \in [0, T] .$$

Bainov and Hristova [75] introduced the following integral inequality of the Gronwall type for piecewise continuous functions that can be used in the sequel.

**Lemma 1.53.** *Let, for $t \geq t_0 \geq 0$, the following inequality hold:*

$$x(t) \leq a(t) + \int_{t_0}^t g(t,s) x(s) ds + \sum_{t_0 < t_k < t} \beta_k(t) x(t_k) ,$$

*where $\beta_k(t) (k \in \mathbb{N})$ are nondecreasing functions for $t \geq t_0$, $a \in PC([t_0, \infty), \mathbb{R}_+)$, $a$ is nondecreasing, and $g(t,s)$ is a continuous nonnegative function for $t, s \geq t_0$ and nondecreasing with respect to $t$ for any fixed $s \geq t_0$. Then, for $t \geq t_0$,*

$$x(t) \leq a(t) \prod_{t_0 < t_k < t} (1 + \beta_k(t)) \exp\left( \int_{t_0}^t g(t,s) ds \right) .$$

**Lemma 1.54** (Ascoli–Arzelà, [155]). *Let $A \subset C(J, \mathbb{R})$; $A$ is relatively compact (i.e., $\overline{A}$ is compact) if*

1. *$A$ is uniformly bounded, i.e., there exists $M > 0$ such that*

$$\|f(x)\| < M \quad \text{for every } f \in A \text{ and } x \in J\,;$$

2. *$A$ is equicontinuous, i.e., for every $\epsilon > 0$ there exists $\delta > 0$ such that for each $x, \overline{x} \in J$, $\|x - \overline{x}\| \le \delta$ implies $\|f(x) - f(\overline{x})\| \le \epsilon$ for every $f \in A$.*

Set $J_0 := \{(x, y, s) : 0 \le s \le x \le a, y \in [0, b]\}$, $J_1 := \{(x, y, s, t) : 0 \le s \le x \le a, 0 \le t \le y \le b\}$, $D_1 := \frac{\partial}{\partial x}$, $D_2 := \frac{\partial}{\partial y}$, and $D_1 D_2 := \frac{\partial^2}{\partial x \partial y}$.

In the sequel we will make use of the following variant of the inequality for two independent variables due to Pachpatte.

**Lemma 1.55** ([211]). *Let $w \in C(J, \mathbb{R}_+)$, $p, D_1 p \in C(J_0, \mathbb{R}_+)$, $q, D_1 q, D_2 q, D_1 D_2 q \in C(J_1, \mathbb{R}_+)$, and $c > 0$ a constant. If*

$$w(x, y) \le c + \int_0^x p(x, y, s) w(s, y) ds + \int_0^x \int_0^y q(x, y, s, t) w(s, t) dt ds$$

*for $(x, y) \in [0, a] \times [0, b]$, then*

$$w(x, y) \le c A(x, y) \exp\left(\int_0^x \int_0^y B(s, t) dt ds\right),$$

*where*

$$A(x, y) = \exp(Q(x, y))\,,$$

$$Q(x, y) = \int_0^x \left[ p(s, y, s) + \int_0^s D_1 p(s, y, \xi) d\xi \right] ds\,,$$

*and*

$$B(x, y) = q(x, y, x, y) A(x, y) + \int_0^x D_1 q(x, y, s, y) A(s, y) ds$$
$$+ \int_0^y D_2 q(x, y, x, t) A(x, t) dt + \int_0^x \int_0^y D_1 D_2 q(x, y, s, t) A(s, t) dt ds\,.$$

From the preceding lemma and with $p \equiv 0$, we get the following lemma.

**Lemma 1.56.** *Let $w \in C(J, \mathbb{R}_+)$, $q, D_1 q, D_2 q, D_1 D_2 q \in C(J_1, \mathbb{R}_+)$, and let $c > 0$ be a constant. If*

$$a w(x, y) \le c + \int_1^x \int_1^y q(x, y, s, t) w(s, t) dt ds$$

*for* $(x, y) \in J$, *then*

$$w(x, y) \leq c \exp\left(\int_1^x \int_1^y B(s, t)dtds\right),$$

*where*

$$B(x, y) = q(x, y, x, y) + \int_1^x D_1 q(x, y, s, y)ds$$
$$+ \int_1^y D_2 q(x, y, x, t)dt + \int_1^x \int_1^y D_1 D_2 q(x, y, s, t)dtds.$$

**Lemma 1.57** ([129]). *Let* $D \subset BC$. *Then* $D$ *is relatively compact in* $BC$ *if the following conditions hold:*

(a) $D$ *is uniformly bounded in* $BC$,

(b) *The functions belonging to* $D$ *are almost equicontinuous on* $[1, \infty) \times [1, b]$, *i.e., equicontinuous on every compact of* $J$.

(c) *The functions from* $D$ *are equiconvergent, that is, given* $\epsilon > 0$ *and* $x \in [1, b]$, *there is a corresponding* $T(\epsilon, x) > 0$ *such that* $|u(t, x) - \lim_{t\to\infty} u(t, x)| < \epsilon$ *for any* $t \geq T(\epsilon, x)$ *and* $u \in D$.

# 2 Nonlinear Implicit Fractional Differential Equations

## 2.1 Introduction

Many techniques have been developed for studying the existence and uniqueness of solutions of initial value problems (IVPs) for fractional differential equations. Several authors have tried to develop techniques that depend on the Darbo or Mönch fixed point theorem with the Hausdorff or Kuratowski measure of noncompactness. The notion of the measure of noncompactness has been defined in many ways. In 1930, Kuratowski [185] defined the measure of noncompactness, $\alpha(A)$, of a bounded subset $A$ of a metric space $(X, d)$, and in 1955, Darbo [132] introduced a new type of fixed point theorem for noncompactness maps.

Recently, fractional differential equations have been studied by Abbas et al. [35, 43], Baleanu et al. [78, 80], Diethelm [137], Kilbas and Marzan [180], Srivastava et al. [181], Lakshmikantham et al. [187], and Samko et al. [239]. The purpose of this chapter is to establish existence and uniqueness results for some classes of implicit fractional differential equations by using fixed point theory (Banach contraction principle, Schauder's fixed point theorem, the nonlinear alternative of a Leray–Schauder type). Two other results are discussed; the first is based on Darbo's fixed point theorem combined with the technique of measures of noncompactness, the second is based on Mönch's fixed point theorem. Some examples are included to show the applicability of our results.

## 2.2 Existence and Stability Results for NIFDE

### 2.2.1 Introduction and Motivations

Recently, some mathematicians have considered boundary value problems (BVPs) for fractional differential equations depending on the fractional derivative. In [89], Benchohra et al. studied the problem involving Caputo's derivative

$$\begin{aligned}
&{}^cD^{\alpha}u(t) = f(t, u(t), {}^cD^{\alpha-1}u(t)), \quad \text{for each } t \in J := [0, \infty),\ 1 < \alpha \le 2,\\
&u(0) = u_0,\ u \text{ is bounded on } J.
\end{aligned}$$

In [203], Murad and Hadid, by means of Schauder fixed point theorem and the Banach contraction principle, considered the BVP for the fractional differential equation

$$\begin{aligned}
&D^{\alpha}y(t) = f(t, y(t), D^{\beta}y(t)), \quad t \in (0, 1),\ 1 < \alpha \le 2,\ 0 < \beta < 1,\ 0 < \gamma \le 1,\\
&y(0) = 0,\ y(1) = I_0^{\gamma}y(s),
\end{aligned}$$

where $D^{\alpha}$ is the Riemann–Liouville fractional derivative and $f\colon [0, 1] \times \mathbb{R} \times \mathbb{R} \to \mathbb{R}$ is a continuous function.

https://doi.org/10.1515/9783110553819-002

In [150], Lakoud and Khaldi studied the following BVP for fractional integral boundary conditions:

$$
\begin{aligned}
&{}^cD^q y(t) = f(t, y(t), {}^cD^p y(t)), \quad t \in (0,1),\ 1 < q \le 2,\ 0 < p < 1,\\
&y(0) = 0,\ y'(1) = \alpha I_0^p y(1),
\end{aligned}
$$

where ${}^cD^\alpha$ is the Caputo fractional derivative and $f\colon [0,1] \times \mathbb{R} \times \mathbb{R} \to \mathbb{R}$ is a continuous function.

The purpose of this section is to establish existence and uniqueness results for the implicit fractional order differential equation

$$ {}^cD^\alpha y(t) = f(t, y(t), {}^cD^\alpha y(t)), \quad \text{for each } t \in J = [0,T],\ T > 0,\ 0 < \alpha \le 1, \tag{2.1} $$

$$ y(0) = y_0, \tag{2.2} $$

where ${}^cD^\alpha$ is the Caputo fractional derivative, $f\colon J \times \mathbb{R} \times \mathbb{R} \to \mathbb{R}$ is a given function, and $y_0 \in \mathbb{R}$.

### 2.2.2 Existence of Solutions

Let us define what we mean by a solution of problem (2.1)–(2.2).

**Definition 2.1.** A function $u \in C^1(J, \mathbb{R})$ is said to be a solution of problem (2.1)–(2.2) if $u$ satisfies equation (2.1) and conditions (2.2) on $J$.

For the existence of solutions for problem (2.1)–(2.2), we need the following auxiliary lemma.

**Lemma 2.2.** *Let a function $f(t,u,v)\colon J \times \mathbb{R} \times \mathbb{R} \to \mathbb{R}$ be continuous. Then the problem (2.1)–(2.2) is equivalent to the problem*

$$ y(t) = y_0 + I^\alpha g(t), \tag{2.3} $$

*where $g \in C(J, \mathbb{R})$ satisfies the functional equation*

$$ g(t) = f(t, y_0 + I^\alpha g(t), g(t)). $$

*Proof.* If ${}^cD^\alpha y(t) = g(t)$, then $I^\alpha\, {}^cD^\alpha y(t) = I^\alpha g(t)$. Thus, we obtain $y(t) = y_0 + I^\alpha g(t)$. □

We are now in a position to state and prove our existence result for problem (2.1)–(2.2) based on Banach's fixed point theorem.

**Theorem 2.3.** *Make the following assumptions:*
(2.3.1) *The function $f\colon J \times \mathbb{R} \times \mathbb{R} \to \mathbb{R}$ is continuous.*

(2.3.2) *There exist constants* $K > 0$ *and* $0 < L < 1$ *such that*

$$|f(t,u,v) - f(t,\bar{u},\bar{v})| \leq K|u-\bar{u}| + L|v-\bar{v}|$$

*for any* $u, v, \bar{u}, \bar{v} \in \mathbb{R}$ *and* $t \in J$.

*If*

$$L + \frac{KT^\alpha}{\Gamma(\alpha+1)} < 1\,, \tag{2.4}$$

*then there exists a unique solution for IVP* (2.1)–(2.2) *on* $J$.

*Proof.* The proof will be given in several steps. Transform problem (2.1)–(2.2) into a fixed point problem. Define the operator $N\colon C(J,\mathbb{R}) \to C(J,\mathbb{R})$ by

$$(Ny)(t) = y_0 + I^\alpha g(t)\,, \tag{2.5}$$

where $g \in C(J,\mathbb{R})$ satisfies the functional equation

$$g(t) = f(t, y(t), g(t))\,.$$

Clearly, the fixed points of operator $N$ are solutions of problem (2.1)–(2.2). Let $u, w \in C(J,\mathbb{R})$. Then for $t \in J$ we have

$$(Nu)(t) - (Nw)(t) = \frac{1}{\Gamma(\alpha)} \int_0^t (t-s)^{\alpha-1}(g(s) - h(s))ds\,,$$

where $g, h \in C(J,\mathbb{R})$ are given by

$$g(t) = f(t, u(t), g(t))\,,$$
$$h(t) = f(t, w(t), h(t))\,.$$

Then for $t \in J$

$$|(Nu)(t) - (Nw)(t)| \leq \frac{1}{\Gamma(\alpha)} \int_0^t (t-s)^{\alpha-1}|g(s) - h(s)|ds\,. \tag{2.6}$$

By (2.3.2) we have

$$\begin{aligned} |g(t) - h(t)| &= |f(t,u(t),g(t)) - f(t,w(t),h(t))| \\ &\leq K|u(t) - w(t)| + L|g(t) - h(t)|\,. \end{aligned}$$

Thus,

$$|g(t) - h(t)| \leq \frac{K}{1-L}|u(t) - w(t)|\,.$$

By (2.6) we have

$$\begin{aligned} |(Nu)(t) - (Nw)(t)| &\leq \frac{K}{(1-L)\Gamma(\alpha)} \int_0^t (t-s)^{\alpha-1}|u(s) - w(s)|ds \\ &\leq \frac{KT^\alpha}{(1-L)\Gamma(\alpha+1)}\|u - w\|_\infty\,. \end{aligned}$$

Then

$$\|Nu - Nw\|_\infty \le \frac{KT^\alpha}{(1-L)\Gamma(\alpha+1)}\|u-w\|_\infty .$$

By (2.4), operator $N$ is a contraction. Hence, by Banach's contraction principle, $N$ has a unique fixed point that is the unique solution of problem (2.1)–(2.2). □

Our next existence result is based on Schauder's fixed point theorem.

**Theorem 2.4.** *Assume* (2.3.1) *and* (2.3.2) *hold and*
(2.4.1) *There exist* $p, q, r \in C(J, \mathbb{R}_+)$ *with* $r^* = \sup_{t\in J} r(t) < 1$ *such that*

$$|f(t,u,w)| \le p(t) + q(t)|u| + r(t)|w| \quad \text{for } t \in J, \text{ and } u, w \in \mathbb{R} .$$

*If*

$$\frac{q^* T^\alpha}{(1-r^*)\Gamma(\alpha+1)} < 1 , \tag{2.7}$$

*where* $p^* = \sup_{t\in J} p(t)$, *and* $q^* = \sup_{t\in J} q(t)$, *then the IVP* (2.1)–(2.2) *has at least one solution.*

*Proof.* Consider operator $N$ defined in (2.5). We will show that $N$ satisfies the assumptions of Schauder's fixed point theorem. The proof will be given in several steps.

*Claim 1: $N$ is continuous.* Let $\{u_n\}$ be a sequence such that $u_n \to u$ in $C(J, \mathbb{R})$. Then for each $t \in J$

$$|N(u_n)(t) - N(u)(t)| \le \frac{1}{\Gamma(\alpha)} \int_0^t (t-s)^{\alpha-1}|g_n(s) - g(s)|ds , \tag{2.8}$$

where $g_n, g \in C(J, \mathbb{R})$ satisfy

$$g_n(t) = f(t, u_n(t), g_n(t))$$

and

$$g(t) = f(t, u(t), g(t)) .$$

By (2.3.2) we have

$$\begin{aligned}|g_n(t) - g(t)| &= |f(t, u_n(t), g_n(t)) - f(t, u(t), g(t))| \\ &\le K|u_n(t) - u(t)| + L|g_n(t) - g(t)| .\end{aligned}$$

Then

$$|g_n(t) - g(t)| \le \frac{K}{1-L}|u_n(t) - u(t)| .$$

Since $u_n \to u$, we have $g_n(t) \to g(t)$ as $n \to \infty$ for each $t \in J$. Let $\eta > 0$ be such that, for each $t \in J$, we have $|g_n(t)| \le \eta$ and $|g(t)| \le \eta$. Then

$$\begin{aligned}(t-s)^{\alpha-1}|g_n(s) - g(s)| &\le (t-s)^{\alpha-1}[|g_n(s)| + |g(s)|] \\ &\le 2\eta(t-s)^{\alpha-1} .\end{aligned}$$

For each $t \in J$ the function $s \to 2\eta(t-s)^{\alpha-1}$ is integrable on $[0, t]$. By the Lebesgue dominated convergence theorem and (2.8),

$$|N(u_n)(t) - N(u)(t)| \to 0 \quad \text{as } n \to \infty .$$

Hence,

$$\|N(u_n) - N(u)\|_\infty \to 0 \quad \text{as } n \to \infty .$$

Consequently, $N$ is continuous.

Let

$$R \geq \frac{M|y_0| + p^* T^\alpha}{M - q^* T^\alpha} ,$$

where $M := (1 - r^*)\Gamma(\alpha + 1)$, and define

$$D_R = \{u \in C(J, \mathbb{R}) : \|u\|_\infty \leq R\} .$$

It is clear that $D_R$ is a bounded, closed, and convex subset of $C(J, \mathbb{R})$.

*Claim 2:* $N(D_R) \subset D_R$. Let $u \in D_R$; we will show that $Nu \in D_R$. For each $t \in J$ we have

$$|Nu(t)| \leq |y_0| + \frac{1}{\Gamma(\alpha)} \int_0^t (t-s)^{\alpha-1} |g(s)| ds . \tag{2.9}$$

By (2.4.1) we have for each $t \in J$

$$\begin{aligned} |g(t)| &= |f(t, u(t), g(t))| \\ &\leq p(t) + q(t)|u(t)| + r(t)|g(t)| \\ &\leq p(t) + q(t)R + r(t)|g(t)| \\ &\leq p^* + q^* R + r^* |g(t)| . \end{aligned}$$

Then

$$|g(t)| \leq \frac{p^* + q^* R}{1 - r^*} := \overline{M} .$$

Thus, (2.9) implies that

$$\begin{aligned} |Nu(t)| &\leq |y_0| + \frac{p^* T^\alpha}{(1 - r^*)\Gamma(\alpha + 1)} + \frac{q^* R T^\alpha}{M} \\ &\leq |y_0| + \frac{p^* T^\alpha}{M} + \frac{q^* R T^\alpha}{M} \\ &\leq R . \end{aligned}$$

Hence, $N(D_R) \subset D_R$.

*Claim 3: $N(D_R)$ is relatively compact.* Let $t_1, t_2 \in J$, $t_1 < t_2$, and let $u \in D_R$. Then

$$\begin{aligned}|N(u)(t_2) - N(u)(t_1)| &= \left|\frac{1}{\Gamma(\alpha)}\int_0^{t_1}[(t_2-s)^{\alpha-1}-(t_1-s)^{\alpha-1}]g(s)ds\right.\\ &\quad\left.+\frac{1}{\Gamma(\alpha)}\int_{t1}^{t_2}[(t_2-s)^{\alpha-1}g(s)ds\right|\\ &\le \frac{\overline{M}}{\Gamma(\alpha+1)}(t_2^\alpha - t_1^\alpha + 2(t_2-t_1)^\alpha)\,.\end{aligned}$$

As $t_1 \to t_2$, the right-hand side of the preceding inequality tends to zero.

As a consequence of Claims 1–3, together with the Ascoli–Arzelà theorem, we conclude that $N\colon C(J,\mathbb{R}) \to C(J,\mathbb{R})$ is continuous and compact. As a consequence of Schauder's fixed point theorem [149], we deduce that $N$ has a fixed point that is a solution of problem (2.1)–(2.2). □

Our next existence result is based on the nonlinear alternative of the Leray–Schauder type.

**Theorem 2.5.** *Assume* (2.3.1), (2.3.2), *and* (2.4.1) *hold. Then IVP* (2.1)–(2.2) *has at least one solution.*

*Proof.* Consider operator $N$ defined in (2.5). We will show that $N$ satisfies the assumptions of the Leray–Schauder fixed point theorem. The proof will be given in several claims.

*Claim 1: Clearly $N$ is continuous.*

*Claim 2: $N$ maps bounded sets to bounded sets in $C(J,\mathbb{R})$.* Indeed, it is enough to show that for any $\rho > 0$ there exist a positive constant $\ell$ such that for each $u \in B_\rho = \{u \in C(J,\mathbb{R})\colon \|u\|_\infty \le \rho\}$ we have $\|N(u)\|_\infty \le \ell$.

For $u \in B_\rho$ we have, for each $t \in J$,

$$|Nu(t)| \le |y_0| + \frac{1}{\Gamma(\alpha)}\int_0^t (t-s)^{\alpha-1}|g(t)|ds\,. \tag{2.10}$$

By (2.4.1), for each $t \in J$, we have

$$\begin{aligned}|g(t)| &= |f(t,u(t),g(t))|\\ &\le p(t) + q(t)|u(t)| + r(t)|g(t)|\\ &\le p(t) + q(t)\rho + r(t)|g(t)|\\ &\le p^* + q^*\rho + r^*|g(t)|\,.\end{aligned}$$

Then

$$|g(t)| \le \frac{p^*+q^*\rho}{1-r^*} := M^*\,.$$

Thus, (2.10) implies that

$$|Nu(t)| \leq |y_0| + \frac{M^* T^\alpha}{\Gamma(\alpha+1)} .$$

Hence,

$$\|Nu\|_\infty \leq |y_0| + \frac{M^* T^\alpha}{\Gamma(\alpha+1)} := l .$$

*Claim 3: Clearly, N maps bounded sets to equicontinuous sets of* $C(J, \mathbb{R})$. We conclude that $N\colon C(J,\mathbb{R}) \longrightarrow C(J,\mathbb{R})$ is continuous and completely continuous.

*Claim 4: A priori bounds.* We now show there exists an open set $U \subseteq C(J,\mathbb{R})$, with $u \neq \lambda N(u)$, for $\lambda \in (0,1)$ and $u \in \partial U$. Let $u \in C(J,\mathbb{R})$ and $u = \lambda N(u)$ for some $0 < \lambda < 1$. Thus, for each $t \in J$ we have

$$u(t) = \lambda y_0 + \frac{\lambda}{\Gamma(\alpha)} \int_0^t (t-s)^{\alpha-1} g(s)ds .$$

This implies by (2.3.2) that for each $t \in J$ we have

$$|u(t)| \leq |y_0| + \frac{1}{\Gamma(\alpha)} \int_0^t (t-s)^{\alpha-1} |g(s)|ds . \tag{2.11}$$

From (2.4.1) we have for each $t \in J$

$$\begin{aligned} |g(t)| &= |f(t, u(t), g(t))| \\ &\leq p(t) + q(t)|u(t)| + r(t)|g(t)| \\ &\leq p^* + q^*|u(t)| + r^*|g(t)| . \end{aligned}$$

Thus,

$$|g(t)| \leq \frac{1}{1-r^*}(p^* + q^*|u(t)|) .$$

Hence,

$$|u(t)| \leq |y_0| + \frac{p^* T^\alpha}{(1-r^*)\Gamma(\alpha+1)} + \frac{q^*}{(1-r^*)\Gamma(\alpha)} \int_0^t (t-s)^{\alpha-1} |u(s)|ds .$$

Then Lemma 1.52 implies that for each $t \in J$

$$|u(t)| \leq \left(|y_0| + \frac{p^* T^\alpha}{(1-r^*)\Gamma(\alpha+1)}\right)\left(1 + \frac{K q^* T^\alpha}{(1-r^*)\Gamma(\alpha+1)}\right) .$$

Thus,

$$\|u\|_\infty \leq \left(|y_0| + \frac{p^* T^\alpha}{(1-r^*)\Gamma(\alpha+1)}\right)\left(1 + \frac{K q^* T^\alpha}{(1-r^*)\Gamma(\alpha+1)}\right) := \overline{M} . \tag{2.12}$$

Let

$$U = \{u \in C(J,\mathbb{R}) \colon \|u\|_\infty < \overline{M} + 1\} .$$

By our choice of $U$, there is no $u \in \partial U$ such that $u = \lambda N(u)$ for $\lambda \in (0,1)$. As a consequence from Leray–Schauder's theorem we deduce that $N$ has a fixed point $u$ in $\overline{U}$ that is a solution of problem (2.1)–(2.2). □

### 2.2.3 Examples

*Example 1.* Consider the Cauchy problem

$$^cD^{\frac{1}{2}}y(t) = \frac{1}{2e^{t+1}\left(1 + |y(t)| + |^cD^{\frac{1}{2}}y(t)|\right)}, \quad \text{for each } t \in [0, 1], \tag{2.13}$$

$$y(0) = 1. \tag{2.14}$$

Set

$$f(t, u, v) = \frac{1}{2e^{t+1}(1 + |u| + |v|)}, \quad t \in [0, 1],\ u, v \in \mathbb{R}.$$

Clearly, the function $f$ is jointly continuous.

For any $u, v, \bar{u}, \bar{v} \in \mathbb{R}$ and $t \in [0, 1]$

$$|f(t, u, v) - f(t, \bar{u}, \bar{v})| \le \frac{1}{2e}(|u - \bar{u}| + |v - \bar{v}|).$$

Hence, condition (2.3.2) is satisfied by $K = L = \frac{1}{2e}$.

It remains to show that condition (2.4) is satisfied. Indeed, we have

$$\frac{KT^\alpha}{(1 - L)\Gamma(\alpha + 1)} = \frac{1}{(2e - 1)\Gamma\left(\frac{3}{2}\right)} < 1.$$

It follows from Theorem 2.3 that problem (2.13)–(2.14) has a unique solution.

*Example 2.* Consider the Cauchy problem

$$^cD^{\frac{1}{2}}y(t) = \frac{\left(2 + |y(t)| + |^cD^{\frac{1}{2}}y(t)|\right)}{2e^{t+1}\left(1 + |y(t)| + |^cD^{\frac{1}{2}}y(t)|\right)}, \quad \text{for each } t \in [0, 1], \tag{2.15}$$

$$y(0) = 1. \tag{2.16}$$

Set

$$f(t, u, v) = \frac{(2 + |u| + |v|)}{2e^{t+1}(1 + |u| + |v|)}, \quad t \in [0, 1],\ u, v \in \mathbb{R}.$$

Clearly, function $f$ is jointly continuous.

For any $u, v, \bar{u}, \bar{v} \in \mathbb{R}$ and $t \in [0, 1]$

$$|f(t, u, v) - f(t, \bar{u}, \bar{v})| \le \frac{1}{2e}(|u - \bar{u}| + |v - \bar{v}|).$$

Hence, condition (2.3.2) is satisfied by $K = L = \frac{1}{2e}$. Also, we have

$$|f(t, u, v)| \le \frac{1}{2e^{t+1}}(2 + |u| + |v|).$$

Thus, condition (2.4.1) is satisfied by $p(t) = \frac{1}{e^{t+1}}$ and $q(t) = r(t) = \frac{1}{2e^{t+1}}$.

Also,

$$\frac{q^* T^\alpha}{(1 - r^*)\Gamma(\alpha + 1)} = \frac{1}{(2e - 1)\Gamma\left(\frac{3}{2}\right)} < 1$$

holds with $T = 1$, $\alpha = \frac{1}{2}$, and $q^* = r^* = \frac{1}{2e}$. It follows from Theorem 2.4 that problem (2.15)–(2.16) has at least one solution.

## 2.3 NIFDE with Nonlocal Conditions

### 2.3.1 Introduction and Motivations

The purpose of this section is to establish the existence, uniqueness, and uniform stability of solutions of the implicit fractional-order differential equation with nonlocal condition:

$$^cD^\alpha y(t) = f(t, y(t), {}^cD^\alpha y(t))\,, \quad \text{for each } t \in J = [0, T],\ T > 0,\ 0 < \alpha \leq 1\,, \tag{2.17}$$

$$y(0) + \varphi(y) = y_0\,, \tag{2.18}$$

where $^cD^\alpha$ is the Caputo fractional derivative, $f : J \times \mathbb{R} \times \mathbb{R} \to \mathbb{R}$ is a given function, $\varphi : C(J, \mathbb{R}) \to \mathbb{R}$ is a continuous function, and $y_0 \in \mathbb{R}$.

### 2.3.2 Existence of Solutions

Let us define what we mean by a solution of problem (2.17)–(2.18).

**Definition 2.6.** A function $u \in C^1(J, \mathbb{R})$ is said to be a solution of problem (2.17)–(2.18) if $u$ satisfies equation (2.17) on $J$ and conditions (2.18).

For the existence of solutions for problem (2.17)–(2.18), we need the following auxiliary lemma.

**Lemma 2.7.** *Let $f : J \times \mathbb{R} \times \mathbb{R} \to \mathbb{R}$ be a continuous function. Then problem* (2.17)–(2.18) *is equivalent to the problem*

$$y(t) = y_0 - \varphi(y) + I^\alpha g(t)\,, \tag{2.19}$$

*where $g \in C(J, \mathbb{R})$ satisfies the functional equation*

$$g(t) = f(t, y_0 - \varphi(y) + I^\alpha g(t), g(t))\,.$$

*Proof.* If $^cD^\alpha y(t) = g(t)$, then $I^\alpha\, {}^cD^\alpha y(t) = I^\alpha g(t)$. Thus, we obtain $y(t) = y_0 - \varphi(y) + I^\alpha g(t)$. □

We are now in a position to state and prove our existence result for problem (2.17)–(2.18) based on Banach's fixed point theorem.

**Theorem 2.8.** *Make the following assumptions:*
(2.8.1)*The function $f : J \times \mathbb{R} \times \mathbb{R} \to \mathbb{R}$ is continuous.*
(2.8.2)*There exist constants $K > 0$ and $0 < L < 1$ such that*

$$|f(t, u, v) - f(t, \bar{u}, \bar{v})| \leq K|u - \bar{u}| + L|v - \bar{v}| \quad \textit{for any } u, v, \bar{u}, \bar{v} \in \mathbb{R}, t \in J\,.$$

(2.8.3)*There exists a constant $0 < \gamma < 1$ such that*

$$|\varphi(u) - \varphi(\bar{u})| \leq \gamma|u - \bar{u}| \quad \textit{for any } u, \bar{u} \in C(J, \mathbb{R})\,.$$

*If*

$$C := \gamma + \frac{KT^{\alpha}}{(1-L)\Gamma(\alpha+1)} < 1\,, \tag{2.20}$$

*then there exists a unique solution for problem* (2.17)–(2.18) *on* $J$.

*Proof.* The proof will be given in several steps. Transform problem (2.17)–(2.18) into a fixed point problem. Define the operator $N\colon C(J,\mathbb{R}) \to C(J,\mathbb{R})$ by

$$N(y)(t) = y_0 - \varphi(y) + I^{\alpha}g(t)\,, \tag{2.21}$$

where $g \in C(J,\mathbb{R})$ satisfies the functional equation

$$g(t) = f(t, y(t), g(t))\,.$$

Clearly, the fixed points of operator $N$ are solutions of problem (2.17)–(2.18). Let $u, w \in C(J,\mathbb{R})$. Then for $t \in J$ we have

$$\begin{aligned}(Nu)(t) - (Nw)(t) &= \varphi(w) - \varphi(u)\\ &\quad + \frac{1}{\Gamma(\alpha)}\int_0^t (t-s)^{\alpha-1}(g(s)-h(s))ds\,,\end{aligned}$$

where $g, h \in C(J,\mathbb{R})$ are given by

$$\begin{aligned}g(t) &= f(t, u(t), g(t))\,,\\ h(t) &= f(t, w(t), h(t))\,.\end{aligned}$$

Then, for $t \in J$,

$$\begin{aligned}|(Nu)(t) - (Nw)(t)| &\le |\varphi(u) - \varphi(w)|\\ &\quad + \frac{1}{\Gamma(\alpha)}\int_0^t (t-s)^{\alpha-1}|g(s)-h(s)|ds\,.\end{aligned} \tag{2.22}$$

By (2.8.2) we have

$$\begin{aligned}|g(t) - h(t)| &= |f(t, u(t), g(t)) - f(t, w(t), h(t))|\\ &\le K|u(t) - w(t)| + L|g(t) - h(t)|\,.\end{aligned}$$

Thus,

$$|g(t) - h(t)| \le \frac{K}{1-L}|u(t) - w(t)|\,.$$

From (2.22) and (2.8.3) we have

$$
\begin{aligned}
|(Nu)(t) - (Nw)(t)| &\le \gamma |u(t) - w(t)| \\
&\quad + \frac{K}{(1-L)\Gamma(\alpha)} \int_0^t (t-s)^{\alpha-1} |u(s) - w(s)| ds \\
&\le \gamma \|u - w\|_\infty \\
&\quad + \sup_{0 \le t \le T} |u(t) - w(t)| \frac{K}{(1-L)\Gamma(\alpha)} \int_0^t (t-s)^{\alpha-1} ds \\
&\le \gamma \|u - w\|_\infty + \frac{KT^\alpha}{(1-L)\Gamma(\alpha+1)} \|u - w\|_\infty .
\end{aligned}
$$

Then

$$
\|Nu - Nw\|_\infty \le \left[ \gamma + \frac{KT^\alpha}{(1-L)\Gamma(\alpha+1)} \right] \|u - w\|_\infty .
$$

By (2.20), operator $N$ is a contraction. Hence, by Banach's contraction principle, $N$ has a unique fixed point that is the unique solution of problem (2.17)–(2.18). □

The second result is based on Krasnosel'skii's fixed point theorem.

Let $\widetilde{M} := \frac{T^\alpha}{(1-L)\Gamma(\alpha+1)}$, $a := |\varphi(0)|$, and $f^* := \sup_{0 \le t \le T} |f(t, 0, 0)|$.

**Theorem 2.9.** *Assume that* (2.8.1)–(2.8.3) *hold. If*

$$
\gamma + \widetilde{M} K < 1 , \tag{2.23}
$$

*then problem* (2.17)–(2.18) *has at least one solution.*

*Proof.* Consider operator $N$ to be defined as in (2.21). We have

$$
N(y)(t) = y_0 - \varphi(y) + \frac{1}{\Gamma(\alpha)} \int_0^t (t-s)^{\alpha-1} g(s) ds ,
$$

where $g \in C(J, \mathbb{R})$ satisfies the functional equation

$$
g(t) = f(t, y(t), g(t)) .
$$

Let

$$
R \ge \frac{|y_0| + a + \widetilde{M} f^*}{1 - \gamma - \widetilde{M} K} ,
$$

and define

$$
D_R = \{u \in C(J, \mathbb{R}) : \|u\|_\infty \le R\} .
$$

It is clear that $D_R$ is a bounded, closed, and convex subset of $C(J, \mathbb{R})$. Define on $D_R$ operators $P$ and $Q$ by

$$
P(u)(t) = \frac{1}{\Gamma(\alpha)} \int_0^t (t-s)^{\alpha-1} g(s) ds , \tag{2.24}
$$

where $g \in C(J, \mathbb{R})$ satisfies the functional equation

$$g(t) = f(t, u(t), g(t))$$

and

$$Q(v)(t) = y_0 - \varphi(v) . \tag{2.25}$$

*Claim 1: For any $u, v \in D_R$, $Pu + Qv \in D_R$.* For any $u, v \in D_R$ and $t \in J$ we have

$$|P(u)(t) + Q(v)(t)| \leq |y_0| + |\varphi(v)| + \frac{1}{\Gamma(\alpha)} \int_0^t (t-s)^{\alpha-1} |g(s)| ds . \tag{2.26}$$

By (2.8.2) we have for each $t \in J$

$$\begin{aligned} |g(t)| &= |f(t, u(t), g(t))| \\ &\leq |f(t, u(t), g(t)) - f(t, 0, 0)| + |f(t, 0, 0)|) \\ &\leq K|u(t)| + L|g(t)| + \sup_{0 \leq t \leq T} |f(t, 0, 0)| \\ &\leq KR + L|g(t)| + f^* . \end{aligned}$$

Then

$$(1 - L)|g(t)| \leq KR + f^* .$$

Thus,

$$|g(t)| \leq \frac{KR + f^*}{1 - L} . \tag{2.27}$$

From (2.8.3) we have

$$\begin{aligned} |\varphi(v)| &\leq |\varphi(v) - \varphi(0)| + |\varphi(0)| \\ &\leq \gamma|v| + a \\ &\leq \gamma R + a . \end{aligned}$$

Then, by (2.26), we get

$$\begin{aligned} |P(u)(t) + Q(v)(t)| &\leq |y_0| + (\gamma R + a) + \frac{KR + f^*}{(1-L)\Gamma(\alpha)} \int_0^t (t-s)^{\alpha-1} ds \\ &\leq |y_0| + (\gamma R + a) + \frac{(KR + f^*)T^\alpha}{(1-L)\Gamma(\alpha+1)} \\ &= |y_0| + \gamma R + a + \widetilde{M}(KR + f^*) \\ &\leq R . \end{aligned}$$

Thus, $Pu + Qv \in D_R$.

*Claim 2: $Q$ is a contraction mapping on $D_R$.* For any $v_1, v_2 \in D_R$, by (2.8.3) we have

$$\begin{aligned} |Q(v_2) - Q(v_1)| &\leq |\varphi(v_2) - \varphi(v_1)| \\ &\leq \gamma|v_2 - v_1| . \end{aligned}$$

Thus,

$$\|Q(v_2) - Q(v_1)\|_\infty \le \gamma \|v_2 - v_1\|_\infty ,$$

and so $Q$ is a contraction mapping.

*Claim 3: $P$ is continuous.* Let $\{u_n\}$ be a sequence such that $u_n \to u$ in $C(J, \mathbb{R})$. Then for each $t \in J$

$$|P(u_n)(t) - P(u)(t)| \le \frac{1}{\Gamma(\alpha)} \int_0^t (t-s)^{\alpha-1} |g_n(s) - g(s)| ds , \tag{2.28}$$

where $g_n, g \in C(J, \mathbb{R})$ satisfy

$$g_n(t) = f(t, u_n(t), g_n(t))$$

and

$$g(t) = f(t, u(t), g(t)) .$$

By (2.8.2) we have

$$\begin{aligned} |g_n(t) - g(t)| &= |f(t, u_n(t), g_n(t)) - f(t, u(t), g(t))| \\ &\le K|u_n(t) - u(t)| + L|g_n(t) - g(t)| , \end{aligned}$$

so

$$|g_n(t) - g(t)| \le \frac{K}{1-L} |u_n(t) - u(t)| .$$

Since $u_n \to u$, we get $g_n(t) \to g(t)$ as $n \to \infty$ for each $t \in J$. Let $\eta > 0$ be such that, for each $t \in J$, we have $|g_n(t)| \le \eta$ and $|g(t)| \le \eta$; then we have

$$\begin{aligned} (t-s)^{\alpha-1} |g_n(s) - g(s)| &\le (t-s)^{\alpha-1} [|g_n(s)| + |g(s)|] \\ &\le 2\eta (t-s)^{\alpha-1} . \end{aligned}$$

For each $t \in J$ the function $s \to 2\eta(t-s)^{\alpha-1}$ is integrable on $[0, t]$; then the Lebesgue dominated convergence theorem and (2.28) imply that

$$|(Pu_n)(t) - (Pu)(t)| \to 0 \quad \text{as } n \to \infty .$$

Hence,

$$\|P(u_n) - P(u)\|_\infty \to 0 \quad \text{as } n \to \infty .$$

Consequently, $P$ is continuous.

*Claim 4: $P$ is compact.* Let $\{u_n\}$ be a sequence on $D_R$. Then, for each $t \in J$, we have

$$|P(u_n)(t)| \le \frac{1}{\Gamma(\alpha)} \int_0^t (t-s)^{\alpha-1} |g_n(s)| ds , \tag{2.29}$$

where $g_n \in C(J, \mathbb{R})$ is given by

$$g_n(t) = f(t, u_n(t), g_n(t)) .$$

By (2.8.2) we have for each $t \in J$

$$\begin{aligned}|g_n(t)| &= |f(t, u_n(t), g(t))| \\ &\le |f(t, u_n(t), g_n(t)) - f(t,0,0)| + |f(t,0,0)|) \\ &\le K|u_n(t)| + L|g_n(t)| + \sup_{0\le t\le T} |f(t,0,0)| \\ &\le KR + L|g_n(t)| + f^* .\end{aligned}$$

Then

$$(1-L)|g_n(t)| \le KR + f^* ,$$

and so

$$|g_n(t)| \le \frac{KR + f^*}{1-L} . \tag{2.30}$$

Thus, (2.29) implies

$$\begin{aligned}|P(u_n)(t)| &\le \frac{KR + f^*}{(1-L)\Gamma(\alpha)} \int_0^t (t-s)^{\alpha-1} ds \\ &\le \frac{(KR+f^*)T^\alpha}{(1-L)\Gamma(\alpha+1)} \\ &\le \widetilde{M}(KR + f^*) ,\end{aligned}$$

and we see that $\{u_n\}$ is uniformly bounded.

Now we prove that $\{P(u_n)\}$ is equicontinuous. Let $t_1, t_2 \in J$, $t_1 < t_2$, and let $u \in D_R$. Then

$$\begin{aligned}|(Pu)(t_2) - (Pu)(t_1)| &= \left| \frac{1}{\Gamma(\alpha)} \int_0^{t_1} [(t_2-s)^{\alpha-1} - (t_1-s)^{\alpha-1}] g(s) ds \right. \\ &\quad \left. + \frac{1}{\Gamma(\alpha)} \int_{t1}^{t_2} (t_2-s)^{\alpha-1} g(s) ds \right| \\ &\le \frac{1}{\Gamma(\alpha)} \int_0^{t_1} |[(t_2-s)^{\alpha-1} - (t_1-s)^{\alpha-1}]| |g(s)| ds \\ &\quad + \frac{1}{\Gamma(\alpha)} \int_{t1}^{t_2} (t_2-s)^{\alpha-1} |g(s)| ds \\ &\le \frac{KR + f^*}{(1-L)\Gamma(\alpha+1)} (t_2^\alpha - t_1^\alpha + 2(t_2-t_1)^\alpha) .\end{aligned}$$

As $t_1 \to t_2$, the right-hand side of the preceding inequality tends to zero. As a consequence of Claims 1–4, together with the Ascoli–Arzelà theorem, we conclude that $N: C(J, \mathbb{R}) \to C(J, \mathbb{R})$ is continuous and compact. As a consequence of Krasnosel'skii's fixed point theorem, we deduce that $N$ has a fixed point that is a solution of problem (2.17)–(2.18). □

### 2.3.3 Stability Results

Here we consider the uniform stability of the solutions of problem (2.17)–(2.18) and adopt the definitions in [138].

**Definition 2.10.** The solution of equation (2.17) is uniformly stable if for any $\epsilon > 0$ there exists $\delta(\epsilon) > 0$ such that for any two solutions $y(t)$ and $\tilde{y}(t)$ corresponding to the initial conditions (2.18) and $\tilde{y}(0) = \tilde{y}_0 - \varphi(\tilde{y})$, respectively, with $|y_0 - \tilde{y}_0| \le \delta$, one has $\|y - \tilde{y}\|_\infty \le \epsilon$.

**Theorem 2.11.** *Assume* (2.8.1)–(2.8.3) *and* (2.20) *hold. Then the solutions of the Cauchy problem* (2.17)–(2.18) *are uniformly stable.*

*Proof.* Let $y$ be a solution of

$$y(t) = y_0 - \varphi(y) + \frac{1}{\Gamma(\alpha)} \int_0^t (t-s)^{\alpha-1} g(s)ds \,, \tag{2.31}$$

where $g \in C(J, \mathbb{R})$ satisfies the functional equation

$$g(t) = f(t, y(t), g(t)) \,,$$

and let $\tilde{y}$ be a solution of equation (2.31) such that

$$\tilde{y}(0) = \tilde{y}_0 - \varphi(\tilde{y}) \,.$$

Then we have

$$\tilde{y}(t) = \tilde{y}_0 - \varphi(\tilde{y}) + \frac{1}{\Gamma(\alpha)} \int_0^t (t-s)^{\alpha-1} h(s)ds \,, \tag{2.32}$$

where $h \in C(J, \mathbb{R})$ satisfies the functional equation

$$h(t) = f(t, \tilde{y}(t), h(t)) \,.$$

By (2.31) and (2.32) we have

$$\begin{aligned} |y(t) - \tilde{y}(t)| &\le |y_0 - \tilde{y}_0| + |\varphi(y) - \varphi(\tilde{y})| \\ &\quad + \frac{1}{\Gamma(\alpha)} \int_0^t (t-s)^{\alpha-1} |g(s) - h(s)| ds \,, \end{aligned} \tag{2.33}$$

and by (2.8.2) we have

$$\begin{aligned} |g(t) - h(t)| &= |f(t, y(t), g(t)) - f(t, \tilde{y}(t), h(t))| \\ &\le K|y(t) - \tilde{y}(t)| + L|g(t) - h(t)| \,, \end{aligned}$$

so

$$|g(t) - h(t)| \le \frac{K}{1-L} |y(t) - \tilde{y}(t)| \,.$$

Thus, (2.33) and (2.8.3) imply that

$$
\begin{aligned}
|y(t)-\tilde{y}(t)| &\le |y_0-\tilde{y}_0| + \gamma|y(t)-\tilde{y}(t)| \\
&\quad + \frac{K}{(1-L)\Gamma(\alpha)}\int_0^t (t-s)^{\alpha-1}|y(s)-\tilde{y}(s)|ds \\
&\le |y_0-\tilde{y}_0| + \gamma\|y-\tilde{y}\|_\infty \\
&\quad + \sup_{0\le t\le T}|y(t)-\tilde{y}(t)|\frac{K}{(1-L)\Gamma(\alpha)}\int_0^t (t-s)^{\alpha-1}ds \\
&\le |y_0-\tilde{y}_0| + \gamma\|y-\tilde{y}\|_\infty \\
&\quad + \|y-\tilde{y}\|_\infty \frac{KT^\alpha}{(1-L)\Gamma(\alpha+1)} .
\end{aligned}
$$

Then

$$\left[1-\gamma-\frac{KT^\alpha}{(1-L)\Gamma(\alpha+1)}\right]\|y-\tilde{y}\|_\infty \le |y_0-\tilde{y}_0| .$$

This implies that

$$\|y-\tilde{y}\|_\infty \le (1-C)^{-1}|y_0-\tilde{y}_0| . \tag{2.34}$$

For $\epsilon > 0$ it suffices to make $(1-C)^{-1}|y_0-\tilde{y}_0| \le \epsilon$. This suggests that we choose $\delta = (1-C)\epsilon$. Therefore, if $|y_0-\tilde{y}_0| \le \delta(\epsilon)$, then $\|y-\tilde{y}\|_\infty \le \epsilon$. This implies that the solution $y$ is uniformly stable. □

### 2.3.4 An Example

Consider the problem with nonlocal conditions

$$ {}^cD^{\frac{1}{2}}y(t) = \frac{1}{2e^{t+1}\left(1+|y(t)|+|{}^cD^{\frac{1}{2}}y(t)|\right)}, \quad \text{for each } t\in[0,1] , \tag{2.35}$$

$$y(0)+\varphi(y) = 1 , \tag{2.36}$$

where

$$\varphi(y) = \frac{|y|}{10+|y|} . \tag{2.37}$$

Set

$$f(t,u,v) = \frac{1}{2e^{t+1}(1+|u|+|v|)}, \quad t\in[0,1],\ u,v\in\mathbb{R} .$$

Clearly, the function $f$ is jointly continuous. For any $u, v, \bar{u}, \bar{v} \in \mathbb{R}$ and $t \in [0,1]$

$$|f(t,u,v)-f(t,\bar{u},\bar{v})| \le \frac{1}{2e}(|u-\bar{u}|+|v-\bar{v}|) .$$

Hence, condition (2.8.2) is satisfied by $K = L = \frac{1}{2e}$. Let

$$\varphi(u) = \frac{u}{10+u}, \qquad u \in [0, \infty),$$

and take $u, v \in [0, \infty)$. Then we have

$$\begin{aligned} |\varphi(u) - \varphi(v)| &= \left|\frac{u}{10+u} - \frac{v}{10+v}\right| = \frac{10|u-v|}{(10+u)(10+v)} \\ &\le \frac{1}{10}|u-v| . \end{aligned}$$

Thus the condition

$$C = \gamma + \frac{KT^\alpha}{(1-L)\Gamma(\alpha+1)} < 1$$

is satisfied by $T = 1$, $\gamma = \frac{1}{10}$, and $\alpha = \frac{1}{2}$. It follows from Theorems 2.9 and 2.11 that problem (2.35)–(2.37) is a unique uniformly stable solution on $J$.

## 2.4 Existence Results for NIFDE in Banach Space

### 2.4.1 Introduction and Motivations

Recently, fractional differential equations have been studied by Abbas et al. [35, 43], Baleanu et al. [78, 80], Diethelm [137], Kilbas and Marzan [180], Srivastava et al. [181], Lakshmikantham et al. [187], and Samko et al. [239]. More recently, some mathematicians have considered BVPs and boundary conditions for implicit fractional differential equations.

In [164], Hu and Wang investigated the existence of solutions of nonlinear fractional differential equations with integral boundary conditions

$$\begin{aligned} &D^\alpha u(t) = f(t, u(t), D^\beta u(t)), \quad t \in (0,1),\ 1 < \alpha \le 2,\ 0 < \beta < 1, \\ &u(0) = u_0,\ u(1) = \int_0^1 g(s)u(s)ds , \end{aligned}$$

where $D^\alpha$ is the Riemann–Liouville fractional derivative, $f\colon [0,1] \times \mathbb{R} \times \mathbb{R} \to \mathbb{R}$ is a continuous function, and $g$ is an integrable function.

In [241], by means of Schauder's fixed point theorem, Su and Liu studied the existence of solutions of nonlinear fractional BVPs involving Caputo's derivative

$$\begin{aligned} &{}^cD^\alpha u(t) = f(t, u(t), {}^cD^\beta u(t)), \quad \text{for each } t \in (0,1),\ 1 < \alpha \le 2,\ 0 < \beta \le 1, \\ &u(0) = u'(1) = 0, \ \text{ or } u'(1) = u(1) = 0, \ \text{ or } u(0) = u(1) = 0 , \end{aligned}$$

where $f\colon [0,1] \times \mathbb{R} \times \mathbb{R} \to \mathbb{R}$ is a given continuous function.

The purpose of this section is to establish existence and uniqueness results for the implicit fractional differential equation

$$^{c}D^{\nu}y(t) = f(t, y(t), {}^{c}D^{\nu}y(t)), \quad \text{for each } t \in J := [0, T],\ T > 0,\ 0 < \nu \leq 1, \tag{2.38}$$

with the initial condition

$$y(0) = y_0, \tag{2.39}$$

where $^{c}D^{\nu}$ is the Caputo fractional derivative, $(E, \|\cdot\|)$ is a real Banach space, $f: J \times E \times E \to E$ is a continuous function, and $y_0 \in E$.

### 2.4.2 Existence of Solutions

Let $(E; \|\cdot\|)$ be a Banach space and $t \in J$. For a given set $V$ of functions $v: J \to E$, let us use the notation

$$V(t) = \{v(t), v \in V\}, \quad t \in J$$

and

$$V(J) = \{v(t): v \in V, t \in J\}.$$

Next we define what we mean by a solution of problem (2.38)–(2.39).

**Definition 2.12.** A function $u \in C^1(J, E)$ is said to be a solution of problem (2.38)–(2.39) if $u$ satisfies equation (2.38) and condition (2.39) on $J$.

For the existence of solutions of problem (2.38)–(2.39), we need the following auxiliary lemma.

**Lemma 2.13.** *Suppose that the function $f(t, u, v): J \times E \times E \to E$ is continuous; then problem* (2.38)–(2.39) *is equivalent to the problem*

$$y(t) = y_0 + I^{\nu}g(t), \tag{2.40}$$

*where $g \in C(J, E)$ satisfies the functional equation*

$$g(t) = f(t, y_0 + I^{\nu}g(t), g(t)).$$

*Proof.* If $^{c}D^{\nu}y(t) = g(t)$, then $I^{\nu}\, {}^{c}D^{\nu}y(t) = I^{\nu}g(t)$. Thus, we obtain $y(t) = y_0 + I^{\nu}g(t)$. □

We list the following conditions:
(2.13.1) The function $f: J \times E \times E \to E$ is continuous.
(2.13.2) There exist constants $K > 0$ and $0 < L < 1$, such that

$$\|f(t, u, v) - f(t, \bar{u}, \bar{v})\| \leq K\|u - \bar{u}\| + L\|v - \bar{v}\|$$

for any $u, v, \bar{u}, \bar{v} \in E$ and $t \in J$.

(2.13.3) There exist $p, q, r \in C(J, \mathbb{R}_+)$, with $r^* = \sup_{t\in J} r(t) < 1$, such that

$$\|f(t, u, w)\| \le p(t) + q(t)|u| + r(t)|w| \quad \text{for } t \in J \text{ and } u, w \in \mathbb{R} .$$

**Remark 2.14** ([66]). If

$$\|f(t, u, v) - f(t, \bar{u}, \bar{v})\| \le K\|u - \bar{u}\| + L\|v - \bar{v}\|$$

for any $u, v, \bar{u}, \bar{v} \in E$ and $t \in J$, then

(2.14.1)

$$\alpha(f(t, B_1, B_2)) \le K\alpha(B_1) + L\alpha(B_2)$$

for each $t \in J$ and bounded sets $B_1, B_2 \subseteq E$.

We are now in a position to state and prove our existence result for problem (2.38)–(2.39) based on the concept of measures of noncompactness and Darbo's fixed point theorem.

**Theorem 2.15.** *Assume* (2.13.1)–(2.13.3). *If*

$$\frac{KT^{\nu}}{(1 - L)\Gamma(\nu + 1)} < 1 , \tag{2.41}$$

*then IVP* (2.38)–(2.39) *has at least one solution on J.*

*Proof.* Transform problem (2.38)–(2.39) into a fixed point problem. Define the operator $N: C(J, E) \to C(J, E)$ by

$$(Ny)(t) = y_0 + I^{\nu} g(t) , \tag{2.42}$$

where $g \in C(J, E)$ satisfies the functional equation

$$g(t) = f(t, y(t), g(t)) .$$

Clearly, the fixed points of operator $N$ are solutions of problem (2.38)–(2.39). We will show that $N$ satisfies the assumptions of Darbo's fixed point theorem. The proof will be given in several steps.

*Claim 1: N is continuous.* Let $u, w \in C(J, E)$, and let $\{u_n\}$ be a sequence such that $u_n \to u$ in $C(J, E)$. Then for each $t \in J$

$$\|N(u_n)(t) - N(u)(t)\| \le \frac{1}{\Gamma(\nu)} \int_0^t (t - s)^{\nu-1} \|g_n(s) - g(s)\| ds , \tag{2.43}$$

where $g_n, g \in C(J, E)$ such that

$$g_n(t) = f(t, u_n(t), g_n(t))$$

and

$$g(t) = f(t, u(t), g(t)) .$$

By (2.13.2), for each $t \in J$, we have

$$\begin{aligned}\|g_n(t) - g(t)\| &= \|f(t, u_n(t), g_n(t)) - f(t, u(t), g(t))\| \\ &\le K\|u_n(t) - u(t)\| + L\|g_n(t) - g(t)\| .\end{aligned}$$

Then

$$\|g_n(t) - g(t)\| \le \frac{K}{1 - L}\|u_n(t) - u(t)\| .$$

Since $u_n \to u$, we get $g_n(t) \to g(t)$ as $n \to \infty$ for each $t \in J$.

Let a positive constant $\eta > 0$ be such that, for each $t \in J$, we have $\|g_n(t)\| \le \eta$ and $\|g(t)\| \le \eta$. Then we have

$$\begin{aligned}(t - s)^{\nu-1}\|g_n(s) - g(s)\| &\le (t - s)^{\nu-1}[\|g_n(s)\| + \|g(s)\|] \\ &\le 2\eta(t - s)^{\nu-1} .\end{aligned}$$

For each $t \in J$, the function $s \to 2\eta(t - s)^{\nu-1}$ is integrable on $[0, t]$, so by the Lebesgue dominated convergence theorem and (7.2),

$$\|N(u_n)(t) - N(u)(t)\| \to 0 \ as \ n \to \infty .$$

Then

$$\|N(u_n) - N(u)\|_\infty \to 0 \ as \ n \to \infty .$$

Consequently, $N$ is continuous.

Let

$$R \ge \frac{M|y_0| + p^* T^\alpha}{M - q^* T^\alpha} , \tag{2.44}$$

where $M := (1 - r^*)\Gamma(\alpha + 1)$, $p^* = \sup_{t\in J} p(t)$, and $q^* = \sup_{t\in J} q(t)$.

Define

$$D_R = \{u \in C(J, E) \colon \|u\|_\infty \le R\} .$$

It is clear that $D_R$ is a bounded, closed, and convex subset of $C(J, E)$.

*Claim 2: $N(D_R) \subset D_R$.* Let $u \in D_R$. We will show that $Nu \in D_R$. We have for each $t \in J$

$$\|Nu(t)\| \le \|y_0\| + \frac{1}{\Gamma(\nu)}\int_0^t (t - s)^{\nu-1}\|g(t)\|ds . \tag{2.45}$$

By (2.13.3) we have

$$\begin{aligned}\|g(t)\| &= \|f(t, u(t), g(t))\| \\ &\le p(t) + q(t)\|u(t)\| + r(t)\|g(t)\| \\ &\le p(t) + q(t)R + r(t)|g(t)| \\ &\le p^* + q^* R + r^*\|g(t)\| .\end{aligned}$$

Then for each $t \in J$

$$\|g(t)\| \leq \frac{p^* + q^* R}{1 - r^*}.$$

Thus, (2.44) and (2.45) imply that

$$\begin{aligned}\|Nu(t)\| &\leq \|y_0\| + \frac{p^* T^\nu}{(1 - r^*)\Gamma(\nu + 1)} + \frac{q^* R T^\nu}{(1 - r^*)\Gamma(\nu + 1)} \\ &\leq \|y_0\| + \frac{p^* T^\nu}{M} + \frac{q^* R T^\nu}{M} \\ &\leq R.\end{aligned}$$

Consequently,

$$N(D_R) \subset D_R.$$

*Claim 3: $N(D_R)$ is bounded and equicontinuous.* By Claim 2 we have $N(D_R) = \{N(u): u \in D_R\} \subset D_R$. Then for each $u \in D_R$ we have $\|N(u)\|_\infty \leq R$. Thus, $N(D_R)$ is bounded. Let $t_1, t_2 \in J$, $t_1 < t_2$, and let $u \in D_R$. Then

$$\begin{aligned}|(Nu)(t_2) - (Nu)(t_1)| &= \left| \frac{1}{\Gamma(\nu)} \int_0^{t_1} [(t_2 - s)^{\nu-1} - (t_1 - s)^{\nu-1}] g(s) ds \right. \\ &\quad \left. + \frac{1}{\Gamma(\nu)} \int_{t1}^{t_2} [(t_2 - s)^{\nu-1} g(s) ds \right| \\ &\leq \frac{M}{\Gamma(\nu + 1)} (t_2^\nu - t_1^\nu + 2(t_2 - t_1)^\nu).\end{aligned}$$

As $t_1 \to t_2$, the right-hand side of the preceding inequality tends to zero, so $N(D_R)$ is equicontinuous.

*Claim 4: The operator $N: B_R \to B_R$ is a strict set contraction.* Let $V \subset B_R$ and $t \in J$; then we have

$$\begin{aligned}\alpha(N(V)(t)) &= \alpha(\{(Ny)(t), y \in V\}) \\ &\leq \frac{1}{\Gamma(\nu)} \left\{ \int_0^t (t - s)^{\nu-1} \alpha(g(s)) ds : y \in V \right\}.\end{aligned}$$

By (2.14.1), Remark 2.14, and Lemma 1.32, for each $s \in J$,

$$\begin{aligned}\alpha(\{g(s): y \in V\}) &= \alpha(\{f(s, y(s), g(s)): y \in V\}) \\ &\leq K\alpha(\{y(s), y \in V\}) + L\alpha(\{g(s): y \in V\}).\end{aligned}$$

Thus,

$$\alpha(\{g(s): y \in V\}) \leq \frac{K}{1 - L} \alpha\{y(s): y \in V\}.$$

Then

$$\begin{aligned}\alpha(N(V)(t)) &\leq \frac{K}{(1-L)\Gamma(\nu)}\int_0^t (t-s)^{\nu-1}\{\alpha(y(s))ds : y \in V\} \\ &\leq \frac{K\alpha_c(V)}{(1-L)\Gamma(\nu)}\int_0^t (t-s)^{\nu-1}ds \\ &\leq \frac{KT^{\nu}}{(1-L)\Gamma(\nu+1)}\alpha_c(V)\,.\end{aligned}$$

Therefore,

$$\alpha_c(NV) \leq \frac{KT^{\nu}}{(1-L)\Gamma(\nu+1)}\alpha_c(V)\,.$$

So, by (2.41), operator $N$ is a set contraction. As a consequence of Theorem 1.45, we deduce that $N$ has a fixed point that is a solution of problem (2.38)–(2.39). □

Our next existence result for problem (2.38)–(2.39) is based on the concept of measures of noncompactness and Mönch's fixed point theorem.

**Theorem 2.16.** *Assume* (2.13.1)–(2.13.3). *Then IVP* (2.38)–(2.39) *has at least one solution.*

*Proof.* Consider operator $N$ defined in (2.42). We will show that $N$ satisfies the assumptions of Mönch's fixed point theorem. We know that $N : B_R \to B_R$ is bounded and continuous; we need to prove that the implication

$$[V = \overline{\text{conv}}N(V) \text{ or } V = N(V) \cup \{0\}] \Rightarrow \alpha(V) = 0$$

holds for every subset $V$ of $B_R$.

Now let $V$ be a subset of $B_R$ such that $V \subset \overline{\text{conv}}(N(V) \cup \{0\})$; now $V$ is bounded and equicontinuous, and therefore the function $t \to v(t) = \alpha(V(t))$ is continuous on $J$. By (2.14.1), Lemma 1.33, and the properties of the measure $\alpha$, we have for each $t \in J$

$$\begin{aligned}v(t) &\leq \alpha(N(V)(t) \cup \{0\}) \\ &\leq \alpha(N(V)(t)) \\ &\leq \alpha\{(Ny)(t) : y \in V\} \\ &\leq \frac{K}{(1-L)\Gamma(\nu)}\int_0^t (t-s)^{\nu-1}\alpha\left(\{(y(s) : y \in V\}\right)ds \\ &\leq \frac{K}{(1-L)\Gamma(\nu)}\int_0^t (t-s)^{\nu-1}v(s)ds\,.\end{aligned}$$

Lemma 1.52 implies that $v(t) = 0$ for each $t \in J$, and so $V(t)$ is relatively compact in $E$. In view of the Ascoli–Arzelà theorem, $V$ is relatively compact in $B_R$. Applying Theorem 1.46 we conclude that $N$ has a fixed point $y \in B_R$. Hence, $N$ has a fixed point that is a solution of the problem (2.38)–(2.39). □

### 2.4.3 An Example

Consider the infinite system

$$ {}^cD^{\frac{1}{2}}y_n(t) = \frac{(3 + \|y_n(t)\| + \|{}^cD^{\frac{1}{2}}y_n(t)\|)}{3e^{t+2}(1 + \|y_n(t)\| + \|{}^cD^{\frac{1}{2}}y_n(t)\|)} \quad \text{for each } t \in [0, 1] , \tag{2.46} $$

$$ y_n(0) = 1 . \tag{2.47} $$

Set

$$ E = l^1 = \{y = (y_1, y_2, \dots, y_n, \dots), \sum_{n=1}^{\infty} |y_n| < \infty\} $$

and

$$ f(t, u, v) = \frac{(3 + \|u\| + \|v\|)}{3e^{t+2}(1 + \|u\| + \|v\|)} , \quad t \in [0, 1], \ u, v \in E , $$

where $E$ is a Banach space with the norm $\|y\| = \sum_{n=1}^{\infty} |y_n|$. Clearly, the function $f$ is jointly continuous. For any $u, v, \bar{u}, \bar{v} \in E$ and $t \in [0, 1]$

$$ \|f(t, u, v) - f(t, \bar{u}, \bar{v})\| \le \frac{2}{3e^2}(\|u - \bar{u}\| + \|v - \bar{v}\|) . $$

Hence condition (2.13.2) is satisfied by $K = L = \frac{2}{3e^2}$. Also,

$$ \|f(t, u, v)\| \le \frac{1}{3e^{t+2}}(3 + \|u\| + \|v\|) . $$

Thus, conditions (2.13.3) and (2.14.1) are satisfied by $p(t) = \frac{1}{e^{t+2}}$, and $q(t) = r(t) = \frac{1}{3e^{t+2}}$. Theorem 2.16 implies that problem (2.46)–(2.47) has at least one solution on $J$.

## 2.5 Existence and Stability Results for Perturbed NIFDE with Finite Delay

### 2.5.1 Introduction

In this section, we establish existence, uniqueness, and stability results for the perturbed functional differential equations of fractional order with finite delay

$$ {}^cD^{\alpha}y(t) = f(t, y_t, {}^cD^{\alpha}y(t)) + g(t, y_t) , \quad t \in J = [0, T], \ T > 0, \ 0 < \alpha \le 1 , \tag{2.48} $$

$$ y(t) = \varphi(t) , \quad t \in [-r, 0], \ r > 0 , \tag{2.49} $$

where $f : J \times C([-r, 0], \mathbb{R}) \times \mathbb{R} \to \mathbb{R}$ and $g : J \times C([-r, 0], \mathbb{R}) \to \mathbb{R}$ are two given functions and $\varphi \in C([-r, 0], \mathbb{R})$. The arguments are based upon the Banach contraction principle and a fixed point theorem of Burton and Kirk.

### 2.5.2 Existence of Solutions

Set

$$Q = \{y\colon [-r, T] \to \mathbb{R}\colon y|_{[-r,0]} \in C([-r, 0], \mathbb{R}) \text{ and } y|_{[0,T]} \in C([0, T], \mathbb{R})\}\,;$$

then $Q$ is a Banach space with the norm

$$\|y\|_Q = \sup_{t\in[-r,T]} |y(t)|\,.$$

**Definition 2.17.** A function $y \in Q$ is called a solution of problem (2.48)–(2.49) if it satisfies equation (2.48) on $J$ and condition (2.49) on $[-r, 0]$.

**Lemma 2.18.** *Let $0 < \alpha \le 1$ and $h\colon [0, T] \to \mathbb{R}$ be a continuous function. The linear problem*

$$\begin{aligned} {}^cD^\alpha y(t) &= h(t)\,, \quad t \in J\,,\\ y(t) &= \varphi(t)\,, \quad t \in [-r, 0]\,, \end{aligned}$$

*has a unique solution given by*

$$y(t) = \begin{cases} \varphi(0) + \dfrac{1}{\Gamma(\alpha)} \displaystyle\int_0^t (t-s)^{\alpha-1} h(s)ds, & t \in J\\ \varphi(t), & t \in [-r, 0]\,. \end{cases}$$

**Lemma 2.19.** *Let $f(t, u, v)\colon J \times C([-r, 0], \mathbb{R}) \times \mathbb{R} \to \mathbb{R}$ be a continuous function. Problem (2.48)–(2.49) is equivalent to the problem*

$$y(t) = \begin{cases} \varphi(0) + I^\alpha K_y(t), & t \in J\,,\\ \varphi(t), & t \in [-r, 0]\,, \end{cases} \tag{2.50}$$

*where $K_y \in C(J, \mathbb{R})$ satisfies the functional equation*

$$K_y(t) = f(t, y_t, K_y(t)) + g(t, y_t)\,.$$

*Proof.* Let $y$ be a solution of problem (2.50); we want to show that $y$ is a solution of (2.48)–(2.49). We have

$$y(t) = \begin{cases} \varphi(0) + I^\alpha K_y(t), & t \in J\\ \varphi(t), & t \in [-r, 0] \end{cases}$$

for $t \in [-r, 0]$, so $y(t) = \varphi(t)$, and we see that condition (2.49) is satisfied.
On the other hand, for $t \in J$ we have

$${}^cD^\alpha y(t) = K_y(t) = f(t, y_t, K_y(t)) + g(t, y_t)\,,$$

so

$${}^cD^\alpha y(t) = f(t, y_t, {}^cD^\alpha y(t)) + g(t, y_t)\,.$$

Then $y$ is a solution of problem (2.48)–(2.49). □

**Lemma 2.20.** *Assume*
(2.20.1)$f: J \times C([-r, 0], \mathbb{R}) \times \mathbb{R} \to \mathbb{R}$ *is a continuous function.*
(2.20.2)*There exist* $K > 0$ *and* $0 < \overline{K} < 1$ *such that*

$$|f(t, u, v) - f(t, \bar{u}, \bar{v})| \leq K \|u - \bar{u}\|_C + \overline{K}|v - \bar{v}|$$

*for any* $u, \bar{u} \in C([-r, 0], \mathbb{R})$, $v, \bar{v} \in \mathbb{R}$ *and* $t \in J$.
(2.20.3)*There exists* $L > 0$ *such that*

$$|g(t, u) - g(t, v)| \leq L\|u - v\|_C$$

*for any* $u, v \in C([-r, 0], \mathbb{R})$ *and* $t \in J$.
*If*

$$\frac{(K + L)T^\alpha}{\left(1 - \overline{K}\right)\Gamma(\alpha + 1)} < 1 , \tag{2.51}$$

*then problem* (2.48)–(2.49) *has a unique solution.*

*Proof.* Consider that the operator $N: Q \to Q$ is defined by

$$(Ny)(t) = \begin{cases} \varphi(0) + I^\alpha K_y(t), & t \in J \\ \varphi(t), & t \in [-r, 0] . \end{cases} \tag{2.52}$$

From Lemma 2.19 it is clear that the fixed points of $N$ are the solutions of problem (2.48)–(2.49). Let $y, \tilde{y} \in Q$. If $t \in [-r, 0]$, then

$$\|Ny(t) - N\tilde{y}(t)\| = 0 ,$$

and for $t \in J$

$$\|Ny(t) - N\tilde{y}(t)\| = \|I^\alpha K_y(t) - I^\alpha K_{\tilde{y}}(t)\| \leq I^\alpha \|K_y(t) - K_{\tilde{y}}(t)\| . \tag{2.53}$$

For any $t \in J$ we have

$$\begin{aligned} \|K_y(t) - K_{\tilde{y}}(t)\| &\leq \|f(t, y_t, K_y(t)) - f(t, \tilde{y}_t, K_{\tilde{y}}(t))\| \\ &\quad + \|g(t, y_t) - g(t, \tilde{y}_t)\| \\ &\leq K \|y_t - \tilde{y}_t\|_C + \overline{K}\|K_y(t) - K_{\tilde{y}}(t)\| \\ &\quad + L \|y_t - \tilde{y}_t\|_C . \end{aligned}$$

Thus,

$$\|K_y(t) - K_{\tilde{y}}(t)\| \leq \frac{K + L}{1 - \overline{K}} \|y_t - \tilde{y}_t\|_C . \tag{2.54}$$

From (2.54) and (2.53) we find that

$$\begin{aligned} \|Ny(t) - N\tilde{y}(t)\| &\leq \frac{K + L}{\left(1 - \overline{K}\right)\Gamma(\alpha)} \int_0^t (t - s)^{\alpha - 1} \|y_s - \tilde{y}_s\|_C \, ds \\ &\leq \frac{(K + L)T^\alpha}{\left(1 - \overline{K}\right)\Gamma(\alpha + 1)} \|y - \tilde{y}\|_Q . \end{aligned}$$

Thus,

$$\|Ny - N\tilde{y}\|_Q \leq \frac{(K+L)T^{\alpha}}{\left(1-\overline{K}\right)\Gamma(\alpha+1)} \|y - \tilde{y}\|_Q .$$

From (2.51) it follows that $N$ has a unique fixed point that is the unique solution of problem (2.48)–(2.49). □

Our next existence result is based on the fixed point theorem of Burton and Kirk.

**Lemma 2.21.** *We consider the operators $F, G\colon Q \to Q$ defined by*

$$F(y)(t) = \begin{cases} \varphi(0) + \dfrac{1}{\Gamma(\alpha)} \displaystyle\int_0^t (t-s)^{\alpha-1} f(s, y_s, K_y(s))ds, & t \in J, \\ \varphi(t), & t \in [-r, 0], \end{cases}$$

$$G(y)(t) = \begin{cases} \dfrac{1}{\Gamma(\alpha)} \displaystyle\int_0^t (t-s)^{\alpha-1} g(s, y_s)ds, & t \in J, \\ 0, & t \in [-r, 0], \end{cases}$$

*where $K_y \in C(J, \mathbb{R})$ satisfies the functional equation*

$$K_y(t) = f(t, y_t, K_y(t)) + g(t, y_t) .$$

To find solutions of (2.48)–(2.49), we must find solutions of the equation

$$y(t) = F(y)(t) + G(y)(t) , \quad \text{for each } t \in [-r, T] .$$

**Remark 2.22.** If $y$ is a fixed point of the operator $F + G$, then $y$ is a solution of problem (2.48)–(2.49).

**Theorem 2.23.** *Assume* (2.20.1) *and* (2.20.3) *hold and*
(2.23.1)*There exist $p, q, r \in C(J, \mathbb{R}_+)$ with $r^* = \sup_{t\in J} r(t) < 1$ such that*

$$|f(t, u, w)| \leq p(t) + q(t)\|u\|_C + r(t)|w|$$

*for $t \in J$, $w \in \mathbb{R}$, and $u \in C([-r, 0], \mathbb{R})$.*

*If*

$$\frac{LT^{\alpha}}{\Gamma(\alpha+1)} < 1 , \tag{2.55}$$

*then IVP* (2.48)–(2.49) *has at least one solution.*

*Proof.* We must show that operators $F$ and $G$ satisfy all conditions of Theorem 1.43. By (2.20.1), (2.20.3), (2.23.1), and the choices of $F$ and $G$, we show that $F$ is completely continuous and $G$ is a contraction. The proof will be given in several claims.

*Claim 1: $F$ is continuous.* Let $\{y_n\}_{n\geq 0}$ be a sequence such that $y_n \to y$ in $Q$; then

$$\|F(y_n)(t) - F(y)(t)\|_Q \leq \frac{T^{\alpha}}{\Gamma(\alpha+1)} \|f(.,y_{n.}, K_{y_n}(.)) - f(.,y_., K_y(.))\|_Q \xrightarrow[n\to\infty]{} 0$$

because $f$ is continuous.

*Claim 2: $F$ maps bounded sets to bounded sets in $Q$.* Let $y^* > 0$, $B_{y^*} = \{y \in Q : \|y\|_Q \leq y^*\}$; we need to show that for each $y^* > 0$ there exists $l > 0$ such that for $y \in B_{y^*}$, $\|Fy\|_Q \leq l$. By (2.23.1) we have for any $t \in J$

$$\begin{aligned}\|F(y)(t)\| &\leq |\varphi(0)| + \frac{1}{\Gamma(\alpha)}\int_0^t (t-s)^{\alpha-1}\|f(s, y_s, K_y(s))\|ds \\ &\leq |\varphi(0)| + \frac{T^\alpha(p^* + q^*y^*)}{\Gamma(\alpha+1)} + \frac{r^*}{\Gamma(\alpha)}\int_0^t (t-s)^{\alpha-1}\|K_y(s)\|ds\,, \end{aligned} \tag{2.56}$$

where $p^* = \sup_{t\in J} p(t)$, and $q^* = \sup_{t\in J} q(t)$. On the other hand, we have for any $t \in J$

$$\begin{aligned}\|K_y(t)\| &\leq \|f(t, y_t, K_y(t))\| + \|g(t, y_t) - g(t, 0)\| + \|g(t, 0)\| \\ &\leq p^* + q^*y^* + r^*\|K_y(t)\| + Ly^* + g^*\,,\end{aligned}$$

where $g^* = \sup_{s\in J}\|g(s, 0)\|$. Thus,

$$\|K_y(t)\| \leq \frac{1}{1-r^*}(p^* + q^*y^* + Ly^* + g^*) := M\,. \tag{2.57}$$

Combining (2.57) and (2.56), we find that

$$\|F(y)(t)\| \leq |\varphi(0)| + \frac{T^\alpha(p^* + q^*y^*)}{\Gamma(\alpha+1)} + \frac{r^*T^\alpha}{\Gamma(\alpha+1)}M := d\,.$$

If $t \in [-r, 0]$, then

$$\|F(y)(t)\| \leq \|\varphi\|_C\,,$$

so

$$\|Fy\|_Q \leq \max\{\|\varphi\|_C, d\} = l\,.$$

*Claim 3: $F$ maps bounded sets to equicontinuous sets of $Q$.* Let $t_1, t_2 \in (0, T]$, $t_1 < t_2$, $B_{y^*}$ be a bounded set of $Q$, which is as defined previously (Claim 2), and let $y \in B_{y^*}$. Then

$$\begin{aligned}\|F(y)(t_2) - F(y)(t_1)\| &\leq \Big\|\frac{1}{\Gamma(\alpha)}\int_0^{t_1}\left[(t_2-s)^{\alpha-1} - (t_1-s)^{\alpha-1}\right]f(s, y_s, K_y(s))ds\Big\| \\ &\quad + \Big\|\frac{1}{\Gamma(\alpha)}\int_{t_1}^{t_2}(t_2-s)^{\alpha-1}f(s, y_s, K_y(s))ds\Big\| \\ &\leq \frac{(p^* + q^*y^* + r^*M)}{\Gamma(\alpha)}\int_0^{t_1}\left[(t_1-s)^{\alpha-1} - (t_2-s)^{\alpha-1}\right]ds \\ &\quad + \frac{(p^* + q^*y^* + r^*M)}{\Gamma(\alpha)}\int_{t_1}^{t_2}(t_2-s)^{\alpha-1}ds\end{aligned}$$

$$\leq \frac{(p^* + q^* y^* + r^* M)}{\Gamma(\alpha+1)} [2(t_2 - t_1)^\alpha - t_1^\alpha - t_2^\alpha]$$
$$\leq \frac{2\,(p^* + q^* y^* + r^* M)}{\Gamma(\alpha+1)} (t_2 - t_1)^\alpha .$$

As $t_1 \to t_2$, the right-hand side of the preceding inequality tends to zero. Consequently, Theorem 1.54 allows us to conclude that the operator $F: Q \to Q$ is relatively compact. Hence, operator $F$ is completely continuous.

*Claim 4: G is a contraction.* Let $y, \tilde{y} \in Q$; then for every $t \in J$

$$\begin{aligned}\|G(y)(t) - G(\tilde{y})(t)\| &\leq \frac{1}{\Gamma(\alpha)} \int_0^t (t-s)^{\alpha-1} \|g(s, y_s) - g(s, \tilde{y}_s)\| ds \\ &\leq \frac{L}{\Gamma(\alpha)} \int_0^t (t-s)^{\alpha-1} \|y_s - \tilde{y}_s\|_C \, ds \\ &\leq \frac{T^\alpha L}{\Gamma(\alpha+1)} \|y - \tilde{y}\|_\infty .\end{aligned}$$

Thus,

$$\|G(y) - G(\tilde{y})\|_Q \leq \frac{T^\alpha L}{\Gamma(\alpha+1)} \|y - \tilde{y}\|_Q .$$

By (2.55) $G$ is a contraction.

*Claim 5: A priori bounds.* We will show that the set

$$\Omega = \left\{ y \in Q : y = \lambda F(y) + \lambda G\left(\frac{y}{\lambda}\right) \quad \text{for } \lambda \in (0,1) \right\}$$

is bounded. In fact, let $y \in \Omega$; then $y = \lambda F(y) + \lambda G(\frac{y}{\lambda})$ for some $0 < \lambda < 1$. Then for each $t \in J$ we have

$$y(t) = \lambda \left[ |\varphi(0)| + \frac{1}{\Gamma(\alpha)} \int_0^t (t-s)^{\alpha-1} f(s, y_s, K_y(s)) ds + \frac{1}{\Gamma(\alpha)} \int_0^t (t-s)^{\alpha-1} g\left(s, \frac{y_s}{\lambda}\right) ds \right] .$$

By (2.20.3) and (2.23.1), for every $t \in J$ we have

$$\begin{aligned}\|y(t)\| &\leq |\varphi(0)| + \frac{q^*}{\Gamma(\alpha)} \int_0^t (t-s)^{\alpha-1} \|y_s\|_C \, ds + \frac{T^\alpha (p^* + r^* M)}{\Gamma(\alpha+1)} \\ &\quad + \frac{\lambda}{\Gamma(\alpha)} \int_0^t (t-s)^{\alpha-1} \left\|g\left(s, \frac{y_s}{\lambda}\right) - g(s, 0)\right\| ds \\ &\quad + \frac{\lambda}{\Gamma(\alpha)} \int_0^t (t-s)^{\alpha-1} \|g(s, 0)\| ds \\ &\leq |\varphi(0)| + \frac{T^\alpha}{\Gamma(\alpha+1)} (p^* + r^* M + g^*) + \frac{(q^* + L)}{\Gamma(\alpha)} \int_0^t (t-s)^{\alpha-1} \|y_s\|_C \, ds .\end{aligned}$$

We consider the function $y$ defined by

$$y(t) = \sup\{\|y(s)\| : -r \le s \le t\}, \quad 0 \le t \le T.$$

There exists $t^* \in [-r, T]$ such that $y(t) = \|y(t^*)\|$.

If $t^* \in [0, T]$, then, by the previous inequality, for $t \in J$ we have

$$y(t) \le |\varphi(0)| + \frac{T^\alpha}{\Gamma(\alpha+1)}(p^* + r^* M + g^*) + \frac{(q^* + L)}{\Gamma(\alpha)} \int_0^t (t-s)^{\alpha-1} y(s) ds.$$

Applying Lemma 1.52, we obtain

$$y(t) \le \left[|\varphi(0)| + \frac{T^\alpha}{\Gamma(\alpha+1)}(p^* + r^* M + g^*)\right]\left[1 + \frac{\overline{\delta}(q^* + L)T^\alpha}{\Gamma(\alpha+1)}\right] := R,$$

where $\overline{\delta} = \overline{\delta}(\alpha)$ is a constant. Thus for any $t \in J$, $\|y\|_\infty \le y(t) \le R$. If $t^* \in [-r, 0]$, then $y(t) = \|\varphi\|_C$. Therefore,

$$\|y\|_Q \le \max\{\|\varphi\|_C, R\} := A.$$

Thus, the set $\Omega$ is bounded. Therefore, problem (2.48)–(2.49) has at least one solution. □

### 2.5.3 Ulam–Hyers Stability Results

For the implicit fractional order differential equation (2.48), we adopt the definition in Rus [224] for Ulam–Hyers stability, generalized Ulam–Hyers stability, Ulam–Hyers–Rassias stability, and generalized Ulam–Hyers–Rassias stability.

**Definition 2.24.** Equation (2.48) is Ulam–Hyers stable if there exists a real number $c_f > 0$ such that for each $\epsilon > 0$ and for each solution $z \in C^1(J, \mathbb{R})$ of the inequality

$$\|{}^cD^\alpha z(t) - f(t, z_t, {}^cD^\alpha z(t)) - g(t, z_t)\| \le \epsilon, \quad t \in J,$$

there exists a solution $y \in C^1(J, \mathbb{R})$ of equation (2.48), with

$$\|z(t) - y(t)\| \le c_f \epsilon, \quad t \in J.$$

**Definition 2.25.** Equation (2.48) is generalized Ulam–Hyers stable if there exists $\psi_f \in C(\mathbb{R}_+, \mathbb{R}_+)$, $\psi_f(0) = 0$, such that for each solution $z \in C^1(J, \mathbb{R})$ of the inequality

$$\|{}^cD^\alpha z(t) - f(t, z_t, {}^cD^\alpha z(t)) - g(t, z_t)\| \le \epsilon, \quad t \in J,$$

there exists a solution $y \in C^1(J, \mathbb{R})$ of equation (2.48), with

$$\|z(t) - y(t)\| \le \psi_f(\epsilon), \quad t \in J.$$

**Definition 2.26.** Equation (2.48) is Ulam–Hyers–Rassias stable with respect to $\phi \in C(J, \mathbb{R}_+)$ if there exists a real number $c_f > 0$ such that for each $\epsilon > 0$ and for each solution $z \in C^1(J, \mathbb{R})$ of the inequality

$$\|{}^cD^\alpha z(t) - f(t, z_t, {}^cD^\alpha z(t)) - g(t, z_t)\| \le \epsilon\phi(t)\,, \quad t \in J\,,$$

there exists a solution $y \in C^1(J, \mathbb{R})$ of equation (2.48), with

$$\|z(t) - y(t)\| \le c_f \epsilon\phi(t), \quad t \in J\,.$$

**Definition 2.27.** Equation (2.48) is generalized Ulam–Hyers–Rassias stable with respect to $\phi \in C(J, \mathbb{R}_+)$ if there exists a real number $c_{f,\phi} > 0$ such that for each solution $z \in C^1(J, \mathbb{R})$ of the inequality

$$\|{}^cD^\alpha z(t) - f(t, z_t, {}^cD^\alpha z(t)) - g(t, z_t)\| \le \phi(t)\,, \quad t \in J\,,$$

there exists a solution $y \in C^1(J, \mathbb{R})$ of equation (2.48), with

$$\|z(t) - y(t)\| \le c_{f,\phi}\phi(t), \quad t \in J\,.$$

**Remark 2.28.** A function $z \in C^1(J, \mathbb{R})$ is a solution of the inequality

$$\|{}^cD^\alpha z(t) - f(t, z_t, {}^cD^\alpha z(t)) - g(t, z_t)\| \le \epsilon\,, \quad t \in J\,,$$

if and only if there exists a function $h \in C(J, \mathbb{R})$ (which depends on $y$) such that
**(i)** $\|h(t)\| \le \epsilon$, $t \in J$,
**(ii)** ${}^cD^\alpha z(t) = f(t, z_t, {}^cD^\alpha z(t)) + g(t, z_t) + h(t)$, $t \in J$.

**Remark 2.29.** Clearly:
**(i)** Definition 2.24 $\Rightarrow$ Definition 2.25.
**(ii)** Definition 2.26 $\Rightarrow$ Definition 2.27.

**Remark 2.30.** A solution of the implicit differential equation

$$\|{}^cD^\alpha z(t) - f(t, z_t, {}^cD^\alpha z(t)) - g(t, z_t)\| \le \epsilon\,, \quad t \in J\,,$$

with fractional order is called a fractional $\epsilon$-solution of the implicit fractional differential equation (2.48).

**Theorem 2.31.** *Assume* (2.20.1)–(2.20.3) *and* (2.51) *hold. Then problem* (2.48)–(2.49) *is Ulam–Hyers stable.*

*Proof.* Let $\epsilon > 0$ and $z \in Q$ be a function such that

$$\|{}^cD^\alpha z(t) - f(t, z_t, {}^cD^\alpha z(t)) - g(t, z_t)\| \le \epsilon \quad \text{for each } t \in J\,.$$

This inequality is equivalent to

$$\|{}^cD^\alpha z(t) - K_z(t)\| \le \epsilon\,. \tag{2.58}$$

Let $y \in Q$ be the unique solution of the problem

$$\begin{cases} {}^cD^\alpha y(t) = f(t, y_t, {}^cD^\alpha y(t)) + g(t, y_t), & t \in J\,, \\ z(t) = y(t) = \varphi(t), & t \in [-r, 0]\,. \end{cases}$$

Integrating inequality (2.58), we obtain

$$\|z(t) - I^\alpha K_z(t)\| \leq \frac{\epsilon T^\alpha}{\Gamma(\alpha+1)}\,.$$

We consider the function $y_2$ defined by

$$y_2(t) = \sup\{\|z(s) - y(s)\| : \ -r \leq s \leq t\}\,, \quad 0 \leq t \leq T\,.$$

Then there exists $t^* \in [-r, T]$ such that $y_2(t) = \|z(t^*) - y(t^*)\|$. If $t^* \in [-r, 0]$, then $y_2(t) = 0$. If $t^* \in [0, T]$, then

$$\begin{aligned} y_2(t) &\leq \|z(t) - I^\alpha K_z(t)\| + I^\alpha\|K_z(t) - K_y(t)\| \\ &\leq \frac{\epsilon T^\alpha}{\Gamma(\alpha+1)} + I^\alpha\|K_z(t) - K_y(t)\|\,. \end{aligned} \tag{2.59}$$

On the other hand, we have

$$\begin{aligned} \|K_z(t) - K_y(t)\| &\leq \|f(t, z_t, K_z(t)) - f(t, y_t, K_y(t)\| \\ &\quad + \|g(t, z_t) - g(t, y_t)\| \\ &\leq (K+L)y_2(t) + \overline{K}\|K_z(t) - K_y(t)\|\,, \end{aligned}$$

so

$$\|K_z(t) - K_y(t)\| \leq \frac{K+L}{1-\overline{K}} y_2(t)\,. \tag{2.60}$$

Substituting (2.60) in inequality (2.59), we get

$$y_2(t) \leq \frac{\epsilon T^\alpha}{\Gamma(\alpha+1)} + \frac{K+L}{\left(1-\overline{K}\right)\Gamma(\alpha)} \int_0^t (t-s)^{\alpha-1} y_2(s)ds\,,$$

and by Gronwall's lemma

$$y_2(t) \leq \frac{\epsilon T^\alpha}{\Gamma(\alpha+1)}\left[1 + \frac{(K+L)T^\alpha \sigma_1}{(1-\overline{K})\Gamma(\alpha+1)}\right] := c\epsilon\,,$$

where $\sigma_1 = \sigma_1(\alpha)$ is a constant. □

If we set $\psi(\epsilon) = c\psi$; $\psi(0) = 0$, then problem (2.48)–(2.49) is generalized Ulam–Hyers stable.

**Theorem 2.32.** *Assume* (2.20.1)–(2.20.3) *and* (2.51) *hold and* (2.32.1) *there exists an increasing function* $\phi \in C(J, \mathbb{R}_+)$ *and there exists* $\lambda_\phi > 0$ *such that for any* $t \in J$

$$I^\alpha \phi(t) \leq \lambda_\phi \phi(t) .$$

*Then problem* (2.48)–(2.49) *is Ulam–Hyers–Rassias stable.*

*Proof.* Let $z \in Q$ be a solution of the inequality

$$\|{}^cD^\alpha z(t) - f(t, z_t, {}^cD^\alpha z(t)) - g(t, z_t)\| \leq \epsilon\phi(t) , \quad t \in J,\ \epsilon > 0 .$$

This inequality is equivalent to

$$\|{}^cD^\alpha z(t) - K_z(t)\| \leq \epsilon\phi(t) . \tag{2.61}$$

Let $y \in Q$ be the unique solution of the Cauchy problem

$$\begin{cases} {}^cD^\alpha y(t) = f(t, y_t, {}^cD^\alpha y(t)) + g(t, y_t), & t \in J , \\ z(t) = y(t) = \varphi(t), & t \in [-r, 0] . \end{cases}$$

Integrating (2.61), we obtain for any $t \in J$

$$\|z(t) - I^\alpha K_z(t)\| \leq \epsilon I^\alpha \phi(t) \leq \epsilon\lambda_\phi \phi(t) .$$

Using the function $y_2$ defined in the proof of Theorem 2.31, we see that if $t^* \in [-r, 0]$, then $y_2(t) = 0$, and if $t^* \in [0, T]$, then we have

$$\begin{aligned} y_2(t) &\leq \|z(t) - I^\alpha K_z(t)\| + I^\alpha \|K_z(t) - K_y(t)\| \\ &\leq \epsilon\lambda_\phi \phi(t) + I^\alpha \|K_z(t) - K_y(t)\| . \end{aligned} \tag{2.62}$$

It follows that

$$\|K_z(t) - K_y(t)\| \leq \frac{K + L}{1 - \overline{K}} y_2(t) . \tag{2.63}$$

Substituting into (2.63) in the inequality (2.62), we obtain

$$y_2(t) \leq \epsilon\lambda_\phi \phi(t) + \frac{K + L}{\left(1 - \overline{K}\right) \Gamma(\alpha)} \int_0^t (t - s)^{\alpha-1} y_2(s) ds ,$$

and by Gronwall's lemma we get

$$\begin{aligned} y_2(t) &\leq \epsilon\lambda_\phi \phi(t) \left[ 1 + \frac{(K + L) T^\alpha \sigma_2}{(1 - \overline{K}) \Gamma(\alpha + 1)} \right] \\ &\leq \left[ \lambda_\phi \left( 1 + \frac{(K + L) T^\alpha \sigma_2}{(1 - \overline{K}) \Gamma(\alpha + 1)} \right) \right] \epsilon\phi(t) = c\epsilon\phi(t) , \end{aligned}$$

where $\sigma_2 = \sigma_2(\alpha)$ is a constant. Thus problem (2.48)–(2.49) is Ulam–Hyers–Rassias stable. □

### 2.5.4 An Example

Consider the problem of the perturbed differential equation of fractional order

$$^{c}D^{\frac{1}{2}}y(t) = \frac{2 + \|y_t\|_C + \|^{c}D^{\frac{1}{2}}y(t)\|}{12e^{t+9}\left(1 + \|y_t\|_C + \|^{c}D^{\frac{1}{2}}y(t)\|\right)} + \frac{e^{-t}\|y_t\|_C}{3\,(1 + \|y_t\|_C)}, \quad t \in [0, 1], \tag{2.64}$$

$$y(t) = \varphi(t), \quad t \in [-r, 0], \tag{2.65}$$

where $\varphi \in C([-r, 0], \mathbb{R})$. Set

$$f(t, u, v) = \frac{2 + u + v}{12e^{t+9}(1 + u + v)}, \quad t \in [0, 1], \; u, v \in [0, +\infty) \times [0, +\infty).$$

It is clear that $f$ is jointly continuous.

For each $u \in C([-r, 0], \mathbb{R})$, $v \in \mathbb{R}$, and $t \in [0, 1]$,

$$f(t, u, v) \le \frac{1}{12e^{t+9}}(2 + \|u\|_C + \|v\|).$$

Hence, condition (2.23.1) is satisfied by $p(t) = \frac{1}{6e^{t+9}}$, $r(t) = q(t) = \frac{1}{12e^{t+9}}$, and $r^* = \frac{1}{12e^9} < 1$. Set

$$g(t, w) = \frac{e^{-t}w}{3(1 + w)}, \, t \in [0, 1], \quad w \in [0, +\infty).$$

It is clear that $g$ is continuous; moreover, we have for any $u, v \in C([-r, 0], \mathbb{R})$ and $t \in J$

$$\|g(t, u) - g(t, v)\| \le \frac{1}{3}\|u - v\|_C.$$

Thus, (2.20.3) is satisfied, and we have

$$\frac{T^{\alpha}L}{\Gamma(\alpha + 1)} = \frac{(1)^{\frac{1}{2}} \times \frac{1}{3}}{\Gamma\left(\frac{3}{2}\right)} = \frac{1}{3\Gamma\left(\frac{3}{2}\right)} = \frac{1}{3 \times \frac{1}{2}\Gamma\left(\frac{1}{2}\right)} = \frac{2}{3\sqrt{\pi}} < 1.$$

Hence, (2.20.1), (2.20.3), (2.23.1), and (2.55) are satisfied, so by Theorem 2.23, problem (2.64)–(2.65) has at least one solution.

## 2.6 Existence and Stability Results for Neutral NIFDE with Finite Delay

### 2.6.1 Introduction

In this section, we establish existence, uniqueness, and stability results for the nonlinear implicit neutral fractional differential equation with finite delay

$$^{c}D^{\alpha}\,[y(t) - g(t, y_t)] = f(t, y_t, {}^{c}D^{\alpha}y(t)), \quad t \in J = [0, T], \; T > 0, \; 0 < \alpha \le 1, \tag{2.66}$$

$$y(t) = \varphi(t), \; t \in [-r, 0], \; r > 0, \tag{2.67}$$

where $f: J \times C([-r,0],\mathbb{R}) \times \mathbb{R} \to \mathbb{R}$ and $g: J \times C([-r,0],\mathbb{R})$ are given functions such that $g(0,\varphi) = 0$ and $\varphi \in C([-r,0],\mathbb{R})$.

Two examples are given to show the applicability of our results.

### 2.6.2 Existence of Solutions

Set

$$\Omega = \{y : [-r,T] \to \mathbb{R} : y|_{[-r,0]} \in C([-r,0],\mathbb{R}) \text{ and } y|_{[0,T]} \in C([0,T],\mathbb{R})\} .$$

Note that $\Omega$ is a Banach space with the norm

$$\|y\|_\Omega = \sup_{t\in[-r,T]} |y(t)| .$$

**Definition 2.33.** A function $y \in \Omega$ is called a solution of problem (2.66)–(2.67) if it satisfies equation (2.66) on $J$ and condition (2.67) on $[-r,0]$.

**Lemma 2.34.** *Let $0 < \alpha \le 1$ and $h: [0,T] \to \mathbb{R}$ be a continuous function. Then the linear problem*

$$\begin{aligned} &{}^cD^\alpha [y(t) - g(t,y_t)] = h(t) , \quad t \in J , \\ &y(t) = \varphi(t) , \quad t \in [-r,0] , \end{aligned}$$

*has a unique solution given by*

$$y(t) = \begin{cases} \varphi(0) + g(t,y_t) + \dfrac{1}{\Gamma(\alpha)} \displaystyle\int_0^t (t-s)^{\alpha-1} h(s)ds, & t \in J , \\ \varphi(t), & t \in [-r,0] . \end{cases}$$

**Lemma 2.35.** *Let $f(t,u,v): J \times C([-r,0],\mathbb{R}) \times \mathbb{R} \to \mathbb{R}$ be a continuous function. Then problem* (2.66)–(2.67) *is equivalent to the problem*

$$y(t) = \begin{cases} \varphi(0) + I^\alpha K_y(t), & t \in J , \\ \varphi(t), & t \in [-r,0] , \end{cases} \tag{2.68}$$

*where $K_y \in C(J,\mathbb{R})$ satisfies the functional equation*

$$K_y(t) = f(t,y_t,K_y(t)) + {}^cD^\alpha g(t,y_t) .$$

*Proof.* Let $y$ be a solution of problem (2.68); we need to show that $y$ is a solution of (2.66)–(2.67). We have

$$y(t) = \begin{cases} \varphi(0) + I^\alpha K_y(t), & t \in J , \\ \varphi(t), & t \in [-r,0] . \end{cases}$$

For $t \in [-r, 0]$ we have $y(t) = \varphi(t)$, so condition (2.67) is satisfied.
On the other hand, for $t \in J$ we have

$$^cD^\alpha y(t) = K_y(t) = f(t, y_t, K_y(t)) + {}^cD^\alpha g(t, y_t) .$$

So

$$^cD^\alpha [y(t) - g(t, y_t)] = f(t, y_t, {}^cD^\alpha y(t)) .$$

Then $y$ is a solution of problem (2.66)–(2.67). □

**Lemma 2.36.** *Assume*

(2.36.1) $f: J \times C([-r, 0], \mathbb{R}) \times \mathbb{R} \to \mathbb{R}$ *is a continuous function.*

(2.36.2) *There exist* $K > 0$ *and* $0 < \overline{K} < 1$ *such that*

$$|f(t, u, v) - f(t, \bar{u}, \bar{v})| \leq K \|u - \bar{u}\|_C + \overline{K}|v - \bar{v}|$$

*for any* $u, \bar{u} \in C([-r, 0], \mathbb{R})$, $v, \bar{v} \in \mathbb{R}$ *and* $t \in J$.

(2.36.3) *There exists* $L > 0$ *such that*

$$|g(t, u) - g(t, v)| \leq L\|u - v\|_C$$

*for any* $u, v \in C([-r, 0], \mathbb{R})$ *and* $t \in J$. *If*

$$\frac{KT^\alpha}{\left(1 - \overline{K}\right)\Gamma(\alpha + 1)} + \frac{L}{\left(1 - \overline{K}\right)} < 1 , \tag{2.69}$$

*then problem* (2.66)–(2.67) *has a unique solution.*

*Proof.* Consider the operator $N: \Omega \to \Omega$ defined by

$$(Ny)(t) = \begin{cases} \varphi(0) + I^\alpha K_y(t), & t \in J \\ \varphi(t), & t \in [-r, 0] . \end{cases} \tag{2.70}$$

By Lemma 2.35, it is clear that the fixed points of $N$ are the solutions of problem (2.66)–(2.67).

Let $y, \tilde{y} \in \Omega$. If $t \in [-r, 0]$, then

$$\|(Ny)(t) - (N\tilde{y})(t)\| = 0 .$$

For $t \in J$ we have

$$\|(Ny)(t) - (N\tilde{y})(t)\| = \|I^\alpha K_y(t) - I^\alpha K_{\tilde{y}}(t)\| \leq I^\alpha \|K_y(t) - K_{\tilde{y}}(t)\| . \tag{2.71}$$

For any $t \in J$

$$\begin{aligned} \|K_y(t) - K_{\tilde{y}}(t)\| &\leq \|f(t, y_t, K_y(t)) - f(t, \tilde{y}_t, K_{\tilde{y}}(t))\| \\ &\quad + {}^cD^\alpha \|g(t, y_t) - g(t, \tilde{y}_t)\| \\ &\leq K \|y_t - \tilde{y}_t\|_C + \overline{K}\|K_y(t) - K_{\tilde{y}}(t)\| \\ &\quad + {}^cD^\alpha \|g(t, y_t) - g(t, \tilde{y}_t)\| . \end{aligned}$$

Thus,

$$\|K_y(t) - K_{\tilde{y}}(t)\| \leq \frac{K}{1-\overline{K}} \|y_t - \tilde{y}_t\|_C + \left(\frac{1}{1-\overline{K}}\right) {}^cD^\alpha \|g(t, y_t) - g(t, \tilde{y}_t)\| . \tag{2.72}$$

Substituting into (2.72) in the inequality (2.71) we have

$$\begin{aligned}
\|Ny(t) - N\tilde{y}(t)\| &\leq \frac{K}{\left(1-\overline{K}\right)\Gamma(\alpha)} \int_0^t (t-s)^{\alpha-1} \|y_s - \tilde{y}_s\|_C \, ds \\
&\quad + \frac{1}{1-\overline{K}} I^\alpha \, {}^cD^\alpha \|g(t, y_t) - g(t, \tilde{y}_t)\| \\
&\leq \frac{KT^\alpha}{\left(1-\overline{K}\right)\Gamma(\alpha+1)} \|y - \tilde{y}\|_\Omega \\
&\quad + \frac{1}{1-\overline{K}} \left(\|g(t, y_t) - g(t, \tilde{y}_t)\| + \|g(0, y_0) - g(0, \tilde{y}_0)\|\right) \\
&\leq \frac{KT^\alpha}{\left(1-\overline{K}\right)\Gamma(\alpha+1)} \|y - \tilde{y}\|_\Omega + \frac{L}{1-\overline{K}} \|y_t - \tilde{y}_t\|_C \\
&\leq \left[\frac{KT^\alpha}{\left(1-\overline{K}\right)\Gamma(\alpha+1)} + \frac{L}{1-\overline{K}}\right] \|y - \tilde{y}\|_\Omega .
\end{aligned}$$

Thus,

$$\|Ny - N\tilde{y}\|_\Omega \leq \left[\frac{KT^\alpha}{\left(1-\overline{K}\right)\Gamma(\alpha+1)} + \frac{L}{\left(1-\overline{K}\right)}\right] \|y - \tilde{y}\|_\Omega .$$

From (2.69) it follows that $N$ has a unique fixed point that is the unique solution of problem (2.66)–(2.67). □

### 2.6.3 Ulam–Hyers Stability Results

A solution of the implicit differential equation

$$\|{}^cD^\alpha z(t) - f(t, z_t, {}^cD^\alpha z(t)) - {}^cD^\alpha g(t, z_t)\| \leq \epsilon , \quad t \in J ,$$

with fractional order is called a fractional $\epsilon$-solution of implicit fractional differential equation (2.66).

**Theorem 2.37.** *Assume* (2.36.1)–(2.36.3) *and* (2.69) *hold. If*

$$\overline{K} + L < 1 , \tag{2.73}$$

*then problem* (2.66)–(2.67) *is Ulam–Hyers stable.*

*Proof.* Let $\epsilon > 0$ and $z \in \Omega$ be a function such that

$$\|{}^cD^\alpha z(t) - f(t, z_t, {}^cD^\alpha z(t)) - {}^cD^\alpha g(t, z_t)\| \leq \epsilon \quad \text{for each } t \in J .$$

This inequality is equivalent to

$$\|^c D^\alpha z(t) - K_z(t)\| \leq \epsilon . \tag{2.74}$$

Let $y \in \Omega$ be the unique solution of the problem

$$\begin{cases} {}^c D^\alpha\,[y(t) - g(t, y_t)] = f(t, y_t, {}^c D^\alpha y(t)), & t \in J \\ z(t) = y(t) = \varphi(t), & t \in [-r, 0] . \end{cases}$$

Integrating the inequality (2.74), we obtain

$$\|z(t) - I^\alpha K_z(t)\| \leq \frac{\epsilon T^\alpha}{\Gamma(\alpha+1)} .$$

We consider the function $y_1$ defined by

$$y_1(t) = \sup\{\|z(s) - y(s)\| : -r \leq s \leq t\} , \quad 0 \leq t \leq T .$$

Then there exists $t^* \in [-r, T]$ such that $y_1(t) = \|z(t^*) - y(t^*)\|$. If $t^* \in [-r, 0]$, then $y_1(t) = 0$. If $t^* \in [0, T]$, then

$$\begin{aligned} y_1(t) &\leq \|z(t) - I^\alpha K_z(t)\| + I^\alpha \|K_z(t) - K_y(t)\| \\ &\leq \frac{\epsilon T^\alpha}{\Gamma(\alpha+1)} + I^\alpha \|K_z(t) - K_y(t)\| . \end{aligned} \tag{2.75}$$

On the other hand, we have

$$\begin{aligned} \|K_z(t) - K_y(t)\| &\leq \|f(t, z_t, K_z(t)) - f(t, y_t, K_y(t)\| \\ &\quad + {}^c D^\alpha \|g(t, z_t) - g(t, y_t)\| \\ &\leq K y_1(t) + \overline{K}\|K_z(t) - K_y(t)\| \\ &\quad + {}^c D^\alpha \|g(t, z_t) - g(t, y_t)\| , \end{aligned}$$

so

$$\|K_z(t) - K_y(t)\| \leq \frac{K}{1-\overline{K}} y_1(t) + \frac{1}{1-\overline{K}}\, {}^c D^\alpha \|g(t, z_t) - g(t, y_t)\| . \tag{2.76}$$

From (2.76) and (2.75) we obtain

$$\begin{aligned} y_1(t) &\leq \frac{\epsilon T^\alpha}{\Gamma(\alpha+1)} + \frac{K}{\left(1-\overline{K}\right)\Gamma(\alpha)} \int_0^t (t-s)^{\alpha-1} y_1(s)\,ds \\ &\quad + \frac{1}{1-\overline{K}} \|g(t, z_t) - g(t, y_t)\| \\ &\leq \frac{\epsilon T^\alpha}{\Gamma(\alpha+1)} + \frac{K}{\left(1-\overline{K}\right)\Gamma(\alpha)} \int_0^t (t-s)^{\alpha-1} y_1(s)\,ds \\ &\quad + \frac{L}{1-\overline{K}} y_1(t) . \end{aligned}$$

Then

$$y_1(t) \le \frac{\epsilon T^{\alpha}\left(1-\overline{K}\right)}{\left[1-\left(\overline{K}+L\right)\right]\Gamma(\alpha+1)} + \frac{K}{\left[1-\left(\overline{K}+L\right)\right]\Gamma(\alpha)}\int_0^t (t-s)^{\alpha-1} y_1(s)ds\,.$$

By Gronwall's lemma

$$y_1(t) \le \frac{\epsilon T^{\alpha}\left(1-\overline{K}\right)}{\left[1-\left(\overline{K}+L\right)\right]\Gamma(\alpha+1)}\left[1+\frac{KT^{\alpha}\sigma_1}{\left[1-\left(\overline{K}_1+L\right)\right]\Gamma(\alpha+1)}\right] := c\epsilon\,,$$

where $\sigma_1 = \sigma_1(\alpha)$ is a constant. This completes the proof of the theorem. Moreover, if we set $\psi(\epsilon) = c\psi$; $\psi(0) = 0$, then problem (2.66)–(2.67) is generalized Ulam–Hyers stable. □

**Theorem 2.38.** *Assume* (2.36.1)–(2.36.3), (2.69), *and* (2.73) *hold and*

(2.45.1) *there exists an increasing function* $\phi \in C(J, \mathbb{R}_+)$, *and there exists* $\lambda_\phi > 0$ *such that for any* $t \in J$

$$I^{\alpha}\phi(t) \le \lambda_\phi \phi(t)\,.$$

*Then problem* (2.66)–(2.67) *is Ulam–Hyers–Rassias stable.*

*Proof.* Let $z \in \Omega$ be a solution of the inequality

$$\|{}^cD^{\alpha} z(t) - f(t, z_t, {}^cD^{\alpha} z(t)) - {}^cD^{\alpha} g(t, z_t)\| \le \epsilon\phi(t)\,, \quad t \in J,\ \epsilon > 0\,.$$

This inequality is equivalent to

$$\|{}^cD^{\alpha} z(t) - K_z(t)\| \le \epsilon\phi(t)\,. \tag{2.77}$$

Let $y \in \Omega$ be the unique solution of the Cauchy problem

$$\begin{cases} {}^cD^{\alpha}\left[y(t) - g(t, y_t)\right] = f(t, y_t, {}^cD^{\alpha} y(t)), & t \in J\,, \\ z(t) = y(t) = \varphi(t), & t \in [-r, 0]\,. \end{cases}$$

Integrating (2.77), we obtain for any $t \in J$

$$\|z(t) - I^{\alpha} K_z(t)\| \le \epsilon I^{\alpha}\phi(t) \le \epsilon\lambda_\phi\phi(t)\,.$$

Using the function $y_1$ defined in the proof of Theorem 2.37, we have that if $t^* \in [-r, 0]$, then $y_1(t) = 0$. If $t^* \in [0, T]$, then we have

$$\begin{aligned} y_1(t) &\le \|z(t) - I^{\alpha} K_z(t)\| + I^{\alpha}\|K_z(t) - K_y(t)\| \\ &\le \epsilon\lambda_\phi\phi(t) + I^{\alpha}\|K_z(t) - K_y(t)\|\,. \end{aligned} \tag{2.78}$$

Thus,

$$\|K_z(t) - K_y(t)\| \le \frac{K}{1-\overline{K}} y_1(t) + \frac{1}{1-\overline{K}}\, {}^cD^{\alpha}\|g(t, z_t) - g(t, y_t)\|\,. \tag{2.79}$$

Substituting into (2.79) in (2.78), we obtain

$$\begin{aligned} y_1(t) &\le \epsilon\lambda_\phi\phi(t) + \frac{K}{\left(1-\overline{K}\right)\Gamma(\alpha)}\int_0^t (t-s)^{\alpha-1}y_1(s)ds \\ &\quad + \frac{1}{1-\overline{K}}\|g(t,z_t)-g(t,y_t)\| \\ &\le \epsilon\lambda_\phi\phi(t) + \frac{K}{\left(1-\overline{K}\right)\Gamma(\alpha)}\int_0^t (t-s)^{\alpha-1}y_1(s)ds + \frac{L}{1-\overline{K}}y_1(t)\,, \end{aligned}$$

so

$$y_1(t) \le \frac{\left(1-\overline{K}\right)\epsilon\lambda_\phi\phi(t)}{1-\left(\overline{K}+L\right)} + \frac{K}{\left[1-\left(\overline{K}+L\right)\right]\Gamma(\alpha)}\int_0^t (t-s)^{\alpha-1}y_1(s)ds\,.$$

By Gronwall's lemma we get

$$\begin{aligned} y_1(t) &\le \frac{\left(1-\overline{K}\right)\epsilon\lambda_\phi\phi(t)}{1-\left(\overline{K}+L\right)}\left[1+\frac{KT^\alpha\sigma_2}{\left[1-\left(\overline{K}+L\right)\right]\Gamma(\alpha+1)}\right] \\ &\le \left[\frac{\left(1-\overline{K}\right)\lambda_\phi}{1-\left(\overline{K}+L\right)}\left(1+\frac{KT^\alpha\sigma_2}{\left[1-\left(\overline{K}+L\right)\right]\Gamma(\alpha+1)}\right)\right]\epsilon\phi(t) = c\epsilon\phi(t)\,, \end{aligned}$$

where $\sigma_2 = \sigma_2(\alpha)$ is a constant. Then problem (2.66)–(2.67) is Ulam–Hyers–Rassias stable. □

### 2.6.4 Examples

*Example 1.* Consider the neutral fractional differential equation

$$^cD^{\frac{1}{2}}\left[y(t) - \frac{te^{-t}\|y_t\|_C}{(9+e^t)(1+\|y_t\|_C)}\right] = \frac{2+\|y_t\|_C+|^cD^{\frac{1}{2}}y(t)|}{12e^{t+9}\left(1+\|y_t\|_C+|^cD^{\frac{1}{2}}y(t)|\right)}\,, \quad t\in[0,1]\,, \tag{2.80}$$

$$y(t) = \varphi(t);\quad t\in[-r,0]\,,\quad r>0\,, \tag{2.81}$$

where $\varphi \in C([-r,0],\mathbb{R})$. Set

$$g(t,w) = \frac{te^{-t}w}{(9+e^t)(1+w)}\,,\quad (t,w)\in[0,1]\times[0,+\infty)$$

and

$$f(t,u,v) = \frac{2+u+v}{12e^{t+9}(1+u+v)}\,,\quad (t,u,v)\in[0,1]\times[0,+\infty)\times[0,+\infty)\,.$$

Observe that $g(0, w) = 0$ for any $w \in [0, +\infty)$. Clearly, the function $f$ is continuous. Hence, (2.36.1) is satisfied. We have

$$|f(t, u, v) - f(t, \bar{u}, \bar{v})| \leq \frac{1}{12e^9}(\|u - \bar{u}\|_C + \|v - \bar{v}\|)$$
$$|g(t, u) - g(t, \bar{u})| \leq \frac{1}{10}\|u - \bar{u}\|_C$$

for any $u, \bar{u} \in C([-r, 0], \mathbb{R})$, $v, \bar{v} \in \mathbb{R}$ and $t \in [0, 1]$. Hence, conditions (2.36.2) and (2.36.3) are satisfied by $K = \overline{K} = \frac{1}{12e^9}$ and $L = \frac{1}{10}$.
The condition

$$\frac{KT^\alpha}{(1 + \overline{K})\Gamma(\alpha + 1)} + \frac{L}{(1 - \overline{K})} = \frac{20 + 12e^9\sqrt{\pi}}{10\sqrt{\pi}(12e^9 - 1)} < 1$$

is satisfied by $T = 1$, $\alpha = \frac{1}{2}$.

By Lemma 2.36, problem (2.80)–(2.81) admits a unique solution.

Since

$$\overline{K} + L = \frac{10 + 12e^9}{120e^9} < 1\,,$$

by Theorem 2.37, problem (2.80)–(2.81) is Ulam–Hyers stable.

*Example 2.* Consider the neutral fractional differential equation

$$^cD^{\frac{1}{2}}\left[y(t) - \frac{t}{5e^{t+2}(1 + \|y_t\|_C)}\right] = \frac{e^{-t}}{7 + e^t}\left[\frac{\|y_t\|_C}{1 + \|y_t\|_C} - \frac{|^cD^{\frac{1}{2}}y(t)|}{1 + |^cD^{\frac{1}{2}}y(t)|}\right], \quad t \in [0, 1]\,, \tag{2.82}$$

$$y(t) = \varphi(t)\,, \quad t \in [-r, 0]\,, \quad r > 0\,, \tag{2.83}$$

where $\varphi \in C([-r, 0], \mathbb{R})$.
Set

$$g(t, w) = \frac{t}{5e^{t+2}(1 + w)}\,, \quad (t, w) \in [0, 1] \times [0, +\infty)\,,$$

and

$$f(t, u, v) = \frac{e^{-t}}{(7 + e^t)}\left(\frac{u}{1 + u} - \frac{v}{1 + v}\right), \quad (t, u, v) \in [0, 1] \times [0, +\infty) \times [0, +\infty)\,.$$

Observe that $g(0, w) = 0$ for any $w \in [0, +\infty)$. Clearly, the function $f$ is continuous, so (2.36.1) is satisfied.

$$|f(t, u, v) - f(t, \bar{u}, \bar{v})| \leq \frac{1}{8}\|u - \bar{u}\|_C + \frac{1}{8}\|v - \bar{v}\|$$
$$|g(t, u) - g(t, \bar{u})| \leq \frac{1}{5e^2}\|u - \bar{u}\|_C$$

for any $u, \bar{u} \in C([-r, 0], \mathbb{R})$, $v, \bar{v} \in \mathbb{R}$ *and* $t \in [0, 1]$.
Hence, conditions (2.36.2) and (2.36.3) are satisfied by $K = \overline{K} = \frac{1}{8}$ and $L = \frac{1}{5e^2}$.
We have

$$\frac{KT^\alpha}{(1 + \overline{K})\Gamma(\alpha + 1)} + \frac{L}{(1 - \overline{K})} = \frac{10e^2 + 8\sqrt{\pi}}{35e^2\sqrt{\pi}} < 1\,,$$

so by Lemma 2.36, problem (2.82)–(2.83) admits a unique solution.
Since

$$\overline{K} + L = \frac{5e^2 + 8}{40e^2} < 1 \,,$$

by Theorem 2.37, problem (2.82)–(2.83) is Ulam–Hyers stable.

## 2.7 Notes and Remarks

The results of Chapter 2 are taken from Benchohra et al. [91, 102, 104]. Other results may be found in [20, 15, 26, 34, 43, 94, 248, 253].

# 3 Impulsive Nonlinear Implicit Fractional Differential Equations

## 3.1 Introduction

Impulsive fractional differential equations are a very important class of fractional differential equations because many phenomena from physics, chemistry, engineering, and biology, for example, can be represented by impulsive fractional differential equations.

Impulsive differential equations describes processes subject to abrupt changes in their states. They have received much attention in the literature and we refer the reader to the books [23, 35, 76, 77, 100, 148, 186, 215, 240], the papers [17, 24, 39, 96, 106, 124, 157, 158, 251], and the references therein.
In this chapter, we establish uniqueness and some Ulam stability and results for several classes of nonlinear implicit fractional differential equations (NIFDEs) with finite delay and fixed time impulses.

## 3.2 Existence and Stability Results for Impulsive NIFDEs with Finite Delay

### 3.2.1 Introduction

In this section, we consider the problem of nonlinear implicit fractional differential equations with finite delay and impulses,

$$^{c}D^{\alpha}_{t_k}y(t) = f(t, y_t, {}^{c}D^{\alpha}_{t_k}y(t)) , \quad \text{for each, } t \in (t_k, t_{k+1}],\ k = 0, \dots, m,\ 0 < \alpha \leq 1 , \tag{3.1}$$

$$\Delta y|_{t_k} = I_k(y_{t_k^-}) , \quad k = 1, \dots, m , \tag{3.2}$$

$$y(t) = \varphi(t) , \quad t \in [-r, 0],\ r > 0 , \tag{3.3}$$

where ${}^{c}D^{\alpha}_{t_k}$ is the Caputo fractional derivative, $f: J \times PC([-r, 0], \mathbb{R}) \times \mathbb{R} \to \mathbb{R}$ is a given function, $I_k: PC([-r, 0], \mathbb{R}) \to \mathbb{R}$, and $\varphi \in PC([-r, 0], \mathbb{R})$, $0 = t_0 < t_1 < \dots < t_m < t_{m+1} = T$. For each function $y_t$ defined on $[-r, T]$ and for any $t \in J$, we denote by $y_t$ the element of $PC([-r, 0], \mathbb{R})$ defined by

$$y_t(\theta) = y(t + \theta) , \quad \theta \in [-r, 0] . \tag{3.4}$$

The arguments are based on the Banach contraction principle and Schaefer's fixed point theorem; here we also present two examples to show the applicability of our results.

https://doi.org/10.1515/9783110553819-003

### 3.2.2 Existence of Solutions

Let $J_0 = [t_0, t_1]$ and $J_k = (t_k, t_{k+1}]$, $k = 1, \dots, m$. Consider the set of functions

$$PC([-r, 0], \mathbb{R}) = \{y : [-r, 0] \to \mathbb{R} : y \in C((\tau_k, \tau_{k+1}], \mathbb{R}), \quad k = 0, \dots, m,$$
$$\text{and there exist } y(\tau_k^-) \text{ and } y(\tau_k^+), \quad k = 1, \dots, m, \text{ with } y(\tau_k^-) = y(\tau_k)\} .$$

$PC([-r, 0], \mathbb{R})$ is a Banach space with the norm

$$\|y\|_{PC} = \sup_{t \in [-r,0]} |y(t)| .$$

Let

$$PC([0, T], \mathbb{R}) = \{y : [0, T] \to \mathbb{R} | y \in C((t_k, t_{k+1}], \mathbb{R}), \quad k = 1, \dots, m,$$
$$\text{and there exist } y(t_k^-) \text{ and } y(t_k^+), \quad k = 1, \dots, m, \text{ with } y(t_k^-) = y(t_k)\} .$$

$PC([0, T], \mathbb{R})$ is a Banach space with the norm

$$\|y\|_C = \sup_{t \in [0,T]} |y(t)| .$$

Notice that

$$\Omega = \{y : [-r, T] \to \mathbb{R} : y|_{[-r,0]} \in PC([-r, 0], \mathbb{R}) \text{ and } y|_{[0,T]} \in PC([0, T], \mathbb{R})\}$$

is a Banach space with the norm

$$\|y\|_\Omega = \sup_{t \in [-r,T]} |y(t)| .$$

**Definition 3.1.** A function $y \in \Omega$ whose $\alpha$-derivative exists on $J_k$ is said to be a solution of (3.1)–(3.3) if $y$ satisfies the equation ${}^cD_{t_k}^\alpha y(t) = f(t, y_t, {}^cD_{t_k}^\alpha y(t))$ on $J_k$ and satisfies the conditions

$$\Delta y|_{t=t_k} = I_k(y_{t_k^-}) , \quad k = 1, \dots, m ,$$
$$y(t) = \varphi(t) , \quad t \in [-r, 0] .$$

To prove the existence of solutions to (3.1)–(3.3), we need the following auxiliary lemma.

**Lemma 3.2.** *Let $0 < \alpha \le 1$, and let $\sigma : J \to \mathbb{R}$ be continuous. A function $y$ is a solution of the fractional integral equation*

$$y(t) = \begin{cases} \varphi(0) + \dfrac{1}{\Gamma(\alpha)} \displaystyle\int_0^t (t-s)^{\alpha-1} \sigma(s) ds, & \text{if } t \in [0, t_1] , \\ \varphi(0) + \displaystyle\sum_{i=1}^k I_i(y_{t_i^-}) + \dfrac{1}{\Gamma(\alpha)} \displaystyle\sum_{i=1}^k \int_{t_{i-1}}^{t_i} (t_i - s)^{\alpha-1} \sigma(s) ds & \\ + \dfrac{1}{\Gamma(\alpha)} \displaystyle\int_{t_k}^t (t-s)^{\alpha-1} \sigma(s) ds, & \text{if } t \in (t_k, t_{k+1}] , \\ \varphi(t), & t \in [-r, 0] , \end{cases} \tag{3.5}$$

*where $k = 1, \dots, m$, if and only if $y$ is a solution of the fractional problem*

$$ {}^cD^\alpha y(t) = \sigma(t)\,, \quad t \in J_k\,, \tag{3.6} $$

$$ \Delta y|_{t=t_k} = I_k(y_{t_k^-})\,, \quad k = 1, \dots, m\,, \tag{3.7} $$

$$ y(t) = \varphi(t)\,, \quad t \in [-r, 0]\,. \tag{3.8} $$

*Proof.* Assume that $y$ satisfies (3.6)–(3.8). If $t \in [0, t_1]$, then

$$ {}^cD^\alpha y(t) = \sigma(t)\,. $$

Lemma 1.9 implies

$$ y(t) = \varphi(0) + I^\alpha \sigma(t) = \varphi(0) + \frac{1}{\Gamma(\alpha)} \int_0^t (t-s)^{\alpha-1} \sigma(s) ds\,. $$

If $t \in (t_1, t_2]$, then Lemma 1.9 implies

$$ \begin{aligned} y(t) &= y(t_1^+) + \frac{1}{\Gamma(\alpha)} \int_{t_1}^t (t-s)^{\alpha-1} \sigma(s) ds \\ &= \Delta y|_{t=t_1} + y(t_1^-) + \frac{1}{\Gamma(\alpha)} \int_{t_1}^t (t-s)^{\alpha-1} \sigma(s) ds \\ &= I_1(y_{t_1^-}) + \left[ \varphi(0) + \frac{1}{\Gamma(\alpha)} \int_0^{t_1} (t_1-s)^{\alpha-1} \sigma(s) ds \right] \\ &\quad + \frac{1}{\Gamma(\alpha)} \int_{t_1}^t (t-s)^{\alpha-1} \sigma(s) ds\,. \\ &= \varphi(0) + I_1(y_{t_1^-}) + \frac{1}{\Gamma(\alpha)} \int_0^{t_1} (t_1-s)^{\alpha-1} \sigma(s) ds \\ &\quad + \frac{1}{\Gamma(\alpha)} \int_{t_1}^t (t-s)^{\alpha-1} \sigma(s) ds\,. \end{aligned} $$

If $t \in (t_2, t_3]$, then from Lemma 1.9 we get

$$
\begin{aligned}
y(t) &= y(t_2^+) + \frac{1}{\Gamma(\alpha)} \int_{t_2}^{t} (t-s)^{\alpha-1} \sigma(s) ds \\
&= \Delta y|_{t=t_2} + y(t_2^-) + \frac{1}{\Gamma(\alpha)} \int_{t_2}^{t} (t-s)^{\alpha-1} \sigma(s) ds \\
&= I_2(y_{t_2^-}) + \left[ \varphi(0) + I_1(y_{t_1^-}) + \frac{1}{\Gamma(\alpha)} \int_0^{t_1} (t_1-s)^{\alpha-1} \sigma(s) ds \right. \\
&\quad \left. + \frac{1}{\Gamma(\alpha)} \int_{t_1}^{t_2} (t_2-s)^{\alpha-1} \sigma(s) ds \right] + \frac{1}{\Gamma(\alpha)} \int_{t_2}^{t} (t-s)^{\alpha-1} \sigma(s) ds . \\
&= \varphi(0) + \left[ I_1(y_{t_1^-}) + I_2(y_{t_2^-}) \right] + \left[ \frac{1}{\Gamma(\alpha)} \int_0^{t_1} (t_1-s)^{\alpha-1} \sigma(s) ds \right. \\
&\quad \left. + \frac{1}{\Gamma(\alpha)} \int_{t_1}^{t_2} (t_2-s)^{\alpha-1} \sigma(s) ds \right] + \frac{1}{\Gamma(\alpha)} \int_{t_2}^{t} (t-s)^{\alpha-1} \sigma(s) ds .
\end{aligned}
$$

Repeating the process in this way, the solution $y(t)$ for $t \in (t_k, t_{k+1}]$, where $k = 1, \dots, m$, can be written

$$
\begin{aligned}
y(t) &= \varphi(0) + \sum_{i=1}^{k} I_i(y_{t_i^-}) + \frac{1}{\Gamma(\alpha)} \sum_{i=1}^{k} \int_{t_{i-1}}^{t_i} (t_i-s)^{\alpha-1} \sigma(s) ds \\
&\quad + \frac{1}{\Gamma(\alpha)} \int_{t_k}^{t} (t-s)^{\alpha-1} \sigma(s) ds .
\end{aligned}
$$

Conversely, assume that $y$ satisfies the impulsive fractional integral equation (3.5). If $t \in [0, t_1]$, then $y(0) = \varphi(0)$. Using the fact that ${}^cD^\alpha$ is the left inverse of $I^\alpha$, we obtain

$$
{}^cD^\alpha y(t) = \sigma(t) , \quad \text{for each } t \in [0, t_1] .
$$

If $t \in (t_k, t_{k+1}]$, $k = 1, \dots, m$, using the fact that ${}^cD^\alpha C = 0$, where $C$ is a constant, we have

$$
{}^cD^\alpha y(t) = \sigma(t) , \quad \text{for each } t \in (t_k, t_{k+1}] .
$$

Also, we can easily show that

$$
\Delta y|_{t=t_k} = I_k(y_{t_k^-}) , \quad k = 1, \dots, m . \qquad \square
$$

We are now in a position to state and prove our existence result for problem (3.1)–(3.3) based on Banach's fixed point.

**Theorem 3.3.** *Make the following assumptions:*
(3.3.1) *The function* $f : J \times PC([-r,0],\mathbb{R}) \times \mathbb{R} \to \mathbb{R}$ *is continuous.*
(3.3.2) *There exist constants* $K > 0$ *and* $0 < L < 1$ *such that*

$$|f(t,u,v) - f(t,\bar{u},\bar{v})| \leq K\|u-\bar{u}\|_{PC} + L|v-\bar{v}|$$

*for any* $u, \bar{u} \in PC([-r,0],\mathbb{R})$, $v, \bar{v} \in \mathbb{R}$ *and* $t \in J$.
(3.3.3) *There exists a constant* $\tilde{l} > 0$ *such that*

$$|I_k(u) - I_k(\overline{u})| \leq \tilde{l}\|u-\overline{u}\|_{PC}$$

*for each* $u, \overline{u} \in PC([-r,0],\mathbb{R})$ *and* $k = 1,\ldots,m$.

*If*

$$m\tilde{l} + \frac{(m+1)KT^{\alpha}}{(1-L)\Gamma(\alpha+1)} < 1\,, \tag{3.9}$$

*then there exists a unique solution for problem* (3.1)–(3.3) *on* $J$.

*Proof.* Transform problem (3.1)–(3.3) into a fixed point problem. Consider the operator $N : \Omega \to \Omega$ defined by

$$(Ny)(t) = \begin{cases} \varphi(0) + \sum\limits_{0<t_k<t} I_k(y_{t_i^-}) + \dfrac{1}{\Gamma(\alpha)} \sum\limits_{0<t_k<t} \displaystyle\int_{t_{k-1}}^{t_k} (t_k-s)^{\alpha-1} g(s)ds \\ +\dfrac{1}{\Gamma(\alpha)} \displaystyle\int_{t_k}^{t} (t-s)^{\alpha-1}\sigma(s)ds, & t \in [0,T]\,, \\ \varphi(t), & t \in [-r,0]\,, \end{cases} \tag{3.10}$$

where $g \in C(J,\mathbb{R})$ is such that

$$g(t) = f(t, y_t, g(t))\,.$$

Clearly, the fixed points of operator $N$ are solutions of problem (3.1)–(3.3).
Let $u, w \in \Omega$. If $t \in [-r,0]$, then

$$|(Nu)(t) - (Nw)(t)| = 0\,.$$

For $t \in J$ we have

$$\begin{aligned} |N(u)(t) - N(w)(t)| \leq{}& \frac{1}{\Gamma(\alpha)} \sum_{0<t_k<t} \int_{t_{k-1}}^{t_k} (t_k-s)^{\alpha-1}|g(s)-h(s)|ds \\ &+ \frac{1}{\Gamma(\alpha)} \int_{t_k}^{t} (t-s)^{\alpha-1}|g(s)-h(s)|ds \\ &+ \sum_{0<t_k<t} |I_k(u_{t_k^-}) - I_k(w_{t_k^-})|\,, \end{aligned}$$

where $g, h \in C(J, \mathbb{R})$ are given by

$$g(t) = f(t, u_t, g(t)) ,$$

and

$$h(t) = f(t, w_t, h(t)) .$$

By (3.3.2) we have

$$\begin{aligned}|g(t) - h(t)| &= |f(t, u_t, g(t)) - f(t, w_t, h(t))|\\ &\le K\|u_t - w_t\|_{PC} + L|g(t) - h(t)| .\end{aligned}$$

Hence,

$$|g(t) - h(t)| \le \frac{K}{1-L}\|u_t - w_t\|_{PC} .$$

Therefore, for each $t \in J$

$$\begin{aligned}|N(u)(t) - N(w)(t)| &\le \frac{K}{(1-L)\Gamma(\alpha)}\sum_{k=1}^{m}\int_{t_{k-1}}^{t_k}(t_k - s)^{\alpha-1}\|u_s - w_s\|_{PC}ds\\ &\quad + \frac{K}{(1-L)\Gamma(\alpha)}\int_{t_k}^{t}(t-s)^{\alpha-1}\|u_s - w_s\|_{PC}ds\\ &\quad + \sum_{k=1}^{m}\tilde{l}\|u_{t_k^-} - w_{t_k^-}\|_{PC} .\\ &\le \left[m\tilde{l} + \frac{mKT^\alpha}{(1-L)\Gamma(\alpha+1)}\right.\\ &\quad \left. + \frac{KT^\alpha}{(1-L)\Gamma(\alpha+1)}\right]\|u - w\|_\Omega .\end{aligned}$$

Thus,

$$\|N(u) - N(w)\|_\Omega \le \left[m\tilde{l} + \frac{(m+1)KT^\alpha}{(1-L)\Gamma(\alpha+1)}\right]\|u - w\|_\Omega .$$

By (3.9), operator $N$ is a contraction. Hence, by Banach's contraction principle, $N$ has a unique fixed point that is the unique solution of problem (3.1)–(3.3). □

Our second result is based on Schaefer's fixed point theorem.

**Theorem 3.4.** *In addition to* (3.3.1), (3.3.2) *assumes that:*

(3.4.1) *There exist* $p, q, r \in C(J, \mathbb{R}_+)$ *with* $r^* = \sup_{t\in J} r(t) < 1$ *such that*

$$|f(t, u, w)| \le p(t) + q(t)\|u\|_{PC} + r(t)|w| \quad \text{for } t \in J,\ u \in PC([-r, 0], \mathbb{R}) \text{ and } w \in \mathbb{R} .$$

(3.4.2) *The functions* $I_k\colon PC([-r, 0], \mathbb{R}) \to \mathbb{R}$ *are continuous, and there exist constants* $M^*, N^* > 0$, *with* $mM^* < 1$, *such that*

$$|I_k(u)| \le M^*\|u\|_{PC} + N^* \quad \text{for each } u \in PC([-r, 0], \mathbb{R}),\ k = 1, \dots, m .$$

*Then problem* (3.1)–(3.3) *has at least one solution.*

*Proof.* Let operator $N$ be defined as in (3.10). We will use Schaefer's fixed point theorem to prove that $N$ has a fixed point. The proof will be given in several steps.

*Step 1: $N$ is continuous.* Let $\{u_n\}$ be a sequence such that $u_n \to u$ in $\Omega$. If $t \in [-r, 0]$, then

$$|(Nu_n)(t) - (Nu)(t)| = 0 .$$

For $t \in J$ we have

$$\begin{aligned} |(Nu_n)(t) - (Nu)(t)| \le & \frac{1}{\Gamma(\alpha)} \sum_{0<t_k<t} \int_{t_{k-1}}^{t_k} (t_k - s)^{\alpha-1} |g_n(s) - g(s)| ds \\ & + \frac{1}{\Gamma(\alpha)} \int_{t_k}^{t} (t - s)^{\alpha-1} |g_n(s) - g(s)| ds \\ & + \sum_{0<t_k<t} |I_k(u_{nt_k^-}) - I_k(u_{t_k^-})| , \end{aligned} \tag{3.11}$$

where $g_n, g \in C(J, \mathbb{R})$ are given by

$$g_n(t) = f(t, u_{nt}, g_n(t))$$

and

$$g(t) = f(t, u_t, g(t)) .$$

From (3.3.2) we have

$$\begin{aligned} |g_n(t) - g(t)| &= |f(t, u_{nt}, g_n(t)) - f(t, u_t, g(t))| \\ &\le K\|u_{nt} - u_t\|_{PC} + L|g_n(t) - g(t)| . \end{aligned}$$

Then

$$|g_n(t) - g(t)| \le \frac{K}{1-L} \|u_{nt} - u_t\|_{PC} .$$

Since $u_n \to u$, $g_n(t) \to g(t)$ as $n \to \infty$ for each $t \in J$. Let $\eta > 0$ be such that for each $t \in J$ we have $|g_n(t)| \le \eta$ and $|g(t)| \le \eta$. Then

$$\begin{aligned} (t - s)^{\alpha-1} |g_n(s) - g(s)| &\le (t - s)^{\alpha-1} [|g_n(s)| + |g(s)|] \\ &\le 2\eta(t - s)^{\alpha-1} \end{aligned}$$

and

$$\begin{aligned} (t_k - s)^{\alpha-1} |g_n(s) - g(s)| &\le (t_k - s)^{\alpha-1} [|g_n(s)| + |g(s)|] \\ &\le 2\eta(t_k - s)^{\alpha-1} . \end{aligned}$$

For each $t \in J$ the functions $s \to 2\eta(t - s)^{\alpha-1}$ and $s \to 2\eta(t_k - s)^{\alpha-1}$ are integrable on $[0, t]$, so by the Lebesgue dominated convergence theorem and (3.11),

$$|(Nu_n)(t) - (Nu)(t)| \to 0 \text{ as } n \to \infty .$$

Hence,

$$\|(Nu_n) - (Nu)\|_\Omega \to 0 \text{ as } n \to \infty .$$

Consequently, $N$ is continuous.

*Step 2: F maps bounded sets to bounded sets in $\Omega$.* It is enough to show that for any $\eta^* > 0$ there exists a positive constant $\ell$ such that for each $u \in B_{\eta^*} = \{u \in \Omega : \|u\|_\Omega \le \eta^*\}$ we have $\|N(u)\|_\Omega \le \ell$. For each $t \in J$ we have

$$\begin{aligned}(Nu)(t) = \varphi(0) + \frac{1}{\Gamma(\alpha)} \sum_{0<t_k<t} \int_{t_{k-1}}^{t_k} (t_k - s)^{\alpha-1} g(s) ds \\ + \frac{1}{\Gamma(\alpha)} \int_{t_k}^{t} (t-s)^{\alpha-1} g(s) ds + \sum_{0<t_k<t} I_k(u_{t_k^-}) ,\end{aligned} \tag{3.12}$$

where $g \in C(J, \mathbb{R})$ is given by

$$g(t) = f(t, u_t, g(t)) .$$

By (3.4.1), for each $t \in J$ we have

$$\begin{aligned}|g(t)| &= |f(t, u_t, g(t))| \\ &\le p(t) + q(t)\|u_t\|_{PC} + r(t)|g(t)| \\ &\le p(t) + q(t)\|u\|_\Omega + r(t)|g(t)| \\ &\le p(t) + q(t)\eta^* + r(t)|g(t)| \\ &\le p^* + q^*\eta^* + r^*|g(t)| ,\end{aligned}$$

where $p^* = \sup_{t\in J} p(t)$ and $q^* = \sup_{t\in J} q(t)$.
Then

$$|g(t)| \le \frac{p^* + q^*\eta^*}{1 - r^*} := M .$$

Thus, (3.12) implies

$$\begin{aligned}|N(u)(t)| &\le |\varphi(0)| + \frac{mMT^\alpha}{\Gamma(\alpha+1)} + \frac{MT^\alpha}{\Gamma(\alpha+1)} + m(M^*\|u_{t_k^-}\|_{PC} + N^*) \\ &\le |\varphi(0)| + \frac{(m+1)MT^\alpha}{\Gamma(\alpha+1)} + m(M^*\|u\|_\Omega + N^*) \\ &\le |\varphi(0)| + \frac{(m+1)MT^\alpha}{\Gamma(\alpha+1)} + m(M^*\eta^* + N^*) := R .\end{aligned}$$

If $t \in [-r, 0]$, then

$$|N(u)(t)| \le \|\varphi\|_{PC} ,$$

so

$$\|N(u)\|_\Omega \le \max\{R, \|\varphi\|_{PC}\} := \ell .$$

*Step 3: F maps bounded sets to equicontinuous sets of $\Omega$.* Let $t_1, t_2 \in (0, T]$, $t_1 < t_2$, $B_{\eta^*}$ be a bounded set of $\Omega$ as in Step 2, and let $u \in B_{\eta^*}$. Then

$$\begin{aligned}
&|N(u)(t_2) - N(u)(t_1)| \\
&\leq \frac{1}{\Gamma(\alpha)} \int_0^{t_1} |(t_2 - s)^{\alpha-1} - (t_1 - s)^{\alpha-1}||g(s)|ds \\
&\quad + \frac{1}{\Gamma(\alpha)} \int_{t_1}^{t_2} |(t_2 - s)^{\alpha-1}||g(s)|ds + \sum_{0<t_k<t_2-t_1} |I_k(u_{t_k^-})| \\
&\leq \frac{M}{\Gamma(\alpha+1)}[2(t_2 - t_1)^\alpha + (t_2^\alpha - t_1^\alpha)] + (t_2 - t_1)(M^*\|u_{t_k^-}\|_{PC} + N^*) \\
&\leq \frac{M}{\Gamma(\alpha+1)}[2(t_2 - t_1)^\alpha + (t_2^\alpha - t_1^\alpha)] + (t_2 - t_1)(M^*\|u\|_\Omega + N^*) \\
&\leq \frac{M}{\Gamma(\alpha+1)}[2(t_2 - t_1)^\alpha + (t_2^\alpha - t_1^\alpha)] + (t_2 - t_1)(M^*\eta^* + N^*) .
\end{aligned}$$

As $t_1 \to t_2$, the right-hand side of the preceding inequality tends to zero. As a consequence of Steps 1–3, together with the Ascoli–Arzelà theorem, we can conclude that $N: \Omega \to \Omega$ is completely continuous.

*Step 4: A priori bounds.* Now it remains to show that the set

$$E = \{u \in \Omega : u = \lambda N(u) \text{ for some } 0 < \lambda < 1\}$$

is bounded. Let $u \in E$; then $u = \lambda N(u)$ for some $0 < \lambda < 1$. Thus, for each $t \in J$ we have

$$\begin{aligned}
u(t) = \lambda\varphi(0) &+ \frac{\lambda}{\Gamma(\alpha)} \sum_{0<t_k<t} \int_{t_{k-1}}^{t_k} (t_k - s)^{\alpha-1} g(s)ds \\
&+ \frac{\lambda}{\Gamma(\alpha)} \int_{t_k}^{t} (t - s)^{\alpha-1} g(s)ds + \lambda \sum_{0<t_k<t} I_k(u_{t_k^-}) ,
\end{aligned} \tag{3.13}$$

and by (3.4.1), for each $t \in J$ we have

$$\begin{aligned}
|g(t)| &= |f(t, u_t, g(t))| \\
&\leq p(t) + q(t)\|u_t\|_{PC} + r(t)|g(t)| \\
&\leq p^* + q^*\|u_t\|_{PC} + r^*|g(t)| .
\end{aligned}$$

Thus,

$$|g(t)| \leq \frac{1}{1 - r^*}(p^* + q^*\|u_t\|_{PC}) .$$

Now (3.13) and (3.4.2) imply that for each $t \in J$ we have

$$|u(t)| \leq |\varphi(0)| + \frac{1}{(1-r^*)\Gamma(\alpha)} \sum_{0<t_k<t} \int_{t_{k-1}}^{t_k} (t_k - s)^{\alpha-1}(p^* + q^*\|u_s\|_{PC})ds$$
$$+ \frac{1}{(1-r^*)\Gamma(\alpha)} \int_{t_k}^{t} (t-s)^{\alpha-1}(p^* + q^*\|u_s\|_{PC})ds$$
$$+ m(M^*\|u_{t_k^-}\|_{PC} + N^*) .$$

Consider the function $\zeta$ defined by

$$\zeta(t) = \sup\{|u(s)| : -r \leq s \leq t\}, 0 \leq t \leq T .$$

Then there exists $t^* \in [-r, T]$ such that $\zeta(t) = |u(t^*)|$. If $t \in [0, T]$, then, by the previous inequality, for $t \in J$ we have

$$\zeta(t) \leq |\varphi(0)| + \frac{1}{(1-r^*)\Gamma(\alpha)} \sum_{0<t_k<t} \int_{t_{k-1}}^{t_k} (t_k - s)^{\alpha-1}(p^* + q^*\zeta(s))ds$$
$$+ \frac{1}{(1-r^*)\Gamma(\alpha)} \int_{t_k}^{t} (t-s)^{\alpha-1}(p^* + q^*\zeta(s))ds$$
$$+ mM^*\zeta(t) + mN^* .$$

Thus, for $t \in J$

$$\zeta(t) \leq \frac{|\varphi(0)| + mN^*}{1-mM^*} + \frac{1}{(1-mM^*)(1-r^*)\Gamma(\alpha)} \sum_{0<t_k<t} \int_{t_{k-1}}^{t_k} (t_k - s)^{\alpha-1}(p^* + q^*\zeta(s))ds$$
$$+ \frac{1}{(1-mM^*)(1-r^*)\Gamma(\alpha)} \int_{t_k}^{t} (t-s)^{\alpha-1}(p^* + q^*\zeta(s))ds$$
$$\leq \frac{|\varphi(0)| + mN^*}{1-mM^*} + \frac{(m+1)p^*T^\alpha}{(1-mM^*)(1-r^*)\Gamma(\alpha+1)}$$
$$+ \frac{(m+1)q^*}{(1-mM^*)(1-r^*)\Gamma(\alpha)} \int_0^t (t-s)^{\alpha-1}\zeta(s)ds .$$

Applying Lemma 1.52, we get

$$\zeta(t) \leq \left[\frac{|\varphi(0)| + mN^*}{1-mM^*} + \frac{(m+1)p^*T^\alpha}{(1-mM^*)(1-r^*)\Gamma(\alpha+1)}\right]$$
$$\times \left[1 + \frac{\delta(m+1)q^*T^\alpha}{(1-mM^*)(1-r^*)\Gamma(\alpha+1)}\right] := A ,$$

where $\delta = \delta(\alpha)$ is a constant. If $t^* \in [-r, 0]$, then $\zeta(t) = \|\varphi\|_{PC}$. Thus, for any $t \in J$ and $\|u\|_\Omega \le \zeta(t)$ we have

$$\|u\|_\Omega \le \max\{\|\varphi\|_{PC}, A\} .$$

This shows that the set $E$ is bounded. As a consequence of Schaefer's fixed point theorem, we deduce that $N$ has a fixed point that is a solution of problem (3.1)–(3.3). □

### 3.2.3 Ulam–Hyers–Rassias stability

Here we adopt the concepts in Wang et al. [252] and introduce Ulam's type stability concepts for problem (3.1)–(3.2).

Let $z \in \Omega$, $\epsilon > 0$, $\psi > 0$, and $\omega \in PC(J, \mathbb{R}_+)$ be nondecreasing. We consider the set of inequalities

$$\begin{cases} |{}^cD^\alpha z(t) - f(t, z_t, {}^cD^\alpha z(t))| \le \epsilon, & t \in (t_k, t_{k+1}],\ k = 1, \dots, m , \\ |\Delta y|_{t_k} - I_k(y_{t_k^-})| \le \epsilon, & k = 1, \dots, m ; \end{cases} \tag{3.14}$$

the set of inequalities

$$\begin{cases} |{}^cD^\alpha z(t) - f(t, z_t, {}^cD^\alpha z(t))| \le \omega(t), & t \in (t_k, t_{k+1}],\ k = 1, \dots, m , \\ |\Delta y|_{t_k} - I_k(y_{t_k^-})| \le \psi, & k = 1, \dots, m ; \end{cases} \tag{3.15}$$

and the set of inequalities

$$\begin{cases} |{}^cD^\alpha z(t) - f(t, z_t, {}^cD^\alpha z(t))| \le \epsilon\omega(t), & t \in (t_k, t_{k+1}],\ k = 1, \dots, m , \\ |\Delta y|_{t_k} - I_k(y_{t_k^-})| \le \epsilon\psi, & k = 1, \dots, m . \end{cases} \tag{3.16}$$

**Definition 3.5.** Problem (3.1)–(3.2) is Ulam–Hyers stable if there exists a real number $c_{f,m} > 0$ such that for each $\epsilon > 0$ and for each solution $z \in \Omega$ of inequality (3.14) there exists a solution $y \in \Omega$ of problem (3.1)–(3.2), with

$$|z(t) - y(t)| \le c_{f,m}\epsilon , \quad t \in J .$$

**Definition 3.6.** Problem (3.1)–(3.2) is generalized Ulam–Hyers stable if there exists $\theta_{f,m} \in C(\mathbb{R}_+, \mathbb{R}_+)$ with $\theta_{f,m}(0) = 0$ such that for each solution $z \in \Omega$ of inequality (3.14) there exists a solution $y \in \Omega$ of problem (3.1)–(3.2), with

$$|z(t) - y(t)| \le \theta_{f,m}(\epsilon) , \quad t \in J .$$

**Definition 3.7.** Problem (3.1)–(3.2) is Ulam–Hyers–Rassias stable with respect to $(\omega, \psi)$ if there exists $c_{f,m,\omega} > 0$ such that for each $\epsilon > 0$ and for each solution $z \in \Omega$ of inequality (3.16) there exists a solution $y \in \Omega$ of problem (3.1)–(3.2), with

$$|z(t) - y(t)| \le c_{f,m,\omega}\epsilon(\omega(t) + \psi) , \quad t \in J .$$

**Definition 3.8.** Problem (3.1)–(3.2) is generalized Ulam–Hyers–Rassias stable with respect to $(\omega, \psi)$ if there exists $c_{f,m,\omega} > 0$ such that for each solution $z \in \Omega$ of inequality (3.15) there exists a solution $y \in \Omega$ of problem (3.1)–(3.2), with

$$|z(t) - y(t)| \leq c_{f,m,\omega}(\omega(t) + \psi), \quad t \in J.$$

**Remark 3.9.** It is clear that (i) Definition 3.5 implies Definition 3.6, (ii) Definition 3.7 implies Definition 3.8, and (iii) Definition 3.7 for $\omega(t) = \psi = 1$ implies Definition 3.5.

**Remark 3.10.** A function $z \in \Omega$ is a solution of inequality (3.16) if and only if there is $\sigma \in PC(J, \mathbb{R})$ and a sequence $\sigma_k$, $k = 1, \ldots, m$ (which depend on $z$) such that
**(i)** $|\sigma(t)| \leq \epsilon\omega(t),\ t \in (t_k, t_{k+1}],\ k = 1, \ldots, m$ and $|\sigma_k| \leq \epsilon\psi$, $k = 1, \ldots, m$;
**(ii)** ${}^cD^\alpha z(t) = f(t, z_t, {}^cD^\alpha z(t)) + \sigma(t),\ t \in (t_k, t_{k+1}], k = 1, \ldots, m$;
**(iii)** $\Delta z|_{t_k} = I_k(z_{t_k^-}) + \sigma_k,\ k = 1, \ldots, m.$

Similar remarks hold for inequalities 3.15 and 3.14.

Now we state the following Ulam–Hyers–Rassias stable result.

**Theorem 3.11.** *Assume* (3.3.1)–(3.3.3) *and* (3.9) *hold and*
(3.11.1) *there exists a nondecreasing function $\omega \in PC(J, \mathbb{R}_+)$, and there exists $\lambda_\omega > 0$ such that, for any $t \in J$,*

$$I^\alpha\omega(t) \leq \lambda_\omega\omega(t).$$

*Then problem* (3.1)–(3.2) *is Ulam–Hyers–Rassias stable with respect to $(\omega, \psi)$.*

*Proof.* Let $z \in \Omega$ be a solution of inequality (3.16). Denote by $y$ the unique solution of the problem

$$\begin{cases} {}^cD^\alpha_{t_k} y(t) = f(t, y_t, {}^cD^\alpha_{t_k} y(t)), & t \in (t_k, t_{k+1}], k = 1, \ldots, m, \\ \Delta y|_{t=t_k} = I_k(y_{t_k^-}), & k = 1, \ldots, m, \\ y(t) = z(t) = \varphi(t), & t \in [-r, 0]. \end{cases}$$

Using Lemma 3.2, we obtain for each $t \in (t_k, t_{k+1}]$

$$y(t) = \varphi(0) + \sum_{i=1}^{k} I_i(y_{t_i^-}) + \frac{1}{\Gamma(\alpha)} \sum_{i=1}^{k} \int_{t_{i-1}}^{t_i} (t_i - s)^{\alpha-1} g(s)ds$$
$$+ \frac{1}{\Gamma(\alpha)} \int_{t_k}^{t} (t - s)^{\alpha-1} g(s)ds,$$

where $g \in C(J, \mathbb{R})$ is given by

$$g(t) = f(t, y_t, g(t)).$$

Since $z$ is a solution of inequality (3.16), by Remark 3.10 we have

$$\begin{cases} {}^cD^\alpha_{t_k} z(t) = f(t, z_t, {}^cD^\alpha_{t_k} z(t)) + \sigma(t), & t \in (t_k, t_{k+1}], k = 1, \ldots, m, \\ \Delta z|_{t=t_k} = I_k(z_{t_k^-}) + \sigma_k, & k = 1, \ldots, m. \end{cases} \tag{3.17}$$

Clearly, the solution of (3.17) is given by

$$z(t) = \varphi(0) + \sum_{i=1}^{k} I_i(z_{t_i^-}) + \sum_{i=1}^{k} \sigma_i + \frac{1}{\Gamma(\alpha)} \sum_{i=1}^{k} \int_{t_{i-1}}^{t_i} (t_i - s)^{\alpha-1} h(s) ds$$
$$+ \frac{1}{\Gamma(\alpha)} \sum_{i=1}^{k} \int_{t_{i-1}}^{t_i} (t_i - s)^{\alpha-1} \sigma(s) ds + \frac{1}{\Gamma(\alpha)} \int_{t_k}^{t} (t - s)^{\alpha-1} h(s) ds$$
$$+ \frac{1}{\Gamma(\alpha)} \int_{t_k}^{t} (t - s)^{\alpha-1} \sigma(s) ds , \quad t \in (t_k, t_{k+1}] ,$$

where $h \in C(J, \mathbb{R})$ is given by

$$h(t) = f(t, z_t, h(t)) .$$

Hence, for each $t \in (t_k, t_{k+1}]$, it follows that

$$|z(t) - y(t)| \le \sum_{i=1}^{k} |\sigma_i| + \sum_{i=1}^{k} |I_i(z_{t_i^-}) - I_i(y_{t_i^-})|$$
$$+ \frac{1}{\Gamma(\alpha)} \sum_{i=1}^{k} \int_{t_{i-1}}^{t_i} (t_i - s)^{\alpha-1} |h(s) - g(s)| ds$$
$$+ \frac{1}{\Gamma(\alpha)} \sum_{i=1}^{k} \int_{t_{i-1}}^{t_i} (t_i - s)^{\alpha-1} |\sigma(s)| ds$$
$$+ \frac{1}{\Gamma(\alpha)} \int_{t_k}^{t} (t - s)^{\alpha-1} |h(s) - g(s)| ds$$
$$+ \frac{1}{\Gamma(\alpha)} \int_{t_k}^{t} (t - s)^{\alpha-1} |\sigma(s)| .$$

Thus,

$$|z(t) - y(t)| \le m\epsilon\psi + (m+1)\epsilon\lambda_\omega \omega(t) + \sum_{i=1}^{k} \tilde{l} \| z_{t_i^-} - y_{t_i^-} \|_{PC}$$
$$+ \frac{1}{\Gamma(\alpha)} \sum_{i=1}^{k} \int_{t_{i-1}}^{t_i} (t_i - s)^{\alpha-1} |h(s) - g(s)| ds$$
$$+ \frac{1}{\Gamma(\alpha)} \int_{t_k}^{t} (t - s)^{\alpha-1} |h(s) - g(s)| ds .$$

By (33.3.2) we have

$$|h(t) - g(t)| = |f(t, z_t, h(t)) - f(t, y_t, g(t))|$$
$$\leq K\|z_t - y_t\|_{PC} + L|g(t) - h(t)| .$$

Then

$$|h(t) - g(t)| \leq \frac{K}{1-L}\|z_t - y_t\|_{PC} .$$

Therefore, for each $t \in J$

$$|z(t) - y(t)| \leq m\epsilon\psi + (m+1)\epsilon\lambda_\omega\omega(t) + \sum_{i=1}^{k}\tilde{l}\|z_{t_i^-} - y_{t_i^-}\|_{PC}$$
$$+ \frac{K}{(1-L)\Gamma(\alpha)}\sum_{i=1}^{k}\int_{t_{i-1}}^{t_i}(t_i - s)^{\alpha-1}\|z_s - y_s\|_{PC}ds$$
$$+ \frac{K}{(1-L)\Gamma(\alpha)}\int_{t_k}^{t}(t-s)^{\alpha-1}\|z_s - y_s\|_{PC}ds .$$

Thus,

$$|z(t) - y(t)| \leq \sum_{0<t_i<t}\tilde{l}\|z_{t_i^-} - y_{t_i^-}\|_{PC} + \epsilon(\psi + \omega(t))(m + (m+1)\lambda_\omega)$$
$$+ \frac{K(m+1)}{(1-L)\Gamma(\alpha)}\int_0^t(t-s)^{\alpha-1}\|z_s - y_s\|_{PC}ds .$$

We consider the function $\zeta_1$ defined by

$$\zeta_1(t) = \sup\{\|z(s) - y(s)\| : \ -r \leq s \leq t\}, 0 \leq t \leq T ;$$

then there exists $t^* \in [-r, T]$ such that $\zeta_1(t) = \|z(t^*) - y(t^*)\|$. If $t^* \in [-r, 0]$, then $\zeta_1(t) = 0$. If $t^* \in [0, T]$, then by the previous inequality we have

$$\zeta_1(t) \leq \sum_{0<t_i<t}\tilde{l}\zeta_1(t_i^-) + \epsilon(\psi + \omega(t))(m + (m+1)\lambda_\omega)$$
$$+ \frac{K(m+1)}{(1-L)\Gamma(\alpha)}\int_0^t(t-s)^{\alpha-1}\zeta_1(s)ds .$$

Applying Lemma 1.53, we get

$$\zeta_1(t) \leq \epsilon(\psi + \omega(t))(m + (m+1)\lambda_\omega)$$
$$\times\left[\prod_{0<t_i<t}(1+\tilde{l})\exp\left(\int_0^t\frac{K(m+1)}{(1-L)\Gamma(\alpha)}(t-s)^{\alpha-1}ds\right)\right]$$
$$\leq c_\omega\epsilon(\psi + \omega(t)) ,$$

where

$$c_\omega = (m + (m+1)\lambda_\omega)\left[\prod_{i=1}^{m}(1+\widetilde{l})\exp\left(\frac{K(m+1)T^\alpha}{(1-L)\Gamma(\alpha+1)}\right)\right]$$
$$= (m + (m+1)\lambda_\omega)\left[(1+\widetilde{l})\exp\left(\frac{K(m+1)T^\alpha}{(1-L)\Gamma(\alpha+1)}\right)\right]^m .$$

Thus, problem (3.1)–(3.2) is Ulam–Hyers–Rassias stable with respect to $(\omega, \psi)$. □

Next, we present the following Ulam–Hyers stability result.

**Theorem 3.12.** *Assume* (3.3.1)–(3.3.3) *and* (3.9) *hold. Then problem* (3.1)–(3.2) *is Ulam–Hyers stable.*

*Proof.* Let $z \in \Omega$ be a solution of inequality (3.14). Denote by $y$ the unique solution of the problem

$$\begin{cases} {}^cD^\alpha_{t_k} y(t) = f(t, y_t, {}^cD^\alpha_{t_k} y(t)), & t \in (t_k, t_{k+1}],\ k = 1, \dots, m\,; \\ \Delta y|_{t=t_k} = I_k(y_{t_k^-}), & k = 1, \dots, m\,; \\ y(t) = z(t) = \varphi(t), & t \in [-r, 0]\,. \end{cases}$$

From the proof of Theorem 3.11 we get the inequality

$$\zeta_1(t) \le \sum_{0<t_i<t} \widetilde{l}\zeta_1(t_i^-) + m\epsilon + \frac{T^\alpha \epsilon(m+1)}{\Gamma(\alpha+1)}$$
$$+ \frac{K(m+1)}{(1-L)\Gamma(\alpha)} \int_0^t (t-s)^{\alpha-1}\zeta_1(s)ds\,.$$

An application of Lemma 1.53 gives

$$\zeta_1(t) \le \epsilon\left(\frac{m\Gamma(\alpha+1) + T^\alpha(m+1)}{\Gamma(\alpha+1)}\right)$$
$$\times \left[\prod_{0<t_i<t}(1+\widetilde{l})\exp\left(\int_0^t \frac{K(m+1)}{(1-L)\Gamma(\alpha)}(t-s)^{\alpha-1}ds\right)\right]$$
$$\le c_\omega \epsilon\,,$$

where

$$c_\omega = \left(\frac{m\Gamma(\alpha+1) + T^\alpha(m+1)}{\Gamma(\alpha+1)}\right)\left[\prod_{i=1}^{m}(1+\widetilde{l})\exp\left(\frac{K(m+1)T^\alpha}{(1-L)\Gamma(\alpha+1)}\right)\right]$$
$$= \left(\frac{m\Gamma(\alpha+1) + T^\alpha(m+1)}{\Gamma(\alpha+1)}\right)\left[(1+\widetilde{l})\exp\left(\frac{K(m+1)T^\alpha}{(1-L)\Gamma(\alpha+1)}\right)\right]^m .$$

Moreover, if we set $\gamma(\epsilon) = c_\omega\epsilon$, $\gamma(0) = 0$, then problem (3.1)–(3.2) is generalized Ulam–Hyers stable. □

### 3.2.4 Examples

*Example 1.* Consider the impulsive problem

$$ {}^cD^{\frac{1}{2}}_{t_k}y(t) = \frac{e^{-t}}{(11+e^t)}\left[\frac{y_t}{1+y_t} - \frac{{}^cD^{\frac{1}{2}}_{t_k}y(t)}{1+{}^cD^{\frac{1}{2}}_{t_k}y(t)}\right] \quad \text{for } t \in J_0 \cup J_1\,, \tag{3.18} $$

$$ \Delta y|_{t=\frac{1}{2}} = \frac{y\left(\frac{1}{2}^-\right)}{10+y\left(\frac{1}{2}^-\right)}\,, \tag{3.19} $$

$$ y(t) = \varphi(t)\,, \quad t \in [-r, 0],\ r > 0\,, \tag{3.20} $$

where $\varphi \in PC([-r, 0], \mathbb{R})$, $J_0 = [0, \frac{1}{2}]$, $J_1 = (\frac{1}{2}, 1]$, $t_0 = 0$, and $t_1 = \frac{1}{2}$.

Set

$$ f(t,u,v) = \frac{e^{-t}}{(11+e^t)}\left[\frac{u}{1+u} - \frac{v}{1+v}\right], \quad t \in [0,1],\ u \in PC([-r,0],\mathbb{R}) \text{ and } v \in \mathbb{R}\,. $$

Clearly, the function $f$ is jointly continuous.

For each $u, \bar{u} \in PC([-r, 0], \mathbb{R})$, $v, \bar{v} \in \mathbb{R}$, and $t \in [0, 1]$:

$$ \begin{aligned} \|f(t,u,v) - f(t,\bar{u},\bar{v})\| &\le \frac{e^{-t}}{(11+e^t)}(\|u-\bar{u}\|_{PC} + \|v-\bar{v}\|) \\ &\le \frac{1}{12}\|u-\bar{u}\|_{PC} + \frac{1}{12}\|v-\bar{v}\|\,. \end{aligned} $$

Hence condition (3.3.2) is satisfied by $K = L = \frac{1}{12}$. Let

$$ I_1(u) = \frac{u}{10+u}\,, \quad u \in PC([-r,0],\mathbb{R})\,, $$

and let $u, v \in PC([-r, 0], \mathbb{R})$. Then we have

$$ |I_1(u) - I_1(v)| = \left|\frac{u}{10+u} - \frac{v}{10+v}\right| \le \frac{1}{10}\|u-v\|_{PC}\,. $$

Thus the condition

$$ \begin{aligned} m\widetilde{l} + \frac{(m+1)KT^\alpha}{(1-L)\Gamma(\alpha+1)} &= \left[\frac{1}{10} + \frac{\frac{2}{12}}{\left(1-\frac{1}{12}\right)\Gamma\left(\frac{3}{2}\right)}\right] \\ &= \frac{4}{11\sqrt{\pi}} + \frac{1}{10} < 1 \end{aligned} $$

is satisfied by $T = 1$, $m = 1$, and $\widetilde{l} = \frac{1}{10}$. It follows from Theorem 3.3 that problem (2.35)–(2.37) has a unique solution on $J$.

For any $t \in [0, 1]$, take $\omega(t) = t$ and $\psi = 1$.

Since

$$ I^{\frac{1}{2}}\omega(t) = \frac{1}{\Gamma\left(\frac{1}{2}\right)}\int_0^t (t-s)^{\frac{1}{2}-1}s\,ds \le \frac{2t}{\sqrt{\pi}}\,, $$

condition (3.11.1) is satisfied with $\lambda_\omega = \frac{2}{\sqrt{\pi}}$. It follows that problem (3.18)–(3.19) is Ulam–Hyers–Rassias stable with respect to $(\omega, \psi)$.

*Example 2.* Consider the impulsive problem

$$^{c}D^{\frac{1}{2}}_{t_k}y(t) = \frac{2 + |y_t| + |^{c}D^{\frac{1}{2}}_{t_k}y(t)|}{108e^{t+3}\left(1 + |y_t| + |^{c}D^{\frac{1}{2}}_{t_k}y(t)|\right)}, \quad \text{for each, } t \in J_0 \cup J_1\,, \tag{3.21}$$

$$\Delta y|_{t=\frac{1}{3}} = \frac{|y\left(\frac{1}{3}^-\right)|}{6 + |y\left(\frac{1}{3}^-\right)|}, \tag{3.22}$$

$$y(t) = \varphi(t)\,, \quad t \in [-r, 0],\ r > 0\,, \tag{3.23}$$

where $\varphi \in PC([-r, 0], \mathbb{R})$ $J_0 = [0, \frac{1}{3}]$, $J_1 = (\frac{1}{3}, 1]$, $t_0 = 0$, and $t_1 = \frac{1}{3}$. Set

$$f(t, u, v) = \frac{2 + |u| + |v|}{108e^{t+3}(1 + |u| + |v|)}, \quad t \in [0, 1],\ u \in PC\left([-r, 0], \mathbb{R}\right), v \in \mathbb{R}\,.$$

Clearly, the function $f$ is jointly continuous. For any $u, \bar{u} \in PC([-r, 0], \mathbb{R})$, $v, \bar{v} \in \mathbb{R}$ and $t \in [0, 1]$

$$|f(t, u, v) - f(t, \bar{u}, \bar{v})| \le \frac{1}{108e^3}(\|u - \bar{u}\|_{PC} + |v - \bar{v}|)\,.$$

Hence condition (3.3.2) is satisfied by $K = L = \frac{1}{108e^3}$. For each $t \in [0, 1]$ we have

$$|f(t, u, v)| \le \frac{1}{108e^{t+3}}(2 + \|u\|_{PC} + |v|)\,.$$

Thus condition (3.4.1) is satisfied by $p(t) = \frac{1}{54e^{t+3}}$ and $q(t) = r(t) = \frac{1}{108e^{t+3}}$. Let

$$I_1(u) = \frac{|u|}{6 + |u|}, \quad u \in PC\left([-r, 0], \mathbb{R}\right)\,.$$

For each $u \in PC([-r, 0], \mathbb{R})$ we have

$$|I_1(u)| \le \frac{1}{6}\|u\|_{PC} + 1\,.$$

Thus, condition (3.4.2) is satisfied by $M^* = \frac{1}{6}$ and $N^* = 1$. It follows from Theorem 3.4 that problem (3.21)–(3.23) has at least one solution on $J$.

## 3.3 Existence Results for Impulsive NIFDE with Finite Delay in Banach Space

### 3.3.1 Introduction

The purpose of this section is to establish existence and uniqueness results for implicit fractional differential equations with finite delay and impulses

$$^{c}D^{\nu}_{t_k}y(t) = f(t, y_t, {}^{c}D^{\nu}_{t_k}y(t)) , \quad \text{for each, } t \in (t_k, t_{k+1}],\ k = 0, \dots, m,\ 0 < \nu \le 1 , \tag{3.24}$$

$$\Delta y|_{t=t_k} = I_k(y_{t_k^-}) , \quad k = 1, \dots, m , \tag{3.25}$$

$$y(t) = \varphi(t) , \quad t \in [-r, 0],\ r > 0 , \tag{3.26}$$

where $^{c}D^{\nu}_{t_k}$ is the Caputo fractional derivative, $(E, \|\cdot\|)$ is a real Banach space, $f\colon J \times PC([-r, 0], E) \times E \to E$ is a given function, $I_k\colon PC([-r, 0], E) \to E$, $\varphi \in PC([-r, 0], E)$, and $0 = t_0 < t_1 < \cdots < t_m < t_{m+1} = T$. For each function $y_t$ defined on $[-r, T]$ and for any $t \in J$ we denote by $y_t$ the element of $PC([-r, 0], E)$ defined by

$$y_t(\theta) = y(t + \theta) , \quad \theta \in [-r, 0] . \tag{3.27}$$

Here, $y_t(.)$ represents the history of the state from time $t - r$ up to time $t$. We have $\Delta y|_{t_k} = y(t_k^+) - y(t_k^-)$, where $y(t_k^+) = \lim_{h\to 0^+} y(t_k + h)$ and $y(t_k^-) = \lim_{h\to 0^-} y(t_k + h)$ represent the right and left limits of $y_t$ at $t = t_k$, respectively.

In this section, two results are discussed: the first is based on Darbo's fixed point theorem combined with the technique of measures of noncompactness; the second uses Mönch's fixed point theorem. Two examples are given to demonstrate the application of our main results.

### 3.3.2 Existence of Solutions

Consider the set of functions

$$\begin{aligned} PC([-r, 0], E) = \{y\colon [-r, 0] \to E\colon y \in C((\tau_k, \tau_{k+1}], E),\ k = 1, \dots, m,\\ \text{and there exist } y(\tau_k^-) \text{ and } y(\tau_k^+),\ k = 1, \dots, m \text{ with } y(\tau_k^-) = y(\tau_k)\} .\end{aligned}$$

Let $PC([-r, 0], E)$ be the Banach space with the norm

$$\|y\|_{PC} = \sup_{t\in[-r,0]} \|y(t)\| .$$

Also, we take

$$\begin{aligned} PC([0, T], E) = \{y\colon [0, T] \to E\colon y \in C((t_k, t_{k+1}], E),\ k = 1, \dots, m,\\ \text{and there exist } y(t_k^-) \text{ and } y(t_k^+),\ k = 1, \dots, m \text{ with } y(t_k^-) = y(t_k)\}\end{aligned}$$

and $PC([0,T],E)$ to be the Banach space with the norm

$$\|y\|_C = \sup_{t\in[0,T]} \|y(t)\| .$$

Let

$$\Omega = \{y\colon [-r,T] \to E\colon y|_{[-r,0]} \in PC([-r,0],E) \text{ and } y|_{[0,T]} \in PC([0,T],E)\} ,$$

and note that $\Omega$ is a Banach space with the norm

$$\|y\|_\Omega = \sup_{t\in[-r,T]} \|y(t)\| .$$

Let us define what we mean by a solution of problem (3.24)–(3.26).

**Definition 3.13.** A function $y \in \Omega$ whose $\nu$-derivative exists on $J_k$ is said to be a solution of (3.24)–(3.26) if $y$ satisfies the equation ${}^cD^\nu_{t_k}y(t) = f(t, y_t, {}^cD^\nu_{t_k}y(t))$ on $J_k$ and satisfies the conditions

$$\Delta y|_{t=t_k} = I_k(y_{t_k^-}) , \quad k = 1,\dots,m ,$$
$$y(t) = \varphi(t) , \quad t \in [-r,0] .$$

To prove the existence of solutions to (3.24)–(3.26), we need the following auxiliary lemma.

**Lemma 3.14.** *Let $0 < \nu \le 1$, and let $\sigma\colon J \to E$ be continuous. A function $y$ is a solution of the fractional integral equation*

$$y(t) = \begin{cases} \varphi(0) + \dfrac{1}{\Gamma(\nu)}\displaystyle\int_0^t (t-s)^{\nu-1}\sigma(s)ds, & \text{if } t \in [0,t_1] , \\ \varphi(0) + \displaystyle\sum_{i=1}^k I_i(y_{t_i^-}) + \dfrac{1}{\Gamma(\nu)}\sum_{i=1}^k \int_{t_{i-1}}^{t_i} (t_i-s)^{\nu-1}\sigma(s)ds & \\ +\dfrac{1}{\Gamma(\nu)}\displaystyle\int_{t_k}^t (t-s)^{\nu-1}\sigma(s)ds, & \text{if } t \in (t_k,t_{k+1}] , \\ \varphi(t), & t \in [-r,0] , \end{cases} \tag{3.28}$$

*where $k = 1,\dots,m$, if and only if $y$ is a solution of the fractional problem*

$$\begin{aligned} {}^cD^\nu y(t) &= \sigma(t), & t &\in J_k , \\ \Delta y|_{t=t_k} &= I_k(y_{t_k^-}), & k &= 1,\dots,m , \\ y(t) &= \varphi(t), & t &\in [-r,0] . \end{aligned}$$

Let us introduce the following conditions:
(3.14.1) The function $f: J \times PC([-r, 0], E) \times E \to E$ is continuous.
(3.14.2) There exist constants $K > 0$ and $0 < L < 1$ such that

$$\|f(t, u, v) - f(t, \bar{u}, \bar{v})\| \le K\|u - \bar{u}\|_{PC} + L\|v - \bar{v}\|$$

for any $u, \bar{u} \in PC([-r, 0], E)$, $v, \bar{v} \in E$ and $t \in J$.
(3.14.3) There exists a constant $\widetilde{l} > 0$ such that

$$\|I_k(u) - I_k(\overline{u})\| \le \widetilde{l}\|u - \overline{u}\|_{PC}$$

for each $u, \overline{u} \in PC([-r, 0], E)$ and $k = 1, \dots, m$.

We are now in a position to state and prove our existence result for problem (3.24)–(3.26) based on the concept of measures of noncompactness and Dafixth's fixed point theorem.

**Remark 3.15** ([66]). Conditions (3.14.2) and (3.14.3) are respectively equivalent to the inequalities

$$\alpha\,(f(t, B_1, B_2)) \le K\alpha(B_1) + L\alpha(B_2)$$

and

$$\alpha(I_k(B_1)) \le \widetilde{l}\alpha(B_1)$$

for any bounded sets $B_1 \subseteq PC([-r, 0], E)$, $B_2 \subseteq E$, for each $t \in J$ and $k = 1, \dots, m$.

**Theorem 3.16.** *Assume* (3.14.1)–(3.14.3). *If*

$$m\widetilde{l} + \frac{(m+1)KT^{\nu}}{(1-L)\Gamma(\nu+1)} < 1\,, \tag{3.29}$$

*then IVP* (3.24)–(3.26) *has at least one solution on* $J$.

*Proof.* Transform problem (3.24)–(3.26) into a fixed point problem. Consider the operator $N: \Omega \to \Omega$ defined by

$$(Ny)(t) = \begin{cases} \varphi(0) + \sum\limits_{0<t_k<t} I_k(y_{t_k^-}) + \dfrac{1}{\Gamma(\nu)} \sum\limits_{0<t_k<t} \displaystyle\int_{t_{k-1}}^{t_k} (t_k - s)^{\nu-1} g(s)ds \\ +\dfrac{1}{\Gamma(\nu)} \displaystyle\int_{t_k}^{t} (t-s)^{\nu-1} g(s)ds, & t \in [0, T]\,, \\ \varphi(t), & t \in [-r, 0]\,, \end{cases} \tag{3.30}$$

where $g \in C(J, E)$ is such that

$$g(t) = f(t, y_t, g(t))\,.$$

Clearly, the fixed points of operator $N$ are solutions of problem (3.24)–(3.26).

We will show that $N$ satisfies the assumptions of Darbo's fixed point theorem. The proof will be given in several claims.

*Claim 1: $N$ is continuous.* Let $\{u_n\}$ be a sequence such that $u_n \to u$ in $\Omega$. If $t \in [-r, 0]$, then

$$\|N(u_n)(t) - N(u)(t)\| = 0 .$$

For $t \in J$ we have

$$\begin{aligned}\|N(u_n)(t) - N(u)(t)\| \leq{}& \frac{1}{\Gamma(\nu)} \sum_{0<t_k<t} \int_{t_{k-1}}^{t_k} (t_k - s)^{\nu-1} \|g_n(s) - g(s)\| ds \\ &+ \frac{1}{\Gamma(\nu)} \int_{t_k}^{t} (t - s)^{\nu-1} \|g_n(s) - g(s)\| ds \\ &+ \sum_{0<t_k<t} \|I_k(u_{nt_k^-}) - I_k(u_{t_k^-})\| ,\end{aligned} \tag{3.31}$$

where $g_n, g \in C(J, E)$ are given by

$$g_n(t) = f(t, u_{nt}, g_n(t))$$

and

$$g(t) = f(t, u_t, g(t)) .$$

By (3.14.2) we have

$$\begin{aligned}\|g_n(t) - g(t)\| &= \|f(t, u_{nt}, g_n(t)) - f(t, u_t, g(t))\| \\ &\leq K\|u_{nt} - u_t\|_{PC} + L\|g_n(t) - g(t)\| .\end{aligned}$$

Then

$$\|g_n(t) - g(t)\| \leq \frac{K}{1 - L} \|u_{nt} - u_t\|_{PC} .$$

Since $u_n \to u$, we get $g_n(t) \to g(t)$ as $n \to \infty$ for each $t \in J$. Let $\eta > 0$ be such that for each $t \in J$ we have $\|g_n(t)\| \leq \eta$ and $\|g(t)\| \leq \eta$. Then we have

$$\begin{aligned}(t - s)^{\nu-1} \|g_n(s) - g(s)\| &\leq (t - s)^{\nu-1} [\|g_n(s)\| + \|g(s)\|] \\ &\leq 2\eta(t - s)^{\nu-1}\end{aligned}$$

and

$$\begin{aligned}(t_k - s)^{\nu-1} \|g_n(s) - g(s)\| &\leq (t_k - s)^{\nu-1} [\|g_n(s)\| + \|g(s)\|] \\ &\leq 2\eta(t_k - s)^{\nu-1} .\end{aligned}$$

For each $t \in J$ the functions $s \to 2\eta(t - s)^{\nu-1}$ and $s \to 2\eta(t_k - s)^{\nu-1}$ are integrable on $[0, t]$. The Lebesgue dominated convergence theorem and (3.31) imply

$$\|N(u_n)(t) - N(u)(t)\| \to 0 \text{ as } n \to \infty .$$

Hence,

$$\|N(u_n) - N(u)\|_\Omega \to 0 \text{ as } n \to \infty .$$

Consequently, $N$ is continuous.

Let $R$ be a constant such that

$$R \geq \max\left\{\frac{(\|\varphi(0)\| + mc_1)\Gamma(\nu+1)(1-L) + (m+1)T^\nu f^*}{\Gamma(\nu+1)(1-L) - \left[(m+1)T^\nu K + m\tilde{l}\Gamma(\nu+1)(1-L)\right]}, \|\varphi\|_{PC}\right\}, \tag{3.32}$$

where

$$c_1 = \max_{1\leq k\leq m}\{\sup\{\|I_k(v)\|, v \in PC([-r,0],E)\}\}$$

and

$$f^* = \sup_{t\in J}\|f(t,0,0)\| .$$

Define

$$D_R = \{u \in \Omega\colon \|u\|_\Omega \leq R\} .$$

It is clear that $D_R$ is a bounded, closed, and convex subset of $\Omega$.

*Claim 2:* $N(D_R) \subset D_R$. Let $u \in D_R$; we show that $Nu \in D_R$. If $t \in [-r, 0]$, then

$$\|N(u)(t)\| \leq \|\varphi\|_{PC} \leq R .$$

If $t \in J$, then we have

$$\begin{aligned}\|N(u)(t)\| \leq \|\varphi(0)\| &+ \frac{1}{\Gamma(\nu)}\sum_{0<t_k<t}\int_{t_{k-1}}^{t_k}(t_k - s)^{\nu-1}\|g(s)\|ds \\ &+ \frac{1}{\Gamma(\nu)}\int_{t_k}^{t}(t-s)^{\nu-1}\|g(s)\|ds + \sum_{0<t_k<t}\|I_k(u_{t_k^-})\| .\end{aligned} \tag{3.33}$$

By (3.14.2), for each $t \in J$ we have

$$\begin{aligned}\|g(t)\| &\leq \|f(t,u_t,g(t)) - f(t,0,0)\| + \|f(t,0,0)\| \\ &\leq K\|u_t\|_{PC} + L\|g(t)\| + f^* \\ &\leq K\|u\|_\Omega + L\|g(t)\| + f^* \\ &\leq KR + L\|g(t)\| + f^* .\end{aligned}$$

Then

$$\|g(t)\| \leq \frac{f^* + KR}{1-L} := M .$$

Thus, (3.32), (3.33), and (3.14.3) imply that

$$\begin{aligned}
\|Nu(t)\| &\le \|\varphi(0)\| + \frac{mMT^{\nu}}{\Gamma(\nu+1)} + \frac{MT^{\nu}}{\Gamma(\nu+1)} + \sum_{k=1}^{m} \|I_k(u_{t_k^-}) - I_k(0)\| + \sum_{k=1}^{m} \|I_k(0)\| \\
&\le \|\varphi(0)\| + \frac{(m+1)MT^{\nu}}{\Gamma(\nu+1)} + m\widetilde{l}\|u_{t_k^-}\|_{PC} + mc_1 \\
&\le \|\varphi(0)\| + \frac{(m+1)MT^{\nu}}{\Gamma(\nu+1)} + m\widetilde{l}\|u\|_{\Omega} + mc_1 \\
&\le \|\varphi(0)\| + \frac{(m+1)MT^{\nu}}{\Gamma(\nu+1)} + m\widetilde{l}R + mc_1 \\
&\le R\,.
\end{aligned}$$

Thus, for each $t \in [-r, T]$ we have $\|Nu(t)\| \le R$. This implies that $\|Nu\|_{\Omega} \le R$. Consequently,

$$N(D_R) \subset D_R\,.$$

*Claim 3: $N(D_R)$ is bounded and equicontinuous.* By Claim 2 we have $N(D_R) = \{N(u) \colon u \in D_R\} \subset D_R$. Thus, for each $u \in D_R$ we have $\|N(u)\|_{\Omega} \le R$. Hence, $N(D_R)$ is bounded. Let $t_1, t_2 \in (0, T]$, $t_1 < t_2$, and let $u \in D_R$. Then

$$\begin{aligned}
&\|N(u)(t_2) - N(u)(t_1)\| \\
&\le \frac{1}{\Gamma(\nu)} \int_0^{t_1} |(t_2-s)^{\nu-1} - (t_1-s)^{\nu-1}| \|g(s)\| ds \\
&\quad + \frac{1}{\Gamma(\nu)} \int_{t_1}^{t_2} |(t_2-s)^{\nu-1}| \|g(s)\| ds + \sum_{0<t_k<t_2-t_1} \|I_k(u_{t_k^-}) - I_k(0)\| + \sum_{0<t_k<t_2-t_1} \|I_k(0)\| \\
&\le \frac{M}{\Gamma(\nu+1)} [2(t_2-t_1)^{\nu} + (t_2^{\nu} - t_1^{\nu})] + (t_2-t_1)(\widetilde{l}\|u_{t_k^-}\|_{PC} + c_1) \\
&\le \frac{M}{\Gamma(\nu+1)} [2(t_2-t_1)^{\nu} + (t_2^{\nu} - t_1^{\nu})] + (t_2-t_1)(\widetilde{l}\|u\|_{\Omega} + c_1) \\
&\le \frac{M}{\Gamma(\nu+1)} [2(t_2-t_1)^{\nu} + (t_2^{\nu} - t_1^{\nu})] + (t_2-t_1)(\widetilde{l}R + c_1)\,.
\end{aligned}$$

As $t_1 \to t_2$, the right-hand side of the preceding inequality tends to zero.

*Claim 4: The operator $N \colon D_R \to D_R$ is a strict set contraction.* Let $V \subset D_R$. If $t \in [-r, 0]$, then

$$\begin{aligned}
\alpha(N(V)(t)) &= \alpha(N(y)(t), y \in V) \\
&= \alpha(\varphi(t), y \in V) \\
&= 0\,.
\end{aligned}$$

If $t \in J$, then we have

$$
\begin{aligned}
\alpha(N(V)(t)) &= \alpha((Ny)(t), y \in V) \\
&\leq \sum_{0<t_k<t} \left\{\alpha(I_k(y_{t_k^-})), y \in V\right\} + \frac{1}{\Gamma(\nu)} \sum_{0<t_k<t} \left\{ \int_{t_{k-1}}^{t_k} (t_k - s)^{\nu-1} \alpha(g(s)) ds, y \in V \right\} \\
&\quad + \frac{1}{\Gamma(\nu)} \left\{ \int_{t_k}^{t} (t - s)^{\nu-1} \alpha(g(s)) ds, y \in V \right\} .
\end{aligned}
$$

Then Remark 3.15 and Lemma 1.32 imply that for each $s \in J$

$$
\begin{aligned}
\alpha(\{g(s), y \in V\}) &= \alpha(\{f(s, y(s), g(s)), y \in V\}) \\
&\leq K\alpha(\{y(s), y \in V\}) + L\alpha(\{g(s), y \in V\}) .
\end{aligned}
$$

Thus,

$$
\alpha\left(\{g(s), y \in V\}\right) \leq \frac{K}{1-L} \alpha\{y(s), y \in V\} .
$$

On the other hand, for each $t \in J$ and $k = 1, \ldots, m$ we have

$$
\sum_{0<t_k<t} \alpha\left(\left\{I_k(y_{t_k^-}), y \in V\right\}\right) \leq m\tilde{l}\alpha(\{y(t), y \in V\}) .
$$

Then

$$
\begin{aligned}
\alpha(N(V)(t)) &\leq m\tilde{l}\alpha(\{y(t), y \in V\}) + \frac{mK}{(1-L)\Gamma(\nu)} \left\{ \int_0^t (t-s)^{\nu-1} \{\alpha(y(s))\} ds, y \in V \right\} \\
&\quad + \frac{K}{(1-L)\Gamma(\nu)} \left\{ \int_0^t (t-s)^{\nu-1} \{\alpha(y(s))\} ds, y \in V \right\} \\
&\leq m\tilde{l}\alpha_c(V) + \left[ \frac{mKT^\nu}{(1-L)\Gamma(\nu+1)} + \frac{KT^\nu}{(1-L)\Gamma(\nu+1)} \right] \alpha_c(V) \\
&= \left[ m\tilde{l} + \frac{(m+1)KT^\nu}{(1-L)\Gamma(\nu+1)} \right] \alpha_c(V) .
\end{aligned}
$$

Therefore,

$$
\alpha_c(NV) \leq \left[ m\tilde{l} + \frac{(m+1)KT^\nu}{(1-L)\Gamma(\nu+1)} \right] \alpha_c(V) .
$$

Thus, by (3.29), operator $N$ is a set contraction. As a consequence of Theorem 1.45, we deduce that $N$ has a fixed point that is a solution of problem (3.24)–(3.26). □

Our next existence result for problem (3.24)–(3.26) is based on the concept of measures of noncompactness and Mönch's fixed point theorem.

**Theorem 3.17.** *Assume* (3.14.1)–(3.14.4) *and* (3.29) *hold. If*

$$m\tilde{l} < 1 ,$$

*then IVP* (3.24)–(3.26) *has at least one solution.*

*Proof.* Consider the operator $N$ defined in (3.30). We will show that $N$ satisfies the assumptions of Mönch's fixed point theorem. We know that $N : D_R \to D_R$ is bounded and continuous, and we need to prove that the implication

$$[V = \overline{\text{conv}} N(V) \text{ or } V = N(V) \cup \{0\}] \text{ implies } \alpha(V) = 0$$

holds for every subset $V$ of $D_R$. Now let $V$ be a subset of $D_R$ such that $V \subset \overline{\text{conv}}(N(V) \cup \{0\})$; $V$ is bounded and equicontinuous, and therefore the function $t \to v(t) = \alpha(V(t))$ is continuous on $[-r, T]$. By Remark 3.15, Lemma 1.33, and the properties of the measure $\alpha$, for each $t \in J$ we have

$$\begin{aligned} v(t) &\le \alpha(N(V)(t) \cup \{0\}) \\ &\le \alpha(N(V)(t)) \\ &\le \alpha\{(Ny)(t), y \in V\} \\ &\le m\tilde{l}\alpha(\{y(t), y \in V\}) + \frac{(m+1)K}{(1-L)\Gamma(\nu)} \left\{ \int_0^t (t-s)^{\nu-1} \{\alpha(y(s))\} ds, y \in V \right\} \\ &= m\tilde{l} v(t) + \frac{(m+1)K}{(1-L)\Gamma(\nu)} \int_0^t (t-s)^{\nu-1} v(s) ds . \end{aligned}$$

Then

$$v(t) \le \frac{(m+1)K}{(1-m\tilde{l})(1-L)\Gamma(\nu)} \int_0^t (t-s)^{\nu-1} v(s) ds .$$

Lemma 1.52 implies that $v(t) = 0$ for each $t \in J$.

For $t \in [-r, 0]$ we have $v(t) = \alpha(\varphi(t)) = 0$, so $V(t)$ is relatively compact in $E$. In view of the Ascoli–Arzelà theorem, $V$ is relatively compact in $D_R$. Applying Theorem 1.46, we conclude that $N$ has a fixed point $y \in D_R$. Hence, $N$ has a fixed point that is a solution of problem (3.24)–(3.26). □

### 3.3.3 Examples

*Example 1.* Consider the infinite system

$$
{}^cD^{\frac{1}{2}}_{t_k} y_n(t) = \frac{e^{-t}}{(11+e^t)}\left[\frac{y_n(t)}{1+y_n(t)} - \frac{{}^cD^{\frac{1}{2}}_{t_k} y_n(t)}{1+{}^cD^{\frac{1}{2}}_{t_k} y_n(t)}\right], \quad \text{for each}, \; t \in J_0 \cup J_1 , \tag{3.34}
$$

$$
\Delta y_n|_{t=\frac{1}{2}} = \frac{y_n\left(\frac{1}{2}^-\right)}{10 + y_n\left(\frac{1}{2}^-\right)}, \tag{3.35}
$$

$$
y_n(t) = \varphi(t), \quad t \in [-r, 0], \; r > 0 , \tag{3.36}
$$

where $\varphi \in PC([-r,0],E)$, $J_0 = [0, \frac{1}{2}]$, $J_1 = (\frac{1}{2}, 1]$, $t_0 = 0$, and $t_1 = \frac{1}{2}$.

Set

$$
E = l^1 = \{y = (y_1, y_2, \dots, y_n, \dots), \sum_{n=1}^{\infty} |y_n| < \infty\} ,
$$

and

$$
f(t,u,v) = \frac{e^{-t}}{(11+e^t)}\left[\frac{u}{1+u} - \frac{v}{1+v}\right], \quad t \in [0,1], \; u \in PC([-r,0],E), \; \text{and } v \in E .
$$

Clearly, the function $f$ is jointly continuous; now $E$ is a Banach space with the norm $\|y\| = \sum_{n=1}^{\infty} |y_n|$. For any $u, \bar{u} \in PC([-r,0],E)$, $v, \bar{v} \in E$ and $t \in [0,1]$

$$
\|f(t,u,v) - f(t,\bar{u},\bar{v})\| \le \frac{1}{12}(\|u - \bar{u}\|_{PC} + \|v - \bar{v}\|) .
$$

Hence, condition (3.14.2) is satisfied by $K = L = \frac{1}{12}$.

Let

$$
I_1(u) = \frac{u}{10+u}, \quad u \in PC([-r,0],E)
$$

and take $u, v \in PC([-r,0],E)$. Then we have

$$
\|I_1(u) - I_1(v)\| = \left\|\frac{u}{10+u} - \frac{v}{10+v}\right\| \le \frac{1}{10}\|u - v\|_{PC} .
$$

Hence, condition (3.14.3) is satisfied by $\widetilde{l} = \frac{1}{10}$.

The conditions

$$
\begin{aligned}
m\widetilde{l} + \frac{(m+1)KT^v}{(1-L)\Gamma(v+1)} &= \left[\frac{1}{10} + \frac{\frac{2}{12}}{\left(1-\frac{1}{12}\right)\Gamma\left(\frac{3}{2}\right)}\right] \\
&= \frac{4}{11\sqrt{\pi}} + \frac{1}{10} < 1
\end{aligned}
$$

are satisfied by $T = m = 1$ and $v = \frac{1}{2}$. It follows from Theorem 3.16 that problem (3.34)–(3.36) has a at least one solution on $J$.

*Example 2.* Consider the impulsive problem

$$ {}^cD^{\frac{1}{2}}_{t_k} y_n(t) = \frac{2 + \|y_n(t)\| + \|{}^cD^{\frac{1}{2}}_{t_k} y_n(t)\|}{108e^{t+3}\left(1 + \|y_n(t)\| + \|{}^cD^{\frac{1}{2}}_{t_k} y_n(t)\|\right)}, \quad \text{for each, } t \in J_0 \cup J_1 , \tag{3.37} $$

$$ \Delta y_n|_{t=\frac{1}{3}} = \frac{\|y_n\left(\frac{1}{3}^-\right)\|}{6 + \|y_n\left(\frac{1}{3}^-\right)\|}, \tag{3.38} $$

$$ y_n(t) = \varphi(t), \quad t \in [-r, 0], \ r > 0 , \tag{3.39} $$

where $\varphi \in PC([-r, 0], E)$, $J_0 = [0, \frac{1}{3}]$, $J_1 = (\frac{1}{3}, 1]$, $t_0 = 0$, and $t_1 = \frac{1}{3}$. Set

$$ E = l^1 = \{y = (y_1, y_2, \dots, y_n, \dots), \sum_{n=1}^{\infty} |y_n| < \infty\} , $$

and

$$ f(t, u, v) = \frac{2 + \|u\| + \|v\|}{108e^{t+3}(1 + \|u\| + \|v\|)}, \quad t \in [0, 1], \ u \in PC([-r, 0], E), \ v \in E . $$

Clearly, the function $f$ is jointly continuous. Now $E$ is a Banach space with the norm $\|y\| = \sum_{n=1}^{\infty} |y_n|$. For any $u, \bar{u} \in PC([-r, 0], E)$, $v, \bar{v} \in E$, and $t \in [0, 1]$,

$$ \|f(t, u, v) - f(t, \bar{u}, \bar{v})\| \le \frac{1}{108e^3}(\|u - \bar{u}\|_{PC} + \|v - \bar{v}\|) . $$

Hence, condition (3.14.2) is satisfied by $K = L = \frac{1}{108e^3}$.

Let

$$ I_1(u) = \frac{\|u\|}{6 + \|u\|}, \quad u \in PC([-r, 0], E) , $$

and take $u, v \in PC([-r, 0], E)$. Then we have

$$ \|I_1(u) - I_1(v)\| = \left\|\frac{u}{6 + u} - \frac{v}{6 + v}\right\| \le \frac{1}{6}\|u - v\|_{PC} . $$

Hence, condition (3.14.3) is satisfied by $\widetilde{l} = \frac{1}{6}$.

The condition

$$ \begin{aligned} m\widetilde{l} + \frac{(m+1)KT^v}{(1-L)\Gamma(v+1)} &= \left[\frac{1}{6} + \frac{\frac{2}{12}}{\left(1 - \frac{1}{12}\right)\Gamma\left(\frac{3}{2}\right)}\right] \\ &= \frac{4}{11\sqrt{\pi}} + \frac{1}{6} < 1 \end{aligned} $$

is satisfied by $T = m = 1$ and $v = \frac{1}{2}$. We also have

$$ m\widetilde{l} = \frac{1}{6} < 1 . $$

It follows from Theorem 3.17 that problem (3.37)–(3.39) has at least one solution on $J$.

## 3.4 Existence and Stability Results for Perturbed Impulsive NIFDE with Finite Delay

### 3.4.1 Introduction

In this section, we establish existence, uniqueness, and stability results for the nonlinear implicit perturbed fractional differential equation with finite delay and impulses

$$^cD^\alpha_{t_k}y(t) = f(t, y_t, {}^cD^\alpha_{t_k}y(t)) + \phi(t, y_t), \quad \text{for } t \in (t_k, t_{k+1}],\ k = 0, \dots, m,\ 0 < \alpha \le 1, \tag{3.40}$$

$$\Delta y|_{t_k} = I_k(y_{t_k^-}), \quad k = 1, \dots, m, \tag{3.41}$$

$$y(t) = \varphi(t), \quad t \in [-r, 0],\ r > 0, \tag{3.42}$$

where $J_0 = [t_0, t_1], J_k = (t_k, t_{k+1}];\ k = 1, \dots, m,\ {}^cD^\alpha_{t_k}$ is the Caputo fractional derivative, $f: J \times PC([-r, 0], \mathbb{R}) \times \mathbb{R} \to \mathbb{R}$ and $\phi: J \times PC([-r, 0], \mathbb{R}) \to \mathbb{R}$ are given functions, $I_k: PC([-r, 0], \mathbb{R}) \to \mathbb{R}$, $\varphi \in PC([-r, 0], \mathbb{R})$, and $0 = t_0 < t_1 < \dots < t_m < t_{m+1} = T$.

The arguments are based on Banach's contraction principle and Schaefer's fixed point theorem. Finally, we present two examples to show the applicability of our results.

### 3.4.2 Existence of Solutions

Consider the Banach space

$$PC([-r, 0], \mathbb{R}) = \{y: [-r, 0] \to \mathbb{R}: y \in C((\tau_k, \tau_{k+1}], \mathbb{R}),\ k = 1, \dots, m, \text{ and there exist } y(\tau_k^-) \text{ and } y(\tau_k^+),\ k = 1, \dots, m \text{ with } y(\tau_k^-) = y(\tau_k)\},$$

with the norm

$$\|y\|_{PC} = \sup_{t\in[-r,0]} |y(t)|;$$

$$PC([0, T], \mathbb{R}) = \{y: [0, T] \to \mathbb{R}: y \in C((t_k, t_{k+1}], \mathbb{R}),\ k = 1, \dots, m, \text{ and there exist } y(t_k^-) \text{ and } y(t_k^+),\ k = 1, \dots, m \text{ with } y(t_k^-) = y(t_k)\},$$

with the norm

$$\|y\|_C = \sup_{t\in[0,T]} |y(t)|;$$

and

$$\Omega = \{y: [-r, T] \to \mathbb{R}: y|_{[-r,0]} \in PC([-r, 0], \mathbb{R}) \text{ and } y|_{[0,T]} \in PC([0, T], \mathbb{R})\},$$

with the norm

$$\|y\|_\Omega = \sup_{t\in[-r,T]} |y(t)|.$$

**Definition 3.18.** A function $y \in \Omega$ whose $\alpha$-derivative exists on $J_k$ is said to be a solution of (3.40)–(3.42) if $y$ satisfies the equation ${}^cD^\alpha_{t_k}y(t) = f(t, y_t, {}^cD^\alpha_{t_k}y(t)) + \phi(t, y_t)$ on $J_k$ and satisfies the conditions

$$\Delta y|_{t=t_k} = I_k(y_{t_k^-}),\ k = 1, \dots, m\,,$$
$$y(t) = \varphi(t)\,, \quad t \in [-r, 0]\,.$$

To prove the existence of solutions to (3.40)–(3.42), we need the following auxiliary lemma.

**Lemma 3.19.** *Let $0 < \alpha \le 1$, and let $\sigma\colon J \to \mathbb{R}$ be continuous. A function $y$ is a solution of the fractional integral equation*

$$y(t) = \begin{cases} \varphi(0) + \dfrac{1}{\Gamma(\alpha)}\displaystyle\int_0^t (t-s)^{\alpha-1}\sigma(s)ds, & \text{if } t \in [0, t_1]\,, \\ \varphi(0) + \displaystyle\sum_{i=1}^{k} I_i(y_{t_i^-}) + \dfrac{1}{\Gamma(\alpha)}\displaystyle\sum_{i=1}^{k}\int_{t_{i-1}}^{t_i} (t_i - s)^{\alpha-1}\sigma(s)ds\,, & \\ +\dfrac{1}{\Gamma(\alpha)}\displaystyle\int_{t_k}^{t} (t-s)^{\alpha-1}\sigma(s)ds, & \text{if } t \in (t_k, t_{k+1}]\,, \\ \varphi(t), & t \in [-r, 0]\,, \end{cases} \tag{3.43}$$

*where $k = 1, \dots, m$, if and only if $y$ is a solution of the fractional problem*

$${}^cD^\alpha y(t) = \sigma(t)\,, \quad t \in J_k\,, \tag{3.44}$$

$$\Delta y|_{t=t_k} = I_k(y_{t_k^-})\,, \quad k = 1, \dots, m\,, \tag{3.45}$$

$$y(t) = \varphi(t)\,, \quad t \in [-r, 0]\,. \tag{3.46}$$

*Proof.* Assume that $y$ satisfies (3.44)–(3.46). If $t \in [0, t_1]$, then

$${}^cD^\alpha y(t) = \sigma(t)\,.$$

Lemma 1.9 implies

$$y(t) = \varphi(0) + I^\alpha\sigma(t) = \varphi(0) + \frac{1}{\Gamma(\alpha)}\int_0^t (t-s)^{\alpha-1}\sigma(s)ds\,.$$

If $t \in (t_1, t_2]$, then Lemma 1.9 implies

$$
\begin{aligned}
y(t) &= y(t_1^+) + \frac{1}{\Gamma(\alpha)} \int_{t_1}^{t} (t-s)^{\alpha-1} \sigma(s) ds \\
&= \Delta y|_{t=t_1} + y(t_1^-) + \frac{1}{\Gamma(\alpha)} \int_{t_1}^{t} (t-s)^{\alpha-1} \sigma(s) ds \\
&= I_1(y_{t_1^-}) + \left[ \varphi(0) + \frac{1}{\Gamma(\alpha)} \int_{0}^{t_1} (t_1-s)^{\alpha-1} \sigma(s) ds \right] \\
&\quad + \frac{1}{\Gamma(\alpha)} \int_{t_1}^{t} (t-s)^{\alpha-1} \sigma(s) ds . \\
&= \varphi(0) + I_1(y_{t_1^-}) + \frac{1}{\Gamma(\alpha)} \int_{0}^{t_1} (t_1-s)^{\alpha-1} \sigma(s) ds \\
&\quad + \frac{1}{\Gamma(\alpha)} \int_{t_1}^{t} (t-s)^{\alpha-1} \sigma(s) ds .
\end{aligned}
$$

If $t \in (t_2, t_3]$, then from Lemma 1.9 we get

$$
\begin{aligned}
y(t) &= y(t_2^+) + \frac{1}{\Gamma(\alpha)} \int_{t_2}^{t} (t-s)^{\alpha-1} \sigma(s) ds \\
&= \Delta y|_{t=t_2} + y(t_2^-) + \frac{1}{\Gamma(\alpha)} \int_{t_2}^{t} (t-s)^{\alpha-1} \sigma(s) ds \\
&= I_2(y_{t_2^-}) + \Bigg[ \varphi(0) + I_1(y_{t_1^-}) + \frac{1}{\Gamma(\alpha)} \int_{0}^{t_1} (t_1-s)^{\alpha-1} \sigma(s) ds \\
&\quad + \frac{1}{\Gamma(\alpha)} \int_{t_1}^{t_2} (t_2-s)^{\alpha-1} \sigma(s) ds \Bigg] + \frac{1}{\Gamma(\alpha)} \int_{t_2}^{t} (t-s)^{\alpha-1} \sigma(s) ds . \\
&= \varphi(0) + \left[ I_1(y_{t_1^-}) + I_2(y_{t_2^-}) \right] + \Bigg[ \frac{1}{\Gamma(\alpha)} \int_{0}^{t_1} (t_1-s)^{\alpha-1} \sigma(s) ds \\
&\quad + \frac{1}{\Gamma(\alpha)} \int_{t_1}^{t_2} (t_2-s)^{\alpha-1} \sigma(s) ds \Bigg] + \frac{1}{\Gamma(\alpha)} \int_{t_2}^{t} (t-s)^{\alpha-1} \sigma(s) ds .
\end{aligned}
$$

Repeating the process in this way, the solution $y(t)$ for $t \in (t_k, t_{k+1}]$, where $k = 1, \dots, m$, can be written

$$y(t) = \varphi(0) + \sum_{i=1}^{k} I_i(y_{t_i^-}) + \frac{1}{\Gamma(\alpha)} \sum_{i=1}^{k} \int_{t_{i-1}}^{t_i} (t_i - s)^{\alpha-1}\sigma(s)ds$$

$$+ \frac{1}{\Gamma(\alpha)} \int_{t_k}^{t} (t - s)^{\alpha-1}\sigma(s)ds .$$

Conversely, assume that $y$ satisfies the impulsive fractional integral equation (3.43). If $t \in [0, t_1]$, then $y(0) = \varphi(0)$. Using the fact that ${}^cD^\alpha$ is the left inverse of $I^\alpha$, we get

$${}^cD^\alpha y(t) = \sigma(t) , \quad \text{for each } t \in [0, t_1] .$$

If $t \in (t_k, t_{k+1}]$, $k = 1, \dots, m$ and using the fact that ${}^cD^\alpha C = 0$, where $C$ is a constant, we get

$${}^cD^\alpha y(t) = \sigma(t) , \quad \text{for each } t \in (t_k, t_{k+1}] .$$

Also, we can easily show that

$$\Delta y|_{t=t_k} = I_k(y_{t_k^-}) , \quad k = 1, \dots, m .$$

□

We are now in a position to state and prove our existence result for problem (3.40)–(3.42) based on Banach's fixed point.

**Theorem 3.20.** *Make the following assumptions:*

(3.20.1) *The functions $f : J \times PC([-r, 0], \mathbb{R}) \times \mathbb{R} \to \mathbb{R}$ and $\phi : J \times PC([-r, 0], \mathbb{R}) \to \mathbb{R}$ are continuous.*

(3.20.2) *There exist constants $K > 0$, $\overline{K} > 0$ and $0 < L < 1$ such that*

$$|f(t, u, v) - f(t, \bar{u}, \bar{v})| \le K\|u - \bar{u}\|_{PC} + L|v - \bar{v}|$$

*and*

$$|\phi(t, u) - \phi(t, \bar{u})| \le \overline{K}\|u - \bar{u}\|_{PC}$$

*for any $u, \bar{u} \in PC([-r, 0], \mathbb{R})$, $v, \bar{v} \in \mathbb{R}$, and $t \in J$.*

(3.20.3) *There exists a constant $\tilde{l} > 0$ such that*

$$|I_k(u) - I_k(\overline{u})| \le \tilde{l}\|u - \overline{u}\|_{PC}$$

*for each $u, \overline{u} \in PC([-r, 0], \mathbb{R})$ and $k = 1, \dots, m$.*

*If*

$$m\tilde{l} + \frac{(m + 1)(K + \overline{K})T^\alpha}{(1 - L)\Gamma(\alpha + 1)} < 1 , \tag{3.47}$$

*then there exists a unique solution for problem (3.40)–(3.42) on $J$.*

*Proof.* Transform problem (3.40)–(3.42) into a fixed point problem. Consider the operator $N: \Omega \to \Omega$ defined by

$$(Ny)(t) = \begin{cases} \varphi(0) + \sum\limits_{0<t_k<t} I_k(y_{t_k^-}) + \dfrac{1}{\Gamma(\alpha)} \sum\limits_{0<t_k<t} \int\limits_{t_{k-1}}^{t_k} (t_k - s)^{\alpha-1} g(s)ds \\ + \dfrac{1}{\Gamma(\alpha)} \int\limits_{t_k}^{t} (t-s)^{\alpha-1} g(s)ds, & t \in [0, T], \\ \varphi(t), & t \in [-r, 0], \end{cases} \tag{3.48}$$

where $g \in C(J, \mathbb{R})$ is given by

$$g(t) = f(t, y_t, g(t)) + \phi(t, y_t).$$

Clearly, the fixed points of operator $N$ are solutions of problem (3.40)–(3.42).

Let $u, w \in \Omega$. If $t \in [-r, 0]$, then

$$|(Nu)(t) - (Nw)(t)| = 0.$$

For $t \in J$ we have

$$\begin{aligned} |(Nu)(t) - (Nw)(t)| \leq &\frac{1}{\Gamma(\alpha)} \sum_{0<t_k<t} \int\limits_{t_{k-1}}^{t_k} (t_k - s)^{\alpha-1} |g(s) - h(s)| ds \\ &+ \frac{1}{\Gamma(\alpha)} \int\limits_{t_k}^{t} (t-s)^{\alpha-1} |g(s) - h(s)| ds \\ &+ \sum_{0<t_k<t} |I_k(u_{t_k^-}) - I_k(w_{t_k^-})|, \end{aligned}$$

where $g, h \in C(J, \mathbb{R})$ are given by

$$g(t) = f(t, u_t, g(t)) + \phi(t, u_t),$$

and

$$h(t) = f(t, w_t, h(t)) + \phi(t, w_t).$$

By (3.20.2) we have

$$\begin{aligned} |g(t) - h(t)| &\leq |f(t, u_t, g(t)) - f(t, w_t, h(t))| + |\phi(t, u_t) - \phi(t, w_t)| \\ &\leq K\|u_t - w_t\|_{PC} + L|g(t) - h(t)| + \overline{K}\|u_t - w_t\|_{PC}. \end{aligned}$$

Then

$$|g(t) - h(t)| \leq \frac{K + \overline{K}}{1 - L} \|u_t - w_t\|_{PC}.$$

Therefore, for each $t \in J$

$$
\begin{aligned}
|(Nu)(t) - (Nw)(t)| &\leq \frac{K + \overline{K}}{(1-L)\Gamma(\alpha)} \sum_{k=1}^{m} \int_{t_{k-1}}^{t_k} (t_k - s)^{\alpha-1} \|u_s - w_s\|_{PC} ds \\
&\quad + \frac{K + \overline{K}}{(1-L)\Gamma(\alpha)} \int_{t_k}^{t} (t - s)^{\alpha-1} \|u_s - w_s\|_{PC} ds \\
&\quad + \sum_{k=1}^{m} \widetilde{l} \|u_{t_k^-} - w_{t_k^-}\|_{PC} . \\
&\leq \left[ m\widetilde{l} + \frac{m(K + \overline{K})T^\alpha}{(1-L)\Gamma(\alpha+1)} \right. \\
&\quad \left. + \frac{(K + \overline{K})T^\alpha}{(1-L)\Gamma(\alpha+1)} \right] \|u - w\|_\Omega .
\end{aligned}
$$

Thus,

$$
\|(Nu) - (Nw)\|_\Omega \leq \left[ m\widetilde{l} + \frac{(m+1)(K + \overline{K})T^\alpha}{(1-L)\Gamma(\alpha+1)} \right] \|u - w\|_\Omega .
$$

By (3.47), operator $N$ is a contraction. Hence, by Banach's contraction principle, $N$ has a unique fixed point that is the unique solution of (3.40)–(3.42). □

Our second result is based on Schaefer's fixed point theorem.

**Theorem 3.21.** *Assume* (3.20.1) *and* (3.20.2) *hold and*

(3.21.1) *There exist* $p, q, r \in C(J, \mathbb{R}_+)$ *with* $r^* = \sup_{t \in J} r(t) < 1$ *such that*

$$
|f(t, u, w)| \leq p(t) + q(t)\|u\|_{PC} + r(t)|w| \quad \text{for } t \in J,\ u \in PC([-r, 0], \mathbb{R}) \text{ and } w \in \mathbb{R} .
$$

(3.21.2) *The functions* $I_k \colon PC([-r, 0], \mathbb{R}) \to \mathbb{R}$ *are continuous and there exist constants* $M^*, N^* > 0$, *with* $mM^* < 1$, *such that*

$$
|I_k(u)| \leq M^* \|u\|_{PC} + N^* \quad \text{for each } u \in PC([-r, 0], \mathbb{R}),\ k = 1, \ldots, m .
$$

*Then problem* (3.40)–(3.42) *has at least one solution.*

*Proof.* Consider operator $N$ defined in (3.48). We will use Schaefer's fixed point theorem to prove that $N$ has a fixed point. The proof will be given in several steps.

*Step 1: $N$ is continuous.* Let $\{u_n\}$ be a sequence such that $u_n \to u$ in $\Omega$. If $t \in [-r, 0]$, then

$$
|(Nu_n)(t) - (Nu)(t)| = 0 .
$$

For $t \in J$ we have

$$\begin{aligned}|(Nu_n)(t) - (Nu)(t)| \le \frac{1}{\Gamma(\alpha)} \sum_{0<t_k<t} \int_{t_{k-1}}^{t_k} (t_k - s)^{\alpha-1} |g_n(s) - g(s)| ds \\ + \frac{1}{\Gamma(\alpha)} \int_{t_k}^{t} (t - s)^{\alpha-1} |g_n(s) - g(s)| ds \\ + \sum_{0<t_k<t} \left| I_k\left(u_n\left(t_k^-\right)\right) - I_k\left(u\left(t_k^-\right)\right)\right| ,\end{aligned} \tag{3.49}$$

where $g_n, g \in C(J, \mathbb{R})$ are given by

$$g_n(t) = f(t, u_{nt}, g_n(t)) + \phi(t, u_{nt})$$

and

$$g(t) = f(t, u_t, g(t)) + \phi(t, u_t) .$$

By (3.21.1) we have

$$\begin{aligned}|g_n(t) - g(t)| &\le |f(t, u_{nt}, g_n(t)) - f(t, u_t, g(t))| + |\phi(t, u_{nt}) - \phi(t, u_t)| \\ &\le K\|u_{nt} - u_t\|_{PC} + L|g_n(t) - g(t)| + \overline{K}\|u_{nt} - u_t\|_{PC} .\end{aligned}$$

Then

$$|g_n(t) - g(t)| \le \frac{K + \overline{K}}{1 - L} \|u_{nt} - u_t\|_{PC} .$$

Since $u_n \to u$, we get $g_n(t) \to g(t)$ as $n \to \infty$ for each $t \in J$. Let $\eta > 0$ be such that for each $t \in J$ we have $|g_n(t)| \le \eta$ and $|g(t)| \le \eta$. Then

$$\begin{aligned}(t - s)^{\alpha-1}|g_n(s) - g(s)| &\le (t - s)^{\alpha-1}[|g_n(s)| + |g(s)|] \\ &\le 2\eta(t - s)^{\alpha-1}\end{aligned}$$

and

$$\begin{aligned}(t_k - s)^{\alpha-1}|g_n(s) - g(s)| &\le (t_k - s)^{\alpha-1}[|g_n(s)| + |g(s)|] \\ &\le 2\eta(t_k - s)^{\alpha-1} .\end{aligned}$$

For each $t \in J$ the functions $s \to 2\eta(t - s)^{\alpha-1}$ and $s \to 2\eta(t_k - s)^{\alpha-1}$ are integrable on $[0, t]$; the Lebesgue dominated convergence theorem and (3.49) imply that

$$|(Nu_n)(t) - (Nu)(t)| \to 0 \text{ as } n \to \infty .$$

Hence

$$\|(Nu_n) - (Nu)\|_\Omega \to 0 \text{ as } n \to \infty ,$$

and so $N$ is continuous.

*Step 2: N maps bounded sets to bounded sets in $\Omega$.* Indeed, it is enough to show that for any $\eta^* > 0$ there exists a positive constant $\ell$ such that for each $u \in B_{\eta^*} = \{u \in \Omega : \|u\|_\Omega \le \eta^*\}$ we have $\|N(u)\|_\Omega \le \ell$. For each $t \in J$ we have

$$(Nu)(t) = \varphi(0) + \frac{1}{\Gamma(\alpha)} \sum_{0<t_k<t} \int_{t_{k-1}}^{t_k} (t_k - s)^{\alpha-1} g(s)ds + \frac{1}{\Gamma(\alpha)} \int_{t_k}^{t} (t-s)^{\alpha-1} g(s)ds + \sum_{0<t_k<t} I_k(u_{t_k^-}) , \tag{3.50}$$

where $g \in C(J, \mathbb{R})$ is given by

$$g(t) = f(t, u_t, g(t)) + \phi(t, y_t) .$$

By (3.21.2), for each $t \in J$ we have

$$\begin{aligned} |g(t)| &\le |f(t, u_t, g(t))| + |\phi(t, y_t)| \\ &\le p(t) + q(t)\|u_t\|_{PC} + r(t)|g(t)| + |\phi(t, y_t) - \phi(t, 0)| + |\phi(t, 0)| \\ &\le p(t) + q(t)\|u_t\|_{PC} + r(t)|g(t)| + \overline{K}\|u_t\|_{PC} + |\phi(t, 0)| \\ &\le p(t) + q(t)\|u\|_\Omega + r(t)|g(t)| + \overline{K}\|u\|_\Omega + |\phi(t, 0)| \\ &\le p(t) + (q(t) + \overline{K})\eta^* + r(t)|g(t)| + |\phi(t, 0)| \\ &\le p^* + (q^* + \overline{K})\eta^* + r^*|g(t)| + \phi^* , \end{aligned}$$

where $p^* = \sup_{t\in J} p(t)$, $q^* = \sup_{t\in J} q(t)$, and $\phi^* = \sup_{t\in J} |g(t, 0)|$.
Then

$$|g(t)| \le \frac{p^* + (q^* + \overline{K})\eta^* + \phi^*}{1 - r^*} := M .$$

Thus, (3.50) implies

$$\begin{aligned} |(Nu)(t)| &\le |\varphi(0)| + \frac{mMT^\alpha}{\Gamma(\alpha+1)} + \frac{MT^\alpha}{\Gamma(\alpha+1)} + m(M^*\|u_{t_k^-}\|_{PC} + N^*) \\ &\le |\varphi(0)| + \frac{(m+1)MT^\alpha}{\Gamma(\alpha+1)} + m(M^*\|u\|_\Omega + N^*) \\ &\le |\varphi(0)| + \frac{(m+1)MT^\alpha}{\Gamma(\alpha+1)} + m(M^*\eta^* + N^*) := R . \end{aligned}$$

If $t \in [-r, 0]$, then

$$|(Nu)(t)| \le \|\varphi\|_{PC} ,$$

so

$$\|N(u)\|_\Omega \le \max\{R, \|\varphi\|_{PC}\} := \ell .$$

*Step 3: N maps bounded sets to equicontinuous sets of $\Omega$.*
Let $t_1, t_2 \in (0, T]$, $t_1 < t_2$, let $B_{\eta^*}$ be a bounded set of $\Omega$ as in Step 2, and let $u \in B_{\eta^*}$.

Then

$$
\begin{aligned}
&|(Nu)(t_2) - (Nu)(t_1)| \\
&\le \frac{1}{\Gamma(\alpha)} \int_0^{t_1} |(t_2 - s)^{\alpha-1} - (t_1 - s)^{\alpha-1}\|g(s)|ds \\
&\quad + \frac{1}{\Gamma(\alpha)} \int_{t_1}^{t_2} |(t_2 - s)^{\alpha-1}\|g(s)|ds + \sum_{0<t_k<t_2-t_1} |I_k(u_{t_k^-})| \\
&\le \frac{M}{\Gamma(\alpha+1)}[2(t_2 - t_1)^\alpha + (t_2^\alpha - t_1^\alpha)] + (t_2 - t_1)(M^*\|u_{t_k^-}\|_{PC} + N^*) \\
&\le \frac{M}{\Gamma(\alpha+1)}[2(t_2 - t_1)^\alpha + (t_2^\alpha - t_1^\alpha)] + (t_2 - t_1)(M^*\|u\|_\Omega + N^*) \\
&\le \frac{M}{\Gamma(\alpha+1)}[2(t_2 - t_1)^\alpha + (t_2^\alpha - t_1^\alpha)] + (t_2 - t_1)(M^*\eta^* + N^*)\,.
\end{aligned}
$$

As $t_1 \to t_2$, the right-hand side of the preceding inequality tends to zero. As a consequence of Steps 1–3, together with the Ascoli–Arzelà theorem, we can conclude that $N\colon \Omega \to \Omega$ is completely continuous.

*Step 4: A priori bounds.* Now it remains to show that the set

$$E = \{u \in \Omega\colon u = \lambda N(u) \quad \text{for some } 0 < \lambda < 1\}$$

is bounded. Let $u \in E$; then $u = \lambda N(u)$ for some $0 < \lambda < 1$. Thus, for each $t \in J$ we have

$$
\begin{aligned}
u(t) = \lambda\varphi(0) &+ \frac{\lambda}{\Gamma(\alpha)} \sum_{0<t_k<t} \int_{t_{k-1}}^{t_k} (t_k - s)^{\alpha-1} g(s)ds \\
&+ \frac{\lambda}{\Gamma(\alpha)} \int_{t_k}^{t} (t - s)^{\alpha-1} g(s)ds + \lambda \sum_{0<t_k<t} I_k(u_{t_k^-})\,.
\end{aligned}
\tag{3.51}
$$

And, by (3.21.1), for each $t \in J$ we have

$$
\begin{aligned}
|g(t)| &\le |f(t, u_t, g(t))| + |\phi(t, y_t)| \\
&\le p(t) + q(t)\|u_t\|_{PC} + r(t)|g(t)| + |\phi(t, y_t) - \phi(t, 0)| + |\phi(t, 0)| \\
&\le p(t) + q(t)\|u_t\|_{PC} + r(t)|g(t)| + \overline{K}\|u_t\|_{PC} + |\phi(t, 0)| \\
&\le p^* + (q^* + \overline{K})\|u_t\|_{PC} + r^*|g(t)| + \phi^*\,.
\end{aligned}
$$

Thus,

$$|g(t)| \le \frac{1}{1 - r^*}(p^* + (q^* + \overline{K})\|u_t\|_{PC} + \phi^*)\,.$$

This implies, by (3.51) and (3.21.2), that for each $t \in J$ we have

$$|u(t)| \leq |\varphi(0)| + \frac{1}{(1-r^*)\Gamma(\alpha)} \sum_{0<t_k<t} \int_{t_{k-1}}^{t_k} (t_k - s)^{\alpha-1}(p^* + \phi^* + (q^* + \overline{K})\|u_s\|_{PC})ds$$
$$+ \frac{1}{(1-r^*)\Gamma(\alpha)} \int_{t_k}^{t} (t-s)^{\alpha-1}(p^* + \phi^* + (q^* + \overline{K})\|u_s\|_{PC})ds$$
$$+ m(M^*\|u_t\|_{PC} + N^*) .$$

Consider the function $v$ defined by

$$v(t) = \sup\{|u(s)| : -r \leq s \leq t\}, \quad 0 \leq t \leq T ;$$

there exists $t^* \in [-r, T]$ such that $v(t) = |u(t^*)|$. If $t \in [0, T]$, then by the previous inequality, for $t \in J$ we have

$$v(t) \leq |\varphi(0)| + \frac{1}{(1-r^*)\Gamma(\alpha)} \sum_{0<t_k<t} \int_{t_{k-1}}^{t_k} (t_k - s)^{\alpha-1}(p^* + \phi^* + (q^* + \overline{K})v(s))ds$$
$$+ \frac{1}{(1-r^*)\Gamma(\alpha)} \int_{t_k}^{t} (t-s)^{\alpha-1}(p^* + \phi^* + (q^* + \overline{K})v(s))ds$$
$$+ mM^*v(t) + mN^* .$$

Thus,

$$v(t) \leq \frac{1}{(1-mM^*)(1-r^*)\Gamma(\alpha)} \sum_{0<t_k<t} \int_{t_{k-1}}^{t_k} (t_k - s)^{\alpha-1}(p^* + \phi^* + (q^* + \overline{K})v(s))ds$$
$$+ \frac{|\varphi(0)| + mN^*}{1-mM^*} + \frac{1}{(1-mM^*)(1-r^*)\Gamma(\alpha)} \int_{t_k}^{t} (t-s)^{\alpha-1}(p^* + \phi^* + (q^* + \overline{K})v(s))ds$$
$$\leq \frac{|\varphi(0)| + mN^*}{1-mM^*} + \frac{(m+1)(p^* + \phi^*)T^\alpha}{(1-mM^*)(1-r^*)\Gamma(\alpha+1)}$$
$$+ \frac{(m+1)(q^* + \overline{K})}{(1-mM^*)(1-r^*)\Gamma(\alpha)} \int_0^t (t-s)^{\alpha-1}v(s)ds .$$

Applying Lemma 1.52, we get

$$v(t) \leq \left[\frac{|\varphi(0)| + mN^*}{1-mM^*} + \frac{(m+1)(p^* + \phi^*)T^\alpha}{(1-mM^*)(1-r^*)\Gamma(\alpha+1)}\right]$$
$$\times \left[1 + \frac{\delta(m+1)(q^* + \overline{K})T^\alpha}{(1-mM^*)(1-r^*)\Gamma(\alpha+1)}\right] := A ,$$

where $\delta = \delta(\alpha)$ is a constant. If $t^* \in [-r, 0]$, then $v(t) = \|\varphi\|_{PC}$; thus, for any $t \in J$, $\|u\|_\Omega \leq v(t)$, and we have

$$\|u\|_\Omega \leq \max\{\|\varphi\|_{PC}, A\} .$$

This shows that the set $E$ is bounded. As a consequence of Schaefer's fixed point theorem, we deduce that $N$ has a fixed point that is a solution of problem (3.40)–(3.42). □

### 3.4.3 Ulam–Hyers Stability Results

Let $z \in PC(J, \mathbb{R})$, $\epsilon > 0$, $\psi > 0$, and $\omega \in PC(J, \mathbb{R}_+)$ is nondecreasing. We consider the sets of inequalities

$$\begin{cases} |{}^cD^\alpha z(t) - f(t, z_t, {}^cD^\alpha z(t)) - \phi(t, z_t)| \leq \epsilon, & t \in (t_k, t_{k+1}],\ k = 1, \ldots, m , \\ |\Delta z|_{t_k} - I_k(z_{t_k^-})| \leq \epsilon, & k = 1, \ldots, m , \end{cases} \tag{3.52}$$

$$\begin{cases} |{}^cD^\alpha z(t) - f(t, z_t, {}^cD^\alpha z(t)) - \phi(t, z_t)| \leq \omega(t), & t \in (t_k, t_{k+1}],\ k = 1, \ldots, m , \\ |\Delta z|_{t_k} - I_k(z_{t_k^-})| \leq \psi, & k = 1, \ldots, m , \end{cases} \tag{3.53}$$

and

$$\begin{cases} |{}^cD^\alpha z(t) - f(t, z_t, {}^cD^\alpha z(t)) - \phi(t, z_t)| \leq \epsilon\omega(t), & t \in (t_k, t_{k+1}],\ k = 1, \ldots, m , \\ |\Delta z|_{t_k} - I_k(z_{t_k^-})| \leq \epsilon\psi, & k = 1, \ldots, m . \end{cases} \tag{3.54}$$

**Definition 3.22.** Problem (3.40)–(3.41) is Ulam–Hyers stable if there exists a real number $c_{f,m} > 0$ such that, for each $\epsilon > 0$ and for each solution $z \in PC(J, \mathbb{R})$ of inequality (3.52), there exists a solution $y \in \Omega$ of problem (3.40)–(3.41), with

$$|z(t) - y(t)| \leq c_{f,m}\epsilon , \quad t \in J .$$

**Definition 3.23.** Problem (3.40)–(3.41) is generalized Ulam–Hyers stable if there exists $\theta_{f,m} \in C(\mathbb{R}_+, \mathbb{R}_+)$, $\theta_{f,m}(0) = 0$ such that, for each solution $z \in PC(J, \mathbb{R})$ of inequality (3.52), there exists a solution $y \in \Omega$ of problem (3.40)–(3.41), with

$$|z(t) - y(t)| \leq \theta_{f,m}(\epsilon) , \quad t \in J .$$

**Definition 3.24.** Problem (3.40)–(3.41) is Ulam–Hyers–Rassias stable with respect to $(\omega, \psi)$ if there exists $c_{f,m,\omega} > 0$ such that, for each $\epsilon > 0$ and for each solution $z \in PC(J, \mathbb{R})$ of inequality (3.54), there exists a solution $y \in \Omega$ of problem (3.40)–(3.41), with

$$|z(t) - y(t)| \leq c_{f,m,\omega}\epsilon(\omega(t) + \psi) , \quad t \in J .$$

**Definition 3.25.** Problem (3.40)–(3.41) is generalized Ulam–Hyers–Rassias stable with respect to $(\omega, \psi)$ if there exists $c_{f,m,\omega} > 0$ such that, for each solution $z \in PC(J, \mathbb{R})$ of inequality (3.53), there exists a solution $y \in \Omega$ of problem (3.40)–(3.41), with

$$|z(t) - y(t)| \leq c_{f,m,\omega}(\omega(t) + \psi) , \quad t \in J .$$

**Remark 3.26.** It is clear that (i) Definition 3.22 implies Definition 3.23, (ii) Definition 3.24 implies Definition 3.25, and (iii) Definition 3.24 for $\omega(t) = \psi = 1$ implies Definition 3.22.

**Remark 3.27.** A function $z \in PC(J, \mathbb{R})$ is a solution of inequality (3.54) if and only if there is $\sigma \in PC(J, \mathbb{R})$ and a sequence $\sigma_k$, $k = 1, \ldots, m$ (which depend on $z$) such that
**(i)** $|\sigma(t)| \leq \epsilon\omega(t)$, $t \in (t_k, t_{k+1}]$, $k = 1, \ldots, m$ and $|\sigma_k| \leq \epsilon\psi$, $k = 1, \ldots, m$;
**(ii)** ${}^cD^\alpha z(t) = f(t, z_t, {}^cD^\alpha z(t)) + \phi(t, z_t) + \sigma(t)$, $t \in (t_k, t_{k+1}]$, $k = 1, \ldots, m$;
**(iii)** $\Delta z|_{t_k} = I_k(z_{t_k^-}) + \sigma_k$, $k = 1, \ldots, m$.

One can have similar remarks for inequalities 3.53 and 3.52.

Now we state the following Ulam–Hyers–Rassias stability results.

**Theorem 3.28.** *Assume* (3.20.1)–(3.20.3) *and* (3.47) *hold and*
(3.28.1) *there exists a nondecreasing function* $\omega \in PC(J, \mathbb{R}_+)$ *and there exists* $\lambda_\omega > 0$ *such that for any* $t \in J$*:*

$$I^\alpha\omega(t) \leq \lambda_\omega\omega(t) .$$

*Then problem* (3.40)–(3.41) *is Ulam–Hyers–Rassias stable with respect to* $(\omega, \psi)$.

*Proof.* Let $z \in \Omega$ be a solution of inequality (3.54). Denote by $y$ the unique solution of the problem

$$\begin{cases} {}^cD^\alpha_{t_k} y(t) = f(t, y_t, {}^cD^\alpha_{t_k} y(t)) + \phi(t, y_t), & t \in (t_k, t_{k+1}], k = 1, \ldots, m , \\ \Delta y|_{t=t_k} = I_k(y_{t_k^-}), & k = 1, \ldots, m , \\ y(t) = z(t) = \varphi(t), & t \in [-r, 0] . \end{cases}$$

Using Lemma 3.19, we obtain for each $t \in (t_k, t_{k+1}]$

$$y(t) = \varphi(0) + \sum_{i=1}^{k} I_i(y_{t_i^-}) + \frac{1}{\Gamma(\alpha)} \sum_{i=1}^{k} \int_{t_{i-1}}^{t_i} (t_i - s)^{\alpha-1} g(s)ds$$

$$+ \frac{1}{\Gamma(\alpha)} \int_{t_k}^{t} (t - s)^{\alpha-1} g(s)ds , \quad t \in (t_k, t_{k+1}] ,$$

where $g \in C(J, \mathbb{R})$ is such that

$$g(t) = f(t, y_t, g(t)) + \phi(t, y_t) .$$

Since $z$ is a solution of inequality (3.54), by Remark 3.27 we have

$$\begin{cases} {}^cD^\alpha_{t_k} z(t) = f(t, z_t, {}^cD^\alpha_{t_k} z(t)) + \phi(t, z_t) + \sigma(t), & t \in (t_k, t_{k+1}], \; k = 1, \ldots, m ; \\ \Delta z|_{t=t_k} = I_k(z_{t_k^-}) + \sigma_k, & k = 1, \ldots, m . \end{cases} \tag{3.55}$$

Clearly, the solution of (3.55) is given by

$$z(t) = \varphi(0) + \sum_{i=1}^{k} I_i(z_{t_i^-}) + \sum_{i=1}^{k} \sigma_i + \frac{1}{\Gamma(\alpha)} \sum_{i=1}^{k} \int_{t_{i-1}}^{t_i} (t_i - s)^{\alpha-1} h(s)ds$$

$$+ \frac{1}{\Gamma(\alpha)} \sum_{i=1}^{k} \int_{t_{i-1}}^{t_i} (t_i - s)^{\alpha-1} \sigma(s)ds + \frac{1}{\Gamma(\alpha)} \int_{t_k}^{t} (t - s)^{\alpha-1} h(s)ds$$

$$+ \frac{1}{\Gamma(\alpha)} \int_{t_k}^{t} (t - s)^{\alpha-1} \sigma(s)ds \,, \quad t \in (t_k, t_{k+1}] \,,$$

where $h \in C(J, \mathbb{R})$ is given by

$$h(t) = f(t, z_t, h(t)) + \phi(t, z_t) \,.$$

Hence, for each $t \in (t_k, t_{k+1}]$ it follows that

$$|z(t) - y(t)| \leq \sum_{i=1}^{k} |\sigma_i| + \sum_{i=1}^{k} |I_i(z_{t_i^-}) - I_i(y_{t_i^-})|$$

$$+ \frac{1}{\Gamma(\alpha)} \sum_{i=1}^{k} \int_{t_{i-1}}^{t_i} (t_i - s)^{\alpha-1} |h(s) - g(s)|ds$$

$$+ \frac{1}{\Gamma(\alpha)} \sum_{i=1}^{k} \int_{t_{i-1}}^{t_i} (t_i - s)^{\alpha-1} |\sigma(s)|ds$$

$$+ \frac{1}{\Gamma(\alpha)} \int_{t_k}^{t} (t - s)^{\alpha-1} |h(s) - g(s)|ds$$

$$+ \frac{1}{\Gamma(\alpha)} \int_{t_k}^{t} (t - s)^{\alpha-1} |\sigma(s)| \,.$$

Thus,

$$|z(t) - y(t)| \leq m\epsilon\psi + (m+1)\epsilon\lambda_\omega \omega(t) + \sum_{i=1}^{k} \widetilde{l} \|z_{t_i^-} - y_{t_i^-}\|_{PC}$$

$$+ \frac{1}{\Gamma(\alpha)} \sum_{i=1}^{k} \int_{t_{i-1}}^{t_i} (t_i - s)^{\alpha-1} |h(s) - g(s)|ds$$

$$+ \frac{1}{\Gamma(\alpha)} \int_{t_k}^{t} (t - s)^{\alpha-1} |h(s) - g(s)|ds \,.$$

By (3.20.2) we have

$$\begin{aligned}|h(t) - g(t)| &\leq |f(t, z_t, h(t)) - f(t, y_t, g(t))| + |\phi(t, z_t) - \phi(t, y_t)| \\ &\leq K\|z_t - y_t\|_{PC} + L|g(t) - h(t)| + \overline{K}\|z_t - y_t\|_{PC} .\end{aligned}$$

Then

$$|h(t) - g(t)| \leq \frac{K + \overline{K}}{1 - L}\|z_t - y_t\|_{PC} .$$

Therefore, for each $t \in J$

$$\begin{aligned}|z(t) - y(t)| \leq{}& m\epsilon\psi + (m+1)\epsilon\lambda_\omega\omega(t) + \sum_{i=1}^{k} \widetilde{l}\|z_{t_i^-} - y_{t_i^-}\|_{PC} \\ &+ \frac{K + \overline{K}}{(1 - L)\Gamma(\alpha)} \sum_{i=1}^{k} \int_{t_{i-1}}^{t_i} (t_i - s)^{\alpha-1}\|z_s - y_s\|_{PC} ds \\ &+ \frac{K + \overline{K}}{(1 - L)\Gamma(\alpha)} \int_{t_k}^{t} (t - s)^{\alpha-1}\|z_s - y_s\|_{PC} ds .\end{aligned}$$

Thus,

$$\begin{aligned}|z(t) - y(t)| \leq{}& \sum_{0<t_i<t} \widetilde{l}\|z_{t_i^-} - y_{t_i^-}\|_{PC} + \epsilon(\psi + \omega(t))(m + (m+1)\lambda_\omega) \\ &+ \frac{(K + \overline{K})(m+1)}{(1 - L)\Gamma(\alpha)} \int_0^t (t - s)^{\alpha-1}\|z_s - y_s\|_{PC} ds .\end{aligned}$$

We consider the function $v_1$ defined by

$$v_1(t) = \sup\{\|z(s) - y(s)\| : -r \leq s \leq t\} , \quad 0 \leq t \leq T .$$

Then there exists $t^* \in [-r, T]$ such that $v_1(t) = \|z(t^*) - y(t^*)\|$. If $t^* \in [-r, 0]$, then $v_1(t) = 0$. If $t^* \in [0, T]$, then, by the previous inequality, we have

$$\begin{aligned}v_1(t) \leq{}& \sum_{0<t_i<t} \widetilde{l} v_1(t_i^-) + \epsilon(\psi + \omega(t))(m + (m+1)\lambda_\omega) \\ &+ \frac{(K + \overline{K})(m+1)}{(1 - L)\Gamma(\alpha)} \int_0^t (t - s)^{\alpha-1} v_1(s) ds .\end{aligned}$$

Applying Lemma 1.53, we get

$$\begin{aligned}v_1(t) &\leq \epsilon(\psi + \omega(t))(m + (m+1)\lambda_\omega) \\ &\quad\times \left[\prod_{0<t_i<t} (1 + \widetilde{l}) \exp\left(\int_0^t \frac{(K + \overline{K})(m+1)}{(1 - L)\Gamma(\alpha)} (t - s)^{\alpha-1} ds\right)\right] \\ &\leq c_\omega \epsilon(\psi + \omega(t)) ,\end{aligned}$$

where

$$c_\omega = (m + (m+1)\lambda_\omega)\left[\prod_{i=1}^{m}(1+\widetilde{l})\exp\left(\frac{(K+\overline{K})(m+1)T^\alpha}{(1-L)\Gamma(\alpha+1)}\right)\right]$$
$$= (m + (m+1)\lambda_\omega)\left[(1+\widetilde{l})\exp\left(\frac{(K+\overline{K})(m+1)T^\alpha}{(1-L)\Gamma(\alpha+1)}\right)\right]^m .$$

Thus, problem (3.40)–(3.41) is Ulam–Hyers–Rassias stable with respect to $(\omega, \psi)$. □

Next we present the following Ulam–Hyers stability result.

**Theorem 3.29.** *Assume* (3.20.1)–(3.20.3) *and* (3.47) *hold. Then problem* (3.40)–(3.41) *is Ulam–Hyers stable.*

*Proof.* Let $z \in \Omega$ be a solution of inequality (3.52). Denote by $y$ the unique solution of the problem

$$\begin{cases} {}^cD^\alpha_{t_k}y(t) = f(t, y_t, {}^cD^\alpha_{t_k}y(t)) + \phi(t, y_t), & t \in (t_k, t_{k+1}],\ k = 1, \dots, m, \\ \Delta y|_{t=t_k} = I_k(y_{t_k^-}), & k = 1, \dots, m, \\ y(t) = z(t) = \varphi(t), & t \in [-r, 0]. \end{cases}$$

From the proof of Theorem 3.28 we get the inequality

$$v_1(t) \le \sum_{0<t_i<t} \widetilde{l}v_1(t_i^-) + m\epsilon + \frac{T^\alpha\epsilon(m+1)}{\Gamma(\alpha+1)}$$
$$+ \frac{(K+\overline{K})(m+1)}{(1-L)\Gamma(\alpha)}\int_0^t (t-s)^{\alpha-1}v_1(s)ds .$$

Applying Lemma 1.53, we obtain

$$v_1(t) \le \epsilon\left(\frac{m\Gamma(\alpha+1) + T^\alpha(m+1)}{\Gamma(\alpha+1)}\right)$$
$$\times\left[\prod_{0<t_i<t}(1+\widetilde{l})\exp\left(\int_0^t \frac{(K+\overline{K})(m+1)}{(1-L)\Gamma(\alpha)}(t-s)^{\alpha-1}ds\right)\right]$$
$$\le c_\omega\epsilon ,$$

where

$$c_\omega = \left(\frac{m\Gamma(\alpha+1) + T^\alpha(m+1)}{\Gamma(\alpha+1)}\right)\left[\prod_{i=1}^{m}(1+\widetilde{l})\exp\left(\frac{(K+\overline{K})(m+1)T^\alpha}{(1-L)\Gamma(\alpha+1)}\right)\right]$$
$$= \left(\frac{m\Gamma(\alpha+1) + T^\alpha(m+1)}{\Gamma(\alpha+1)}\right)\left[(1+\widetilde{l})\exp\left(\frac{(K+\overline{K})(m+1)T^\alpha}{(1-L)\Gamma(\alpha+1)}\right)\right]^m .$$

Moreover, if we set $\gamma(\epsilon) = c_\omega\epsilon; \gamma(0) = 0$, then problem (3.40)–(3.41) is generalized Ulam–Hyers stable. □

### 3.4.4 Examples

*Example 1.* Consider the impulsive problem

$$^{c}D^{\frac{1}{2}}_{t_k}y(t) = \frac{e^{-t}}{(11+e^t)}\left[\frac{|y_t|}{1+|y_t|} - \frac{|^{c}D^{\frac{1}{2}}_{t_k}y(t)|}{1+|^{c}D^{\frac{1}{2}}_{t_k}y(t)|}\right] + \frac{|y_t|}{6(1+|y_t|)}\,, \quad \text{for each, } t \in J_0 \cup J_1\,, \tag{3.56}$$

$$\Delta y|_{t=\frac{1}{2}} = \frac{|y\left(\frac{1}{2}^-\right)|}{10+|y\left(\frac{1}{2}^-\right)|}\,, \tag{3.57}$$

$$y(t) = \varphi(t)\,, \quad t \in [-r,0],\ r > 0\,, \tag{3.58}$$

where $\varphi \in PC([-r,0],\mathbb{R})$, $J_0 = [0,\frac{1}{2}]$, $J_1 = (\frac{1}{2},1]$, $t_0 = 0$, and $t_1 = \frac{1}{2}$. Set

$$f(t,u,v) = \frac{e^{-t}}{(11+e^t)}\left[\frac{|u|}{1+|u|} - \frac{|v|}{1+|v|}\right]$$

and

$$\phi(t,u) = \frac{|u|}{6(1+|u|)}$$

for any $t \in [0,1]$, $u \in PC([-r,0],\mathbb{R})$, and $v \in \mathbb{R}$. Clearly, the functions $f$, $\phi$ are jointly continuous. For each $u, \bar{u} \in PC([-r,0],\mathbb{R})$, $v, \bar{v} \in \mathbb{R}$, and $t \in [0,1]$,

$$\begin{aligned}\|f(t,u,v) - f(t,\bar{u},\bar{v})\| &\le \frac{e^{-t}}{(11+e^t)}(\|u-\bar{u}\|_{PC} + \|v-\bar{v}\|)\\ &\le \frac{1}{12}\|u-\bar{u}\|_{PC} + \frac{1}{12}\|v-\bar{v}\|\end{aligned}$$

and

$$|\phi(t,u) - \phi(t,\bar{u})| \le \frac{1}{6}\|u-\bar{u}\|_{PC}\,.$$

Hence, condition (3.20.2) is satisfied by $K = L = \frac{1}{12}$, $\overline{K} = \frac{1}{6}$.

Let

$$I_1(u) = \frac{|u|}{10+|u|}\,, \quad u \in PC([-r,0],\mathbb{R})\,,$$

and $u, v \in PC([-r,0],\mathbb{R})$. Then we have

$$|I_1(u) - I_1(v)| = \left|\frac{|u|}{10+|u|} - \frac{|v|}{10+|v|}\right| \le \frac{1}{10}\|u-v\|_{PC}\,.$$

Thus,

$$\begin{aligned}m\tilde{l} + \frac{(m+1)(K+\overline{K})T^{\alpha}}{(1-L)\Gamma(\alpha+1)} &= \frac{1}{10} + \frac{\frac{3}{6}}{\left(1-\frac{1}{12}\right)\Gamma\left(\frac{3}{2}\right)}\\ &= \frac{12}{11\sqrt{\pi}} + \frac{1}{10} < 1\,.\end{aligned}$$

It follows from Theorem 3.20 that problem (3.56)–(3.58) has a unique solution on $J$.

Set, for any $t \in [0, 1]$, $\omega(t) = t$, $\psi = 1$. Since

$$I^{\frac{1}{2}} \omega(t) = \frac{1}{\Gamma\left(\frac{1}{2}\right)} \int_0^t (t-s)^{\frac{1}{2}-1} s ds \leq \frac{2t}{\sqrt{\pi}},$$

condition (3.28.1) is satisfied by $\lambda_\omega = \frac{2}{\sqrt{\pi}}$. It follows that problem (3.56)–(3.57) is Ulam–Hyers–Rassias stable with respect to $(\omega, \psi)$.

*Example 2.* Consider the impulsive problem

$$^cD_{t_k}^{\frac{1}{2}} y(t) = \frac{2 + |y_t| + |^cD_{t_k}^{\frac{1}{2}} y(t)|}{108e^{t+3}\left(1 + |y_t| + |^cD_{t_k}^{\frac{1}{2}} y(t)|\right)} + \frac{e^{-t}|y_t|}{(3+e^t)(1+|y_t|)} \quad \text{for each} \quad t \in J_0 \cup J_1, \tag{3.59}$$

$$\Delta y|_{t=\frac{1}{3}} = \frac{|y\left(\frac{1}{3}^-\right)|}{6 + |y\left(\frac{1}{3}^-\right)|}, \tag{3.60}$$

$$y(t) = \varphi(t), \quad t \in [-r, 0], r > 0, \tag{3.61}$$

where $\varphi \in PC([-r, 0], \mathbb{R})$ $J_0 = [0, \frac{1}{3}]$, $J_1 = (\frac{1}{3}, 1]$, $t_0 = 0$, and $t_1 = \frac{1}{3}$. Set

$$f(t, u, v) = \frac{2 + |u| + |v|}{108e^{t+3}(1 + |u| + |v|)}$$

and

$$\phi(t, u) = \frac{e^{-t}|u|}{(3 + e^t)(1 + |u|)}$$

for any $t \in [0, 1]$, $u \in PC([-r, 0], \mathbb{R})$, $v \in \mathbb{R}$. Clearly, the functions $f, \phi$ are jointly continuous.

For any $u, \bar{u} \in PC([-r, 0], \mathbb{R})$, $v, \bar{v} \in \mathbb{R}$, and $t \in [0, 1]$,

$$|f(t, u, v) - f(t, \bar{u}, \bar{v})| \leq \frac{1}{108e^3}(\|u - \bar{u}\|_{PC} + |v - \bar{v}|)$$

and

$$|\phi(t, u) - \phi(t, \bar{u})| \leq \frac{1}{4}\|u - \bar{u}\|_{PC}.$$

Hence, condition (3.20.2) is satisfied by $K = L = \frac{1}{108e^3}$, $\overline{K} = \frac{1}{4}$.

For each $t \in [0, 1]$ we have

$$|f(t, u, v)| \leq \frac{1}{108e^{t+3}}(2 + \|u\|_{PC} + |v|).$$

Thus, condition (3.21.1) is satisfied by $p(t) = \frac{1}{54e^{t+3}}$ and $q(t) = r(t) = \frac{1}{108e^{t+3}}$. Let

$$I_1(u) = \frac{|u|}{6 + |u|}, \quad u \in PC([-r, 0], \mathbb{R}).$$

For each $u \in PC([-r,0],\mathbb{R})$ we have

$$|I_1(u)| \leq \frac{1}{6}\|u\|_{PC} + 1 .$$

Since $mM^* < 1$, condition (3.21.2) is satisfied. It follows from Theorem 3.21 that problem (3.59)–(3.61) has at least one solution on $J$.

## 3.5 Existence and Stability Results for Neutral Impulsive NIFDE with Finite Delay

### 3.5.1 Introduction

The purpose of this section is to establish some existence, uniqueness, and stability results for the following implicit neutral differential equations of fractional order with finite delay and impulses:

$$^cD^{\alpha}_{t_k}[y(t) - \phi(t, y_t)] = f(t, y_t, {^cD^{\alpha}_{t_k}}y(t)) , \qquad \text{for each } t \in (t_k, t_{k+1}] , \quad k = 0, \ldots, m, 0 < \alpha \leq 1 , \tag{3.62}$$

$$\Delta y|_{t_k} = I_k(y_{t_k^-}) \qquad k = 1, \ldots, m , \tag{3.63}$$

$$y(t) = \varphi(t) , \qquad t \in [-r, 0],\ r > 0 , \tag{3.64}$$

where $f: J \times PC([-r,0],\mathbb{R}) \times \mathbb{R} \to \mathbb{R}$ and $\phi: J \times PC([-r,0],\mathbb{R}) \to \mathbb{R}$ are given functions with $\phi(0,\varphi) = 0$, $I_k: PC([-r,0],\mathbb{R}) \to \mathbb{R}$ and $\varphi \in PC([-r,0],\mathbb{R})$, $0 = t_0 < t_1 < \cdots < t_m < t_{m+1} = T$, and $PC([-r,0],\mathbb{R})$ is a space to be specified later. Here, $\Delta y|_{t_k} = y(t_k^+) - y(t_k^-)$, where $y(t_k^+) = \lim_{h\to 0^+} y(t_k + h)$ and $y(t_k^-) = \lim_{h\to 0^-} y(t_k + h)$ represent the right and left limits of $y_t$ at $t = t_k$, respectively.

The arguments are based upon the Banach contraction principle and Schaefer's fixed point theorem. An example is included to show the applicability of our results.

### 3.5.2 Existence of Solutions

Consider the Banach space

$$PC([-r,0],\mathbb{R}) = \{y: [-r,0] \to \mathbb{R}: y \in C((\tau_k, \tau_{k+1}],\mathbb{R}),\ k=1,\ldots,l, \text{ and there exist } y(\tau_k^-) \text{ and } y(\tau_k^+),\ k=1,\ldots,l \text{ with } y(\tau_k^-)=y(\tau_k)\} ,$$

with the norm

$$\|y\|_{PC} = \sup_{t\in[-r,0]} |y(t)| .$$

Take

$$PC([0,T],\mathbb{R}) = \{y: [0,T] \to \mathbb{R}: y \in C((t_k, t_{k+1}],\mathbb{R}),\ k = 1,\ldots,m, \text{ and there exist } y(t_k^-) \text{ and } y(t_k^+),\ k = 1,\ldots,m \text{ with } y(t_k^-) = y(t_k)\}$$

to be our Banach space with the norm

$$\|y\|_C = \sup_{t\in[0,T]} |y(t)| .$$

Also,

$$\Omega = \{y\colon [-r, T] \to \mathbb{R}\colon y|_{[-r,0]} \in PC([-r, 0], \mathbb{R}) \text{ and } y|_{[0,T]} \in PC([0, T], \mathbb{R})\}$$

is a Banach space with the norm

$$\|y\|_\Omega = \sup_{t\in[-r,T]} |y(t)| .$$

**Definition 3.30.** A function $y \in \Omega$ whose $\alpha$-derivative exists on $J_k$ is said to be a solution of (3.62)–(3.64) if $y$ satisfies the equation ${}^cD^{\alpha}_{t_k}(y(t) - \phi(t, y_t)) = f(t, y_t, {}^cD^{\alpha}_{t_k}y(t))$ on $J_k$ and satisfies the conditions

$$\Delta y|_{t=t_k} = I_k(y_{t_k^-}) , \quad k = 1, \dots, m ,$$
$$y(t) = \varphi(t) , \quad t \in [-r, 0] .$$

To prove the existence of solutions to (3.62)–(3.64), we need the following auxiliary lemma.

**Lemma 3.31.** *Let $0 < \alpha \le 1$, and let $\sigma\colon J \to \mathbb{R}$ be continuous. A function $y$ is a solution of the fractional integral equation*

$$y(t) = \begin{cases} \varphi(0) + \phi(t, y_t) + \dfrac{1}{\Gamma(\alpha)}\displaystyle\int_0^t (t-s)^{\alpha-1}\sigma(s)ds, & \text{if } t \in [0, t_1] , \\ \varphi(0) + \phi(t, y_t) + \displaystyle\sum_{i=1}^{k} I_i(y_{t_i^-}) + \dfrac{1}{\Gamma(\alpha)}\displaystyle\sum_{i=1}^{k}\int_{t_{i-1}}^{t_i} (t_i - s)^{\alpha-1}\sigma(s)ds & \\ +\dfrac{1}{\Gamma(\alpha)}\displaystyle\int_{t_k}^{t} (t-s)^{\alpha-1}\sigma(s)ds, & \text{if } t \in (t_k, t_{k+1}] , \\ \varphi(t), & t \in [-r, 0] , \end{cases} \tag{3.65}$$

*where $k = 1, \dots, m$, if and only if $y$ is a solution of the following fractional problem:*

$${}^cD^{\alpha}(y(t) - \phi(t, y_t)) = \sigma(t) , \quad t \in J_k , \tag{3.66}$$
$$\Delta y|_{t=t_k} = I_k(y_{t_k^-}) , \quad k = 1, \dots, m , \tag{3.67}$$
$$y(t) = \varphi(t) , \quad t \in [-r, 0] . \tag{3.68}$$

*Proof.* Assume that $y$ satisfies (3.66)–(3.68). If $t \in [0, t_1]$, then

$${}^cD^{\alpha}(y(t) - \phi(t, y_t)) = \sigma(t) .$$

Lemma 1.9 implies

$$y(t) - \phi(t, y_t) = \varphi(0) + I^\alpha \sigma(t) = \varphi(0) + \frac{1}{\Gamma(\alpha)} \int_0^t (t-s)^{\alpha-1} \sigma(s) ds .$$

If $t \in (t_1, t_2]$, then Lemma 1.9 implies

$$\begin{aligned}
y(t) - \phi(t, y_t) &= y(t_1^+) - \phi(t_1, y_{t_1}) + \frac{1}{\Gamma(\alpha)} \int_{t_1}^t (t-s)^{\alpha-1} \sigma(s) ds \\
&= \Delta y|_{t=t_1} + y(t_1^-) - \phi(t_1, y_{t_1}) + \frac{1}{\Gamma(\alpha)} \int_{t_1}^t (t-s)^{\alpha-1} \sigma(s) ds \\
&= I_1(y_{t_1^-}) + \left[ \varphi(0) + \frac{1}{\Gamma(\alpha)} \int_0^{t_1} (t_1 - s)^{\alpha-1} \sigma(s) ds \right] \\
&\quad + \frac{1}{\Gamma(\alpha)} \int_{t_1}^t (t-s)^{\alpha-1} \sigma(s) ds . \\
&= \varphi(0) + I_1(y_{t_1^-}) + \frac{1}{\Gamma(\alpha)} \int_0^{t_1} (t_1 - s)^{\alpha-1} \sigma(s) ds \\
&\quad + \frac{1}{\Gamma(\alpha)} \int_{t_1}^t (t-s)^{\alpha-1} \sigma(s) ds .
\end{aligned}$$

If $t \in (t_2, t_3]$, then from Lemma 1.9 we get

$$\begin{aligned}
y(t) - \phi(t, y_t) &= y(t_2^+) - \phi(t_2, y_{t_2}) + \frac{1}{\Gamma(\alpha)} \int_{t_2}^t (t-s)^{\alpha-1} \sigma(s) ds \\
&= \Delta y|_{t=t_2} + y(t_2^-) - \phi(t_2, y_{t_2}) + \frac{1}{\Gamma(\alpha)} \int_{t_2}^t (t-s)^{\alpha-1} \sigma(s) ds \\
&= I_2(y_{t_2^-}) + \left[ \varphi(0) + I_1(y_{t_1^-}) + \frac{1}{\Gamma(\alpha)} \int_0^{t_1} (t_1 - s)^{\alpha-1} \sigma(s) ds \right. \\
&\quad \left. + \frac{1}{\Gamma(\alpha)} \int_{t_1}^{t_2} (t_2 - s)^{\alpha-1} \sigma(s) ds \right] + \frac{1}{\Gamma(\alpha)} \int_{t_2}^t (t-s)^{\alpha-1} \sigma(s) ds
\end{aligned}$$

$$= \varphi(0) + \left[I_1(y_{t_1^-}) + I_2(y_{t_2^-})\right] + \left[\frac{1}{\Gamma(\alpha)}\int_0^{t_1}(t_1 - s)^{\alpha-1}\sigma(s)ds + \frac{1}{\Gamma(\alpha)}\int_{t_1}^{t_2}(t_2 - s)^{\alpha-1}\sigma(s)ds\right] + \frac{1}{\Gamma(\alpha)}\int_{t_2}^{t}(t - s)^{\alpha-1}\sigma(s)ds .$$

Repeating the process in this way, the solution $y(t)$ for $t \in (t_k, t_{k+1}]$, $k = 1, \ldots, m$ can be written

$$y(t) = \varphi(0) + \phi(t, y_t) + \sum_{i=1}^{k} I_i(y_{t_i^-}) + \frac{1}{\Gamma(\alpha)}\sum_{i=1}^{k}\int_{t_{i-1}}^{t_i}(t_i - s)^{\alpha-1}\sigma(s)ds + \frac{1}{\Gamma(\alpha)}\int_{t_k}^{t}(t - s)^{\alpha-1}\sigma(s)ds .$$

Conversely, assume that $y$ satisfies impulsive fractional integral equation (3.65). If $t \in [0, t_1]$, then $y(0) = \varphi(0)$. Using the fact that ${}^cD^\alpha$ is the left inverse of $I^\alpha$, we get

$${}^cD^\alpha(y(t) - \phi(t, y_t)) = \sigma(t) , \quad \text{for each } t \in [0, t_1] .$$

If $t \in (t_k, t_{k+1}]$, $k = 1, \ldots, m$, and using the fact that ${}^cD^\alpha C = 0$, where $C$ is a constant, we get

$${}^cD^\alpha(y(t) - \phi(t, y_t)) = \sigma(t) \quad \text{for each } t \in (t_k, t_{k+1}] .$$

Also, we can easily show that

$$\Delta y|_{t=t_k} = I_k(y_{t_k^-}) , \quad k = 1, \ldots, m .$$

□

We are now in a position to state and prove our existence result for problem (3.62)–(3.64) based on Banach's fixed point.

**Theorem 3.32.** *Make the following assumptions:*

(3.32.1) *The function $f : J \times PC([-r, 0], \mathbb{R}) \times \mathbb{R} \to \mathbb{R}$ is continuous.*

(3.32.2) *There exist constants $K > 0$, $\overline{L} > 0$ and $0 < L < 1$ such that*

$$|f(t, u, v) - f(t, \bar{u}, \bar{v})| \le K\|u - \bar{u}\|_{PC} + L|v - \bar{v}|$$

*and*

$$|\phi(t, u) - \phi(t, \bar{u})| \le \overline{L}\|u - \bar{u}\|_{PC}$$

*for any $u, \bar{u} \in PC([-r, 0], \mathbb{R})$, $v, \bar{v} \in \mathbb{R}$, and $t \in J$.*

(3.32.3) *There exists a constant $\tilde{l} > 0$ such that*

$$|I_k(u) - I_k(\overline{u})| \le \tilde{l}\|u - \overline{u}\|_{PC}$$

*for each $u, \overline{u} \in PC([-r, 0], \mathbb{R})$ and $k = 1, \ldots, m$.*

*If*

$$m\tilde{l} + \overline{L} + \frac{(m+1)KT^{\alpha}}{(1-L)\Gamma(\alpha+1)} < 1\,, \tag{3.69}$$

*then there exists a unique solution to problem (3.62)–(3.64) on J.*

*Proof.* Consider the operator $N\colon \Omega \to \Omega$ defined by

$$(Ny)(t) = \begin{cases} \varphi(0) + \phi(t, y_t) + \sum\limits_{0<t_k<t} I_k(y_{t_k^-}) + \frac{1}{\Gamma(\alpha)} \sum\limits_{0<t_k<t} \int\limits_{t_{k-1}}^{t_k} (t_k - s)^{\alpha-1} g(s)ds \\ + \frac{1}{\Gamma(\alpha)} \int\limits_{t_k}^{t} (t-s)^{\alpha-1} g(s)ds, & t \in [0, T]\,, \\ \varphi(t), & t \in [-r, 0]\,, \end{cases} \tag{3.70}$$

where $g \in C(J, \mathbb{R})$ is such that

$$g(t) = f(t, y_t, g(t))\,.$$

Clearly, the fixed points of operator $N$ are solutions of problem (3.62)–(3.64). Let $u, w \in \Omega$. If $t \in [-r, 0]$, then

$$|(Nu)(t) - (Nw)(t)| = 0\,.$$

For $t \in J$ we have

$$\begin{aligned} |(Nu)(t) - (Nw)(t)| \le{}& \frac{1}{\Gamma(\alpha)} \sum_{0<t_k<t} \int\limits_{t_{k-1}}^{t_k} (t_k - s)^{\alpha-1} |g(s) - h(s)| ds \\ &+ \frac{1}{\Gamma(\alpha)} \int\limits_{t_k}^{t} (t-s)^{\alpha-1} |g(s) - h(s)| ds + |\phi(t, u_t) - \phi(t, w_t)| \\ &+ \sum_{0<t_k<t} |I_k(u_{t_k^-}) - I_k(w_{t_k^-})|\,, \end{aligned}$$

where $g, h \in C(J, \mathbb{R})$ are given by

$$g(t) = f(t, u_t, g(t))$$

and

$$h(t) = f(t, w_t, h(t))\,.$$

By (3.32.2) we have

$$\begin{aligned} |g(t) - h(t)| &= |f(t, u_t, g(t)) - f(t, w_t, h(t))| \\ &\le K\|u_t - w_t\|_{PC} + L|g(t) - h(t)|\,. \end{aligned}$$

Then

$$|g(t) - h(t)| \le \frac{K}{1-L} \|u_t - w_t\|_{PC}\,.$$

Therefore, for each $t \in J$

$$
\begin{aligned}
|(Nu)(t) - (Nw)(t)| &\leq \frac{K}{(1-L)\Gamma(\alpha)} \sum_{k=1}^{m} \int_{t_{k-1}}^{t_k} (t_k - s)^{\alpha-1} \|u_s - w_s\|_{PC} ds \\
&+ \frac{K}{(1-L)\Gamma(\alpha)} \int_{t_k}^{t} (t-s)^{\alpha-1} \|u_s - w_s\|_{PC} ds \\
&+ \sum_{k=1}^{m} \widetilde{l} \|u_{t_k^-} - w_{t_k^-}\|_{PC} + \overline{L} \|u_t - w_t\|_{PC} \\
&\leq \left[ m\widetilde{l} + \overline{L} + \frac{mKT^\alpha}{(1-L)\Gamma(\alpha+1)} \right. \\
&\left. + \frac{KT^\alpha}{(1-L)\Gamma(\alpha+1)} \right] \|u - w\|_\Omega .
\end{aligned}
$$

Thus,

$$
\|N(u) - N(w)\|_\Omega \leq \left[ m\widetilde{l} + \overline{L} + \frac{(m+1)KT^\alpha}{(1-L)\Gamma(\alpha+1)} \right] \|u - w\|_\Omega .
$$

By (3.69), operator $N$ is a contraction. Hence, by Banach's contraction principle, $N$ has a unique fixed point that is the unique solution of problem (3.62)–(3.64). □

Our second result is based on Schaefer's fixed point theorem.

**Theorem 3.33.** *Assume* (3.32.1) *and* (3.32.2) *hold and*

(3.33.1) *There exist* $p, q, r \in C(J, \mathbb{R}_+)$ *with* $r^* = \sup_{t \in J} r(t) < 1$ *such that*

$$
|f(t, u, w)| \leq p(t) + q(t)\|u\|_{PC} + r(t)|w| \quad \text{for } t \in J,\ u \in PC([-r, 0], \mathbb{R}) \text{ and } w \in \mathbb{R} ;
$$

(3.33.2) *The functions* $I_k : PC([-r, 0], \mathbb{R}) \to \mathbb{R}$ *are continuous and there exist constants* $M^*, N^* > 0$ *such that*

$$
|I_k(u)| \leq M^* \|u\|_{PC} + N^* \quad \text{for each } u \in PC([-r, 0], \mathbb{R}),\ k = 1, \ldots, m ;
$$

(3.33.3) *The function* $\phi$ *is completely continuous, and for each bounded set* $B_{\eta^*}$ *in* $\Omega$ *the set* $\{t \to \phi(t, y_t) : y \in B_{\eta^*}\}$ *is equicontinuous in* $PC(J, \mathbb{R})$ *and there exist two constants* $d_1 > 0$, $d_2 > 0$ *with* $mM^* + d_1 < 1$ *such that*

$$
|\phi(t, u)| \leq d_1 \|u\|_{PC} + d_2 , \quad t \in J,\ u \in PC([-r, 0], \mathbb{R}) .
$$

*Then problem* (3.62)–(3.64) *has at least one solution.*

*Proof.* We consider the operator $N_1 : \Omega \to \Omega$ defined by

$$
N_1 y(t) = \begin{cases}
\varphi(0) + \sum_{0<t_k<t} I_k(y_{t_k^-}) + \frac{1}{\Gamma(\alpha)} \sum_{0<t_k<t} \int_{t_{k-1}}^{t_k} (t_k - s)^{\alpha-1} g(s) ds \\
+ \frac{1}{\Gamma(\alpha)} \int_{t_k}^{t} (t-s)^{\alpha-1} g(s) ds, & t \in [0, T] , \\
\varphi(t), & t \in [-r, 0] .
\end{cases}
$$

Operator $N$ defined in (3.70) can be written

$$(Ny)(t) = \phi(t, y_t) + N_1 y(t), \quad \text{for each } t \in J .$$

We will use Schaefer's fixed point theorem to prove that $N$ has a fixed point. So we must show that $N$ is completely continuous. Since $\phi$ is completely continuous by (3.33.3), we will show that $N_1$ is completely continuous. The proof will be given in several steps.

*Step 1: $N_1$ is continuous.* Let $\{u_n\}$ be a sequence such that $u_n \to u$ in $\Omega$. If $t \in [-r, 0]$, then

$$|N_1(u_n)(t) - N_1(u)(t)| = 0 .$$

For $t \in J$ we have

$$\begin{aligned}
|N_1(u_n)(t) - N_1(u)(t)| \le{}& \frac{1}{\Gamma(\alpha)} \sum_{0<t_k<t} \int_{t_{k-1}}^{t_k} (t_k - s)^{\alpha-1} |g_n(s) - g(s)| ds \\
&+ \frac{1}{\Gamma(\alpha)} \int_{t_k}^{t} (t - s)^{\alpha-1} |g_n(s) - g(s)| ds \\
&+ \sum_{0<t_k<t} |I_k(u_{nt_k^-}) - I_k(u_{t_k^-})| \\
\le{}& \frac{1}{\Gamma(\alpha)} \sum_{0<t_k<t} \int_{t_{k-1}}^{t_k} (t_k - s)^{\alpha-1} |g_n(s) - g(s)| ds \\
&+ \frac{1}{\Gamma(\alpha)} \int_{t_k}^{t} (t - s)^{\alpha-1} |g_n(s) - g(s)| ds \\
&+ \sum_{0<t_k<t} \tilde{l} \|u_{nt_k^-} - u_{t_k^-}\|_{PC} ,
\end{aligned}$$

and so

$$\begin{aligned}
|N_1(u_n)(t) - N_1(u)(t)| \le{}& \frac{1}{\Gamma(\alpha)} \sum_{0<t_k<t} \int_{t_{k-1}}^{t_k} (t_k - s)^{\alpha-1} |g_n(s) - g(s)| ds \\
&+ \frac{1}{\Gamma(\alpha)} \int_{t_k}^{t} (t - s)^{\alpha-1} |g_n(s) - g(s)| ds \\
&+ m\tilde{l} \|u_n - u\|_{\Omega} ,
\end{aligned} \tag{3.71}$$

where $g_n, g \in C(J, \mathbb{R})$ are given by

$$g_n(t) = f(t, u_{nt}, g_n(t)) ,$$

and

$$g(t) = f(t, u_t, g(t)) .$$

By (3.32.2) we have

$$\begin{aligned}|g_n(t) - g(t)| &= |f(t, u_{nt}, g_n(t)) - f(t, u_t, g(t))| \\ &\leq K\|u_{nt} - u_t\|_{PC} + L|g_n(t) - g(t)| .\end{aligned}$$

Then

$$|g_n(t) - g(t)| \leq \frac{K}{1-L}\|u_{nt} - u_t\|_{PC} .$$

Since $u_n \to u$, we get $g_n(t) \to g(t)$ as $n \to \infty$ for each $t \in J$. Let $\eta > 0$ be such that for each $t \in J$ we have $|g_n(t)| \leq \eta$ and $|g(t)| \leq \eta$. Then we have

$$\begin{aligned}(t-s)^{\alpha-1}|g_n(s) - g(s)| &\leq (t-s)^{\alpha-1}[|g_n(s)| + |g(s)|] \\ &\leq 2\eta(t-s)^{\alpha-1}\end{aligned}$$

and

$$\begin{aligned}(t_k-s)^{\alpha-1}|g_n(s) - g(s)| &\leq (t_k-s)^{\alpha-1}[|g_n(s)| + |g(s)|] \\ &\leq 2\eta(t_k-s)^{\alpha-1} .\end{aligned}$$

For each $t \in J$ the functions $s \to 2\eta(t-s)^{\alpha-1}$ and $s \to 2\eta(t_k-s)^{\alpha-1}$ are integrable on $[0, t]$, then the Lebesgue dominated convergence theorem and (3.71) imply that

$$|N_1(u_n)(t) - N_1(u)(t)| \to 0 \quad \text{as } n \to \infty .$$

Hence,

$$\|N_1(u_n) - N_1(u)\|_\Omega \to 0 \quad \text{as } n \to \infty .$$

Consequently, $N_1$ is continuous.

*Step 2: $N_1$ maps bounded sets to bounded sets in $\Omega$.* Indeed, it is enough to show that for any $\eta^* > 0$ there exists a positive constant $\ell$ such that for each $u \in B_{\eta^*} = \{u \in \Omega\colon \|u\|_\Omega \leq \eta^*\}$ we have $\|N_1(u)\|_\Omega \leq \ell$. For each $t \in J$ we have

$$\begin{aligned}N_1(u)(t) = \varphi(0) &+ \frac{1}{\Gamma(\alpha)} \sum_{0<t_k<t} \int_{t_{k-1}}^{t_k} (t_k - s)^{\alpha-1} g(s)ds \\ &+ \frac{1}{\Gamma(\alpha)} \int_{t_k}^{t} (t-s)^{\alpha-1} g(s)ds , \\ &+ \sum_{0<t_k<t} I_k(u_{t_k^-}) ,\end{aligned} \tag{3.72}$$

where $g \in C(J, \mathbb{R})$ is such that

$$g(t) = f(t, u_t, g(t)) .$$

By (3.33.1), for each $t \in J$ we have

$$\begin{aligned}|g(t)| &= |f(t, u_t, g(t))| \\ &\leq p(t) + q(t)\|u_t\|_{PC} + r(t)|g(t)| \\ &\leq p(t) + q(t)\|u\|_{\Omega} + r(t)|g(t)| \\ &\leq p(t) + q(t)\eta^* + r(t)|g(t)| \\ &\leq p^* + q^*\eta^* + r^*|g(t)|\,,\end{aligned}$$

where $p^* = \sup_{t\in J} p(t)$, and $q^* = \sup_{t\in J} q(t)$.

Then

$$|g(t)| \leq \frac{p^* + q^*\eta^*}{1 - r^*} := M\,.$$

Thus, (3.72) implies

$$\begin{aligned}|N_1(u)(t)| &\leq |\varphi(0)| + \frac{mMT^\alpha}{\Gamma(\alpha+1)} + \frac{MT^\alpha}{\Gamma(\alpha+1)} + \sum_{k=1}^{m}\left(M^*\|u_{t_k^-}\|_{PC} + N^*\right) \\ &\leq |\varphi(0)| + \frac{(m+1)MT^\alpha}{\Gamma(\alpha+1)} + m\left(M^*\|u_{t_k^-}\|_{\Omega} + N^*\right) \\ &\leq |\varphi(0)| + \frac{(m+1)MT^\alpha}{\Gamma(\alpha+1)} + m\left(M^*\eta^* + N^*\right) := R\,.\end{aligned}$$

If $t \in [-r, 0]$, then

$$|N_1(u)(t)| \leq \|\varphi\|_{PC}\,,$$

so

$$\|N_1(u)\|_{\Omega} \leq \max\left\{R, \|\varphi\|_{PC}\right\} := \ell\,.$$

*Step 3: $N_1$ maps bounded sets to equicontinuous sets of $\Omega$.* Let $\tau_1, \tau_2 \in (0, T]$, $\tau_1 < \tau_2$, $B_{\eta^*}$ be a bounded set of $\Omega$ as in Step 2, and let $u \in B_{\eta^*}$. Then

$$\begin{aligned}&|N_1(u)(\tau_2) - N_1(u)(\tau_1)| \\ &\leq \frac{1}{\Gamma(\alpha)}\int_0^{\tau_1} |(\tau_2 - s)^{\alpha-1} - (\tau_1 - s)^{\alpha-1}||g(s)|ds \\ &\quad + \frac{1}{\Gamma(\alpha)}\int_{\tau_1}^{\tau_2} |(\tau_2 - s)^{\alpha-1}||g(s)|ds + \sum_{0<t_k<\tau_2-\tau_1} |I_k(u_{t_k^-})| \\ &\leq \frac{M}{\Gamma(\alpha+1)}[2(\tau_2-\tau_1)^\alpha + (\tau_2^\alpha - \tau_1^\alpha)] + (\tau_2 - \tau_1)\left(M^*\|u_{t_k^-}\|_{\Omega} + N^*\right) \\ &\leq \frac{M}{\Gamma(\alpha+1)}[2(\tau_2-\tau_1)^\alpha + (\tau_2^\alpha - \tau_1^\alpha)] + (\tau_2 - \tau_1)\left(M^*\eta^* + N^*\right)\,.\end{aligned}$$

As $\tau_1 \to \tau_2$, the right-hand side of the preceding inequality tends to zero. As a consequence of Steps 1–3, together with the Ascoli–Arzelà theorem, we can conclude that $N_1 : \Omega \to \Omega$ is completely continuous.

*Step 4: A priori bounds.* Now it remains to show that the set

$$E = \{u \in \Omega : u = \lambda N(u) \text{ for some } 0 < \lambda < 1\}$$

is bounded. Let $u \in E$. Then $u = \lambda N(u)$ for some $0 < \lambda < 1$. Thus, for each $t \in J$ we have

$$u(t) = \lambda\varphi(0) + \lambda\phi(t, y_t) + \frac{\lambda}{\Gamma(\alpha)} \sum_{0<t_k<t} \int_{t_{k-1}}^{t_k} (t_k - s)^{\alpha-1} g(s)ds + \frac{\lambda}{\Gamma(\alpha)} \int_{t_k}^{t} (t-s)^{\alpha-1} g(s)ds + \lambda \sum_{0<t_k<t} I_k(u_{t_k^-}) . \tag{3.73}$$

From (3.33.1), for each $t \in J$ we have

$$\begin{aligned} |g(t)| &= |f(t, u_t, g(t))| \\ &\le p(t) + q(t)\|u_t\|_{PC} + r(t)|g(t)| \\ &\le p^* + q^*\|u_t\|_{PC} + r^*|g(t)| . \end{aligned}$$

Thus,

$$|g(t)| \le \frac{1}{1-r^*}(p^* + q^*\|u_t\|_{PC}) .$$

This implies, by (3.73), (3.33.2), and (3.33.3), that for each $t \in J$ we have

$$\begin{aligned} |u(t)| \le |\varphi(0)| &+ d_1\|u_t\|_{PC} + d_2 \\ &+ \frac{1}{(1-r^*)\Gamma(\alpha)} \sum_{0<t_k<t} \int_{t_{k-1}}^{t_k} (t_k - s)^{\alpha-1}(p^* + q^*\|u_s\|_{PC})ds \\ &+ \frac{1}{(1-r^*)\Gamma(\alpha)} \int_{t_k}^{t} (t-s)^{\alpha-1}(p^* + q^*\|u_s\|_{PC})ds \\ &+ m\left(M^*\|u_{t_k^-}\|_{PC} + N^*\right) . \end{aligned}$$

Consider the function $v$ defined by

$$v(t) = \sup\{|u(s)| : -r \le s \le t\} , \quad 0 \le t \le T .$$

Then there exists $t^* \in [-r, T]$ such that $v(t) = |u(t^*)|$. If $t \in [0, T]$, then, by the previous inequality, for $t \in J$ we have

$$\begin{aligned} v(t) \le |\varphi(0)| &+ \frac{1}{(1-r^*)\Gamma(\alpha)} \sum_{0<t_k<t} \int_{t_{k-1}}^{t_k} (t_k - s)^{\alpha-1}(p^* + q^* v(s))ds \\ &+ \frac{1}{(1-r^*)\Gamma(\alpha)} \int_{t_k}^{t} (t-s)^{\alpha-1}(p^* + q^* v(s))ds \\ &+ (mM^* + d_1)v(t) + (mN^* + d_2) . \end{aligned}$$

Thus,

$$\begin{aligned} v(t) &\leq \frac{1}{(1-(mM^*+d_1))(1-r^*)\Gamma(\alpha)} \sum_{0<t_k<t} \int_{t_{k-1}}^{t_k} (t_k-s)^{\alpha-1}(p^*+q^*v(s))ds \\ &\quad + \frac{|\varphi(0)|+mN^*+d_2}{1-(mM^*+d_1)} \\ &\quad + \frac{1}{(1-(mM^*+d_1))(1-r^*)\Gamma(\alpha)} \int_{t_k}^{t} (t-s)^{\alpha-1}(p^*+q^*v(s))ds \\ &\leq \frac{|\varphi(0)|+mN^*+d_2}{1-(mM^*+d_1)} + \frac{(m+1)p^*T^\alpha}{(1-(mM^*+d_1))(1-r^*)\Gamma(\alpha+1)} \\ &\quad + \frac{(m+1)q^*}{(1-(mM^*+d_1))(1-r^*)\Gamma(\alpha)} \int_0^t (t-s)^{\alpha-1}v(s)ds\,. \end{aligned}$$

Applying Lemma 1.52, we get

$$\begin{aligned} v(t) &\leq \left[\frac{|\varphi(0)|+mN^*+d_2}{1-(mM^*+d_1)} + \frac{(m+1)p^*T^\alpha}{(1-(mM^*+d_1))(1-r^*)\Gamma(\alpha+1)}\right] \\ &\quad \times \left[1+\frac{\delta(m+1)q^*T^\alpha}{(1-(mM^*+d_1))(1-r^*)\Gamma(\alpha+1)}\right] := A\,, \end{aligned}$$

where $\delta = \delta(\alpha)$ is a constant. If $t^* \in [-r, 0]$, then $v(t) = \|\varphi\|_{PC}$; thus, for any $t \in J$, $\|u\|_\Omega \leq v(t)$ we have

$$\|u\|_\Omega \leq \max\{\|\varphi\|_{PC}, A\}\,.$$

This shows that set $E$ is bounded. As a consequence of Schaefer's fixed point theorem, we deduce that $N$ has a fixed point that is a solution of the problem (3.62)–(3.64). □

### 3.5.3 Ulam–Hyers Stability Results

Here we adopt the concepts in Wang et al. [252] and introduce Ulam's type stability concepts for problem (3.62)–(3.63).

Let $z \in PC(J, \mathbb{R})$, $\epsilon > 0$, $\psi > 0$, and let $\omega \in PC(J, \mathbb{R}_+)$ be nondecreasing. We consider the sets of inequalities

$$\begin{cases} |{}^cD^\alpha(z(t)-\phi(t,z_t)) - f(t,z_t,{}^cD^\alpha z(t))| \leq \epsilon, & t \in (t_k, t_{k+1}], k = 1, \dots, m\,, \\ |\Delta z|_{t=t_k} - I_k(z_{t_k^-})| \leq \epsilon, & k = 1, \dots, m\,, \end{cases} \tag{3.74}$$

$$\begin{cases} |{}^cD^\alpha(z(t)-\phi(t,z_t)) - f(t,z_t,{}^cD^\alpha z(t))| \leq \omega(t), & t \in (t_k, t_{k+1}], k = 1, \dots, m\,, \\ |\Delta z|_{t=t_k} - I_k(z_{t_k^-})| \leq \psi, & k = 1, \dots, m\,, \end{cases} \tag{3.75}$$

and

$$\begin{cases} |{}^cD^\alpha(z(t)-\phi(t,z_t))-f(t,z_t,{}^cD^\alpha z(t))|\le\epsilon\omega(t), & t\in(t_k,t_{k+1}], k=1,\dots,m, \\ |\Delta z|_{t=t_k}-I_k(z_{t_k^-})|\le\epsilon\psi, & k=1,\dots,m. \end{cases} \tag{3.76}$$

**Remark 3.34.** A function $z\in PC(J,\mathbb{R})$ is a solution of inequality (3.76) if and only if there is $\sigma\in PC(J,\mathbb{R})$ and a sequence $\sigma_k$, $k=1,\dots,m$ (which depend on $z$) such that

**(i)** $|\sigma(t)|\le\epsilon\omega(t),\ t\in(t_k,t_{k+1}],\ k=1,\dots,m$ and $|\sigma_k|\le\epsilon\psi$, $k=1,\dots,m$;

**(ii)** ${}^cD^\alpha(z(t)-\phi(t,z_t))=f(t,z_t,{}^cD^\alpha z(t))+\sigma(t),\ t\in(t_k,t_{k+1}], k=1,\dots,m$;

**(iii)** $\Delta z|_{t_k}=I_k(z_{t_k^-})+\sigma_k,\ k=1,\dots,m$.

One can provide remarks for inequalities 3.75 and 3.74.

**Theorem 3.35.** *Assume* (3.32.1)–(3.32.3) *and* (3.69) *hold and*

(3.35.1) *there exists a nondecreasing function* $\omega\in PC(J,\mathbb{R}_+)$, *and there exists* $\lambda_\omega>0$ *such that for any* $t\in J$

$$I^\alpha\omega(t)\le\lambda_\omega\omega(t).$$

*If* $\overline{L}<1$, *then problem* (3.62)–(3.63) *is Ulam–Hyers–Rassias stable with respect to* $(\omega,\psi)$.

*Proof.* Let $z\in\Omega$ be a solution of inequality (3.76). Denote by $y$ the unique solution of the problem

$$\begin{cases} {}^cD^\alpha_{t_k}[y(t)-\phi(t,y_t)]=f(t,y_t,{}^cD^\alpha_{t_k}y(t)), & t\in(t_k,t_{k+1}],\ k=1,\dots,m, \\ \Delta y|_{t=t_k}=I_k(y_{t_k^-}), & k=1,\dots,m, \\ y(t)=z(t)=\varphi(t), & t\in[-r,0]. \end{cases}$$

Using Lemma 3.31, for each $t\in(t_k,t_{k+1}]$ we obtain

$$y(t)=\varphi(0)+\phi(t,y_t)+\sum_{i=1}^k I_i(y_{t_i^-})+\frac{1}{\Gamma(\alpha)}\sum_{i=1}^k\int_{t_{i-1}}^{t_i}(t_i-s)^{\alpha-1}g(s)ds$$
$$+\frac{1}{\Gamma(\alpha)}\int_{t_k}^t(t-s)^{\alpha-1}g(s)ds,\quad t\in(t_k,t_{k+1}],$$

where $g\in C(J,\mathbb{R})$ is given by

$$g(t)=f(t,y_t,g(t)).$$

Since $z$ is a solution of inequality (3.76), by Remark 3.34 we have

$$\begin{cases} {}^cD^\alpha_{t_k}[z(t)-\phi(t,z_t)]=f(t,z_t,{}^cD^\alpha_{t_k}z(t))+\sigma(t), & t\in(t_k,t_{k+1}], k=1,\dots,m, \\ \Delta z|_{t=t_k}=I_k(z_{t_k^-})+\sigma_k, & k=1,\dots,m. \end{cases} \tag{3.77}$$

Clearly, the solution of (3.77) is given by

$$z(t) = \varphi(0) + \phi(t, z_t) + \sum_{i=1}^{k} I_i(z_{t_i^-}) + \sum_{i=1}^{k} \sigma_i + \frac{1}{\Gamma(\alpha)} \sum_{i=1}^{k} \int_{t_{i-1}}^{t_i} (t_i - s)^{\alpha-1} h(s)ds$$
$$+ \frac{1}{\Gamma(\alpha)} \sum_{i=1}^{k} \int_{t_{i-1}}^{t_i} (t_i - s)^{\alpha-1} \sigma(s)ds + \frac{1}{\Gamma(\alpha)} \int_{t_k}^{t} (t - s)^{\alpha-1} h(s)ds$$
$$+ \frac{1}{\Gamma(\alpha)} \int_{t_k}^{t} (t - s)^{\alpha-1} \sigma(s)ds \,, \quad t \in (t_k, t_{k+1}] \,,$$

where $h \in C(J, \mathbb{R})$ is given by

$$h(t) = f(t, z_t, h(t)) \,.$$

Hence, for each $t \in (t_k, t_{k+1}]$ it follows that

$$|z(t) - y(t)| \le \sum_{i=1}^{k} |\sigma_i| + |\phi(t, z_t) - \phi(t, y_t)| + \sum_{i=1}^{k} |I_i(z_{t_i^-}) - I_i(y_{t_i^-})|$$
$$+ \frac{1}{\Gamma(\alpha)} \sum_{i=1}^{k} \int_{t_{i-1}}^{t_i} (t_i - s)^{\alpha-1} |\sigma(s)|ds$$
$$+ \frac{1}{\Gamma(\alpha)} \sum_{i=1}^{k} \int_{t_{i-1}}^{t_i} (t_i - s)^{\alpha-1} |h(s) - g(s)|ds$$
$$+ \frac{1}{\Gamma(\alpha)} \int_{t_k}^{t} (t - s)^{\alpha-1} |h(s) - g(s)|ds$$
$$+ \frac{1}{\Gamma(\alpha)} \int_{t_k}^{t} (t - s)^{\alpha-1} |\sigma(s)| \,.$$

Thus,

$$|z(t) - y(t)| \le m\epsilon\psi + (m+1)\epsilon\lambda_\omega \omega(t) + \overline{L}\|z_t - y_t\|_{PC} + \sum_{i=1}^{k} \tilde{l}\|z_{t_i^-} - y_{t_i^-}\|_{PC}$$
$$+ \frac{1}{\Gamma(\alpha)} \sum_{i=1}^{k} \int_{t_{i-1}}^{t_i} (t_i - s)^{\alpha-1} |h(s) - g(s)|ds$$
$$+ \frac{1}{\Gamma(\alpha)} \int_{t_k}^{t} (t - s)^{\alpha-1} |h(s) - g(s)|ds \,.$$

By (3.32.2) we have

$$|h(t) - g(t)| = |f(t, z_t, h(t)) - f(t, y_t, g(t))|$$
$$\leq K\|z_t - y_t\|_{PC} + L|g(t) - h(t)| .$$

Then

$$|h(t) - g(t)| \leq \frac{K}{1-L}\|z_t - y_t\|_{PC} .$$

Therefore, for each $t \in J$

$$|z(t) - y(t)| \leq m\epsilon\psi + (m+1)\epsilon\lambda_\omega\omega(t) + \overline{L}\|z_t - y_t\|_{PC} + \sum_{i=1}^{k}\widetilde{l}\|z_{t_i^-} - y_{t_i^-}\|_{PC}$$
$$+ \frac{K}{(1-L)\Gamma(\alpha)}\sum_{i=1}^{k}\int_{t_{i-1}}^{t_i}(t_i - s)^{\alpha-1}\|z_s - y_s\|_{PC}ds$$
$$+ \frac{K}{(1-L)\Gamma(\alpha)}\int_{t_k}^{t}(t-s)^{\alpha-1}\|z_s - y_s\|_{PC}ds .$$

Thus,

$$|z(t) - y(t)| \leq \sum_{0<t_i<t}\widetilde{l}\|z_{t_i^-} - y_{t_i^-}\|_{PC} + \epsilon(\psi + \omega(t))(m + (m+1)\lambda_\omega)$$
$$+ \overline{L}\|z_t - y_t\|_{PC} + \frac{K(m+1)}{(1-L)\Gamma(\alpha)}\int_0^t (t-s)^{\alpha-1}\|z_s - y_s\|_{PC}ds .$$

We consider the function $v_1$ defined by

$$v_1(t) = \sup\{\|z(s) - y(s)\| : \ -r \leq s \leq t\} , \quad 0 \leq t \leq T .$$

Then there exists $t^* \in [-r, T]$ such that $v_1(t) = \|z(t^*) - y(t^*)\|$. If $t^* \in [-r, 0]$, then $v_1(t) = 0$. If $t^* \in [0, T]$, then, by the previous inequality, we have

$$v_1(t) \leq \sum_{0<t_i<t}\frac{\widetilde{l}}{1-\overline{L}}v_1(t_i^-) + \frac{\epsilon(\psi + \omega(t))(m + (m+1)\lambda_\omega)}{1-\overline{L}}$$
$$+ \frac{K(m+1)}{(1-\overline{L})(1-L)\Gamma(\alpha)}\int_0^t (t-s)^{\alpha-1}v_1(s)ds .$$

Applying Lemma 1.53, we get

$$v_1(t) \leq \frac{\epsilon(\psi + \omega(t))(m + (m+1)\lambda_\omega)}{1-\overline{L}}$$
$$\times \left[\prod_{0<t_i<t}\left(1 + \frac{\widetilde{l}}{1-\overline{L}}\right)\exp\left(\int_0^t \frac{K(m+1)}{(1-\overline{L})(1-L)\Gamma(\alpha)}(t-s)^{\alpha-1}ds\right)\right]$$
$$\leq c_\omega\epsilon(\psi + \omega(t)) ,$$

where

$$c_\omega = \frac{(m+(m+1)\lambda_\omega)}{1-\overline{L}}\left[\prod_{i=1}^{m}\left(1+\frac{\widetilde{l}}{1-\overline{L}}\right)\exp\left(\frac{K(m+1)T^\alpha}{(1-\overline{L})(1-L)\Gamma(\alpha+1)}\right)\right]$$
$$= \frac{(m+(m+1)\lambda_\omega)}{1-\overline{L}}\left[\left(1+\frac{\widetilde{l}}{1-\overline{L}}\right)\exp\left(\frac{K(m+1)T^\alpha}{(1-\overline{L})(1-L)\Gamma(\alpha+1)}\right)\right]^m .$$

Thus, problem (3.62)–(3.63) is Ulam–Hyers–Rassias stable with respect to $(\omega, \psi)$. □

Next we present the following Ulam–Hyers stability result.

**Theorem 3.36.** *Assume* (3.32.1)–(3.32.3) *and* (3.69) *hold. If* $\overline{L} < 1$, *then problem* (3.62)–(3.63) *is Ulam–Hyers stable.*

*Proof.* Let $z \in \Omega$ be a solution of inequality (3.74). Denote by $y$ the unique solution of the problem

$$\begin{cases} {}^cD^\alpha_{t_k}[y(t)-\phi(t,y_t)] = f(t,y_t,{}^cD^\alpha_{t_k}y(t)), & t \in (t_k,t_{k+1}], k=1,\dots,m , \\ \Delta y|_{t=t_k} = I_k(y_{t_k^-}), & k=1,\dots,m , \\ y(t)=z(t)=\varphi(t), & t\in[-r,0] . \end{cases}$$

From the proof of Theorem 3.35 we get the inequality

$$v_1(t) \le \sum_{0<t_i<t}\frac{\widetilde{l}}{1-\overline{L}}v_1(t_i^-) + \frac{m\epsilon}{1-\overline{L}} + \frac{T^\alpha\epsilon(m+1)}{(1-\overline{L})\Gamma(\alpha+1)}$$
$$+ \frac{K(m+1)}{(1-\overline{L})(1-L)\Gamma(\alpha)}\int_0^t (t-s)^{\alpha-1}v_1(s)ds .$$

Applying Lemma 1.53, we get

$$v_1(t) \le \epsilon\left(\frac{m\Gamma(\alpha+1)+T^\alpha(m+1)}{(1-\overline{L})\Gamma(\alpha+1)}\right)$$
$$\times\left[\prod_{0<t_i<t}\left(1+\frac{\widetilde{l}}{1-\overline{L}}\right)\exp\left(\int_0^t\frac{K(m+1)}{(1-\overline{L})(1-L)\Gamma(\alpha)}(t-s)^{\alpha-1}ds\right)\right]$$
$$\le c\epsilon ,$$

where

$$c = \left(\frac{m\Gamma(\alpha+1)+T^\alpha(m+1)}{(1-\overline{L})\Gamma(\alpha+1)}\right)\left[\prod_{i=1}^{m}\left(1+\frac{\widetilde{l}}{1-\overline{L}}\right)\exp\left(\frac{K(m+1)T^\alpha}{(1-\overline{L})(1-L)\Gamma(\alpha+1)}\right)\right]$$
$$= \left(\frac{m\Gamma(\alpha+1)+T^\alpha(m+1)}{(1-\overline{L})\Gamma(\alpha+1)}\right)\left[\left(1+\frac{\widetilde{l}}{1-\overline{L}}\right)\exp\left(\frac{K(m+1)T^\alpha}{(1-\overline{L})(1-L)\Gamma(\alpha+1)}\right)\right]^m .$$

Moreover, if we set $\gamma(\epsilon) = c\epsilon; \gamma(0) = 0$, then problem (3.62)–(3.63) is generalized Ulam–Hyers stable. □

### 3.5.4 An Example

Consider the impulsive problem, for each $t \in J_0 \cup J_1$,

$$ {}^c D_{t_k}^{\frac{1}{2}} \left[ y(t) - \frac{te^{-t}|y_t|}{(9+e^t)(1+|y_t|)} \right] = \frac{e^{-t}}{(11+e^t)} \left[ \frac{|y_t|}{1+|y_t|} - \frac{|{}^c D_{t_k}^{\frac{1}{2}} y(t)|}{1+|{}^c D_{t_k}^{\frac{1}{2}} y(t)|} \right], \tag{3.78} $$

$$ \Delta y|_{t=\frac{1}{2}} = \frac{|y(\frac{1}{2}^-)|}{10+|y(\frac{1}{2}^-)|}, \tag{3.79} $$

$$ y(t) = \varphi(t), \quad t \in [-r, 0], \ r > 0, \tag{3.80} $$

where $\varphi \in PC([-r,0],\mathbb{R})$, $J_0 = [0, \frac{1}{2}]$, $J_1 = (\frac{1}{2}, 1]$, $t_0 = 0$, and $t_1 = \frac{1}{2}$.

For $t \in [0,1]$, $u \in PC([-r,0],\mathbb{R})$, and $v \in \mathbb{R}$, set

$$ f(t,u,v) = \frac{e^{-t}}{(11+e^t)} \left[ \frac{|u|}{1+|u|} - \frac{|v|}{1+|v|} \right] $$

and

$$ \phi(t,u) = \frac{te^{-t}|u|}{(9+e^t)(1+|u|)}. $$

Notice that $\phi(0,\varphi) = 0$ for any $\varphi \in PC([-r,0],\mathbb{R})$. Clearly, the function $f$ is jointly continuous. For each $u, \bar{u} \in PC([-r,0],\mathbb{R})$, $v, \bar{v} \in \mathbb{R}$ and $t \in [0,1]$, and we have

$$ \begin{aligned} \|f(t,u,v) - f(t,\bar{u},\bar{v})\| &\le \frac{e^{-t}}{(11+e^t)} (\|u-\bar{u}\|_{PC} + \|v-\bar{v}\|) \\ &\le \frac{1}{12}\|u-\bar{u}\|_{PC} + \frac{1}{12}\|v-\bar{v}\| \end{aligned} $$

and

$$ \|\phi(t,u) - \phi(t,\bar{u})\| \le \frac{1}{10}\|u-\bar{u}\|_{PC}. $$

Hence, condition (3.32.2) is satisfied by $K = L = \frac{1}{12}$, $\overline{L} = \frac{1}{10}$.

Let

$$ I_1(u) = \frac{|u|}{10+|u|}, \quad u \in PC([-r,0],\mathbb{R}). $$

For each $u, v \in PC([-r,0],\mathbb{R})$ we have

$$ |I_1(u) - I_1(v)| = \left| \frac{|u|}{10+|u|} - \frac{|v|}{10+|v|} \right| \le \frac{1}{10}\|u-v\|_{PC}. $$

Thus condition

$$ \begin{aligned} m\tilde{l} + \overline{L} + \frac{(m+1)KT^\alpha}{(1-L)\Gamma(\alpha+1)} &= \frac{2}{10} + \frac{\frac{1}{6}}{\left(1-\frac{1}{12}\right)\Gamma\left(\frac{3}{2}\right)} \\ &= \frac{4}{11\sqrt{\pi}} + \frac{2}{10} < 1 \end{aligned} $$

is satisfied. From Theorem 3.32, problem (3.78)–(3.80) has a unique solution on $J$.

Set, for any $t \in [0, 1]$, $\omega(t) = t$ and $\psi = 1$. Since

$$I^{\frac{1}{2}} \omega(t) = \frac{1}{\Gamma\left(\frac{1}{2}\right)} \int_0^t (t-s)^{\frac{1}{2}-1} s ds \leq \frac{2t}{\sqrt{\pi}},$$

(3.35.1) is satisfied by $\lambda_\omega = \frac{2}{\sqrt{\pi}}$. Since $\overline{L} < 1$, it follows that problem (3.78)–(3.79) is Ulam–Hyers–Rassias stable with respect to $(\omega, \psi)$.

## 3.6 Notes and Remarks

The results of Chapter 3 are taken from Benchohra et al. [90, 92]. Other results may be found in [17, 14, 41, 53, 57, 106, 124, 158].

# 4 Boundary Value Problems for Nonlinear Implicit Fractional Differential Equations

## 4.1 Introduction

In this chapter, we establish the existence and uniqueness of solutions to some boundary value problem (BVPs) for implicit fractional differential equations with $0 < \alpha \leq 1$ and $1 < \alpha \leq 2$. Then we consider the stability of solutions of other classes of BVP of implicit fractional differential equations with local and nonlocal conditions in Banach spaces.

## 4.2 BVP for NIFDE with $0 < \alpha \leq 1$

### 4.2.1 Introduction and Motivations

The purpose of this section is to establish existence and uniqueness results of solutions for a class of boundary value problem (BVP) for the implicit fractional order differential equation

$$^cD^\alpha y(t) = f(t, y(t), {}^cD^\alpha y(t))\,, \quad \text{for each } t \in J = [0, T],\, T > 0,\ 0 < \alpha \leq 1\,, \tag{4.1}$$

$$ay(0) + by(T) = c\,, \tag{4.2}$$

where $f : J \times \mathbb{R} \times \mathbb{R} \to \mathbb{R}$ is a given function and $a$, $b$, and $c$ are real constants, with $a + b \neq 0$.

We present three results for problem (4.1)–(4.2). The first one is based on the Banach contraction principle, the second one on Schauder's fixed point theorem, and the last one on the nonlinear alternative of Leray–Schauder type.

### 4.2.2 Existence of Solutions

Let us define what we mean by a solution of problem (4.1)–(4.2).

**Definition 4.1.** A function $u \in C(J)$ is said to be a solution of problem (4.1)–(4.2) if $u$ satisfies equation (4.1) and conditions (4.2) on $J$.

For the existence of solutions to problem (4.1)–(4.2), we need the following auxiliary lemma.

**Lemma 4.2** ([95]). *Let $0 < \alpha \leq 1$ and $g : J \to \mathbb{R}$ be continuous. A function $y$ is a solution of the implicit fractional boundary value problem*

$$^cD^\alpha y(t) = g(t)\,, \quad \textit{for each } t \in J,\ 0 < \alpha \leq 1\,,$$

$$ay(0) + b(T) = c\,,$$

https://doi.org/10.1515/9783110553819-004

*where $a$, $b$, and $c$ are real constants with $a + b \neq 0$, if and only if $y$ is a solution of the fractional integral equation*

$$y(t) = \frac{c}{a+b} + \frac{1}{\Gamma(\alpha)} \int_0^t (t-s)^{\alpha-1} g(s)ds - \frac{b}{(a+b)\Gamma(\alpha)} \int_0^T (T-s)^{\alpha-1} g(s)ds .$$

**Theorem 4.3.** *Make the following assumptions:*
(4.3.1) *The function $f : J \times \mathbb{R} \times \mathbb{R} \to \mathbb{R}$ is continuous.*
(4.3.2) *There exist constants $K > 0$ and $0 < L < 1$ such that*

$$|f(t, u, v) - f(t, \bar{u}, \bar{v})| \leq K|u - \bar{u}| + L|v - \bar{v}|$$

*for any $u, v, \bar{u}, \bar{v} \in \mathbb{R}$ and $t \in J$.*

*If*

$$\frac{KT^\alpha}{(1-L)\Gamma(\alpha+1)} \left(1 + \frac{|b|}{|a+b|}\right) < 1 , \tag{4.3}$$

*then there exists a unique solution for BVP (4.1)–(4.2) on $J$.*

*Proof.* Define the operator $N : C(J, \mathbb{R}) \to C(J, \mathbb{R})$ by

$$N(y)(t) = \frac{c}{a+b} + \frac{1}{\Gamma(\alpha)} \int_0^t (t-s)^{\alpha-1} g(s)ds - \frac{b}{(a+b)\Gamma(\alpha)} \int_0^T (T-s)^{\alpha-1} g(s)ds , \tag{4.4}$$

where $g \in C(J, \mathbb{R})$ satisfies the functional equation

$$g(t) = f(t, y(t), g(t)) .$$

Clearly, the fixed points of the operator $N$ are solutions of problem (4.1)–(4.2). Let $u, w \in C(J, \mathbb{R})$. Then for $t \in J$ we have

$$(Nu)(t) - (Nw)(t) = \frac{1}{\Gamma(\alpha)} \int_0^t (t-s)^{\alpha-1}(g(s) - h(s))ds - \frac{b}{(a+b)\Gamma(\alpha)} \int_0^T (T-s)^{\alpha-1}(g(s) - h(s))ds ,$$

where $g, h \in C(J, \mathbb{R})$ is such that

$$g(t) = f(t, u(t), g(t)) ,$$

and

$$h(t) = f(t, w(t), h(t)) .$$

Then, for $t \in J$,

$$|(Nu)(t) - (Nw)(t)| \leq \frac{1}{\Gamma(\alpha)} \int_0^t (t-s)^{\alpha-1} |g(s) - h(s)| ds + \frac{|b|}{|a+b|\Gamma(\alpha)} \int_0^T (T-s)^{\alpha-1} |g(s) - h(s)| ds . \quad (4.5)$$

By (4.3.2) we have

$$\begin{aligned} |g(t) - h(t)| &= |f(t, u(t), g(t)) - f(t, w(t), h(t))| \\ &\leq K|u(t) - w(t)| + L|g(t) - h(t)| . \end{aligned}$$

Thus,

$$|g(t) - h(t)| \leq \frac{K}{1-L} |u(t) - w(t)| .$$

By (4.5), for $t \in J$ we have

$$\begin{aligned} |(Nu)(t) - (Nw)(t)| &\leq \frac{K}{(1-L)\Gamma(\alpha)} \int_0^t (t-s)^{\alpha-1} |u(s) - w(s)| ds \\ &\quad + \frac{|b|K}{|a+b|(1-L)\Gamma(\alpha)} \int_0^T (T-s)^{\alpha-1} |u(s) - w(s)| ds \\ &\leq \frac{KT^\alpha}{(1-L)\Gamma(\alpha+1)} \left(1 + \frac{|b|}{|a+b|}\right) \|u - w\|_\infty . \end{aligned}$$

Then

$$\|Nu - Nw\|_\infty \leq \frac{KT^\alpha}{(1-L)\Gamma(\alpha+1)} \left(1 + \frac{|b|}{|a+b|}\right) \|u - w\|_\infty .$$

By (4.3), the operator $N$ is a contraction. Hence, by Banach's contraction principle, $N$ has a unique fixed point that is the unique solution of problem (4.1)–(4.2). □

Our next existence result is based on Schauder's fixed point theorem.

**Theorem 4.4.** *Assume* (4.3.1) *and* (4.3.2) *hold and*
(4.4.1) *There exist* $p, q, r \in C(J, \mathbb{R}_+)$ *with* $r^* = \sup_{t \in J} r(t) < 1$ *such that*

$$|f(t, u, w)| \leq p(t) + q(t)|u| + r(t)|w| \text{ for } t \in J, \text{ and } u, w \in \mathbb{R} .$$

*If*

$$q^* M \left(1 + \frac{|b|}{|a+b|}\right) < 1 , \quad (4.6)$$

*where* $q^* = \sup_{t \in J} q(t)$, *and* $M = \frac{T^\alpha}{(1-r^*)\Gamma(\alpha+1)}$, *then BVP* (4.1)–(4.2) *has at least one solution.*

*Proof.* We will show that the operator $N$ defined in (4.4) satisfies the assumptions of Schauder's fixed point theorem. The proof will be given in several steps.

*Claim 1: $N$ is continuous.* Let $\{u_n\}$ be a sequence such that $u_n \to u$ in $C(J, \mathbb{R})$. Then for each $t \in J$

$$|N(u_n)(t) - N(u)(t)| \leq \frac{1}{\Gamma(\alpha)} \int_0^t (t-s)^{\alpha-1} |g_n(s) - g(s)| ds + \frac{|b|}{|a+b|\Gamma(\alpha)} \int_0^T (T-s)^{\alpha-1} |g_n(s) - g(s)| ds\,, \tag{4.7}$$

where $g_n, g \in C(J, \mathbb{R})$ such that

$$g_n(t) = f(t, u_n(t), g_n(t))$$

and

$$g(t) = f(t, u(t), g(t))\,.$$

By (4.3.2) we have

$$\begin{aligned}|g_n(t) - g(t)| &= |f(t, u_n(t), g_n(t)) - f(t, u(t), g(t))| \\ &\leq K|u_n(t) - u(t)| + L|g_n(t) - g(t)|\,.\end{aligned}$$

Then

$$|g_n(t) - g(t)| \leq \frac{K}{1-L} |u_n(t) - u(t)|\,.$$

Since $u_n \to u$, we get $g_n(t) \to g(t)$ as $n \to \infty$ for each $t \in J$. Let $\eta > 0$ be such that, for each $t \in J$, we have $|g_n(t)| \leq \eta$ and $|g(t)| \leq \eta$; then we have

$$\begin{aligned}(t-s)^{\alpha-1}|g_n(s) - g(s)| &\leq (t-s)^{\alpha-1}[|g_n(s)| + |g(s)|] \\ &\leq 2\eta(t-s)^{\alpha-1}\,.\end{aligned}$$

For each $t \in J$, the function $s \to 2\eta(t-s)^{\alpha-1}$ is integrable on $[0, t]$; then the Lebesgue dominated convergence theorem and (4.7) imply that

$$|N(u_n)(t) - N(u)(t)| \to 0 \text{ as } n \to \infty\,.$$

Hence,

$$\|N(u_n) - N(u)\|_\infty \to 0 \text{ as } n \to \infty\,.$$

Consequently, $N$ is continuous.

Let $p^* = \sup_{t\in J} p(t)$ and

$$R \geq \frac{\frac{|c|}{|a+b|} + \left(1 + \frac{|b|}{|a+b|}\right) p^* M}{1 - \left(1 + \frac{|b|}{|a+b|}\right) q^* M}\,,$$

and define

$$D_R = \{u \in C(J, \mathbb{R}) : \|u\|_\infty \le R\} .$$

It is clear that $D_R$ is a bounded, closed, and convex subset of $C(J, \mathbb{R})$.

*Claim 2: $N(D_R) \subset D_R$.* Let $u \in D_R$; we show that $Nu \in D_R$. For each $t \in J$ we have

$$|Nu(t)| \le \frac{|c|}{|a+b|} + \frac{1}{\Gamma(\alpha)} \int_0^t (t-s)^{\alpha-1} |g(s)| ds + \frac{|b|}{|a+b|\Gamma(\alpha)} \int_0^T (T-s)^{\alpha-1} |g(s)| ds . \tag{4.8}$$

By (4.4.1), for each $t \in J$ we have

$$\begin{aligned} |g(t)| &= |f(t, u(t), g(t))| \\ &\le p(t) + q(t)|u(t)| + r(t)|g(t)| \\ &\le p(t) + q(t)R + r(t)|g(t)| \\ &\le p^* + q^*R + r^*|g(t)| . \end{aligned}$$

Then

$$|g(t)| \le \frac{p^* + q^*R}{1 - r^*} := M_1 .$$

Thus, (4.8) implies that

$$\begin{aligned} |Nu(t)| &\le \frac{|c|}{|a+b|} + \frac{(p^* + q^*R)T^\alpha}{(1-r^*)\Gamma(\alpha+1)} + \frac{|b|(p^* + q^*R)T^\alpha}{|a+b|(1-r^*)\Gamma(\alpha+1)} \\ &\le \frac{|c|}{|a+b|} + (p^* + q^*R)M + \frac{|b|(p^* + q^*R)M}{|a+b|} \\ &\le \frac{|c|}{|a+b|} + p^*M\left(1 + \frac{|b|}{|a+b|}\right) + q^*M\left(1 + \frac{|b|}{|a+b|}\right)R \\ &\le R . \end{aligned}$$

Then $N(D_R) \subset D_R$.

*Claim 3: $N(D_R)$ is relatively compact.* Let $t_1, t_2 \in J$, $t_1 < t_2$, and let $u \in D_R$. Then

$$\begin{aligned} |N(u)(t_2) - N(u)(t_1)| &= \left| \frac{1}{\Gamma(\alpha)} \int_0^{t_1} [(t_2-s)^{\alpha-1} - (t_1-s)^{\alpha-1}] g(s) ds \right. \\ &\quad \left. + \frac{1}{\Gamma(\alpha)} \int_{t1}^{t_2} [(t_2-s)^{\alpha-1} g(s) ds \right| \\ &\le \frac{M_1}{\Gamma(\alpha+1)} (t_2^\alpha - t_1^\alpha + 2(t_2-t_1)^\alpha) . \end{aligned}$$

As $t_1 \to t_2$, the right-hand side of the preceding inequality tends to zero.

As a consequence of Claims 1–3, together with the Ascoli–Arzelà theorem, we conclude that $N: C(J,\mathbb{R}) \to C(J,\mathbb{R})$ is continuous and compact. As a consequence of Schauder's fixed point theorem [149], we deduce that $N$ has a fixed point that is a solution of problem (4.1)–(4.2). □

Our next existence result is based on a nonlinear alternative of the Leray–Schauder type.

**Theorem 4.5.** *Assume* (4.3.1), (4.3.2), (4.4.1), *and* (4.6) *hold. Then BVP* (4.1)–(4.2) *has at least one solution.*

*Proof.* Consider the operator $N$ defined in (4.4). We will show that $N$ satisfies the assumptions of the Leray–Schauder fixed point theorem. The proof will be given in several claims.

*Claim 1: Clearly N is continuous.*

*Claim 2: N maps bounded sets to bounded sets in* $C(J,\mathbb{R})$. Indeed, it is enough to show that for any $\rho > 0$ there exist a positive constant $\ell$ such that for each $u \in B_\rho = \{u \in C(J,\mathbb{R}) : \|u\|_\infty \le \rho\}$ we have $\|N(u)\|_\infty \le \ell$.

For $u \in B_\rho$, we have, for each $t \in J$,

$$\begin{aligned}|(Nu)(t)| \le \frac{|c|}{|a+b|} + \frac{1}{\Gamma(\alpha)}\int_0^t (t-s)^{\alpha-1}|g(s)|ds \\ + \frac{|b|}{|a+b|\Gamma(\alpha)}\int_0^T (T-s)^{\alpha-1}|g(s)|ds\,.\end{aligned} \tag{4.9}$$

By (4.4.1), for each $t \in J$ we have

$$\begin{aligned}|g(t)| &= |f(t,u(t),g(t))| \\ &\le p(t) + q(t)|u(t)| + r(t)|g(t)| \\ &\le p(t) + q(t)\rho + r(t)|g(t)| \\ &\le p^* + q^*\rho + r^*|g(t)|\,.\end{aligned}$$

Then

$$|g(t)| \le \frac{p^* + q^*\rho}{1-r^*} := M^*\,.$$

Thus, (4.9) implies that

$$|(Nu)(t)| \le \frac{|c|}{|a+b|} + \frac{M^* T^\alpha}{\Gamma(\alpha+1)} + \frac{|b|M^* T^\alpha}{|a+b|\Gamma(\alpha+1)}\,.$$

Consequently,

$$\|N(u)\|_\infty \le \frac{|c|}{|a+b|} + \frac{M^* T^\alpha}{\Gamma(\alpha+1)} + \frac{|b|M^* T^\alpha}{|a+b|\Gamma(\alpha+1)} := l\,.$$

*Claim 3: Clearly, N maps bounded sets to equicontinuous sets of* $C(J, \mathbb{R})$.

We conclude that $N\colon C(J, \mathbb{R}) \longrightarrow C(J, \mathbb{R})$ is continuous and completely continuous.

*Claim 4: A priori bounds.* We now show that there exists an open set $U \subseteq C(J, \mathbb{R})$, with $u \neq \lambda N(u)$, for $\lambda \in (0, 1)$ and $u \in \partial U$. Let $u \in C(J, \mathbb{R})$ and $u = \lambda N(u)$ for some $0 < \lambda < 1$. Thus, for each $t \in J$ we have

$$u(t) = \lambda \frac{c}{a+b} + \frac{\lambda}{\Gamma(\alpha)} \int_0^t (t-s)^{\alpha-1} g(s)ds + \frac{\lambda b}{(a+b)\Gamma(\alpha)} \int_0^T (T-s)^{\alpha-1} |g(s)|ds \,.$$

This implies by (4.3.2) that for each $t \in J$ we have

$$\begin{aligned} |u(t)| \le \frac{|c|}{|a+b|} &+ \frac{1}{\Gamma(\alpha)} \int_0^t (t-s)^{\alpha-1} g(s)ds \\ &+ \frac{|b|}{|a+b|\Gamma(\alpha)} \int_0^T (T-s)^{\alpha-1} |g(s)|ds \,. \end{aligned} \tag{4.10}$$

Additionally, by (4.4.1), for each $t \in J$ we have

$$\begin{aligned} |g(t)| &= |f(t, u(t), g(t))| \\ &\le p(t) + q(t)|u(t)| + r(t)|g(t)| \\ &\le p^* + q^*|u(t)| + r^*|g(t)| \,. \end{aligned}$$

Thus,

$$|g(t)| \le \frac{1}{1-r^*}(p^* + q^*|u(t)|) \,.$$

Hence,

$$\begin{aligned} |u(t)| &\le \frac{|c|}{|a+b|} + \frac{p^* T^\alpha}{(1-r^*)\Gamma(\alpha+1)} \left(1 + \frac{|b|}{|a+b|}\right) \\ &\quad + \frac{q^*}{(1-r^*)\Gamma(\alpha)} \int_0^t (t-s)^{\alpha-1} |u(s)|ds \\ &\quad + \frac{|b|q^*}{(1-r^*)|a+b|\Gamma(\alpha)} \int_0^T (T-s)^{\alpha-1} |u(s)|ds \\ &\le \frac{|c|}{|a+b|} + \frac{p^* T^\alpha}{(1-r^*)\Gamma(\alpha+1)} \left(1 + \frac{|b|}{|a+b|}\right) \\ &\quad + \frac{q^* \|u\|_\infty}{(1-r^*)\Gamma(\alpha)} \int_0^t (t-s)^{\alpha-1} ds \\ &\quad + \frac{|b|q^* \|u\|_\infty}{(1-r^*)|a+b|\Gamma(\alpha)} \int_0^T (T-s)^{\alpha-1} ds \end{aligned}$$

$$\leq \frac{|c|}{|a+b|} + \frac{p^* T^\alpha}{(1-r^*)\Gamma(\alpha+1)}\left(1+\frac{|b|}{|a+b|}\right) + \frac{q^* T^\alpha}{(1-r^*)\Gamma(\alpha+1)}\left(1+\frac{|b|}{|a+b|}\right)\|u\|_\infty .$$

Then for each $t \in J$ we have

$$\|u\|_\infty \leq \frac{|c|}{|a+b|} + \frac{p^* T^\alpha}{(1-r^*)\Gamma(\alpha+1)}\left(1+\frac{|b|}{|a+b|}\right) + \frac{q^* T^\alpha}{(1-r^*)\Gamma(\alpha+1)}\left(1+\frac{|b|}{|a+b|}\right)\|u\|_\infty .$$

Thus, for each $t \in J$,

$$\|u\|_\infty \left[1-\left(1+\frac{|b|}{|a+b|}\right)q^* M\right] \leq \frac{|c|}{|a+b|} + \left(1+\frac{|b|}{|a+b|}\right)p^* M .$$

Consequently,

$$\|u\|_\infty \leq \frac{\frac{|c|}{|a+b|} + \left(1+\frac{|b|}{|a+b|}\right)p^* M}{1-\left(1+\frac{|b|}{|a+b|}\right)q^* M} := \overline{M} . \tag{4.11}$$

Let

$$U = \{u \in C(J,\mathbb{R}) : \|u\|_\infty < \overline{M}+1\} .$$

By our choice of $U$, there is no $u \in \partial U$ such that $u = \lambda N(u)$ for $\lambda \in (0,1)$. As a consequence of Leray–Schauder's theorem ([149]), we deduce that $N$ has a fixed point $u$ in $\overline{U}$ that is a solution of problem (4.1)–(4.2). □

### 4.2.3 Examples

*Example 1.* Consider the BVP

$$^cD^{\frac{1}{2}}y(t) = \frac{1}{10e^{t+2}\left(1+|y(t)|+|^cD^{\frac{1}{2}}y(t)|\right)}, \quad \text{for each } t \in [0,1] , \tag{4.12}$$

$$y(0)+y(1) = 0 . \tag{4.13}$$

Set

$$f(t,u,v) = \frac{1}{10e^{t+2}(1+|u|+|v|)}, \quad t \in [0,1],\ u,v \in \mathbb{R} .$$

Clearly, the function $f$ is jointly continuous.

For any $u, v, \bar{u}, \bar{v} \in \mathbb{R}$ and $t \in [0,1]$,

$$|f(t,u,v) - f(t,\bar{u},\bar{v})| \leq \frac{1}{10e^2}(|u-\bar{u}| + |v-\bar{v}|) .$$

Hence condition (4.3.2) is satisfied by $K = L = \frac{1}{10e^2}$. Thus, the condition

$$\frac{KT^\alpha}{(1-L)\Gamma(\alpha+1)}\left(1+\frac{|b|}{|a+b|}\right) = \frac{3}{2(10e^2-1)\Gamma\left(\frac{3}{2}\right)} = \frac{3}{(10e^2-1)\sqrt{\pi}} < 1$$

is satisfied by $a = b = T = 1$, $c = 0$, and $\alpha = \frac{1}{2}$. It follows from Theorem 4.3 that problem (4.12)–(4.13) as a unique solution on $J$.

*Example 2.* Consider the BVP

$$ {}^{c}D^{\frac{1}{2}}y(t) = \frac{\left(2 + |y(t)| + |{}^{c}D^{\frac{1}{2}}y(t)|\right)}{12e^{t+9}\left(1 + |y(t)| + |{}^{c}D^{\frac{1}{2}}y(t)|\right)}, \quad \text{for each } t \in [0,1], \tag{4.14} $$

$$ \frac{1}{2}y(0) + \frac{1}{2}y(1) = 1. \tag{4.15} $$

Set

$$ f(t,u,v) = \frac{(2 + |u| + |v|)}{12e^{t+9}(1 + |u| + |v|)}, \quad t \in [0,1],\ u, v \in \mathbb{R}. $$

Clearly, the function $f$ is jointly continuous. For any $u, v, \bar{u}, \bar{v} \in \mathbb{R}$ and $t \in [0,1]$

$$ |f(t,u,v) - f(t,\bar{u},\bar{v})| \leq \frac{1}{12e^{9}}(|u-\bar{u}| + |v-\bar{v}|). $$

Hence, condition (4.3.2) is satisfied by $K = L = \frac{1}{12e^{9}}$. Also, we have

$$ |f(t,u,v)| \leq \frac{1}{12e^{t+9}}(2 + |u| + |v|). $$

Thus, condition (4.34.1) is satisfied by $p(t) = \frac{1}{6e^{t+9}}$ and $q(t) = r(t) = \frac{1}{12e^{t+9}}$. The condition

$$ q^{*}M\left(1 + \frac{|b|}{|a+b|}\right) = \frac{3}{2(12e^{9}-1)\Gamma\left(\frac{3}{2}\right)} = \frac{3}{(12e^{9}-1)\sqrt{\pi}} < 1 $$

is satisfied by $a = b = \frac{1}{2}$, $c = T = 1$, $\alpha = \frac{1}{2}$, and $q^{*} = r^{*} = \frac{1}{12e^{9}}$.

It follows from Theorem 4.4 that problem (4.14)–(4.15) has at least one solution on $J$.

## 4.3 BVP for NIFDE with $1 < \alpha \leq 2$

### 4.3.1 Introduction and Motivations

The purpose of this section is to establish existence and uniqueness results to the following implicit fractional order differential equation:

$$ {}^{c}D^{\alpha}y(t) = f(t, y(t), {}^{c}D^{\alpha}y(t)), \quad \text{for each } t \in J = [0,T],\ T > 0,\ 1 < \alpha \leq 2, \tag{4.16} $$

$$ y(0) = y_0, \quad y(T) = y_1, \tag{4.17} $$

where $f : J \times \mathbb{R} \times \mathbb{R} \to \mathbb{R}$ is a given function and $y_0, y_1 \in \mathbb{R}$.

We present three results for problem (4.16)–(4.17). The first one is based on the Banach contraction principle, the second one on Schauder's fixed point theorem, and the last one on the nonlinear alternative of the Leray–Schauder type.

### 4.3.2 Existence of Solutions

Let us define what we mean by a solution of problem (4.16)–(4.17).

**Definition 4.6.** A function $u \in C^1(J)$ is said to be a solution of problem (4.16)–(4.17) if $u$ satisfies equation (4.16) on $J$ and conditions (4.17).

#### 4.3.2.1 Preparatory Lemmas

For the existence of solutions of problem (4.16)–(4.17), we need the following auxiliary lemma.

**Lemma 4.7.** *Let $1 < \alpha \leq 2$ and $g: J \to \mathbb{R}$ be continuous. A function $y$ is a solution of the fractional BVP*

$$\begin{aligned} {}^cD^\alpha y(t) &= f(t, y(t), {}^cD^\alpha y(t)), \quad \textit{for each, } t \in J,\ 1 < \alpha \leq 2, \\ y(0) &= y_0, \quad y(T) = y_1, \end{aligned}$$

*if and only if $y$ is a solution of the fractional integral equation*

$$y(t) = l(t) + \int_0^T G(t,s) f\left(s, l(s) + \int_0^T G(t,\tau) g(\tau) d\tau, g(s)\right) ds, \tag{4.18}$$

*where*

$$\begin{aligned} l(t) &= \left(1 - \frac{t}{T}\right) y_0 + \frac{t}{T} y_1 = y_0 + \frac{(y_1 - y_0)}{T} t, \\ {}^cD^\alpha y(t) &= g(t), \end{aligned} \tag{4.19}$$

*and*

$$G(t,s) = \frac{1}{\Gamma(\alpha)} \begin{cases} (t-s)^{\alpha-1} - \frac{t}{T}(T-s)^{\alpha-1} & \textit{if } 0 \leq s \leq t, \\ -\frac{t}{T}(T-s)^{\alpha-1} & \textit{if } t \leq s \leq T. \end{cases} \tag{4.20}$$

*Proof.* By Lemma 1.9 we reduce (4.16)–(4.17) to the equation

$$y(t) = I^\alpha g(t) + c_0 + c_1 t = \frac{1}{\Gamma(\alpha)} \int_0^t (t-s)^{\alpha-1} g(s) ds + c_0 + c_1 t$$

for some constants $c_0$, and $c_1 \in \mathbb{R}$. Conditions (4.17) give

$$c_0 = y_0, \quad c_1 = \frac{1}{T} y_T - \frac{1}{T} y_0 - \frac{1}{T\Gamma(\alpha)} \int_0^T (T-s)^{\alpha-1} g(s) ds.$$

Then the solution of (4.16)–(4.17) is given by

$$\begin{aligned} y(t) &= \frac{1}{\Gamma(\alpha)} \int_0^t (t-s)^{\alpha-1} g(s)ds - \frac{t}{T\Gamma(\alpha)} \int_0^T (T-s)^{\alpha-1} g(s)ds \\ &\quad + \left(1 - \frac{t}{T}\right) y_0 + \frac{t}{T} y_1 \\ &= \frac{1}{\Gamma(\alpha)} \Bigg[ \int_0^t [(t-s)^{\alpha-1} - \frac{t}{T}(T-s)^{\alpha-1}] g(s)ds \\ &\quad - \frac{t}{T} \int_t^T (T-s)^{\alpha-1} g(s)ds \Bigg] + \left(1 - \frac{t}{T}\right) y_0 + \frac{t}{T} y_1 . \end{aligned}$$

Hence, we get (4.18). Inversely, if $y$ satisfies (4.18), then equations (4.16)–(4.17) hold. □

**Remark 4.8.** From the expression of $G(t,s)$, it is obvious that $G(t,s)$ is continuous on $[0,T] \times [0,T]$. Use the notation

$$G^* := \sup\{|G(t,s)|,\ (t,s) \in J \times J\} .$$

We are now in a position to state and prove our existence result for problem (4.16)–(4.17) based on Banach's fixed point.

**Theorem 4.9.** *Make the following assumptions:*
(4.9.1) *The function* $f : J \times \mathbb{R} \times \mathbb{R} \to \mathbb{R}$ *is continuous.*
(4.9.2) *There exist constants* $K > 0$ *and* $0 < L < 1$ *such that*

$$|f(t,u,v) - f(t,\bar{u},\bar{v})| \leq K|u - \bar{u}| + L|v - \bar{v}|$$

*for any* $u, v, \bar{u}, \bar{v} \in \mathbb{R}$ *and* $t \in J$.

*If*

$$\frac{KTG^*}{1-L} < 1 , \tag{4.21}$$

*then there exists a unique solution for BVP* (4.16)–(4.17).

*Proof.* The proof will be given in several steps. Transform problem (4.16)–(4.17) into a fixed point problem. Define the operator $N : C(J, \mathbb{R}) \to C(J, \mathbb{R})$ by

$$N(y)(t) = l(t) + \int_0^T G(t,s)k(s)ds , \tag{4.22}$$

where $k \in C(J)$ satisfies the implicit functional equation

$$k(t) = f(t, y(t), k(t)) ,$$

and $l$ and $G$ are the functions defined by (4.19) and (4.20), respectively.

Clearly, the fixed points of the operator $N$ are solutions of problem (4.16)–(4.17). Let $u, w \in C(J)$. Then for $t \in J$ we have

$$(Nu)(t) - (Nw)(t) = \int_0^T G(t,s)(g(s) - h(s))ds\,,$$

where $g, h \in C(J)$ are such that

$$g(t) = f(t, u(t), g(t))$$

and

$$h(t) = f(t, w(t), h(t))\,.$$

Then, for $t \in J$,

$$|(Nu)(t) - (Nw)(t)| \leq \int_0^T |G(t,s)||g(s) - h(s)|ds\,. \tag{4.23}$$

By (4.9.2) we have

$$\begin{aligned}|g(t) - h(t)| &= |f(t, u(t), g(t)) - f(t, w(t), h(t))| \\ &\leq K|u(t) - w(t)| + L|g(t) - h(t)|\,.\end{aligned}$$

Thus,

$$|g(t) - h(t)| \leq \frac{K}{1-L}|u(t) - w(t)|\,.$$

By (4.23) we have

$$\begin{aligned}|(Nu)(t) - (Nw)(t)| &\leq \frac{K}{(1-L)}\int_0^T |G(t,s)||u(s) - w(s)|ds \\ &\leq \frac{KTG^*}{1-L}\|u - w\|_\infty\,.\end{aligned}$$

Then

$$\|Nu - Nw\|_\infty \leq \frac{KTG^*}{1-L}\|u - w\|_\infty\,.$$

By (4.21), the operator $N$ is a contraction. Hence, by Banach's contraction principle, $N$ has a unique fixed point that is the unique solution of problem (4.16)–(4.17). □

Our next existence result is based on Schauder's fixed point theorem.

**Theorem 4.10.** *Assume* (4.9.1) *and* (4.9.2) *hold and*

(4.10.1) *There exist* $p, q, r \in C(J, \mathbb{R}_+)$ *with* $r^* = \sup_{t\in J} r(t) < 1$ *such that*

$$|f(t,u,w)| \leq p(t) + q(t)|u| + r(t)|w| \quad \text{for } t \in J, \text{ and } u, w \in \mathbb{R}\,.$$

*If*

$$\frac{q^* T G^*}{1 - r^*} < 1\,, \tag{4.24}$$

*where* $q^* = \sup_{t \in J} q(t)$, *then BVP* (4.16)–(4.17) *has at least one solution.*

*Proof.* Consider the operator $N$ defined in (4.22). We will show that $N$ satisfies the assumptions of Schauder's fixed point theorem. The proof will be given in several steps.

*Claim 1: $N$ is continuous.* Let $\{u_n\}$ be a sequence such that $u_n \to u$ in $C(J)$. Then for each $t \in J$

$$|N(u_n)(t) - N(u)(t)| \leq \int_0^T |G(t,s)||g_n(s) - g(s)|ds\,, \tag{4.25}$$

where $g_n, g \in C(J)$ such that

$$g_n(t) = f(t, u_n(t), g_n(t))$$

and

$$g(t) = f(t, u(t), g(t))\,.$$

By (4.9.2), we have

$$\begin{aligned} |g_n(t) - g(t)| &= |f(t, u_n(t), g_n(t)) - f(t, u(t), g(t))| \\ &\leq K|u_n(t) - u(t)| + L|g_n(t) - g(t)|\,. \end{aligned}$$

Then

$$|g_n(t) - g(t)| \leq \frac{K}{1-L}|u_n(t) - u(t)|\,.$$

Since $u_n \to u$, we get $g_n(t) \to g(t)$ as $n \to \infty$ for each $t \in J$. Let $\eta > 0$ be such that, for each $t \in J$, we have $|g_n(t)| \leq \eta$ and $|g(t)| \leq \eta$. Then we have

$$\begin{aligned} |G(t,s)||g_n(s) - g(s)| &\leq |G(t,s)|[|g_n(s)| + |g(s)|] \\ &\leq 2\eta|G(t,s)|\,. \end{aligned}$$

For each $t \in J$ the function $s \to 2\eta|G(t,s)|$ is integrable on $J$. Then the Lebesgue dominated convergence theorem and (4.25) imply that

$$|N(u_n)(t) - N(u)(t)| \to 0 \text{ as } n \to \infty\,.$$

Hence,

$$\|N(u_n) - N(u)\|_\infty \to 0 \text{ as } n \to \infty\,.$$

Consequently, $N$ is continuous.

Let

$$R \geq \frac{(2|y_0| + |y_1|)(1 - r^*) + G^* T p^*}{M}\,,$$

where $M := 1 - r^* - G^* T q^*$ and $p^* = \sup_{t \in J} p(t)$. Define

$$D_R = \{u \in C(J) \colon \|u\|_\infty \leq R\}\,.$$

It is clear that $D_R$ is a bounded, closed, and convex subset of $C(J)$.

*Claim 2:* $N(D_R) \subset D_R$. Let $u \in D_R$; we show that $Nu \in D_R$. For each $t \in J$ we have

$$\begin{aligned}|Nu(t)| &\leq |l(t)| + \int_0^T |G(t,s)||g(s)|ds \\ &\leq |y_0| + |y_1 - y_0| + G^* \int_0^T |g(s)|ds \\ &\leq 2|y_0| + |y_1| + G^* \int_0^T |g(s)|ds\,, \end{aligned} \tag{4.26}$$

where $g(t) = f(t, u(t), g(t))$.

From (4.10.1), for each $t \in J$ we have

$$\begin{aligned}|g(t)| &= |f(t, u(t), g(t))| \\ &\leq p(t) + q(t)|u(t)| + r(t)|g(t)| \\ &\leq p(t) + q(t)R + r(t)|g(t)| \\ &\leq p^* + q^*R + r^*|g(t)|\,. \end{aligned}$$

Then

$$|g(t)| \leq \frac{p^* + q^*R}{1 - r^*}\,.$$

Thus, (4.26) implies that, for each $t \in J$,

$$\begin{aligned}|Nu(t)| &\leq 2|y_0| + |y_1| + \frac{p^* + q^*R}{1 - r^*} G^* T \\ &\leq R\,. \end{aligned}$$

Then $N(D_R) \subset D_R$.

*Claim 3: $N(D_R)$ is relatively compact.* Let $t_1, t_2 \in J$, $t_1 < t_2$, and let $u \in D_R$. Then

$$\begin{aligned}|N(u)(t_2) - N(u)(t_1)| &= \left| l(t_2) - l(t_1) + \int_0^T [G(t_2,s) - G(t_1,s)]g(s)ds \right| \\ &= \left| \frac{(y_1 - y_0)}{T}(t_2 - t_1) + \int_0^T [G(t_2,s) - G(t_1,s)]g(s)ds \right| \\ &\leq \left| \frac{(y_1 - y_0)}{T}(t_2 - t_1) \right| + \frac{p^* + q^*R}{1 - r^*} \left| \int_0^T [G(t_2,s) - G(t_1,s)]ds \right|\,. \end{aligned}$$

As $t_1 \to t_2$, the right-hand side of the preceding inequality tends to zero.

As a consequence of Claims 1–3, together with the Ascoli–Arzelà theorem, we conclude that $N\colon C(J) \to C(J)$ is continuous and compact. As a consequence of Schauder's fixed point theorem, we deduce that $N$ has a fixed point that is a solution of problem (4.16)–(4.17). □

Our next existence result is based on a nonlinear alternative of the Leray–Schauder type.

**Theorem 4.11.** *Assume* (4.9.1), (4.9.2), (4.10.1), *and* (4.24) *hold. Then the initial value problem (IVP)* (4.16)–(4.17) *has at least one solution.*

*Proof.* Consider the operator $N$ defined in (4.22). We will show that $N$ satisfies the assumptions of the Leray–Schauder fixed point theorem. The proof will be given in several claims.

*Claim 1: Clearly $N$ is continuous.*

*Claim 2: $N$ maps bounded sets to bounded sets in $C(J)$.* Indeed, it is enough to show that for any $\rho > 0$ there exist a positive constant $\ell$ such that for each $u \in B_\rho = \{u \in C(J, \mathbb{R}) : \|u\|_\infty \leq \rho\}$ we have $\|N(u)\|_\infty \leq \ell$.

For $u \in B_\rho$ we have, for each $t \in J$,

$$\begin{aligned}|Nu(t)| &\leq |l(t)| + \int_0^T |G(t,s)||g(s)|ds \\ &\leq |y_0| + |y_1 - y_0| + G^* \int_0^T |g(s)|ds\,.\end{aligned}$$

Then

$$|Nu(t)| \leq 2|y_0| + |y_1| + G^* \int_0^T |g(t)|ds\,. \tag{4.27}$$

By (4.10.1), for each $t \in J$ we have

$$\begin{aligned}|g(t)| &= |f(t, u(t), g(t))| \\ &\leq p(t) + q(t)|u(t)| + r(t)|g(t)| \\ &\leq p(t) + q(t)\rho + r(t)|g(t)| \\ &\leq p^* + q^*\rho + r^*|g(t)|\,.\end{aligned}$$

Then

$$|g(t)| \leq \frac{p^* + q^*\rho}{1 - r^*} := M^*\,.$$

Thus, (4.27) implies that

$$|Nu(t)| \leq 2|y_0| + |y_1| + G^* M^* T\,.$$

Thus,

$$\|Nu\|_\infty \leq 2|y_0| + |y_1| + G^* M^* T := l\,.$$

*Claim 3: Clearly, $N$ maps bounded sets to equicontinuous sets of $C(J)$.*

We conclude that $N : C(J) \longrightarrow C(J)$ is continuous and completely continuous.

*Claim 4: A priori bounds.* We now show there exists an open set $U \subseteq C(J)$ with $u \neq \lambda N(u)$ for $\lambda \in (0, 1)$ and $u \in \partial U$. Let $u \in C(J)$ and $u = \lambda N(u)$ for some $0 < \lambda < 1$. Thus, for each $t \in J$ we have

$$u(t) = \lambda l(t) + \lambda \int_0^T G(t, s)g(s)ds .$$

This implies by (3.9.2) that, for each $t \in J$, we have

$$|u(t)| \leq 2|y_0| + |y_1| + \int_0^T |G(t, s)||g(s)|ds . \tag{4.28}$$

By (4.10.1), for each $t \in J$, we have

$$\begin{aligned} |g(t)| &= |f(t, u(t), g(t))| \\ &\leq p(t) + q(t)|u(t)| + r(t)|g(t)| \\ &\leq p^* + q^*|u(t)| + r^*|g(t)| . \end{aligned}$$

Thus,

$$|g(t)| \leq \frac{1}{1 - r^*}(p^* + q^*|u(t)|) .$$

Hence,

$$\begin{aligned} |u(t)| &\leq \left(2|y_0| + |y_1| + \frac{p^* TG^*}{1 - r^*}\right) + \frac{q^* G^*}{1 - r^*} \int_0^T |u(s)|ds \\ &\leq \left(2|y_0| + |y_1| + \frac{p^* TG^*}{1 - r^*}\right) + \frac{q^* TG^*}{1 - r^*}\|u\|_\infty . \end{aligned}$$

Then

$$\|u\|_\infty \leq \left(2|y_0| + |y_1| + \frac{p^* TG^*}{1 - r^*}\right) + \frac{q^* TG^*}{1 - r^*}\|u\|_\infty .$$

Thus,

$$\|u\|_\infty \leq \frac{M_1}{1 - \frac{q^* TG^*}{1 - r^*}} := \overline{M} ,$$

where $M_1 = 2|y_0| + |y_1| + \frac{p^* TG^*}{1 - r^*}$.

Let

$$U = \{u \in C(J) : \|u\|_\infty < \overline{M} + 1\} .$$

By our choice of $U$, there is no $u \in \partial U$ such that $u = \lambda N(u)$ for $\lambda \in (0, 1)$. As a consequence of Leray–Schauder's theorem, we deduce that $N$ has a fixed point $u$ in $\overline{U}$ that is a solution of (4.16)–(4.17). □

### 4.3.3 Examples

*Example 1.* Consider the BVP

$$^{c}D^{\frac{3}{2}}y(t) = \frac{1}{3e^{t+2}\left(1 + |y(t)| + |^{c}D^{\frac{3}{2}}y(t)|\right)}, \quad \text{for each } t \in [0, 1], \tag{4.29}$$

$$y(0) = 1, \quad y(1) = 2. \tag{4.30}$$

Set

$$f(t, u, v) = \frac{1}{3e^{t+2}(1 + |u| + |v|)}, \quad t \in [0, 1], \; u, v \in \mathbb{R}.$$

Clearly, the function $f$ is jointly continuous.

For any $u, v, \bar{u}, \bar{v} \in \mathbb{R}$ and $t \in [0, 1]$

$$|f(t, u, v) - f(t, \bar{u}, \bar{v})| \leq \frac{1}{3e^2}(|u - \bar{u}| + |v - \bar{v}|).$$

Hence, condition (4.9.2) is satisfied by $K = \frac{1}{3e^2}$ and $L = \frac{1}{3e^2} < 1$.

From (4.20) the function $G$ is given by

$$G(t, s) = \frac{1}{\Gamma\left(\frac{3}{2}\right)} \begin{cases} (t - s)^{\frac{1}{2}} - t(1 - s)^{\frac{1}{2}} & \text{if } 0 \leq s \leq t \\ -t(1 - s)^{\frac{1}{2}} & \text{if } t \leq s \leq 1. \end{cases}$$

Clearly, $G^* < \frac{2}{\Gamma(\frac{3}{2})}$. Thus, condition

$$\frac{KTG^*}{1 - L} < 1$$

is satisfied by $T = 1$ and $\alpha = \frac{3}{2}$. It follows from Theorem 4.9 that problem (4.29)–(4.30) has a unique solution on $J$.

*Example 2.* Consider the BVP

$$^{c}D^{\frac{3}{2}}y(t) = \frac{\left(6 + |y(t)| + |^{c}D^{\frac{3}{2}}y(t)|\right)}{10e^{t+1}\left(1 + |y(t)| + |^{c}D^{\frac{3}{2}}y(t)|\right)}, \quad \text{for each } t \in [0, 1], \tag{4.31}$$

$$y(0) = 1, \quad y(1) = 2. \tag{4.32}$$

Set

$$f(t, u, v) = \frac{6 + |u| + |v|}{10e^{t+1}(1 + |u| + |v|)}, \quad t \in [0, 1], \; u, v \in \mathbb{R}.$$

Clearly, the function $f$ is jointly continuous.

For each $u, v, \bar{u}, \bar{v} \in \mathbb{R}$ and $t \in [0, 1]$,

$$|f(t, u, v) - f(t, \bar{u}, \bar{v})| \leq \frac{1}{2e}(|u - \bar{u}| + |v - \bar{v}|).$$

Hence, condition (4.9.2) is satisfied by $K = L = \frac{1}{2e}$. Also, for each $u, v, \in \mathbb{R}$ and $t \in [0, 1]$ we have

$$|f(t, u, v)| \leq \frac{1}{10e^{t+1}}(6 + |u| + |v|).$$

Thus, condition (4.10.1) is satisfied by $p(t) = \frac{3}{5e^{t+1}}$ and $q(t) = r(t) = \frac{1}{10e^{t+1}}$. Clearly, $p^* = \frac{3}{5e}$, $q^* = \frac{1}{10e}$, and $r^* = \frac{1}{10e} < 1$.

From (4.20) the function $G$ is given by

$$G(t,s) = \frac{1}{\Gamma\left(\frac{3}{2}\right)} \begin{cases} (t-s)^{\frac{1}{2}} - t(1-s)^{\frac{1}{2}} & \text{if } 0 \le s \le t \\ -t(1-s)^{\frac{1}{2}} & \text{if } t \le s \le 1 . \end{cases}$$

Clearly, $G^* < \frac{2}{\Gamma(\frac{3}{2})}$. Thus, condition

$$\frac{q^* T G^*}{1 - r^*} < 1$$

is satisfied by $T = 1$ and $\alpha = \frac{3}{2}$. It follows from Theorems 4.10 and 4.11 that problem (4.31)–(4.32) has at least one solution on $J$.

## 4.4 Stability Results for BVP for NIFDE

### 4.4.1 Introduction and Motivations

In this section, we establish some existence, uniqueness, and stability results for the implicit fractional order differential equations

$$^{c}D^{\alpha}y(t) = f(t, y(t), {}^{c}D^{\alpha}y(t)) \quad \text{for each } t \in J = [0,T],\ T > 0,\ 0 < \alpha \le 1 , \tag{4.33}$$

$$ay(0) + by(T) = c , \tag{4.34}$$

where $f : J \times \mathbb{R} \times \mathbb{R} \longrightarrow \mathbb{R}$ is a continuous function, and $a, b, c$ are real constants, with $a + b \neq 0$ and

$$^{c}D^{\alpha}y(t) = f(t, y(t), {}^{c}D^{\alpha}y(t)) , \quad \text{for each } t \in J = [0,T],\ T > 0,\ 0 < \alpha \le 1 , \tag{4.35}$$

$$y(0) + g(y) = y_0 , \tag{4.36}$$

where $f : J \times \mathbb{R} \times \mathbb{R} \to \mathbb{R}$ is a given function, $g : C(J, \mathbb{R}) \to \mathbb{R}$ is a continuous function, and $y_0 \in \mathbb{R}$.

### 4.4.2 Existence of solutions

Let us define what we mean by a solution of problem (4.33)–(4.34) and (4.35)–(4.36).

**Definition 4.12.** A function $u \in C^1(J, \mathbb{R})$ is said to be a solution of problem (4.33)–(4.34) if $u$ satisfies equation (4.33) and conditions (4.34) on $J$, and a function $y \in C^1(J, \mathbb{R})$ is called a solution of problem (4.35)–(4.36) if $y$ satisfies equation (4.35) and conditions (4.36) on $J$.

For the existence of solutions to problems (4.33)–(4.34) and (4.35)–(4.36), we need the following auxiliary lemma.

**Lemma 4.13.** *Let* $0 < \alpha \leq 1$, *and let* $h\colon [0, T] \longrightarrow \mathbb{R}$ *be a continuous function. The linear problem*

$$^{c}D^{\alpha}y(t) = h(t)\,, \quad t \in J\,, \tag{4.37}$$

$$ay(0) + by(T) = c \tag{4.38}$$

*has a unique solution given by*

$$y(t) = \frac{1}{\Gamma(\alpha)}\int_0^t (t-s)^{\alpha-1}h(s)ds - \frac{1}{a+b}\left[\frac{b}{\Gamma(\alpha)}\int_0^T (T-s)^{\alpha-1}h(s)ds - c\right]. \tag{4.39}$$

*Proof.* By the integration of formula (4.37) we obtain

$$y(t) = y_0 + \frac{1}{\Gamma(\alpha)}\int_0^t (t-s)^{\alpha-1}h(s)ds\,. \tag{4.40}$$

We use condition (4.38) to compute the constant $y_0$, so we have

$$ay(0) = ay_0 \text{ and by } y(T) = by_0 + \frac{b}{\Gamma(\alpha)}\int_0^T (T-s)^{\alpha-1}h(s)ds\,;$$

then $ay(0) + by(T) = c$. Since

$$y_0 = \frac{-1}{(a+b)}\left[\frac{b}{\Gamma(\alpha)}\int_0^T (T-s)^{\alpha-1}h(s)ds - c\right],$$

we can use this in (4.40) to obtain (4.39). □

**Lemma 4.14.** *Let* $f(t, u, v)\colon J \times \mathbb{R} \times \mathbb{R} \longrightarrow \mathbb{R}$ *be a continuous function; then problem (4.33)–(4.34) is equivalent to the problem*

$$y(t) = \tilde{A} + I^{\alpha}g(t)\,, \tag{4.41}$$

*where* $g \in C(J, \mathbb{R})$ *satisfies the functional equation*

$$g(t) = f(t, \tilde{A} + I^{\alpha}g(t), g(t))$$

*and*

$$\tilde{A} = \frac{1}{a+b}\left[c - \frac{b}{\Gamma(\alpha)}\int_0^T (T-s)^{\alpha-1}g(s)ds\right].$$

*Proof.* Let $y$ be a solution of (4.41). We will show that $y$ is a solution of (4.33)–(4.34). We have

$$y(t) = \tilde{A} + I^\alpha g(t)\,.$$

Thus, $y(0) = \tilde{A}$ and $y(T) = \tilde{A} + \frac{1}{\Gamma(\alpha)}\int_0^T (T-s)^{\alpha-1} g(s)ds$, so

$$\begin{aligned} ay(0) + by(T) &= \frac{-ab}{(a+b)\Gamma(\alpha)}\int_0^T (T-s)^{\alpha-1} g(s)ds \\ &\quad + \frac{ac}{a+b} - \frac{b^2}{(a+b)\Gamma(\alpha)}\int_0^T (T-s)^{\alpha-1} g(s)ds \\ &\quad + \frac{bc}{a+b} + \frac{b}{\Gamma(\alpha)}\int_0^T (T-s)^{\alpha-1} g(s)ds\,. \\ &= c\,. \end{aligned}$$

Then

$$ay(0) + by(T) = c\,.$$

On the other hand, we have

$$\begin{aligned} {}^cD^\alpha y(t) &= {}^cD^\alpha(\tilde{A} + I^\alpha g(t)) = g(t) \\ &= f(t, y(t), {}^cD^\alpha y(t))\,. \end{aligned}$$

Thus, $y$ is a solution of problem (4.33)–(4.34). □

**Lemma 4.15.** *Let $0 < \alpha \le 1$, and let $h\colon [0, T] \longrightarrow \mathbb{R}$ be a continuous function. The linear problem*

$$\begin{aligned} {}^cD^\alpha y(t) &= h(t)\,, \quad t \in J \\ y(0) + g(y) &= y_0 \end{aligned}$$

*has a unique solution given by*

$$y(t) = y_0 - g(y) + \frac{1}{\Gamma(\alpha)}\int_0^t (t-s)^{\alpha-1} h(s)ds\,.$$

**Lemma 4.16.** *Let $f\colon J\times\mathbb{R}\times\mathbb{R} \longrightarrow \mathbb{R}$ be a continuous function; then problem (4.35)–(4.36) is equivalent to the problem*

$$y(t) = y_0 - g(y) + I^\alpha K_y(t)\,,$$

*where $K_y(t) = f(t, y(t), K_y(t))$.*

**Theorem 4.17.** *Make the following assumption:*
(4.17.1) *There exist two constants* $K > 0$ *and* $0 < L < 1$ *such that*

$$\|f(t,u,v) - f(t,\overline{u},\overline{v})\| \le K\|u-\overline{u}\| + L\|v-\overline{v}\| \quad \text{for each } t \in J \text{ and } u,\overline{u},v,\overline{v} \in \mathbb{R}\,.$$

*If*

$$\frac{KT^\alpha}{(1-L)\Gamma(\alpha+1)}\left(1+\frac{\|b\|}{\|a+b\|}\right) < 1\,, \tag{4.42}$$

*then problem (4.33)–(4.34) has a unique solution.*

*Proof.* Let $N$ be the operator defined by

$$N\colon C(J,\mathbb{R}) \longrightarrow C(J,\mathbb{R})$$

$$Ny(t) = \tilde{A}_y + \frac{1}{\Gamma(\alpha)}\int_0^t (t-s)^{\alpha-1} g_y(s)ds\,,$$

where

$$g_y(t) = f(t, \tilde{A}_y + I^\alpha g_y(t), g_y(t))$$

and

$$\tilde{A}_y = \frac{1}{a+b}\left[c - \frac{b}{\Gamma(\alpha)}\int_0^T (T-s)^{\alpha-1} g_y(s)ds\right].$$

By Lemma 4.14, it is clear that the fixed points of $N$ are solutions of (4.33)–(4.34). Let $y_1, y_2 \in C(J,\mathbb{R})$ and $t \in J$; then we have

$$\begin{aligned}\|Ny_1(t) - Ny_2(t)\| &\le \frac{1}{\Gamma(\alpha)}\int_0^t (t-s)^{\alpha-1}\|g_{y_1}(s) - g_{y_2}(s)\|ds \\ &\quad + \frac{\|b\|}{\|a+b\|\Gamma(\alpha)}\int_0^T (T-s)^{\alpha-1}\|g_{y_1}(s) - g_{y_2}(s)\|ds\end{aligned} \tag{4.43}$$

and

$$\begin{aligned}\|g_{y_1}(t) - g_{y_2}(t)\| &= \|f(t, y_1(t), {}^cD^\alpha y_1(t)) - f(t, y_2(t), {}^cD^\alpha y_2(t))\| \\ &\le K\|y_1(t) - y_2(t)\| + L\|g_{y_1}(t) - g_{y_2}(t)\|\,.\end{aligned}$$

Then

$$\|g_{y_1}(t) - g_{y_2}(t)\| \le \frac{K}{1-L}\|y_1(t) - y_2(t)\|\,. \tag{4.44}$$

By replacing (4.44) in inequality (4.43), we obtain

$$\begin{aligned}\|Ny_1(t) - Ny_2(t)\| &\le \frac{K}{(1-L)\Gamma(\alpha)}\int_0^t (t-s)^{\alpha-1}\|y_1(s) - y_2(s)\|ds \\ &\quad + \frac{\|b\|K}{(1-L)\|a+b\|\Gamma(\alpha)}\int_0^T (T-s)^{\alpha-1}\|y_1(s) - y_2(s)\|ds \\ &\le \frac{KT^\alpha}{(1-L)\Gamma(\alpha+1)}\|y_1 - y_2\|_\infty \\ &\quad + \frac{\|b\|KT^\alpha}{(1-L)\|a+b\|\Gamma(\alpha+1)}\|y_1 - y_2\|_\infty\,.\end{aligned}$$

Then

$$\|Ny_1 - Ny_2\|_\infty \le \left[\frac{KT^\alpha}{(1-L)\Gamma(\alpha+1)}\left(1 + \frac{\|b\|}{\|a+b\|}\right)\right]\|y_1 - y_2\|_\infty\,.$$

From (4.42), the operator $N$ has a unique fixed point that is the unique solution. □

**Theorem 4.18.** *Make the following assumption:*
(4.18.1) *There exist* $K > 0$, $0 < \overline{K} < 1$, *and* $0 < L < 1$ *such that*

$$\|f(t,u,v) - f(t,\overline{u},\overline{v})\| \le K\|u - \overline{u}\| + \overline{K}\|v - \overline{v}\| \quad \text{for any } u,\overline{u},v,\overline{v} \in \mathbb{R}$$

*and*

$$\|g(y) - g(\overline{y})\| \le L\|y - \overline{y}\| \quad \text{for any } y,\overline{y} \in C(J,\mathbb{R})\,.$$

*If*

$$L + \frac{KT^\alpha}{(1-\overline{K})\Gamma(\alpha+1)} < 1\,, \tag{4.45}$$

*then BVP* (4.35)–(4.36) *has a unique solution on J.*

*Proof.* Let

$$N\colon C(J,\mathbb{R}) \to C(J,\mathbb{R})$$

$$Ny(t) = y_0 - g(y) + \frac{1}{\Gamma(\alpha)}\int_0^t (t-s)^{\alpha-1}K_y(s)ds\,,$$

where

$$K_y(t) = f(t, y(t), K_y(t))\,.$$

By Lemma 4.16, it is easy to see that the fixed points of $N$ are the solutions of problem (4.35)–(4.36). Let $y_1, y_2 \in C(J,\mathbb{R})$; for any $t \in J$ we have

$$\|Ny_1(t) - Ny_2(t)\| \le \|g(y_1) - g(y_2)\| + \frac{1}{\Gamma(\alpha)}\int_0^t (t-s)^{\alpha-1}\|K_{y_1}(s) - K_{y_2}(s).\|ds\,.$$

Then

$$\|Ny_1(t) - Ny_2(t)\| \le L\|y_1(t) - y_2(t)\| + \frac{1}{\Gamma(\alpha)} \int_0^t (t-s)^{\alpha-1} \|K_{y_1}(s) - K_{y_2}(s)\| ds . \tag{4.46}$$

On the other hand, for every $t \in J$ we have

$$\begin{aligned} \|K_{y_1}(t) - K_{y_2}(t)\| &= \|f(t, y_1(t), K_{y_1}(t)) - f(t, y_2(t), K_{y_2}(t))\| \\ &\le K\|y_1(t) - y_2(t)\| + \overline{K}\|K_{y_1}(t) - K_{y_2}(t)\| . \end{aligned}$$

Thus,

$$\|K_{y_1}(t) - K_{y_2}(t)\| \le \frac{K}{1-\overline{K}} \|y_1(t) - y_2(t)\| . \tag{4.47}$$

Replacing (4.47) in inequality (4.46), we obtain

$$\begin{aligned} \|Ny_1(t) - Ny_2(t)\| &\le L\|y_1(t) - y_2(t)\| + \frac{K}{\left(1-\overline{K}\right)\Gamma(\alpha)} \int_0^t (t-s)^{\alpha-1} \|y_1(s) - y_2(s)\| \\ &\le \left[L + \frac{KT^\alpha}{\left(1-\overline{K}\right)\Gamma(\alpha+1)}\right] \|y_1 - y_2\|_\infty . \end{aligned}$$

Then

$$\|Ny_1 - Ny_2\|_\infty \le \left[L + \frac{KT^\alpha}{\left(1-\overline{K}\right)\Gamma(\alpha+1)}\right] \|y_1 - y_2\|_\infty .$$

Thus, $N$ is a contraction. Hence, the operator $N$ has a unique fixed point that is the unique solution of problem (4.35)–(4.36). □

### 4.4.3 Ulam–Hyers–Rassias stability

**Definition 4.19.** A solution of the implicit differential inequality

$$\|{}^cD^\alpha z(t) - f(t, z(t), {}^cD^\alpha z(t))\| \le \epsilon , \quad t \in J ,$$

with fractional order is called a fractional $\epsilon$-solution of the implicit fractional differential equation (4.33).

**Theorem 4.20.** *Assume* (4.17.1) *and* (4.42) *hold; then problem* (4.33)–(4.34) *is Ulam–Hyers stable.*

*Proof.* Let $\epsilon > 0$ and $z \in C^1(J, \mathbb{R})$ be a function that satisfies the inequality

$$\|^c D^\alpha z(t) - f(t, z(t), {}^c D^\alpha z(t))\| \le \epsilon \text{ for any } t \in J\,, \tag{4.48}$$

and let $y \in C(J, \mathbb{R})$ be the unique solution of the Cauchy problem

$$\begin{cases} {}^c D^\alpha y(t) = f(t, y(t), {}^c D^\alpha y(t)); & t \in J;\ 0 < \alpha \le 1 \\ y(0) = z(0)\,, \quad y(T) = z(T)\,. \end{cases}$$

Using Lemma 4.14, we obtain

$$y(t) = \tilde{A}_y + \frac{1}{\Gamma(\alpha)} \int_0^t (t-s)^{\alpha-1} g_y(s) ds\,.$$

On the other hand, if $y(T) = z(T)$ and $y(0) = z(0)$, then $\tilde{A}_y = \tilde{A}_z$. Indeed,

$$\|\tilde{A}_y - \tilde{A}_z\| \le \frac{\|b\|}{\|a+b\|\Gamma(\alpha)} \int_0^T (T-s)^{\alpha-1} \|g_y(s) - g_z(s)\| ds\,,$$

and by inequality (4.44) we find

$$\begin{aligned} \|\tilde{A}_y - \tilde{A}_z\| &\le \frac{\|b\|K}{(1-L)\|a+b\|\Gamma(\alpha)} \int_0^T (T-s)^{\alpha-1} \|y(s) - z(s)\| ds \\ &= \frac{\|b\|K}{(1-L)\|a+b\|} I^\alpha \|y(T) - z(T)\| = 0\,. \end{aligned}$$

Thus,

$$\tilde{A}_y = \tilde{A}_z\,.$$

Hence, we have

$$y(t) = \tilde{A}_z + \frac{1}{\Gamma(\alpha)} \int_0^t (t-s)^{\alpha-1} g_y(s) ds\,.$$

By integration of inequality (4.48), we obtain

$$\|z(t) - \tilde{A}_z - \frac{1}{\Gamma(\alpha)} \int_0^t (t-s)^{\alpha-1} g_z(s) ds\| \le \frac{\epsilon t^\alpha}{\Gamma(\alpha+1)} \le \frac{\epsilon T^\alpha}{\Gamma(\alpha+1)}\,,$$

with

$$g_z(t) = f(t, \tilde{A}_z + I^\alpha g_z(t), g_z(t))\,.$$

We have for any $t \in J$

$$\begin{aligned}\|z(t)-y(t)\| &= \left| z(t)-\tilde{A}_z - \frac{1}{\Gamma(\alpha)}\int_0^t (t-s)^{\alpha-1} g_z(s)ds \right. \\ &\quad \left. + \frac{1}{\Gamma(\alpha)}\int_0^t (t-s)^{\alpha-1}\left(g_z(s)-g_y(s)\right) ds \right| \\ &\le \|z(t)-\tilde{A}_z - \frac{1}{\Gamma(\alpha)}\int_0^t (t-s)^{\alpha-1} g_z(s)ds\| \\ &\quad + \frac{1}{\Gamma(\alpha)}\int_0^t (t-s)^{\alpha-1}\|g_z(s)-g_y(s)\|ds\,.\end{aligned}$$

Using (4.44), we obtain

$$\|z(t)-y(t)\| \le \frac{\epsilon T^\alpha}{\Gamma(\alpha+1)} + \frac{K}{(1-L)\Gamma(\alpha)}\int_0^t (t-s)^{\alpha-1}\|z(s)-y(s)\|ds\,,$$

and by Gronwall's lemma we get

$$\|z(t)-y(t)\| \le \frac{\epsilon T^\alpha}{\Gamma(\alpha+1)}\left[1+\frac{\gamma K T^\alpha}{(1-L)\Gamma(\alpha+1)}\right] := c\epsilon\,,$$

where $\gamma = \gamma(\alpha)$ is a constant. Moreover, if we set $\psi(\epsilon) = c\epsilon$, $\psi(0) = 0$, then problem (4.33)–(4.34) is generalized Ulam–Hyers stable. □

**Theorem 4.21.** *Assume* (4.17.1) *and* (4.42) *hold and*

(4.27.1) *there exists an increasing function* $\varphi \in C(J, \mathbb{R}_+)$, *and there exists* $\lambda_\varphi > 0$ *such that for any* $t \in J$,

$$I^\alpha \varphi(t) \le \lambda_\varphi \varphi(t)\,.$$

*Then problem* (4.33)–(4.34) *is Ulam–Hyers–Rassias stable.*

*Proof.* Let $z \in C^1(J, \mathbb{R})$ be a solution of the inequality

$$\|{}^cD^\alpha z(t) - f(t, z(t), {}^cD^\alpha z(t))\| \le \epsilon\varphi(t)\,, \quad t \in J,\ \epsilon > 0\,, \tag{4.49}$$

and let $y \in C(J, \mathbb{R})$ be the unique solution of the Cauchy problem

$$\begin{cases} {}^cD^\alpha y(t) = f(t, y(t), {}^cD^\alpha y(t)); & t \in J;\ 0 < \alpha \le 1 \\ y(0) = z(0)\,, \quad y(T) = z(T)\,. \end{cases}$$

It follows from the proof of the previous theorem that

$$y(t) = \tilde{A}_z + \frac{1}{\Gamma(\alpha)}\int_0^t (t-s)^{\alpha-1} g_y(s)ds\,.$$

By integration of (4.49), we obtain

$$\|z(t) - \tilde{A}_z - \frac{1}{\Gamma(\alpha)}\int_0^t (t-s)^{\alpha-1} g_z(s)ds\| \le \frac{\epsilon}{\Gamma(\alpha)}\int_0^t (t-s)^{\alpha-1}\varphi(s)ds$$
$$\le \epsilon\lambda_\varphi \varphi(t)\,.$$

On the other hand, we have

$$\begin{aligned}\|z(t) - y(t)\| &= \Bigg| z(t) - \tilde{A}_z - \frac{1}{\Gamma(\alpha)}\int_0^t (t-s)^{\alpha-1} g_z(s)ds \\ &\quad + \frac{1}{\Gamma(\alpha)}\int_0^t (t-s)^{\alpha-1}\left(g_z(s) - g_y(s)\right) ds\Bigg| \\ &\le \|z(t) - \tilde{A}_z - \frac{1}{\Gamma(\alpha)}\int_0^t (t-s)^{\alpha-1} g_z(s)ds\| \\ &\quad + \frac{1}{\Gamma(\alpha)}\int_0^t (t-s)^{\alpha-1}\|g_z(s) - g_y(s)\|ds\,.\end{aligned}$$

Using (4.44), we have

$$\|z(t) - y(t)\| \le \epsilon\lambda_\varphi\varphi(t) + \frac{K}{(1-L)\Gamma(\alpha)}\int_0^t (t-s)^{\alpha-1}\|z(s) - y(s)\|ds\,.$$

By applying Gronwall's lemma, we get that for any $t \in J$:

$$\|z(t) - y(t)\| \le \epsilon\lambda_\varphi\varphi(t) + \frac{\gamma_1\epsilon K\lambda_\varphi}{(1-L)\Gamma(\alpha)}\int_0^t (t-s)^{\alpha-1}\varphi(s)ds\,,$$

where $\gamma_1 = \gamma_1(\alpha)$ is constant, and by (2.27.2) we have

$$\|z(t) - y(t)\| \le \epsilon\lambda_\varphi\varphi(t) + \frac{\gamma_1\epsilon K\lambda_\varphi^2\varphi(t)}{(1-L)} = \left(1 + \frac{\gamma_1 K\lambda_\varphi}{(1-L)}\right)\epsilon\lambda_\varphi\varphi(t)\,.$$

Then for any $t \in J$

$$\|z(t) - y(t)\| \le \left[\left(1 + \frac{\gamma_1 K\lambda_\varphi}{1-L}\right)\lambda_\varphi\right]\epsilon\varphi(t) = c\epsilon\varphi(t)\,.$$ □

**Theorem 4.22.** *Assume* (4.27.1) *and* (4.45) *hold; then problem* (4.35)–(4.36) *is Ulam–Hyers stable.*

*Proof.* Let $\epsilon > 0$, and let $z \in C^1(J, \mathbb{R})$, satisfying the inequality

$$\|{}^cD^\alpha z(t) - f(t, z(t), {}^cD^\alpha z(t))\| \le \epsilon \quad \text{for every } t \in J\,, \tag{4.50}$$

and let $y \in C(J, \mathbb{R})$ be the unique solution of the Cauchy problem

$$\begin{cases} {}^cD^\alpha y(t) = f(t, y(t), {}^cD^\alpha y(t)), & t \in J,\ 0 < \alpha \le 1 \\ z(0) + g(y) = y_0 . \end{cases}$$

Thus,

$$y(t) = y_0 - g(y) + \frac{1}{\Gamma(\alpha)} \int_0^t (t-s)^{\alpha-1} K_y(s) ds ,$$

where $K_y(t) = f(t, y(t), K_y(t))$. By integration of inequality (4.50), we find

$$\|z(t) - y_0 + g(z) - \frac{1}{\Gamma(\alpha)} \int_0^t (t-s)^{\alpha-1} K_z(s) ds\| \le \frac{\epsilon T^\alpha}{\Gamma(\alpha+1)} ,$$

where $K_z(t) = f(t, z(t), K_z(t))$. For every $t \in J$ we have

$$\begin{aligned} \|z(t) - y(t)\| &\le \|z(t) - y_0 + g(z) - \frac{1}{\Gamma(\alpha)} \int_0^t (t-s)^{\alpha-1} K_z(s) ds\| \\ &\quad + \|g(y) - g(z) + \frac{1}{\Gamma(\alpha)} \int_0^t (t-s)^{\alpha-1} \left(K_z(s) - K_y(s)\right) ds\| \\ &\le \frac{\epsilon T^\alpha}{\Gamma(\alpha+1)} + \|g(z) - g(y)\| + \frac{1}{\Gamma(\alpha)} \int_0^t (t-s)^{\alpha-1} \|K_z(s) - K_y(s)\| ds . \end{aligned}$$

Using (4.47), we obtain

$$\|z(t) - y(t)\| \le \frac{\epsilon T^\alpha}{\Gamma(\alpha+1)} + L\|z(t) - y(t)\| + \frac{K}{\left(1-\overline{K}\right)\Gamma(\alpha)} \int_0^t (t-s)^{\alpha-1} \|z(s) - y(s)\| ds .$$

Thus,

$$\|z(t) - y(t)\| \le \frac{\epsilon T^\alpha}{(1-L)\Gamma(\alpha+1)} + \frac{K}{(1-L)\left(1-\overline{K}\right)\Gamma(\alpha)} \int_0^t (t-s)^{\alpha-1} \|z(s) - y(s)\| ds .$$

Using Gronwall's lemma, for every $t \in J$ we obtain

$$\|z(t) - y(t)\| \le \frac{\epsilon T^\alpha}{(1-L)\Gamma(\alpha+1)} \left[ 1 + \frac{\gamma K T^\alpha}{(1-L)\left(1-\overline{K}\right)\Gamma(\alpha+1)} \right] := c\epsilon ,$$

where $\gamma = \gamma(\alpha)$ is a constant, so problem (4.35)–(4.36) is Ulam–Hyers stable. If we set $\psi(\epsilon) = c\epsilon$; $\psi(0) = 0$, then problem (4.35)–(4.36) is generalized Ulam–Hyers stable. □

**Theorem 4.23.** *Assume that* (4.27.1) *and inequality* (4.45) *and*

(4.23.1) *there exists an increasing function* $\varphi \in C(J, \mathbb{R}_+)$, *and there exists* $\lambda_\varphi > 0$ *such that for any* $t \in J$

$$I^\alpha \varphi(t) \le \lambda_\varphi \varphi(t)$$

*are satisfied;*

*then problem* (4.35)–(4.36) *is Ulam–Hyers–Rassias stable.*

### 4.4.4 Examples

*Example 1.* Consider the BVP

$$^cD^{\frac{1}{2}}y(t) = \frac{1}{10e^{t+2}\left(1 + |y(t)| + |^cD^{\frac{1}{2}}y(t)|\right)} \quad \text{for each } t \in [0,1]\,, \tag{4.51}$$

$$y(0) + y(1) = 0\,. \tag{4.52}$$

Set

$$f(t,u,v) = \frac{1}{10e^{t+2}(1 + |u| + |v|)}\,, \quad t \in [0,1],\ u,v \in \mathbb{R}\,.$$

Clearly, the function $f$ is jointly continuous.

For any $u, v, \bar{u}, \bar{v} \in \mathbb{R}$ and $t \in [0,1]$

$$|f(t,u,v) - f(t,\bar{u},\bar{v})| \le \frac{1}{10e^2}(|u-\bar{u}| + |v-\bar{v}|)\,.$$

Hence, condition (4.27.1) is satisfied by $K = L = \frac{1}{10e^2}$.
Thus, condition

$$\frac{KT^\alpha}{(1-L)\Gamma(\alpha+1)}\left(1 + \frac{|b|}{|a+b|}\right) = \frac{3}{2(10e^2-1)\Gamma\left(\frac{3}{2}\right)} = \frac{3}{(10e^2-1)\sqrt{\pi}} < 1$$

is satisfied by $a = b = T = 1$, $c = 0$, and $\alpha = \frac{1}{2}$. From Theorem 4.17, problem (4.51)–(4.52) has a unique solution on $J$, and Theorem 4.20 implies that problem (4.51)–(4.52) is Ulam–Hyers stable.

*Example 2.* Consider the BVP

$$^cD^{\frac{1}{2}}y(t) = \frac{e^{-t}}{(9+e^t)}\left[\frac{\|y(t)\|}{1+\|y(t)\|} - \frac{\|^cD^{\frac{1}{2}}y(t)\|}{1+\|^cD^{\frac{1}{2}}y(t)\|}\right]\,, \quad t \in J = [0,1]\,, \tag{4.53}$$

$$y(0) + \sum_{i=1}^{n} c_i y(t_i) = 1\,, \tag{4.54}$$

where $0 < t_1 < t_2 < \cdots < t_n < 1$ and $c_i = 1, \ldots, n$ are positive constants, with

$$\sum_{i=1}^{n} c_i < \frac{1}{3}\,.$$

Set

$$f(t,u,v) = \frac{e^{-t}}{(9+e^t)}\left[\frac{u}{1+u} - \frac{v}{1+v}\right]\,, \quad t \in [0,1],\ u,v \in [0,+\infty)\,.$$

Clearly, the function $f$ is continuous. For each $u, \bar{u}, v, \bar{v} \in \mathbb{R}$ and $t \in [0,1]$

$$\begin{aligned}\|f(t,u,v) - f(t,\bar{u},\bar{v})\| &\le \frac{e^{-t}}{(9+e^t)}(\|u-\bar{u}\| + \|v-\bar{v}\|)\\ &\le \frac{1}{10}\|u-\bar{u}\| + \frac{1}{10}\|v-\bar{v}\|\,.\end{aligned}$$

On the other hand, we have

$$\begin{aligned}\|g(u)-g(\bar{u})\| &= \|\sum_{i=1}^{n} c_i u - \sum_{i=1}^{n} c_i \bar{u}\| \\ &\le \sum_{i=1}^{n} c_i \|u-\bar{u}\| \\ &< \frac{1}{3}\|u-\bar{u}\| .\end{aligned}$$

Hence, condition (4.27.1) is satisfied by $K = \overline{K} = \frac{1}{10}$ and $L = \frac{1}{3}$. We have

$$L + \frac{KT^{\alpha}}{\left(1-\overline{K}\right)\Gamma(\alpha+1)} = \frac{1}{3} + \frac{1}{9\Gamma\left(\frac{3}{2}\right)} = \frac{9\sqrt{\pi}+6}{27\sqrt{\pi}} < 1 .$$

From Theorem 4.18, problem (4.53)–(4.54) has a unique solution on $J$, and Theorem 4.22 implies that problem (4.53)–(4.54) is Ulam–Hyers stable.

## 4.5 BVP for NIFDE in Banach Space

### 4.5.1 Introduction and Motivations

Recently, fractional differential equations have been studied by Abbas et al. [35, 43], Baleanu et al. [78, 80], Diethelm [137], Kilbas and Marzan [180], Srivastava et al. [181], Lakshmikantham et al. [187], and Samko et al. [239]. More recently, some mathematicians have considered BVPs and boundary conditions for implicit fractional differential equations.

In [164], Hu and Wang investigated the existence of a solutions to a nonlinear fractional differential equation with an integral boundary condition:

$$\begin{aligned}&D^{\alpha}u(t) = f(t, u(t), D^{\beta}u(t)) , \quad t \in (0,1),\ 1<\alpha\le 2,\ 0<\beta<1 , \\ &u(0) = u_0 , \quad u(1) = \int_0^1 g(s)u(s)ds ,\end{aligned}$$

where $f\colon [0,1]\times\mathbb{R}\times\mathbb{R} \to \mathbb{R}$ is a continuous function and $g$ is an integrable function. In [241], by means of Schauder's fixed point theorem, Su and Liu studied the existence of nonlinear fractional BVPs involving Caputo's derivative:

$$\begin{aligned}&{}^{c}D^{\alpha}u(t) = f(t, u(t), {}^{c}D^{\beta}u(t)) , \quad \text{for each } t \in (0,1),\ 1<\alpha\le 2,\ 0<\beta\le 1 , \\ &u(0) = u'(1) = 0, \ \text{ or } \ u'(1) = u(1) = 0, \ \text{ or } \ u(0) = u(1) = 0 ,\end{aligned}$$

where $f\colon [0,1]\times\mathbb{R}\times\mathbb{R} \to \mathbb{R}$ is a continuous function.

Many techniques have been developed for studying the existence and uniqueness of solutions of initial and BVPs for fractional differential equations. Several authors tried to develop a technique that depends on the Darbo or the Mönch fixed point theorem with the Hausdorff or Kuratowski measure of noncompactness. The notion of the measure of noncompactness was defined in many ways. In 1930, Kuratowski [185] defined the measure of noncompactness, $\alpha(A)$, of a bounded subset $A$ of a metric space $(X, d)$, and in 1955 Darbo [132] introduced a new type of fixed point theorem for set contractions.

The purpose of this section is to establish existence and uniqueness results for problems of implicit fractional differential equations in Banach space:

$$
\begin{gathered}
{}^cD^\nu y(t) = f(t, y(t), {}^cD^\nu y(t)), \quad \text{for each, } t \in J := [0, T],\ T > 0,\ 0 < \nu \leq 1\,, \\
ay(0) + by(T) = c\,,
\end{gathered}
$$

where $(E, \|\cdot\|)$ is a real Banach space, $f: J \times E \times E \to E$ is a given function, and $a, b$ are real, with $a + b \neq 0$, $c \in E$, and

$$
\begin{gathered}
{}^cD^\nu y(t) = f(t, y(t), {}^cD^\nu y(t)), \quad \text{for every } t \in J := [0, T], T > 0,\ 0 < \nu \leq 1\,, \\
y(0) + g(y) = y_0\,,
\end{gathered}
$$

where $f: J \times E \times E \to E$ is a given function, $g: C(J, E) \to E$ is a continuous function, and $y_0 \in E$. The results of this section are based on Darbo's fixed point theorem combined with the technique of measures of noncompactness and on Mönch's fixed point theorem.

### 4.5.2 Existence Results for BVPs in Banach Space

The purpose of this section is to establish sufficient conditions for the existence of solutions to the problem of implicit fractional differential equations with a Caputo fractional derivative:

$$
{}^cD^\nu y(t) = f(t, y(t), {}^cD^\nu y(t)), \quad \text{for each, } t \in J := [0, T],\ T > 0,\ 0 < \nu \leq 1\,, \tag{4.55}
$$

$$
ay(0) + by(T) = c\,, \tag{4.56}
$$

where $f: J \times E \times E \to E$ is a given function and $a, b$ are real, with $a + b \neq 0$ and $c \in E$.

For a given set $V$ of functions $v: J \to E$ let us use the notation

$$
V(t) = \{v(t), v \in V\}\,, \quad t \in J
$$

and

$$
V(J) = \{v(t) : v \in V,\ t \in J\}\,.
$$

Let us define what we mean by a solution of problem (4.55)–(4.56).

**Definition 4.24.** A function $y \in C^1(J, E)$ is said to be a solution of problem (4.55)–(4.56) if $y$ satisfies equation (4.55) on $J$ and conditions (4.56).

For the existence of solutions of problem (4.55)–(4.56), we need the following auxiliary lemma.

**Lemma 4.25** ([79]). *Let $0 < \nu \le 1$, and let $h\colon [0,T] \longrightarrow E$ be a continuous function. The linear problem*

$$
\begin{aligned}
{}^cD^\nu y(t) &= h(t)\,, \quad t \in J\,,\\
ay(0) + by(T) &= c\,,
\end{aligned}
$$

*has a unique solution given by*

$$
\begin{aligned}
y(t) = {} & \frac{1}{\Gamma(\nu)} \int_0^t (t-s)^{\nu-1} h(s)ds \\
& - \frac{1}{a+b}\left[\frac{b}{\Gamma(\nu)} \int_0^T (T-s)^{\nu-1} h(s)ds - c\right].
\end{aligned}
$$

**Lemma 4.26.** *Let $f(t,u,v)\colon J \times E \times E \longrightarrow E$ be a continuous function; then problem (4.55)–(4.56) is equivalent to the problem*

$$y(t) = \tilde{A} + I^\nu g(t)\,, \tag{4.57}$$

*where $g \in C(J,E)$ satisfies the functional equation*

$$g(t) = f(t, \tilde{A} + I^\nu g(t), g(t))$$

*and*

$$\tilde{A} = \frac{1}{a+b}\left[c - \frac{b}{\Gamma(\nu)} \int_0^T (T-s)^{\nu-1} g(s)ds\right].$$

*Proof.* Let $y$ be a solution of (4.57). We will show that $y$ is a solution of (4.55)–(4.56). We have

$$y(t) = \tilde{A} + I^\nu g(t)\,.$$

Thus, $y(0) = \tilde{A}$ and $y(T) = \tilde{A} + \frac{1}{\Gamma(\nu)} \int_0^T (T-s)^{\nu-1} g(s)ds$, so

$$
\begin{aligned}
ay(0) + by(T) = {} & \frac{-ab}{(a+b)\Gamma(\nu)} \int_0^T (T-s)^{\alpha-1} g(s)ds \\
& + \frac{ac}{a+b} - \frac{b^2}{(a+b)\Gamma(\nu)} \int_0^T (T-s)^{\nu-1} g(s)ds \\
& + \frac{bc}{a+b} + \frac{b}{\Gamma(\nu)} \int_0^T (T-s)^{\nu-1} g(s)ds\,. \\
= {} & c\,.
\end{aligned}
$$

Then

$$ay(0) + by(T) = c .$$

On the other hand, we have

$$\begin{aligned} {}^cD^\nu y(t) &= {}^cD^\nu(\tilde{A} + I^\nu g(t)) = g(t) \\ &= f(t, y(t), {}^cD^\nu y(t)) . \end{aligned}$$

Thus, $y$ is a solution of problem (4.55)–(4.56). □

Let us list the conditions:

(4.33.1) The function $f: J \times E \times E \to E$ is continuous.

(4.33.2) There exist constants $K > 0$ and $0 < L < 1$ such that

$$\|f(t, u, v) - f(t, \bar{u}, \bar{v})\| \le K\|u - \bar{u}\| + L\|v - \bar{v}\|$$

for any $u, v, \bar{u}, \bar{v} \in E$, and $t \in J$.

We are now in a position to state and prove our existence result for problem (4.55)–(4.56) based on the concept of measures of noncompactness and Darbo's fixed point theorem.

**Remark 4.27** ([66]). Condition (4.33.2) is equivalent to the inequality

$$\alpha\,(f(t, B_1, B_2)) \le K\alpha(B_1) + L\alpha(B_2)$$

for any bounded sets $B_1, B_2 \subseteq E$ and for each $t \in J$.

**Theorem 4.28.** *Assume that* (4.33.1) *and* (4.33.2) *hold. If*

$$\frac{(|b| + |a + b|)T^\nu K}{|a + b|\Gamma(\nu + 1)(1 - L)} < 1 , \tag{4.58}$$

*and*

$$\frac{KT^\nu}{(1 - L)\Gamma(\nu + 1)} < 1 , \tag{4.59}$$

*then IVP* (4.55)–(4.56) *has at least one solution on* $J$.

*Proof.* Transform problem (4.55)–(4.56) into a fixed point problem. Define the operator $N: C(J, E) \to C(J, E)$ by

$$N(y)(t) = \tilde{A} + I^\nu g(t) , \tag{4.60}$$

where $g \in C(J, E)$ satisfies the functional equation

$$g(t) = f(t, y(t), g(t))$$

and

$$\tilde{A} = \frac{1}{a + b}\left[c - \frac{b}{\Gamma(\nu)}\int_0^T (T - s)^{\nu - 1} g(s)ds\right] .$$

Clearly, the fixed points of the operator $N$ are solutions of problem (4.55)–(4.56). We will show that $N$ satisfies the assumptions of Darbo's fixed point theorem. The proof will be given in several claims.

*Claim 1: N is continuous.* Let $\{u_n\}$ be a sequence such that $u_n \to u$ in $C(J, E)$. Then for each $t \in J$

$$\|N(u_n)(t) - N(u)(t)\| \leq \frac{|b|}{|a+b|\Gamma(\nu)} \int_0^T (T-s)^{\nu-1} \|g_n(s) - g(s)\| ds + \frac{1}{\Gamma(\nu)} \int_0^t (t-s)^{\nu-1} \|g_n(s) - g(s)\| ds , \tag{4.61}$$

where $g_n, g \in C(J, E)$ such that

$$g_n(t) = f(t, u_n(t), g_n(t))$$

and

$$g(t) = f(t, u(t), g(t)) .$$

By (4.33.2), for each $t \in J$ we have

$$\begin{aligned} \|g_n(t) - g(t)\| &= \|f(t, u_n(t), g_n(t)) - f(t, u(t), g(t))\| \\ &\leq K\|u_n(t) - u(t)\| + L\|g_n(t) - g(t)\| . \end{aligned}$$

Then

$$\|g_n(t) - g(t)\| \leq \frac{K}{1-L} \|u_n(t) - u(t)\| .$$

Since $u_n \to u$, we get $g_n(t) \to g(t)$, as $n \to \infty$ for each $t \in J$.

Let $\eta > 0$ be such that, for each $t \in J$, we have $\|g_n(t)\| \leq \eta$ and $\|g(t)\| \leq \eta$. Then we have

$$\begin{aligned} (t-s)^{\nu-1} \|g_n(s) - g(s)\| &\leq (t-s)^{\nu-1} [\|g_n(s)\| + \|g(s)\|] \\ &\leq 2\eta(t-s)^{\nu-1} . \end{aligned}$$

For each $t \in J$, the function $s \to 2\eta(t-s)^{\nu-1}$ is integrable on $[0, t]$; then the Lebesgue dominated convergence theorem and (4.61) imply that

$$\|N(u_n)(t) - N(u)(t)\| \to 0 \text{ as } n \to \infty .$$

Thus,

$$\|N(u_n) - N(u)\|_\infty \to 0 \text{ as } n \to \infty .$$

Hence, $N$ is continuous.

Let $R$ be a constant such that

$$R \geq \frac{\|c\|\Gamma(\nu+1)(1-L) + (|b| + |a+b|)T^\nu f^*}{|a+b|\Gamma(\nu+1)(1-L) - (|b| + |a+b|)T^\nu K}, \quad \text{where } f^* = \sup_{t \in J} \|f(t, 0, 0)\| . \tag{4.62}$$

Define

$$D_R = \{u \in C(J, E) : \|u\|_\infty \le R\} .$$

It is clear that $D_R$ is a bounded, closed, and convex subset of $C(J, E)$.

*Claim 2:* $N(D_R) \subset D_R$. Let $u \in D_R$; we show that $Nu \in D_R$. For each $t \in J$ we have

$$\|Nu(t)\| \le \frac{\|c\|}{|a+b|} + \frac{|b|}{|a+b|\Gamma(\nu)} \int_0^T (T-s)^{\nu-1}\|g(s)\|ds + \frac{1}{\Gamma(\nu)} \int_0^t (t-s)^{\nu-1}\|g(s)\|ds . \tag{4.63}$$

By (4.33.2), for each $t \in J$ we have

$$\begin{aligned}\|g(t)\| &= \|f(t, u(t), g(t)) - f(t, 0, 0) + f(t, 0, 0)\| \\ &\le \|f(t, u(t), g(t)) - f(t, 0, 0)\| + \|f(t, 0, 0)\| \\ &\le K\|u(t)\| + L\|g(t)\| + f^* \\ &\le KR + L\|g(t)\| + f^* .\end{aligned}$$

Then

$$\|g(t)\| \le \frac{f^* + KR}{1-L} := M .$$

Thus, (4.62) and (4.63) imply that

$$\begin{aligned}\|Nu(t)\| &\le \frac{\|c\|}{|a+b|} + \left[\frac{|b|}{|a+b|} + 1\right] \frac{T^\nu}{\Gamma(\nu+1)} \left(\frac{f^* + KR}{1-L}\right) \\ &\le \frac{\|c\|}{|a+b|} + \frac{(|b| + |a+b|)T^\nu f^*}{|a+b|\Gamma(\nu+1)(1-L)} \\ &\quad + \frac{(|b| + |a+b|)T^\nu KR}{|a+b|\Gamma(\nu+1)(1-L)} \\ &\le R .\end{aligned}$$

Consequently,

$$N(D_R) \subset D_R .$$

*Claim 3: $N(D_R)$ is bounded and equicontinuous.* By Claim 2 we have $N(D_R) = \{N(u) : u \in D_R\} \subset D_R$. Thus, for each $u \in D_R$ we have $\|N(u)\|_\infty \le R$. Thus, $N(D_R)$ is bounded. Let $t_1, t_2 \in J$, $t_1 < t_2$, and let $u \in D_R$. Then

$$\begin{aligned}\|N(u)(t_2) - N(u)(t_1)\| &= \left\| \frac{1}{\Gamma(\nu)} \int_0^{t_1} [(t_2-s)^{\nu-1} - (t_1-s)^{\nu-1}]g(s)ds \right. \\ &\quad \left. + \frac{1}{\Gamma(\nu)} \int_{t1}^{t_2} (t_2-s)^{\nu-1}g(s)ds \right\| \\ &\le \frac{M}{\Gamma(\nu+1)}(t_2^\nu - t_1^\nu + 2(t_2-t_1)^\nu) .\end{aligned}$$

As $t_1 \to t_2$, the right-hand side of the preceding inequality tends to zero.

*Claim 4: The operator $N\colon D_R \to D_R$ is a strict set contraction.* Let $V \subset D_R$ and $t \in J$; then we have

$$\begin{aligned}\alpha(N(V)(t)) &= \alpha((Ny)(t), y \in V)\\ &\le \frac{1}{\Gamma(\nu)}\left\{\int_0^t (t-s)^{\nu-1}\alpha(g(s))ds, y \in V\right\}.\end{aligned}$$

Then Remark 4.27 implies that, for each $s \in J$,

$$\begin{aligned}\alpha(\{g(s), y \in V\}) &= \alpha(\{f(s, y(s), g(s)), y \in V\})\\ &\le K\alpha(\{y(s), y \in V\}) + L\alpha(\{g(s), y \in V\}).\end{aligned}$$

Thus,

$$\alpha(\{g(s), y \in V\}) \le \frac{K}{1-L}\alpha\{y(s), y \in V\}.$$

Then

$$\begin{aligned}\alpha(N(V)(t)) &\le \frac{K}{(1-L)\Gamma(\nu)}\left\{\int_0^t (t-s)^{\nu-1}\{\alpha(y(s))\}ds, y \in V\right\}\\ &\le \frac{K\alpha_c(V)}{(1-L)\Gamma(\nu)}\int_0^t (t-s)^{\nu-1}ds\\ &\le \frac{KT^\nu}{(1-L)\Gamma(\nu+1)}\alpha_c(V).\end{aligned}$$

Therefore,

$$\alpha_c(NV) \le \frac{KT^\nu}{(1-L)\Gamma(\nu+1)}\alpha_c(V).$$

So, by (4.59), the operator $N$ is a set contraction. As a consequence of Theorem 1.45, we deduce that $N$ has a fixed point that is a solution of problem (4.55)–(4.56). □

Our next existence result for problem (4.55)–(4.56) is based on the concept of measures of noncompactness and Mönch's fixed point theorem.

**Theorem 4.29.** *Assume* (4.33.1), (4.33.2), *and* (4.58) *hold. Then IVP* (4.55)–(4.56) *has at least one solution.*

*Proof.* Consider the operator $N$ defined in (4.60). We will show that $N$ satisfies the assumptions of Mönch's fixed point theorem. We know that $N\colon D_R \to D_R$ is bounded and continuous, and we need to prove that the implication

$$[V = \overline{\text{conv}}N(V) \text{ or } V = N(V) \cup \{0\}] \text{ implies } \alpha(V) = 0$$

holds for every subset $V$ of $D_R$. Now let $V$ be a subset of $D_R$ such that $V \subset \overline{\mathrm{conv}}(N(V) \cup \{0\})$. $V$ is bounded and equicontinuous, and therefore the function $t \to v(t) = \alpha(V(t))$ is continuous on $J$. By Remark 4.27, Lemma 1.33, and the properties of the measure $\alpha$ we have for each $t \in J$

$$\begin{aligned} v(t) &\leq \alpha(N(V)(t) \cup \{0\}) \\ &\leq \alpha(N(V)(t)) \\ &\leq \alpha\{(Ny)(t), y \in V\} \\ &\leq \frac{K}{(1-L)\Gamma(v)} \int_0^t (t-s)^{v-1}\{\alpha(y(s))ds, y \in V\} \\ &\leq \frac{K}{(1-L)\Gamma(v)} \int_0^t (t-s)^{v-1} v(s)ds\,. \end{aligned}$$

Lemma 1.52 implies that $v(t) = 0$ for each $t \in J$, $V(t)$ is relatively compact in $E$. In view of the Ascoli–Arzelà theorem, $V$ is relatively compact in $D_R$. Applying now Theorem 1.46, we conclude that $N$ has a fixed point $y \in D_R$. Hence, $N$ has a fixed point that is a solution of problem (4.55)–(4.56). □

### 4.5.3 Existence Results for Nonlocal BVP in Banach Space

The purpose of this section is to establish sufficient conditions for the existence of solutions to the BVP for implicit fractional differential equations with a Caputo fractional derivative:

$$^cD^v y(t) = f(t, y(t), {}^cD^v y(t))\,, \quad \text{for every } t \in J := [0, T],\, T > 0,\ 0 < v \leq 1\,, \tag{4.64}$$

$$y(0) + g(y) = y_0\,, \tag{4.65}$$

where $f : J \times E \times E \to E$ is a given function, $g : C(J, E) \to E$ is a continuous function, and $y_0 \in E$. Finally, an example is given to demonstrate the application of our main results.

Let $(E; \|\cdot\|)$ be a Banach space, and $t \in J$. We denote by $C(J, E)$ the space of $E$ valued continuous functions on $J$ with the usual supremum norm

$$\|y\|_\infty = \sup\{\|y(t)\| : t \in J\}$$

for any $y \in C(J, E)$.

**Definition 4.30.** A function $y \in C^1(J, E)$ is called a solution of problem (4.64)–(4.65) if it satisfies equation (4.64) on $J$ and condition (4.65).

**Lemma 4.31.** *Let $0 < \nu \le 1$, and let $h\colon [0, T] \longrightarrow E$ be a continuous function. The linear problem*

$$ {}^cD^\nu y(t) = h(t)\,, \quad t \in J\,, $$
$$ y(0) + g(y) = y_0 $$

*has a unique solution given by*

$$ y(t) = y_0 - g(y) + \frac{1}{\Gamma(\nu)} \int_0^t (t-s)^{\nu-1} h(s)ds\,. $$

**Lemma 4.32.** *Let $f\colon J \times E \times E \longrightarrow E$ be a continuous function; then problem (4.64)–(4.65) is equivalent to the problem*

$$ y(t) = y_0 - g(y) + I^\nu H(t) $$

*where $H(t) = f(t, y(t), H(t))$.*

Introduce the following condition:
(4.39.1) There exists $0 < \overline{K}$ such that

$$ \|g(u) - g(\overline{u})\| \le \overline{K}\|u - \overline{u}\| \quad \text{for any } u, \overline{u} \in C(J, E)\,. $$

**Remark 4.33** ([66]). Condition (4.39.1) is equivalent to the inequality

$$ \alpha(g(B)) \le \overline{K}\alpha(B) $$

for any bounded sets $B \subseteq E$.

**Theorem 4.34.** *Assume (4.33.11), (4.33.2), and (4.39.1). If*

$$ \overline{K} + \frac{KT^\nu}{(1-L)\Gamma(\nu+1)} < 1\,, \tag{4.66} $$

*then IVP (4.64)–(4.65) has at least one solution on $J$.*

**Theorem 4.35.** *Assume (4.33.11), (4.33.2), (4.39.1), and (4.66) hold. If $\overline{K} < 1$, then IVP (4.64)–(4.65) has at least one solution.*

### 4.5.4 Examples

*Example 1.* Consider the infinite system

$$ {}^cD^{\frac{1}{2}} y_n(t) = \frac{\left(3 + \|y_n(t)\| + \|{}^cD^{\frac{1}{2}} y_n(t)\|\right)}{3e^{t+2}\left(1 + \|y_n(t)\| + \|{}^cD^{\frac{1}{2}} y_n(t)\|\right)} \quad \text{for each } t \in [0, 1]\,, \tag{4.67} $$

$$ y_n(0) + y_n(1) = 0\,. \tag{4.68} $$

Set

$$E = l^1 = \{y = (y_1, y_2, \dots, y_n, \dots), \sum_{n=1}^{\infty} |y_n| < \infty\},$$

and

$$f(t, u, v) = \frac{(3 + \|u\| + \|v\|)}{3e^{t+2}(1 + \|u\| + \|v\|)}, \quad t \in [0, 1], \ u, v \in E.$$

$E$ is a Banach space with the norm $\|y\| = \sum_{n=1}^{\infty} |y_n|$.

Clearly, the function $f$ is jointly continuous. For any $u, v, \bar{u}, \bar{v} \in E$ and $t \in [0, 1]$

$$\|f(t, u, v) - f(t, \bar{u}, \bar{v})\| \le \frac{1}{3e^2}(\|u - \bar{u}\| + \|v - \bar{v}\|).$$

Hence, condition (4.33.2) is satisfied by $K = L = \frac{1}{3e^2}$. The conditions

$$\frac{(|b| + |a + b|)T^{\nu}K}{|a + b|\Gamma(\nu + 1)(1 - L)} = \frac{1}{\sqrt{\pi}\left(e^2 - \frac{1}{3}\right)} < 1$$

and

$$\frac{KT^{\nu}}{(1 - L)\Gamma(\nu + 1)} = \frac{2}{(3e^2 - 1)\sqrt{\pi}} < 1$$

are satisfied by $a = b = T = 1, c = 0$, and $\nu = \frac{1}{2}$. From Theorem 4.28, problem (4.67)–(4.68) has at least one solution on $J$.

*Example 2.* Consider the BVP

$$^cD^{\frac{1}{2}} y_n(t) = \frac{e^{-t}}{(9 + e^t)}\left[1 + \frac{\|y_n(t)\|}{1 + \|y_n(t)\|} - \frac{\|^cD^{\frac{1}{2}} y_n(t)\|}{1 + \|^cD^{\frac{1}{2}} y_n(t)\|}\right], \quad t \in J = [0, 1], \tag{4.69}$$

$$y_n(0) + \sum_{i=1}^{m} c_i y_n(t_i) = 1, \tag{4.70}$$

where $0 < t_1 < t_2 < \cdots < t_m < 1$ and $c_i = 1, \dots, m$ are positive constants, with

$$\sum_{i=1}^{m} c_i < \frac{1}{3}.$$

Set

$$E = l^1 = \{y = (y_1, y_2, \dots, y_n, \dots), \quad \sum_{n=1}^{\infty} |y_n| < \infty\},$$

and

$$f(t, u, v) = \frac{e^{-t}}{(9 + e^t)}\left[1 + \frac{\|u\|}{1 + \|u\|} - \frac{\|v\|}{1 + \|v\|}\right], \quad t \in [0, 1], \ u, v \in E.$$

$E$ is a Banach space with the norm $\|y\| = \sum_{n=1}^{\infty} |y_n|$.

Clearly, the function $f$ is continuous. For each $u, \bar{u}, v, \bar{v} \in E$ and $t \in [0, 1]$

$$\begin{aligned}\|f(t, u, v) - f(t, \bar{u}, \bar{v})\| &\le \frac{e^{-t}}{9 + e^t}(\|u - \bar{u}\| + \|v - \bar{v}\|) \\ &\le \frac{1}{10}\|u - \bar{u}\| + \frac{1}{10}\|v - \bar{v}\|.\end{aligned}$$

Hence, condition (4.33.2) is satisfied by $K = L = \frac{1}{10}$.
On the other hand, we have for any $u, \bar{u} \in E$

$$\|g(u) - g(\bar{u})\| \leq \frac{1}{3}\|u - \bar{u}\| .$$

Hence, condition (4.39.1) is satisfied by $\overline{K} = \frac{1}{3}$. Also, the condition

$$\overline{K} + \frac{KT^{\nu}}{(1-L)\Gamma(\nu+1)} = \frac{9\sqrt{\pi}+6}{27\sqrt{\pi}} < 1$$

is satisfied by $T = 1$ and $\nu = \frac{1}{2}$. It follows from Theorem 4.35 that problem (4.69)–(4.70) has at least one solution on $J$.

## 4.6 $L^1$-Solutions of BVP for NIFDE

### 4.6.1 Introduction and Motivations

More recently, considerable attention has been paid to the existence of solutions of BVPs and boundary conditions for implicit fractional differential equations and integral equations with a Caputo fractional derivative. See, for example, [47, 53, 94, 164, 260] and references therein.

In [203], Murad and Hadid, by means of Schauder's fixed-point theorem and the Banach contraction principle, considered the BVP for the fractional differential equation

$$\begin{aligned} &D^{\alpha}y(t) = f(t, y(t), D^{\beta}y(t)), \quad t \in J := (0,1),\ 1 < \alpha \leq 2,\ 0 < \beta < 1,\ 0 < \gamma \leq 1, \\ &y(0) = 0, \quad y(1) = I_0^{\gamma}y(s), \end{aligned}$$

where $f\colon [0,1] \times \mathbb{R} \times \mathbb{R} \to \mathbb{R}$ is a continuous function, and $D^{\alpha}$ is the Riemann–Liouville fractional derivative.

In [150], Guezane-Lakoud and Khaldi studied the BVP of the fractional integral boundary conditions

$$\begin{aligned} &{}^cD^{q}y(t) = f(t, y(t), {}^cD^{p}y(t)), \quad t \in J := (0,1),\ 1 < q \leq 2,\ 0 < p < 1, \\ &y(0) = 0, \quad y'(1) = \alpha I_0^{p}y(1), \end{aligned}$$

where $f\colon [0,1] \times \mathbb{R} \times \mathbb{R} \to \mathbb{R}$ is a continuous function, and $D^{\alpha}$ is the Caputo fractional derivative.

In [241], by means of Schauder's fixed-point theorem, Su and Liu studied the existence of nonlinear fractional BVPs involving Caputo's derivative:

$$\begin{aligned} &{}^cD^{\alpha}u(t) = f(t, u(t), {}^cD^{\beta}u(t)), \quad t \in J := (0,1),\ 1 < \alpha \leq 2,\ 0 < \beta < 1, \\ &u(0) = 0 = u'(1) = 0 \text{ or } u'(1) = u(1) = 0 \text{ or } u(0) = u(1) = 0, \end{aligned}$$

where $f\colon [0,1] \times \mathbb{R} \times \mathbb{R} \to \mathbb{R}$ is a continuous function.

In [103], Benchohra and Lazreg studied the existence of continuous solutions of problem (4.71)–(4.72) and the implicit fractional order differential equation

$$^cD^\alpha y(t) = f(t, y(t), {}^cD^\alpha y(t))\,, \quad t \in J := [0, T],\ 0 < \alpha \le 1\,,$$

with boundary condition

$$ay(0) = y_0 + By(T) = c\,,$$

where $f : J \times \mathbb{R} \times \mathbb{R} \to \mathbb{R}$ is a given function, $^cD^\alpha$ is the Caputo fractional derivative, and $a, b, c$ are real constants, with $a + b \neq 0$.

The purpose of this section is to establish existence and uniqueness of integrable solutions to BVPs for the fractional order implicit differential equation

$$^cD^\alpha y(t) = f(t, y(t), {}^cD^\alpha y(t))\,, \quad t \in J := [0, T],\ 1 < \alpha \le 2\,, \tag{4.71}$$

$$y(0) = y_0\,, \quad y(T) = y_T\,, \tag{4.72}$$

where $f : J \times \mathbb{R} \times \mathbb{R} \to \mathbb{R}$ is a given function, $y_0, y_T \in \mathbb{R}$.

### 4.6.2 Existence of solutions

**Definition 4.36.** A function $y \in L^1(J, \mathbb{R})$ is said to be a solution of BVP (4.71)–(4.72) if $y$ satisfies (4.71) and (4.72).

For the existence of solutions to problem (4.71)–(4.72), we need the following auxiliary lemma.

**Lemma 4.37.** *Let $1 < \alpha \le 2$ and let $x \in L^1(J, \mathbb{R})$. The BVP (4.71)–(4.72) is equivalent to the integral equation*

$$y(t) = \frac{1}{\Gamma(\alpha)} \int_0^T G(t, s)x(s)ds + y_0 + \frac{(y_T - y_0)t}{T}\,, \tag{4.73}$$

*where $x$ is the solution of the functional integral equation*

$$x(t) = f\left(t, \frac{1}{\Gamma(\alpha)} \int_0^T G(t, s)x(s)ds + y_0 + \frac{(y_T - y_0)t}{T}, x(t)\right)\,, \tag{4.74}$$

*and $G(t, s)$ is the Green's function defined by*

$$G(t, s) := \begin{cases} (t - s)^{\alpha-1} - \frac{t(T-s)^{\alpha-1}}{T}, & 0 \le s \le t \le T\,, \\ \frac{-t(T-s)^{\alpha-1}}{T}, & 0 \le t \le s \le T\,. \end{cases} \tag{4.75}$$

*Proof.* Let $^cD^\alpha y(t) = x(t)$ in equation (4.71); then

$$x(t) = f(t, y(t), x(t))\,, \tag{4.76}$$

and Lemma 1.9 implies that

$$y(t) = c_0 + c_1 t + \frac{1}{\Gamma(\alpha)} \int_0^t (t-s)^{\alpha-1} x(s) ds \, .$$

From (4.72), a simple calculation gives

$$c_0 = y_0$$

and

$$c_1 = -\frac{1}{T\Gamma(\alpha)} \int_0^T (T-s)^{\alpha-1} x(s) ds + \frac{(y_T - y_0)}{T} \, .$$

Hence, we get equation (4.73).

Inversely, we prove that equation (4.73) satisfies BVP (4.71)–(4.72). Differentiating (4.73), we get

$${}^cD^\alpha y(t) = x(t) = f(t, y(t), {}^cD^\alpha y(t)) \, .$$

By (4.73) and (4.75) we have

$$y(t) = \frac{1}{\Gamma(\alpha)} \int_0^t (t-s)^{\alpha-1} x(s) ds - \frac{t}{T\Gamma(\alpha)} \int_0^T (T-s)^{\alpha-1} x(s) ds + y_0 + \frac{(y_T - y_0)t}{T} \, . \quad (4.77)$$

A simple calculation gives $y(0) = y_0$ and $y(T) = y_T$. □

Let us introduce the following conditions:

**(4.44.1)** $f\colon [0, T] \times \mathbb{R}^2 \longrightarrow \mathbb{R}$ is measurable in $t \in [0, T]$, for any $(u_1, u_2) \in \mathbb{R}^2$ and continuous in $(u_1, u_2) \in \mathbb{R}^2$ for almost all $t \in [0, T]$.

**(4.44.2)** There exist a positive function $a \in L^1[0, T]$ and constants $b_i > 0$, $i = 1, 2$, such that

$$|f(t, u_1, u_2)| \le |a(t)| + b_1|u_1| + b_2|u_2|, \forall (t, u_1, u_2) \in [0, T] \times \mathbb{R}^2 \, .$$

Our first result is based on Schauder's fixed point theorem.

**Theorem 4.38.** *Assume* (4.44.1) *and* (4.44.2) *hold. If*

$$\frac{b_1 G_0 T}{\Gamma(\alpha)} + b_2 < 1 \, , \quad (4.78)$$

*then BVP* (4.71)–(4.72) *has at least one solution* $y \in L^1(J, \mathbb{R})$.

*Proof.* Transform problem (4.71)–(4.72) into a fixed point problem. Consider the operator

$$H\colon L^1(J, \mathbb{R}) \longrightarrow L^1(J, \mathbb{R})$$

defined by

$$(Hx)(t) = f\left(t, \frac{1}{\Gamma(\alpha)}\int_0^T G(t,s)x(s)ds + y_0 + \frac{(y_T - y_0)t}{T}, x(t)\right), \tag{4.79}$$

where $G$ is given by 4.75. Let

$$G_0 := \max[|G(t,s)|, (t,s) \in J \times J]$$

and

$$r = \frac{b_1(|y_0| + |y_T|)T + \|a\|_{L_1}}{1 - \left(\frac{b_1 G_0 T}{\Gamma(\alpha)} + b_2\right)}.$$

Consider the set

$$B_r = \{x \in L^1([0,T], \mathbb{R}) : \|x\|_{L_1} \leq r\}.$$

Clearly, $B_r$ is nonempty, bounded, convex, and closed.

We will now show that $HB_r \subset B_r$; indeed, for each $x \in B_r$, from conditions (4.44.2) and (4.78) we get

$$\begin{aligned}
\|Hx\|_{L_1} &= \int_0^T |Hx(t)|dt \\
&= \int_0^T \left| f\left(t, \frac{1}{\Gamma(\alpha)}\int_0^T G(t,s)x(s)ds + y_0 + \frac{(y_T - y_0)t}{T}, x(t)\right)\right| dt \\
&\leq \int_0^T \left[ |a(t)| + b_1 \left| \frac{1}{\Gamma(\alpha)}\int_0^T G(t,s)x(s)ds - \left(\frac{t}{T} - 1\right)y_0 + \frac{t}{T}y_T \right| + b_2|x(t)| \right] dt \\
&\leq \|a\|_{L_1} + \frac{b_1 G_0 T}{\Gamma(\alpha)}\|x\|_{L_1} + b_1(|y_0| + |y_T|)T + b_2\|x\|_{L_1} \\
&\leq b_1(|y_0| + |y_T|)T + \|a\|_{L_1} + \left(\frac{b_1 G_0 T}{\Gamma(\alpha)} + b_2\right) r \\
&\leq r.
\end{aligned}$$

Then $HB_r \subset B_r$. Assumption (4.44.1) implies that $H$ is continuous. We will now show that $H$ is compact, that is, $HB_r$ is relatively compact. Clearly, $HB_r$ is bounded in $L^1(J, \mathbb{R})$, i.e., condition (i) of Kolmogorov's compactness criterion is satisfied. It remains to show that $(Hx)_h \longrightarrow (Hx)$ in $L^1(J, \mathbb{R})$ for each $x \in B_r$.

Let $x \in B_r$; then we have

$$\begin{aligned}
&\|(Hx)_h - (Hx)\|_{L^1} \\
&= \int_0^T |(Hx)_h(t) - (Hx)(t)| dt \\
&= \int_0^T \left| \frac{1}{h} \int_t^{t+h} (Hx)(s) ds - (Hx)(t) \right| dt \\
&\le \int_0^T \left( \frac{1}{h} \int_t^{t+h} |(Hx)(s) - (Hx)(t)| ds \right) dt \\
&\le \int_0^T \left( \frac{1}{h} \int_t^{t+h} \left| f\left( s, \frac{1}{\Gamma(\alpha)} \int_0^T G(s,\tau)x(\tau)d\tau + y_0 + \frac{(y_T - y_0)s}{T}, x(s) \right) \right. \right. \\
&\quad \left. \left. - f\left( t, \frac{1}{\Gamma(\alpha)} \int_0^T G(t,s)x(s)ds + y_0 + \frac{(y_T - y_0)t}{T}, x(t) \right) \right| ds \right) dt .
\end{aligned}$$

Since $x \in B_r \subset L^1(J, \mathbb{R})$, condition (4.44.2) implies that $f \in L^1(J, \mathbb{R})$. Thus, we have

$$\begin{aligned}
&\frac{1}{h} \int_t^{t+h} \left| f\left( s, \frac{1}{\Gamma(\alpha)} \int_0^T G(s,\tau)x(\tau)d\tau + y_0 + \frac{(y_T - y_0)s}{T}, x(s) \right) \right. \\
&\quad \left. - f\left( t, \frac{1}{\Gamma(\alpha)} \int_0^T G(t,s)x(s)ds + y_0 + \frac{(y_T - y_0)t}{T}, x(t) \right) \right| ds \longrightarrow 0 , \quad \text{as } h \longrightarrow 0,\ t \in J .
\end{aligned}$$

Hence,

$$(Hx)_h \longrightarrow (Hx) \text{ uniformly as } h \longrightarrow 0 .$$

Then by Kolmogorov's compactness criterion, $HB_r$ is relatively compact. As a consequence of Schauder's fixed point theorem, BVP (4.71)–(4.72) has at least one solution in $B_r$. □

The next result is based on the Banach contraction principle.

**Theorem 4.39.** *Assume* (4.44.1) *holds and*

**(4.46.1)** *There exist constants $k_1, k_2 > 0$ such that*

$$|f(t, x_1, y_1) - f(t, x_2, y_2)| \le k_1|x_1 - x_2| + k_2|y_1 - y_2|, \quad t \in [0, T],\ x_1, x_2, y_1, y_2 \in \mathbb{R}.$$

*If*

$$\frac{k_1 T G_0}{\Gamma(\alpha)} + k_2 < 1 , \tag{4.80}$$

*then BVP* (4.71)–(4.72) *has a unique solution $y \in L^1([0, T], \mathbb{R})$.*

*Proof.* We will use the Banach contraction principle to prove that $H$ defined by (4.79) has a fixed point. Let $x, y \in L^1(J, \mathbb{R})$, and $t \in J$. Then we have

$$\begin{aligned}|(Hx)(t) - (Hy)(t)| &= \left| f\left(t, \frac{1}{\Gamma(\alpha)} \int_0^T G(t,s)x(s)ds + y_0 + \frac{(y_T - y_0)t}{T}, x(t)\right)\right. \\ &\quad \left. - f\left(t, \frac{1}{\Gamma(\alpha)} \int_0^T G(t,s)y(s)ds + y_0 + \frac{(y_T - y_0)t}{T}, y(t)\right)\right| . \\ &\leq \frac{k_1}{\Gamma(\alpha)} \int_0^T |G(t,s)(x(s) - y(s))|ds + k_2|x(t) - y(t)| \\ &\leq \frac{k_1 G_0}{\Gamma(\alpha)} \int_0^T |x(s) - y(s)|ds + k_2|x(t) - y(t)| .\end{aligned}$$

Thus,

$$\begin{aligned}\|(Hx) - (Hy)\|_{L_1} &\leq \frac{k_1 T G_0}{\Gamma(\alpha)} \|x - y\|_{L_1} + k_2 \int_0^T |x(t) - y(t)|dt \\ &\leq \frac{k_1 T G_0}{\Gamma(\alpha)} \|x - y\|_{L_1} + k_2\|x - y\|_{L_1} \\ &\leq \left(\frac{k_1 T G_0}{\Gamma(\alpha)} + k_2\right) \|x - y\|_{L_1} .\end{aligned}$$

Consequently, by (4.80), $H$ is a contraction. As a consequence of the Banach contraction principle, the operator $H$ has a fixed point that is a solution of problem (4.71)–(4.72). □

### 4.6.3 Nonlocal problem

This section is devoted to some existence and uniqueness results for the class of the nonlocal problem

$$^cD^\alpha y(t) = f(t, y(t), {}^cD^\alpha y(t)) , \quad t \in J := [0, T], \ 1 < \alpha \leq 2 , \tag{4.81}$$

$$y(0) = g(y) , \quad y(T) = y_T , \tag{4.82}$$

where $g\colon L^1(J, \mathbb{R}) \to \mathbb{R}$ a continuous function. The nonlocal condition can be applied in physics with better effect than the classical initial condition $y(0) = y_0$. For example, $g(y)$ may be given by

$$g(y) = \sum_{i=1}^{p} c_i y(t_i) ,$$

where $c_i, i = 1, 2, \ldots, p$ are given constants and $0 < \cdots < t_p < T$. Nonlocal conditions were initiated by Byszewski [117] when he proved the existence and uniqueness of mild and classical solutions of nonlocal Cauchy problems. As remarked by Byszewski [117, 118], the nonlocal condition can be more useful than the standard initial condition to describe some physical phenomena.

Let us introduce the following set of conditions on the function $g$.

**(4.46.2)** There exists a constant $\tilde{k} > 0$ such that

$$|g(y) - g(\tilde{y}| \leq \tilde{k}|y - \tilde{y}| \quad \text{for each } y, \tilde{y} \in L^1(J, \mathbb{R})\,.$$

**Theorem 4.40.** *Assume* (4.44.1), (4.46.1), *and* (4.46.2) *hold. If*

$$\frac{2k_1 T^\alpha}{\Gamma(\alpha+1)} + k_1\tilde{k} + k_2 < 1\,, \tag{4.83}$$

*then BVP* (4.81)–(4.82) *has a unique solution* $y \in L^1(J, \mathbb{R})$.

Transform problem (4.81)–(4.82) into a fixed point problem. Consider the operator

$$\tilde{H}: L^1(J, \mathbb{R}) \longrightarrow L^1(J, \mathbb{R})$$

defined by

$$\begin{aligned}(\tilde{H}x)(t) &= f\Bigg(t, \frac{1}{\Gamma(\alpha)}\int_0^t (t-s)^{\alpha-1}x(s)ds - \frac{t}{T\Gamma(\alpha)}\int_0^T (T-s)^{\alpha-1}x(s)ds \\ &\quad - \left(\frac{t}{T} - 1\right)g(y) + \frac{t}{T}y_T, x(t)\Bigg)\,.\end{aligned} \tag{4.84}$$

*Proof.* We will use the Banach contraction principle to prove that $\tilde{H}$ defined by (4.84) has a fixed point. Let $x, y \in L^1(J, \mathbb{R})$, and $t \in J$. Then we have

$$\begin{aligned}&|(\tilde{H}x)(t) - (\tilde{H}y)(t)| \\ &= \Bigg|f\Bigg(t, \frac{1}{\Gamma(\alpha)}\int_0^t (t-s)^{\alpha-1}x(s)ds - \frac{t}{T\Gamma(\alpha)}\int_0^T (T-s)^{\alpha-1}x(s)ds \\ &\quad - \left(\frac{t}{T} - 1\right)g(x) + \frac{t}{T}y_T, x(t)\Bigg) \\ &\quad - f\Bigg(t, \frac{1}{\Gamma(\alpha)}\int_0^t (t-s)^{\alpha-1}y(s)ds - \frac{t}{T\Gamma(\alpha)}\int_0^T (T-s)^{\alpha-1}y(s)ds \\ &\quad - \left(\frac{t}{T} - 1\right)g(y) + \frac{t}{T}y_T, y(t)\Bigg)\Bigg| \\ &\leq \frac{k_1}{\Gamma(\alpha)}\int_0^t (t-s)^{\alpha-1}|(x(s) - y(s))|ds + \frac{k_1}{\Gamma(\alpha)}\int_0^T (T-s)^{\alpha-1}|(x(s) - y(s))|ds \\ &\quad + k_1|g(x) - g(y)| + k_2|x(t) - y(t)|\,.\end{aligned}$$

Thus,

$$
\begin{aligned}
\|(\tilde{H}x)-(\tilde{H}y)\|_{L_1} &\le \frac{k_1\|x-y\|_{L_1}}{\Gamma(\alpha)}\int_0^t (t-s)^{\alpha-1}ds + \frac{k_1\|x-y\|_{L_1}}{\Gamma(\alpha)}\int_0^T (T-s)^{\alpha-1}ds \\
&\quad + k_1\tilde{k}\|x-y\|_{L_1} + k_2\|x-y\|_{L_1} \\
&\le \frac{2k_1T^\alpha}{\Gamma(\alpha+1)}\|x-y\|_{L_1} + k_1\tilde{k}\|x-y\|_{L_1} + k_2\|x-y\|_{L_1} \\
&\le \left(\frac{2k_1T^\alpha}{\Gamma(\alpha+1)} + k_1\tilde{k} + k_2\right)\|x-y\|_{L_1} .
\end{aligned}
$$

Consequently, by (4.83), $\tilde{H}$ is a contraction. As a consequence of the Banach contraction principle, we deduce that $\tilde{H}$ has a fixed point that is a solution of problem (4.81)–(4.82). □

### 4.6.4 Examples

*Example 1.* Let us consider the BVP

$$
{}^cD^\alpha y(t) = \frac{e^{-t}}{(e^t+6)(1+|y(t)|+|{}^cD^\alpha y(t)|)}, \quad t \in J := [0,1],\ 1<\alpha\le 2, \tag{4.85}
$$

$$
y(0)=1, \quad y(1)=2. \tag{4.86}
$$

Set

$$
f(t,y,z) = \frac{e^{-t}}{(e^t+6)(1+y+z)}, \quad (t,y,z) \in J\times[0,+\infty)\times[0,+\infty).
$$

Let $y, z \in [0,+\infty)$ and $t \in J$. Then we have

$$
\begin{aligned}
|f(t,y_1,z_1)-f(t,y_2,z_2)| &= \left|\frac{e^{-t}}{e^t+6}\left(\frac{1}{1+y_1+z_1}-\frac{1}{1+y_2+z_2}\right)\right| \\
&\le \frac{e^{-t}(|y_1-y_2|+|z_1-z_2|)}{(e^t+6)(1+y_1+z_1)(1+y_2+z_2)} \\
&\le \frac{e^{-t}}{(e^t+6)}(|y_1-y_2|+|z_1-z_2|) \\
&\le \frac{1}{7}|y_1-y_2| + \frac{1}{7}|z_1-z_2| .
\end{aligned}
$$

Hence, condition (4.46.1) holds, with $k_1 = k_2 = \frac{1}{7}$. Condition (4.80) is satisfied by $T = 1$. Indeed,

$$
\frac{k_1TG_0}{\Gamma(\alpha)} + k_2 = \frac{G_0}{7\Gamma(\alpha)} + \frac{1}{7} < 1 . \tag{4.87}
$$

Then, by Theorem 4.39, problem (4.85)–(4.86) has a unique integrable solution on $[0,1]$ for values of $\alpha$ satisfying condition (4.87).

*Example 2.* Let us consider the nonlocal BVP

$$^{c}D^{\alpha}y(t) = \frac{e^{-t}}{(e^{t}+9)(1+|y(t)|+|^{c}D^{\alpha}y(t)|)}, \quad t \in J := [0,1],\ 1 < \alpha \le 2, \tag{4.88}$$

$$y(0) = \sum_{i=1}^{n} c_i y(t_i), \quad y(1) = 0, \tag{4.89}$$

where $0 < \cdots < t_n < 1$, $c_i$, $i = 1, 2, \dots, n$, are given positive constants with $\sum_{i=1}^{n} c_i < \frac{4}{5}$. Set

$$f(t, y, z) = \frac{e^{-t}}{(e^{t}+9)(1+y+z)}, \quad (t, y, z) \in J \times [0, +\infty) \times [0, +\infty),$$

and

$$g(y) = \sum_{i=1}^{n} c_i y(t_i).$$

Let $y, z \in [0, +\infty)$ and $t \in J$. Then we have

$$\begin{aligned}
|f(t, y_1, z_1) - f(t, y_2, z_2)| &= \left|\frac{e^{-t}}{e^{t}+9}\left(\frac{1}{1+y_1+z_1} - \frac{1}{1+y_2+z_2}\right)\right| \\
&\le \frac{e^{-t}(|y_1 - y_2| + |z_1 - z_2|)}{(e^{t}+9)(1+y_1+z_1)(1+y_2+z_2)} \\
&\le \frac{e^{-t}}{(e^{t}+9)}(|y_1 - y_2| + |z_1 - z_2|) \\
&\le \frac{1}{10}|y_1 - y_2| + \frac{1}{10}|z_1 - z_2|.
\end{aligned}$$

Hence, condition (4.46.1) holds, with $k_1 = k_2 = \frac{1}{10}$. Also, we have

$$|g(x) - g(y)| \le \sum_{i=1}^{n} c_i |x - y|.$$

Hence, (4.46.2) is satisfied by $\tilde{k} = \sum_{i=1}^{n} c_i$. Condition (4.83) is satisfied by $T = 1$. Indeed,

$$\frac{2k_1 T^{\alpha}}{\Gamma(\alpha+1)} + k_1\tilde{k} + k_2 = \frac{1}{5\Gamma(\alpha+1)} + \frac{1}{10}\sum_{i=1}^{n} c_i + \frac{1}{10} < 1 \Longleftrightarrow \Gamma(\alpha+1) > \frac{10}{41}. \tag{4.90}$$

Then by Theorem 4.40, problem (4.88)–(4.89) has a unique integrable solution on $[0, 1]$ for values of $\alpha$ satisfying condition (4.90).

## 4.7 Notes and Remarks

The results of Chapter 4 are taken from Benchohra et al. [91, 103, 109]. Other results may be found in [95, 97, 202].

# 5 Boundary Value Problems for Impulsive NIFDE

## 5.1 Introduction and Motivations

The theory of impulsive differential equations of integer order has found extensive application in realistic mathematical modeling of a wide variety of practical situations and has emerged as an important area of investigation in recent years. See [76, 77, 100, 148, 186, 215, 240], as well as [124, 157, 158, 251], and references therein.

Very recently, antiperiodic boundary value problems (BVPs) of fractional differential equations have received considerable attention because they occur in the mathematical modeling of a variety of physical processes. See for example [45, 53, 54, 52, 55, 88, 98, 125, 248, 249, 250, 259].

In this chapter, we establish existence, uniqueness, and stability results for some classes of BVPs for impulsive nonlinear implicit fractional differential equations (NIFDEs). Next, we present other results of existence and uniqueness for BVPs for NIFDEs with impulses in Banach spaces.

## 5.2 Existence Results for Impulsive NIFDEs

### 5.2.1 Introduction

In this section, we establish existence, uniqueness, and stability results of solutions for the BVP for NIFDEs with impulse and Caputo fractional derivatives:

$$ {}^cD^{\alpha}_{t_k}y(t) = f(t, y, {}^cD^{\alpha}_{t_k}y(t))\,, \quad \text{for each, } t \in (t_k, t_{k+1}],\ k = 0, \dots, m,\ 0 < \alpha \le 1\,, \tag{5.1}$$

$$ \Delta y|_{t=t_k} = I_k(y(t_k^-))\,, \quad k = 1, \dots, m\,, \tag{5.2}$$

$$ ay(0) + by(T) = c\,, \tag{5.3}$$

where ${}^cD^{\alpha}_{t_k}$ is the Caputo fractional derivative, $f\colon J \times \mathbb{R} \times \mathbb{R} \to \mathbb{R}$ is a given function, $I_k\colon \mathbb{R} \to \mathbb{R}$, and $a, b, c$ are real constants, with $a+b \neq 0, 0 = t_0 < t_1 < \dots < t_m < t_{m+1} = T$, $\Delta y|_{t=t_k} = y(t_k^+) - y(t_k^-)$, and $y(t_k^+) = \lim_{h\to 0^+} y(t_k + h)$ and $y(t_k^-) = \lim_{h\to 0^-} y(t_k + h)$ represent the right and left limits of $y(t)$ at $t = t_k$, respectively.

The arguments are based upon the Banach contraction principle and Schaefer's fixed point theorem.

### 5.2.2 Existence of Solutions

Consider the Banach space

$$ PC(J, \mathbb{R}) = \{y\colon J \to \mathbb{R}\colon y \in C((t_k, t_{k+1}], \mathbb{R}),\ k = 0, \dots, m \text{ and there exist } y(t_k^-) \text{ and } y(t_k^+),\ k = 1, \dots, m \text{ with } y(t_k^-) = y(t_k)\}\,,$$

https://doi.org/10.1515/9783110553819-005

with the norm

$$\|y\|_{PC} = \sup_{t\in J} |y(t)| .$$

**Definition 5.1.** A function $y \in PC(J, \mathbb{R})$ whose $\alpha$-derivative exists on $J_k$ is said to be a solution of (5.1)–(5.3) if $y$ satisfies the equation ${}^cD^\alpha_{t_k} y(t) = f(t, y(t), {}^cD^\alpha_{t_k} y(t))$ on $J_k$ and satisfies the conditions

$$\Delta y|_{t=t_k} = I_k(y(t_k^-)) , \quad k = 1, \dots, m ,$$
$$ay(0) + by(T) = c .$$

To prove the existence of solutions to (5.1)–(5.3), we need the following auxiliary lemma.

**Lemma 5.2.** *Let $0 < \alpha \le 1$, and let $\sigma\colon J \to \mathbb{R}$ be continuous. A function $y$ is a solution of the fractional integral equation*

$$y(t) = \begin{cases} \frac{-1}{a+b}\left[ b\sum_{i=1}^m I_i(y(t_i^-)) + \frac{b}{\Gamma(\alpha)}\sum_{i=1}^m \int_{t_{i-1}}^{t_i} (t_i - s)^{\alpha-1}\sigma(s)ds \right. \\ \left. + \frac{b}{\Gamma(\alpha)}\int_{t_m}^{T} (T-s)^{\alpha-1}\sigma(s)ds - c \right] + \frac{1}{\Gamma(\alpha)}\int_0^t (t-s)^{\alpha-1}\sigma(s)ds \quad \text{if } t \in [0, t_1] \\ \frac{-1}{a+b}\left[ b\sum_{i=1}^m I_i(y(t_i^-)) + \frac{b}{\Gamma(\alpha)}\sum_{i=1}^m \int_{t_{i-1}}^{t_i} (t_i - s)^{\alpha-1}\sigma(s)ds \right. \\ \left. + \frac{b}{\Gamma(\alpha)}\int_{t_m}^{T} (T-s)^{\alpha-1}\sigma(s)ds - c \right] + \sum_{i=1}^k I_i(y(t_i^-)) + \frac{1}{\Gamma(\alpha)}\sum_{i=1}^k \int_{t_{i-1}}^{t_i} (t_i-s)^{\alpha-1}\sigma(s)ds \\ + \frac{1}{\Gamma(\alpha)}\int_{t_k}^{t} (t-s)^{\alpha-1}\sigma(s)ds \quad \text{if } t \in (t_k, t_{k+1}] , \end{cases} \tag{5.4}$$

*where $k = 1, \dots, m$ if and only if $y$ is a solution of the fractional BVP*

$${}^cD^\alpha y(t) = \sigma(t) , \quad t \in J_k , \tag{5.5}$$
$$\Delta y|_{t=t_k} = I_k(y(t_k^-)) , \quad k = 1, \dots, m , \tag{5.6}$$
$$ay(0) + by(T) = c . \tag{5.7}$$

*Proof.* Assume that $y$ satisfies (5.5)–(5.7). If $t \in [0, t_1]$, then

$${}^cD^\alpha y(t) = \sigma(t) .$$

Lemma 1.9 implies

$$y(t) = c_0 + I^\alpha\sigma(t) = c_0 + \frac{1}{\Gamma(\alpha)}\int_0^t (t-s)^{\alpha-1}\sigma(s)ds$$

for $c_0 \in \mathbb{R}$. If $t \in (t_1, t_2]$, then Lemma 1.9 implies

$$\begin{aligned}
y(t) &= y(t_1^+) + \frac{1}{\Gamma(\alpha)} \int_{t_1}^{t} (t-s)^{\alpha-1}\sigma(s)ds \\
&= \Delta y|_{t=t_1} + y(t_1^-) + \frac{1}{\Gamma(\alpha)} \int_{t_1}^{t} (t-s)^{\alpha-1}\sigma(s)ds \\
&= I_1(y(t_1^-)) + \left[ c_0 + \frac{1}{\Gamma(\alpha)} \int_{0}^{t_1} (t_1-s)^{\alpha-1}\sigma(s)ds \right] \\
&\quad + \frac{1}{\Gamma(\alpha)} \int_{t_1}^{t} (t-s)^{\alpha-1}\sigma(s)ds\,. \\
&= c_0 + I_1(y(t_1^-)) + \frac{1}{\Gamma(\alpha)} \int_{0}^{t_1} (t_1-s)^{\alpha-1}\sigma(s)ds \\
&\quad + \frac{1}{\Gamma(\alpha)} \int_{t_1}^{t} (t-s)^{\alpha-1}\sigma(s)ds\,.
\end{aligned}$$

If $t \in (t_2, t_3]$, then from Lemma 1.9 we get

$$\begin{aligned}
y(t) &= y(t_2^+) + \frac{1}{\Gamma(\alpha)} \int_{t_2}^{t} (t-s)^{\alpha-1}\sigma(s)ds \\
&= \Delta y|_{t=t_2} + y(t_2^-) + \frac{1}{\Gamma(\alpha)} \int_{t_2}^{t} (t-s)^{\alpha-1}\sigma(s)ds \\
&= I_2(y(t_2^-)) + \left[ c_0 + I_1(y(t_1^-)) + \frac{1}{\Gamma(\alpha)} \int_{0}^{t_1} (t_1-s)^{\alpha-1}\sigma(s)ds \right. \\
&\quad \left. + \frac{1}{\Gamma(\alpha)} \int_{t_1}^{t_2} (t_2-s)^{\alpha-1}\sigma(s)ds \right] + \frac{1}{\Gamma(\alpha)} \int_{t_2}^{t} (t-s)^{\alpha-1}\sigma(s)ds\,. \\
&= c_0 + [I_1(y(t_1^-)) + I_2(y(t_2^-))] + \left[ \frac{1}{\Gamma(\alpha)} \int_{0}^{t_1} (t_1-s)^{\alpha-1}\sigma(s)ds \right. \\
&\quad \left. + \frac{1}{\Gamma(\alpha)} \int_{t_1}^{t_2} (t_2-s)^{\alpha-1}\sigma(s)ds \right] + \frac{1}{\Gamma(\alpha)} \int_{t_2}^{t} (t-s)^{\alpha-1}\sigma(s)ds\,.
\end{aligned}$$

Repeating the process in this way, the solution $y(t)$ for $t \in (t_k, t_{k+1}]$, where $k = 1, \ldots, m$, can be written

$$y(t) = c_0 + \sum_{i=1}^{k} I_i(y(t_i^-)) + \frac{1}{\Gamma(\alpha)} \sum_{i=1}^{k} \int_{t_{i-1}}^{t_i} (t_i - s)^{\alpha-1} \sigma(s) ds$$
$$+ \frac{1}{\Gamma(\alpha)} \int_{t_k}^{t} (t - s)^{\alpha-1} \sigma(s) ds .$$

Applying the boundary conditions $ay(0) + by(T) = c$, we get

$$c = c_0(a + b) + b \sum_{i=1}^{m} I_i(y(t_i^-)) + \frac{b}{\Gamma(\alpha)} \sum_{i=1}^{m} \int_{t_{i-1}}^{t_i} (t_i - s)^{\alpha-1} \sigma(s) ds$$
$$+ \frac{b}{\Gamma(\alpha)} \int_{t_m}^{T} (T - s)^{\alpha-1} \sigma(s) ds .$$

Then

$$c_0 = \frac{-1}{a + b} \left[ b \sum_{i=1}^{m} I_i(y(t_i^-)) + \frac{b}{\Gamma(\alpha)} \sum_{i=1}^{m} \int_{t_{i-1}}^{t_i} (t_i - s)^{\alpha-1} \sigma(s) ds \right.$$
$$\left. + \frac{b}{\Gamma(\alpha)} \int_{t_m}^{T} (T - s)^{\alpha-1} \sigma(s) ds - c \right] .$$

Thus, if $t \in (t_k, t_{k+1}]$, where $k = 1, \ldots, m$, then

$$y(t) = \frac{-1}{a + b} \left[ b \sum_{i=1}^{m} I_i(y(t_i^-)) + \frac{b}{\Gamma(\alpha)} \sum_{i=1}^{m} \int_{t_{i-1}}^{t_i} (t_i - s)^{\alpha-1} \sigma(s) ds \right.$$
$$\left. + \frac{b}{\Gamma(\alpha)} \int_{t_m}^{T} (T - s)^{\alpha-1} \sigma(s) ds - c \right] + \sum_{i=1}^{k} I_i(y(t_i^-))$$
$$+ \frac{1}{\Gamma(\alpha)} \sum_{i=1}^{k} \int_{t_{i-1}}^{t_i} (t_i - s)^{\alpha-1} \sigma(s) ds + \frac{1}{\Gamma(\alpha)} \int_{t_k}^{t} (t - s)^{\alpha-1} \sigma(s) ds .$$

Conversely, assume that $y$ satisfies impulsive fractional integral equation (5.4). If $t \in [0, t_1]$, then $ay(0) + by(T) = c$. Using the fact that ${}^cD^\alpha$ is the left inverse of $I^\alpha$, we get

$${}^cD^\alpha y(t) = \sigma(t) \quad \text{for each} \quad t \in [0, t_1] .$$

If $t \in (t_k, t_{k+1}]$, $k = 1, \ldots, m$, and using the fact that ${}^cD^\alpha C = 0$, where $C$ is a constant, we get

$${}^cD^\alpha y(t) = \sigma(t) , \quad \text{for each} \quad t \in (t_k, t_{k+1}] .$$

Also, we can easily show that

$$\Delta y|_{t=t_k} = I_k(y(t_k^-)), \quad k = 1, \dots, m.$$ □

We are now in a position to state and prove our existence result for problem (5.1)–(5.3) based on Banach's fixed point.

**Theorem 5.3.** *Make the following assumptions:*
(5.3.1) *The function $f: J \times \mathbb{R} \times \mathbb{R} \to \mathbb{R}$ is continuous.*
(5.3.2) *There exist constants $K > 0$ and $0 < L < 1$ such that*

$$|f(t, u, v) - f(t, \bar{u}, \bar{v})| \leq K|u - \bar{u}| + L|v - \bar{v}|$$

*for any $u, v, \bar{u}, \bar{v} \in \mathbb{R}$ and $t \in J$.*
(5.3.3) *There exists a constant $\tilde{l} > 0$ such that*

$$|I_k(u) - I_k(\overline{u})| \leq \tilde{l}|u - \overline{u}|$$

*for each $u, \overline{u} \in \mathbb{R}$ and $k = 1, \dots, m$.*

*If*

$$\left(\frac{|b|}{|a+b|} + 1\right)\left[m\tilde{l} + \frac{(m+1)KT^\alpha}{(1-L)\Gamma(\alpha+1)}\right] < 1, \tag{5.8}$$

*then there exists a unique solution for BVP (5.1)–(5.3) on $J$.*

*Proof.* Transform problem (5.1)–(5.3) into a fixed point problem. Consider the operator $N: PC(J, \mathbb{R}) \to PC(J, \mathbb{R})$ defined by

$$\begin{aligned} N(y)(t) = {} & \frac{-1}{a+b}\left[b\sum_{i=1}^{m} I_i(y(t_i^-)) + \frac{b}{\Gamma(\alpha)}\sum_{i=1}^{m}\int_{t_{i-1}}^{t_i}(t_i - s)^{\alpha-1}g(s)ds\right. \\ & \left. + \frac{b}{\Gamma(\alpha)}\int_{t_m}^{T}(T-s)^{\alpha-1}g(s)ds - c\right] + \frac{1}{\Gamma(\alpha)}\sum_{0<t_k<t}\int_{t_{k-1}}^{t_k}(t_k - s)^{\alpha-1}g(s)ds \\ & + \frac{1}{\Gamma(\alpha)}\int_{t_k}^{t}(t-s)^{\alpha-1}g(s)ds + \sum_{0<t_k<t} I_k(y(t_k^-)), \end{aligned} \tag{5.9}$$

where $g \in C(J, \mathbb{R})$ is such that

$$g(t) = f(t, y(t), g(t)).$$

Clearly, the fixed points of operator $N$ are solutions of problem (5.1)–(5.3).

Let $u, w \in PC(J, \mathbb{R})$. Then for $t \in J$ we have

$$\begin{aligned}|N(u)(t) - N(w)(t)| \leq \frac{|b|}{|a+b|} \Bigg[ &\sum_{i=1}^{m} |I_i(u(t_i^-)) - I_i(w(t_i^-))| \\ &+ \frac{1}{\Gamma(\alpha)} \sum_{i=1}^{m} \int_{t_{i-1}}^{t_i} (t_i - s)^{\alpha-1} |g(s) - h(s)| ds \\ &+ \frac{1}{\Gamma(\alpha)} \int_{t_m}^{T} (T-s)^{\alpha-1} |g(s) - h(s)| ds \Bigg] \\ &+ \frac{1}{\Gamma(\alpha)} \sum_{0<t_k<t} \int_{t_{k-1}}^{t_k} (t_k - s)^{\alpha-1} |g(s) - h(s)| ds \\ &+ \frac{1}{\Gamma(\alpha)} \int_{t_k}^{t} (t-s)^{\alpha-1} |g(s) - h(s)| ds \\ &+ \sum_{0<t_k<t} |I_k(u(t_k^-)) - I_k(w(t_k^-))| \,,\end{aligned}$$

where $g, h \in C(J, \mathbb{R})$ is such that

$$g(t) = f(t, u(t), g(t)) \,,$$

and

$$h(t) = f(t, w(t), h(t)) \,.$$

By (5.3.2) we have

$$\begin{aligned}|g(t) - h(t)| &= |f(t, u(t), g(t)) - f(t, w(t), h(t))| \\ &\leq K|u(t) - w(t)| + L|g(t) - h(t)| \,.\end{aligned}$$

Then

$$|g(t) - h(t)| \leq \frac{K}{1-L} |u(t) - w(t)| \,.$$

Therefore, for each $t \in J$

$$\begin{aligned}|N(u)(t) - N(w)(t)| \leq \frac{|b|}{|a+b|} \Bigg[ &\sum_{k=1}^{m} \tilde{l} |u(t_k^-) - w(t_k^-)| \\ &+ \frac{K}{(1-L)\Gamma(\alpha)} \sum_{k=1}^{m} \int_{t_{k-1}}^{t_k} (t_k - s)^{\alpha-1} |u(s) - w(s)| ds \\ &+ \frac{K}{(1-L)\Gamma(\alpha)} \int_{t_m}^{T} (T-s)^{\alpha-1} |u(s) - w(s)| ds \Bigg]\end{aligned}$$

$$+\frac{K}{(1-L)\Gamma(\alpha)}\sum_{k=1}^{m}\int_{t_{k-1}}^{t_k}(t_k-s)^{\alpha-1}|u(s)-w(s)|ds$$
$$+\frac{K}{(1-L)\Gamma(\alpha)}\int_{t_k}^{t}(t-s)^{\alpha-1}|u(s)-w(s)|ds$$
$$+\sum_{k=1}^{m}\tilde{l}|u(t_k^-)-w(t_k^-)|\,.$$
$$\leq\left(\frac{|b|}{|a+b|}+1\right)\left[m\tilde{l}+\frac{mKT^{\alpha}}{(1-L)\Gamma(\alpha+1)}\right.$$
$$\left.+\frac{KT^{\alpha}}{(1-L)\Gamma(\alpha+1)}\right]\|u-w\|_{PC}\,.$$

Thus,

$$\|N(u)-N(w)\|_{PC}\leq\left(\frac{|b|}{|a+b|}+1\right)\left[m\tilde{l}+\frac{(m+1)KT^{\alpha}}{(1-L)\Gamma(\alpha+1)}\right]\|u-w\|_{PC}\,.$$

By (5.8), operator $N$ is a contraction. Hence, by Banach's contraction principle, $N$ has a unique fixed point that is the unique solution of problem (5.1)–(5.3). □

Our second result is based on Schaefer's fixed point theorem.

**Theorem 5.4.** *Assume* (5.3.1) *and* (5.3.2) *hold and*
(5.4.1) *There exist* $p, q, r \in C(J, \mathbb{R}_+)$ *with* $r^* = \sup_{t\in J} r(t) < 1$ *such that*

$$|f(t,u,w)|\leq p(t)+q(t)|u|+r(t)|w|\quad \text{for } t\in J \text{ and } u,w\in\mathbb{R}\,;$$

(5.4.2) *The functions* $I_k : \mathbb{R} \to \mathbb{R}$ *are continuous, and there exist constants* $M^*$, $N^* > 0$ *such that*

$$|I_k(u)|\leq M^*|u|+N^*\quad \text{for each } u\in\mathbb{R},\ k=1,\dots,m\,.$$

*If*

$$\left(\frac{|b|}{|a+b|}+1\right)\left(mM^*+\frac{(m+1)q^*T^{\alpha}}{(1-r^*)\Gamma(\alpha+1)}\right)<1\,, \tag{5.10}$$

*then BVP* (5.1)–(5.3) *has at least one solution on* $J$.

*Proof.* Consider operator $N$ defined in (5.9). We will use Schaefer's fixed point theorem to prove that $N$ has a fixed point. The proof will be given in several steps.

*Step 1: N is continuous.* Let $\{u_n\}$ be a sequence such that $u_n \to u$ in $PC(J, \mathbb{R})$. Then for each $t \in J$

$$\begin{aligned}|N(u_n)(t) - N(u)(t)| \le \frac{|b|}{|a+b|}\Bigg[&\sum_{i=1}^{m} |I_k(u_n(t_k^-)) - I_k(u(t_k^-))| \\ &+ \frac{1}{\Gamma(\alpha)}\sum_{i=1}^{m}\int_{t_{i-1}}^{t_i} (t_i - s)^{\alpha-1}|g_n(s) - g(s)|ds \\ &+ \frac{1}{\Gamma(\alpha)}\int_{t_m}^{T}(T-s)^{\alpha-1}|g_n(s) - g(s)|ds\Bigg] \\ &+ \frac{1}{\Gamma(\alpha)}\sum_{0<t_k<t}\int_{t_{k-1}}^{t_k}(t_k - s)^{\alpha-1}|g_n(s) - g(s)|ds \\ &+ \frac{1}{\Gamma(\alpha)}\int_{t_k}^{t}(t-s)^{\alpha-1}|g_n(s) - g(s)|ds \\ &+ \sum_{0<t_k<t}|I_k(u_n(t_k^-)) - I_k(u(t_k^-))|\,,\end{aligned} \tag{5.11}$$

where $g_n, g \in C(J, \mathbb{R})$ such that

$$g_n(t) = f(t, u_n(t), g_n(t))$$

and

$$g(t) = f(t, u(t), g(t))\,.$$

By (5.3.2) we have

$$\begin{aligned}|g_n(t) - g(t)| &= |f(t, u_n(t), g_n(t)) - f(t, u(t), g(t))| \\ &\le K|u_n(t) - u(t)| + L|g_n(t) - g(t)|\,.\end{aligned}$$

Then

$$|g_n(t) - g(t)| \le \frac{K}{1-L}|u_n(t) - u(t)|\,.$$

Since $u_n \to u$, we get $g_n(t) \to g(t)$ as $n \to \infty$ for each $t \in J$. Let $\eta > 0$ be such that, for each $t \in J$, we have $|g_n(t)| \le \eta$ and $|g(t)| \le \eta$. Then we have

$$\begin{aligned}(t-s)^{\alpha-1}|g_n(s) - g(s)| &\le (t-s)^{\alpha-1}[|g_n(s)| + |g(s)|] \\ &\le 2\eta(t-s)^{\alpha-1}\end{aligned}$$

and

$$\begin{aligned}(t_k-s)^{\alpha-1}|g_n(s) - g(s)| &\le (t_k-s)^{\alpha-1}[|g_n(s)| + |g(s)|] \\ &\le 2\eta(t_k-s)^{\alpha-1}\,.\end{aligned}$$

For each $t \in J$ the functions $s \to 2\eta(t-s)^{\alpha-1}$ and $s \to 2\eta(t_k-s)^{\alpha-1}$ are integrable on $[0,t]$, then the Lebesgue dominated convergence theorem and (5.11) imply that

$$|N(u_n)(t) - N(u)(t)| \to 0 \text{ as } n \to \infty .$$

Hence

$$\|N(u_n) - N(u)\|_{PC} \to 0 \text{ as } n \to \infty .$$

Consequently, $N$ is continuous.

*Step 2: $N$ maps bounded sets to bounded sets in $PC(J,\mathbb{R})$.* Indeed, it is enough to show that for any $\eta^* > 0$ there exists a positive constant $\ell$ such that for each $u \in B_{\eta^*} = \{u \in PC(J,\mathbb{R}) : \|u\|_{PC} \le \eta^*\}$ we have $\|N(u)\|_{PC} \le \ell$. For each $t \in J$ we have

$$\begin{aligned} N(u)(t) = \frac{-1}{a+b}&\left[b\sum_{i=1}^{m} I_i(u(t_i^-)) + \frac{b}{\Gamma(\alpha)}\sum_{i=1}^{m}\int_{t_{i-1}}^{t_i}(t_i-s)^{\alpha-1}g(s)ds\right.\\ &\left.+\frac{b}{\Gamma(\alpha)}\int_{t_m}^{T}(T-s)^{\alpha-1}g(s)ds - c\right] + \frac{1}{\Gamma(\alpha)}\sum_{0<t_k<t}\int_{t_{k-1}}^{t_k}(t_k-s)^{\alpha-1}g(s)ds\\ &+\frac{1}{\Gamma(\alpha)}\int_{t_k}^{t}(t-s)^{\alpha-1}g(s)ds + \sum_{0<t_k<t} I_k(u(t_k^-)) , \end{aligned} \tag{5.12}$$

where $g \in C(J,\mathbb{R})$ is such that

$$g(t) = f(t, u(t), g(t)) .$$

By (5.4.1), for each $t \in J$ we have

$$\begin{aligned} |g(t)| &= |f(t,u(t),g(t))| \\ &\le p(t) + q(t)|u(t)| + r(t)|g(t)| \\ &\le p(t) + q(t)\eta^* + r(t)|g(t)| \\ &\le p^* + q^*\eta^* + r^*|g(t)| , \end{aligned}$$

where $p^* = \sup_{t\in J} p(t)$, and $q^* = \sup_{t\in J} q(t)$.

Then

$$|g(t)| \le \frac{p^* + q^*\eta^*}{1-r^*} := M .$$

Thus, (5.12) implies

$$\begin{aligned} |N(u)(t)| &\le \frac{|b|}{|a+b|}\left[m(M^*|u| + N^*) + \frac{mMT^\alpha}{\Gamma(\alpha+1)} + \frac{MT^\alpha}{\Gamma(\alpha+1)}\right] \\ &\quad + \frac{|c|}{|a+b|} + \frac{mMT^\alpha}{\Gamma(\alpha+1)} + \frac{MT^\alpha}{\Gamma(\alpha+1)} + m(M^*|u| + N^*) \\ &\le \left(\frac{|b|}{|a+b|} + 1\right)\left[m(M^*|u| + N^*) + \frac{(m+1)MT^\alpha}{\Gamma(\alpha+1)}\right] + \frac{|c|}{|a+b|} . \end{aligned}$$

Then

$$\|N(u)\|_{PC} \le \left(\frac{|b|}{|a+b|}+1\right)\left[m(M^*\eta^*+N^*)+\frac{(m+1)MT^\alpha}{\Gamma(\alpha+1)}\right]+\frac{|c|}{|a+b|}:=\ell\,.$$

*Step 3: N maps bounded sets to equicontinuous sets of $PC(J,\mathbb{R})$.* Let $\tau_1, \tau_2 \in J$, $\tau_1 < \tau_2$, $B_{\eta^*}$ be a bounded set of $PC(J,\mathbb{R})$ as in Step 2, and let $u \in B_{\eta^*}$. Then

$$\begin{aligned}
&|N(u)(\tau_2)-N(u)(\tau_1)|\\
&\le \frac{1}{\Gamma(\alpha)}\int_0^{\tau_1}|(\tau_2-s)^{\alpha-1}-(\tau_1-s)^{\alpha-1}\|g(s)|ds\\
&\quad+\frac{1}{\Gamma(\alpha)}\int_{\tau_1}^{\tau_2}|(\tau_2-s)^{\alpha-1}\|g(s)|ds+\sum_{0<t_k<\tau_2-\tau_1}|I_k(u(t_k^-))|\\
&\le \frac{M}{\Gamma(\alpha+1)}[2(\tau_2-\tau_1)^\alpha+(\tau_2^\alpha-\tau_1^\alpha)]+(\tau_2-\tau_1)(M^*|u|+N^*)\\
&\le \frac{M}{\Gamma(\alpha+1)}[2(\tau_2-\tau_1)^\alpha+(\tau_2^\alpha-\tau_1^\alpha)]+(\tau_2-\tau_1)(M^*\eta^*+N^*)\,.
\end{aligned}$$

As $\tau_1 \to \tau_2$, the right-hand side of the preceding inequality tends to zero. As a consequence of Steps 1–3, together with the Ascoli–Arzelà theorem, we can conclude that $N\colon PC(J,\mathbb{R}) \to PC(J,\mathbb{R})$ is completely continuous.

*Step 4: A priori bounds.* Now it remains to show that the set

$$E=\{u\in PC(J,\mathbb{R})\colon u=\lambda N(u)\ \text{ for some }\ 0<\lambda<1\}$$

is bounded. Let $u \in E$; then $u = \lambda N(u)$ for some $0 < \lambda < 1$. Thus, for each $t \in J$ we have

$$\begin{aligned}
u(t)&=\frac{-1}{a+b}\left[b\lambda\sum_{i=1}^m I_i(u(t_i^-))+\frac{b\lambda}{\Gamma(\alpha)}\sum_{i=1}^m\int_{t_{i-1}}^{t_i}(t_i-s)^{\alpha-1}g(s)ds\right.\\
&\quad\left.+\frac{b\lambda}{\Gamma(\alpha)}\int_{t_m}^{T}(T-s)^{\alpha-1}g(s)ds-c\lambda\right]+\frac{\lambda}{\Gamma(\alpha)}\sum_{0<t_k<t}\int_{t_{k-1}}^{t_k}(t_k-s)^{\alpha-1}g(s)ds\\
&\quad+\frac{\lambda}{\Gamma(\alpha)}\int_{t_k}^{t}(t-s)^{\alpha-1}g(s)ds+\lambda\sum_{0<t_k<t}I_k(u(t_k^-))\,.
\end{aligned}\tag{5.13}$$

From (5.4.1), for each $t \in J$ we have

$$\begin{aligned}
|g(t)|&=|f(t,u(t),g(t))|\\
&\le p(t)+q(t)|u(t)|+r(t)|g(t)|\\
&\le p^*+q^*|u(t)|+r^*|g(t)|\,.
\end{aligned}$$

Thus,

$$\begin{aligned}
|g(t)|&\le\frac{1}{1-r^*}(p^*+q^*|u(t)|)\\
&\le\frac{1}{1-r^*}(p^*+q^*\|u\|_{PC})\,.
\end{aligned}$$

This implies, by (5.13) and (5.4.2), that for each $t \in J$ we have

$$|u(t)| \leq \frac{|b|}{|a+b|}\left[m(M^*\|u\|_{PC}+N^*)+\frac{mT^\alpha(p^*+q^*\|u\|_{PC})}{(1-r^*)\Gamma(\alpha+1)}+\frac{T^\alpha(p^*+q^*\|u\|_{PC})}{(1-r^*)\Gamma(\alpha+1)}\right]$$
$$+\frac{|c|}{|a+b|}+\frac{mT^\alpha(p^*+q^*\|u\|_{PC})}{(1-r^*)\Gamma(\alpha+1)}+\frac{T^\alpha(p^*+q^*\|u\|_{PC})}{(1-r^*)\Gamma(\alpha+1)}+m(M^*\|u(t)\|_{PC}+N^*)\,.$$

Then

$$\|u\|_{PC} \leq \left(\frac{|b|}{|a+b|}+1\right)\left[m(M^*\|u(t)\|_{PC}+N^*)+\frac{(m+1)\,(p^*+q^*\|u\|_{PC})\,T^\alpha}{(1-r^*)\Gamma(\alpha+1)}\right]$$
$$+\frac{|c|}{|a+b|}$$
$$\leq \left(\frac{|b|}{|a+b|}+1\right)\left(mN^*+\frac{(m+1)p^*T^\alpha}{(1-r^*)\Gamma(\alpha+1)}\right)+\frac{|c|}{|a+b|}$$
$$+\left(\frac{|b|}{|a+b|}+1\right)\left(mM^*+\frac{(m+1)q^*T^\alpha}{(1-r^*)\Gamma(\alpha+1)}\right)\|u\|_{PC}\,.$$

Thus,

$$\left[1-\left(\frac{|b|}{|a+b|}+1\right)\left(mM^*+\frac{(m+1)q^*T^\alpha}{(1-r^*)\Gamma(\alpha+1)}\right)\right]\|u\|_{PC} \leq \left(\frac{|b|}{|a+b|}+1\right)\left[\frac{|c|}{|a+b|}\right.$$
$$\left.+mN^*+\frac{(m+1)p^*T^\alpha}{(1-r^*)\Gamma(\alpha+1)}\right].$$

Finally, by (5.10) we have

$$\|u\|_{PC} \leq \frac{\left(\frac{|b|}{|a+b|}+1\right)\left[mN^*+\frac{(m+1)p^*T^\alpha}{(1-r^*)\Gamma(\alpha+1)}+\frac{|c|}{|a+b|}\right]}{\left[1-\left(\frac{|b|}{|a+b|}+1\right)\left(mM^*+\frac{(m+1)q^*T^\alpha}{(1-r^*)\Gamma(\alpha+1)}\right)\right]} := R\,.$$

This shows that set $E$ is bounded. As a consequence of Schaefer's fixed point theorem, we deduce that $N$ has a fixed point that is a solution of problem (5.1)–(5.3). □

### 5.2.3 Ulam–Hyers Rassias stability

Here we adopt the concepts in Wang et al. [252] and introduce Ulam's type stability concepts for problem (5.1)–(5.2).

Let $z \in PC^1(J,\mathbb{R})$, $\epsilon > 0$, $\psi > 0$, and $\varphi \in PC(J,\mathbb{R}_+)$ is nondecreasing. We consider the set of inequalities

$$\begin{cases} |{}^cD^\alpha z(t) - f(t,z(t),{}^cD^\alpha z(t))| \leq \epsilon\,, & t \in (t_k,t_{k+1}],\ k=1,\dots,m \\ |\Delta z(t_k) - I_k(z(t_k^-))| \leq \epsilon\,, & k=1,\dots,m\,; \end{cases} \tag{5.14}$$

the set of inequalities

$$\begin{cases} |{}^cD^\alpha z(t) - f(t,z(t),{}^cD^\alpha z(t))| \leq \varphi(t)\,, & t \in (t_k,t_{k+1}],\ k=1,\dots,m \\ |\Delta z(t_k) - I_k(z(t_k^-))| \leq \psi\,, & k=1,\dots,m\,; \end{cases} \tag{5.15}$$

and the set of inequalities

$$\begin{cases} |{}^cD^\alpha z(t) - f(t, z(t), {}^cD^\alpha z(t))| \leq \epsilon\varphi(t)\,, & t \in (t_k, t_{k+1}],\ k = 1, \dots, m \\ |\Delta z(t_k) - I_k(z(t_k^-))| \leq \epsilon\psi\,, & k = 1, \dots, m\,. \end{cases} \tag{5.16}$$

**Definition 5.5.** Problem (5.1)–(5.2) is Ulam–Hyers stable if there exists a real number $c_{f,m} > 0$ such that for each $\epsilon > 0$ and for each solution $z \in PC^1(J, \mathbb{R})$ of inequality (5.14) there exists a solution $y \in PC^1(J, \mathbb{R})$ of problem (5.1)–(5.2), with

$$|z(t) - y(t)| \leq c_{f,m}\epsilon\,, \quad t \in J\,.$$

**Definition 5.6.** Problem (5.1)–(5.2) is generalized Ulam–Hyers stable if there exists $\theta_{f,m} \in C(\mathbb{R}_+, \mathbb{R}_+)$, $\theta_{f,m}(0) = 0$ such that for each solution $z \in PC^1(J, \mathbb{R})$ of inequality (5.14) there exists a solution $y \in PC^1(J, \mathbb{R})$ of problem (5.1)–(5.2), with

$$|z(t) - y(t)| \leq \theta_{f,m}(\epsilon)\,, \quad t \in J\,.$$

**Definition 5.7.** Problem (5.1)–(5.2) is Ulam–Hyers–Rassias stable with respect to $(\varphi, \psi)$ if there exists $c_{f,m,\varphi} > 0$ such that for each $\epsilon > 0$ and for each solution $z \in PC^1(J, \mathbb{R})$ of inequality (5.16) there exists a solution $y \in PC^1(J, \mathbb{R})$ of problem (5.1)–(5.2), with

$$|z(t) - y(t)| \leq c_{f,m,\varphi}\epsilon(\varphi(t) + \psi)\,, \quad t \in J\,.$$

**Definition 5.8.** Problem (5.1)–(5.2) is generalized Ulam–Hyers–Rassias stable with respect to $(\varphi, \psi)$ if there exists $c_{f,m,\varphi} > 0$ such that for each solution $z \in PC^1(J, \mathbb{R})$ of inequality (5.15) there exists a solution $y \in PC^1(J, \mathbb{R})$ of problem (5.1)–(5.2), with

$$|z(t) - y(t)| \leq c_{f,m,\varphi}(\varphi(t) + \psi)\,, \quad t \in J\,.$$

**Remark 5.9.** It is clear that (i) Definition 5.5 implies Definition 5.6, (ii) Definition 5.7 implies Definition 5.8, and (iii) Definition 5.7 for $\varphi(t) = \psi = 1$ implies Definition 5.5.

**Remark 5.10.** A function $z \in PC^1(J, \mathbb{R})$ is a solution of inequality (5.16) if and only if there is $\sigma \in PC(J, \mathbb{R})$ and a sequence $\sigma_k$, $k = 1, \dots, m$ (which depend on $z$) such that
**(i)** $|\sigma(t)| \leq \epsilon\varphi(t)$, $t \in (t_k, t_{k+1}]$, $k = 1, \dots, m$ and $|\sigma_k| \leq \epsilon\psi$, $k = 1, \dots, m$;
**(ii)** ${}^cD^\alpha z(t) = f(t, z(t), {}^cD^\alpha z(t)) + \sigma(t)$, $t \in (t_k, t_{k+1}]$, $k = 1, \dots, m$;
**(iii)** $\Delta z(t_k) = I_k(z(t_k^-)) + \sigma_k$, $k = 1, \dots, m$.

One could make similar remarks for inequalities 5.15 and 5.14. Now we state the Ulam–Hyers–Rassias stability result.

**Theorem 5.11.** *Assume* (5.3.1)–(5.3.3) *and* (5.8) *hold and*
(5.11.1) *there exists a nondecreasing function* $\varphi \in PC(J, \mathbb{R}_+)$ *and there exists* $\lambda_\varphi > 0$ *such that for any* $t \in J$,

$$I^\alpha\varphi(t) \leq \lambda_\varphi\varphi(t)\,.$$

*Then problem* (5.1)–(5.2) *is Ulam–Hyers–Rassias stable with respect to* $(\varphi, \psi)$.

*Proof.* Let $z \in PC^1(J, \mathbb{R})$ be a solution of inequality (5.16). Denote by $y$ the unique solution of the BVP

$$\begin{cases} {}^cD^\alpha_{t_k} y(t) = f(t, y(t), {}^cD^\alpha_{t_k} y(t)), & t \in (t_k, t_{k+1}], k = 1, \dots, m\,; \\ \Delta y(t_k) = I_k(y(t_k^-)), & k = 1, \dots, m\,; \\ ay(0) + by(T) = c\,; \\ y(0) = z(0)\,. \end{cases}$$

Using Lemma 5.2, we obtain for each $t \in (t_k, t_{k+1}]$

$$y(t) = y(0) + \sum_{i=1}^{k} I_i(y(t_i^-)) + \frac{1}{\Gamma(\alpha)} \sum_{i=1}^{k} \int_{t_{i-1}}^{t_i} (t_i - s)^{\alpha-1} g(s)ds$$
$$+ \frac{1}{\Gamma(\alpha)} \int_{t_k}^{t} (t - s)^{\alpha-1} g(s)ds\,, \quad t \in (t_k, t_{k+1}]\,,$$

where $g \in C(J, \mathbb{R})$ is such that

$$g(t) = f(t, y(t), g(t))\,.$$

Since $z$ is a solution of inequality (5.16) and by Remark 5.10, we have

$$\begin{cases} {}^cD^\alpha_{t_k} z(t) = f(t, z(t), {}^cD^\alpha_{t_k} z(t)) + \sigma(t), & t \in (t_k, t_{k+1}], k = 1, \dots, m\,; \\ \Delta z(t_k) = I_k(z(t_k^-)) + \sigma_k, & k = 1, \dots, m\,. \end{cases} \tag{5.17}$$

Clearly, the solution of (5.17) is given by

$$z(t) = z(0) + \sum_{i=1}^{k} I_i(z(t_i^-)) + \sum_{i=1}^{k} \sigma_i + \frac{1}{\Gamma(\alpha)} \sum_{i=1}^{k} \int_{t_{i-1}}^{t_i} (t_i - s)^{\alpha-1} h(s)ds$$
$$+ \frac{1}{\Gamma(\alpha)} \sum_{i=1}^{k} \int_{t_{i-1}}^{t_i} (t_i - s)^{\alpha-1} \sigma(s)ds + \frac{1}{\Gamma(\alpha)} \int_{t_k}^{t} (t - s)^{\alpha-1} h(s)ds$$
$$+ \frac{1}{\Gamma(\alpha)} \int_{t_k}^{t} (t - s)^{\alpha-1} \sigma(s)ds\,, \quad t \in (t_k, t_{k+1}]\,,$$

where $h \in C(J, \mathbb{R})$ is such that

$$h(t) = f(t, z(t), h(t))\,.$$

Hence, for each $t \in (t_k, t_{k+1}]$, it follows that

$$\begin{aligned}|z(t) - y(t)| &\le \sum_{i=1}^{k} |\sigma_i| + \sum_{i=1}^{k} |I_i(z(t_i^-)) - I_i(y(t_i^-))| \\ &+ \frac{1}{\Gamma(\alpha)} \sum_{i=1}^{k} \int_{t_{i-1}}^{t_i} (t_i - s)^{\alpha-1} |h(s) - g(s)| ds \\ &+ \frac{1}{\Gamma(\alpha)} \sum_{i=1}^{k} \int_{t_{i-1}}^{t_i} (t_i - s)^{\alpha-1} |\sigma(s)| ds \\ &+ \frac{1}{\Gamma(\alpha)} \int_{t_k}^{t} (t - s)^{\alpha-1} |h(s) - g(s)| ds \\ &+ \frac{1}{\Gamma(\alpha)} \int_{t_k}^{t} (t - s)^{\alpha-1} |\sigma(s)| .\end{aligned}$$

Thus,

$$\begin{aligned}|z(t) - y(t)| &\le m\epsilon\psi + (m+1)\epsilon\lambda_\varphi \varphi(t) + \sum_{i=1}^{k} \tilde{l} |z(t_i^-) - y(t_i^-)| \\ &+ \frac{1}{\Gamma(\alpha)} \sum_{i=1}^{k} \int_{t_{i-1}}^{t_i} (t_i - s)^{\alpha-1} |h(s) - g(s)| ds \\ &+ \frac{1}{\Gamma(\alpha)} \int_{t_k}^{t} (t - s)^{\alpha-1} |h(s) - g(s)| ds .\end{aligned}$$

By (5.3.2) we have

$$\begin{aligned}|h(t) - g(t)| &= |f(t, z(t), h(t)) - f(t, y(t), g(t))| \\ &\le K|z(t) - y(t)| + L|g(t) - h(t)| .\end{aligned}$$

Then

$$|h(t) - g(t)| \le \frac{K}{1 - L} |z(t) - y(t)| .$$

Therefore, for each $t \in J$

$$\begin{aligned}|z(t) - y(t)| &\le m\epsilon\psi + (m+1)\epsilon\lambda_\varphi \varphi(t) + \sum_{i=1}^{k} \tilde{l} |z(t_i^-) - y(t_i^-)| \\ &+ \frac{K}{(1 - L)\Gamma(\alpha)} \sum_{i=1}^{k} \int_{t_{i-1}}^{t_i} (t_i - s)^{\alpha-1} |z(s) - y(s)| ds \\ &+ \frac{K}{(1 - L)\Gamma(\alpha)} \int_{t_k}^{t} (t - s)^{\alpha-1} |z(s) - y(s)| ds .\end{aligned}$$

Thus,

$$|z(t) - y(t)| \leq \sum_{i=1}^{k} \widetilde{l}|z(t_i^-) - y(t_i^-)| + \epsilon(\psi + \varphi(t))(m + (m+1)\lambda_\varphi) + \frac{K(m+1)}{(1-L)\Gamma(\alpha)} \int_0^t (t-s)^{\alpha-1}|z(s) - y(s)|ds .$$

Applying Lemma 1.53, we get

$$\begin{aligned} |z(t) - y(t)| &\leq \epsilon(\psi + \varphi(t))(m + (m+1)\lambda_\varphi) \\ &\quad \times \left[ \prod_{0<t_k<t} (1 + \widetilde{l}) \exp\left( \int_0^t \frac{K(m+1)}{(1-L)\Gamma(\alpha)} (t-s)^{\alpha-1} ds \right) \right] \\ &\leq c_\varphi \epsilon(\psi + \varphi(t)) , \end{aligned}$$

where

$$\begin{aligned} c_\varphi &= (m + (m+1)\lambda_\varphi) \left[ \prod_{k=1}^{m} (1 + \widetilde{l}) \exp\left( \frac{K(m+1)T^\alpha}{(1-L)\Gamma(\alpha+1)} \right) \right] \\ &= (m + (m+1)\lambda_\varphi) \left[ (1 + \widetilde{l}) \exp\left( \frac{K(m+1)T^\alpha}{(1-L)\Gamma(\alpha+1)} \right) \right]^m . \end{aligned}$$

Thus, problem (5.1)–(5.2) is Ulam–Hyers–Rassias stable with respect to $(\varphi, \psi)$. The proof is complete. □

Next we present the Ulam–Hyers stability result.

**Theorem 5.12.** *Assume* (5.3.1)–(5.3.3) *and* (5.8) *hold. Then problem* (5.1)–(5.2) *is Ulam–Hyers stable.*

*Proof.* Let $z \in PC^1(J, \mathbb{R})$ be a solution of inequality (5.14). Denote by $y$ the unique solution of the BVP

$$\begin{cases} {}^cD_{t_k}^\alpha y(t) = f(t, y(t), {}^cD_{t_k}^\alpha y(t)), & t \in (t_k, t_{k+1}], k = 1, \ldots, m ; \\ \Delta y(t_k) = I_k(y(t_k^-)), & k = 1, \ldots, m ; \\ ay(0) + by(T) = c ; \\ y(0) = z(0) . \end{cases}$$

From the proof of Theorem 5.11 we get the inequality

$$|z(t) - y(t)| \leq \sum_{i=1}^{k} \widetilde{l}|(z(t_i^-)) - (y(t_i^-))| + m\epsilon + \frac{T^\alpha \epsilon(m+1)}{\Gamma(\alpha+1)} + \frac{K(m+1)}{(1-L)\Gamma(\alpha)} \int_0^t (t-s)^{\alpha-1}|z(s) - y(s)|ds .$$

Applying Lemma 1.53, we get

$$|z(t)-y(t)| \le \epsilon\left(\frac{m\Gamma(\alpha+1)+T^{\alpha}(m+1)}{\Gamma(\alpha+1)}\right)$$
$$\times\left[\prod_{0<t_k<t}(1+\tilde{l})\exp\left(\int_0^t \frac{K(m+1)}{(1-L)\Gamma(\alpha)}(t-s)^{\alpha-1}ds\right)\right]$$
$$\le c_{\varphi}\epsilon\,,$$

where

$$\begin{aligned} c_{\varphi} &= \left(\frac{m\Gamma(\alpha+1)+T^{\alpha}(m+1)}{\Gamma(\alpha+1)}\right)\left[\prod_{k=1}^{m}(1+\tilde{l})\exp\left(\frac{K(m+1)T^{\alpha}}{(1-L)\Gamma(\alpha+1)}\right)\right] \\ &= \left(\frac{m\Gamma(\alpha+1)+T^{\alpha}(m+1)}{\Gamma(\alpha+1)}\right)\left[(1+\tilde{l})\exp\left(\frac{K(m+1)T^{\alpha}}{(1-L)\Gamma(\alpha+1)}\right)\right]^{m}. \end{aligned}$$

Moreover, if we set $y(\epsilon)=c\epsilon;\, y(0)=0$, then problem (5.1)–(5.2) is generalized Ulam–Hyers stable. □

### 5.2.4 Examples

*Example 1.* Consider the impulsive BVP

$$^{c}D^{\frac{1}{2}}_{t_k}y(t) = \frac{1}{5e^{t+2}\left(1+|y(t)|+|^{c}D^{\frac{1}{2}}_{t_k}y(t)|\right)}, \quad \text{for each, } t\in J_0\cup J_1\,, \tag{5.18}$$

$$\Delta y|_{t=\frac{1}{2}} = \frac{\left|y\left(\frac{1}{2}^{-}\right)\right|}{10+\left|y\left(\frac{1}{2}^{-}\right)\right|}, \tag{5.19}$$

$$2y(0)-y(1)=3\,, \tag{5.20}$$

where $J_0=[0,\frac{1}{2}]$, $J_1=(\frac{1}{2},1]$, $t_0=0$, and $t_1=\frac{1}{2}$.

Set

$$f(t,u,v)=\frac{1}{5e^{t+2}(1+|u|+|v|)}, \quad t\in[0,1],\ u,v\in\mathbb{R}\,.$$

Clearly, the function $f$ is jointly continuous.

For each $u, v, \bar{u}, \bar{v}\in\mathbb{R}$ and $t\in[0,1]$

$$|f(t,u,v)-f(t,\bar{u},\bar{v})| \le \frac{1}{5e^2}(|u-\bar{u}|+|v-\bar{v}|)\,.$$

Hence, condition (5.3.2) is satisfied by $K=L=\frac{1}{5e^2}$. Let

$$I_1(u)=\frac{u}{10+u}, \quad u\in[0,\infty)\,.$$

Let $u, v \in [0, \infty)$. Then we have

$$|I_1(u) - I_1(v)| = \left|\frac{u}{10+u} - \frac{v}{10+v}\right| = \frac{10|u-v|}{(10+u)(10+v)} \le \frac{1}{10}|u-v| .$$

Thus, the condition

$$\left(\frac{|b|}{|a+b|} + 1\right)\left[m\tilde{l} + \frac{(m+1)KT^\alpha}{(1-L)\Gamma(\alpha+1)}\right] = 2\left[\frac{1}{10} + \frac{\frac{2}{5e^2}}{\left(1-\frac{1}{5e^2}\right)\Gamma\left(\frac{3}{2}\right)}\right]$$
$$= 2\left[\frac{4}{(5e^2-1)\sqrt{\pi}} + \frac{1}{10}\right] < 1$$

is satisfied by $T = 1$, $a = 2$, $b = -1$, $c = 3$, $m = 1$, and $\tilde{l} = \frac{1}{10}$. It follows from Theorem 5.3 that problem (5.18)–(5.20) has a unique solution on $J$.

Set, for any $t \in [0, 1]$, $\varphi(t) = t$, $\psi = 1$.

Since

$$I^{\frac{1}{2}}\varphi(t) = \frac{1}{\Gamma\left(\frac{1}{2}\right)}\int_0^t (t-s)^{\frac{1}{2}-1} s ds \le \frac{2t}{\sqrt{\pi}} ,$$

(5.11.1) is satisfied by $\lambda_\varphi = \frac{2}{\sqrt{\pi}}$. Thus, problem (5.18)–(5.19) is Ulam–Hyers–Rassias stable with respect to $(\varphi, \psi)$.

*Example 2.* Consider the impulsive antiperiodic problem

$$^cD_{t_k}^{\frac{1}{2}} y(t) = \frac{2 + |y(t)| + |^cD_{t_k}^{\frac{1}{2}} y(t)|}{108e^{t+3}\left(1 + |y(t)| + |^cD_{t_k}^{\frac{1}{2}} y(t)|\right)} , \quad \text{for each, } t \in J_0 \cup J_1 , \tag{5.21}$$

$$\Delta y|_{t=\frac{1}{3}} = \frac{|y\left(\frac{1}{3}^-\right)|}{6 + |y\left(\frac{1}{3}^-\right)|} , \tag{5.22}$$

$$y(0) = -y(1) , \tag{5.23}$$

where $J_0 = [0, \frac{1}{3}]$, $J_1 = (\frac{1}{3}, 1]$, $t_0 = 0$, and $t_1 = \frac{1}{3}$. Set

$$f(t, u, v) = \frac{2 + |u| + |v|}{108e^{t+3}(1 + |u| + |v|)} , \quad t \in [0, 1],\ u, v \in \mathbb{R} .$$

Clearly, the function $f$ is jointly continuous.

For any $u, v, \bar{u}, \bar{v} \in \mathbb{R}$ and $t \in [0, 1]$

$$|f(t, u, v) - f(t, \bar{u}, \bar{v})| \le \frac{1}{108e^3}(|u - \bar{u}| + |v - \bar{v}|) .$$

Hence, condition (5.3.2) is satisfied by $K = L = \frac{1}{108e^3}$.

For each $t \in [0, 1]$ we have

$$|f(t, u, v)| \le \frac{1}{108e^{t+3}}(2 + |u| + |v|) .$$

Thus, condition (5.4.1) is satisfied by $p(t) = \frac{1}{54e^{t+3}}$ and $q(t) = r(t) = \frac{1}{108e^{t+3}}$. Let

$$I_1(u) = \frac{u}{6+u}, \quad u \in [0, \infty).$$

For each $u \in [0, \infty)$ we have

$$|I_1(u)| \leq \frac{1}{6}u + 1.$$

Thus, condition (5.4.2) is satisfied by $M^* = \frac{1}{6}$ and $N^* = 1$. Thus, the condition

$$\left(\frac{|b|}{|a+b|} + 1\right)\left(mM^* + \frac{(m+1)q^*T^\alpha}{(1-r^*)\Gamma(\alpha+1)}\right) = \frac{3}{2}\left(\frac{1}{6} + \frac{4}{(108e^3 - 1)\sqrt{\pi}}\right) < 1$$

is satisfied by $T = 1$, $a = 1$, $b = 1$, $c = 0$, $m = 1$, and $q^*(t) = r^*(t) = \frac{1}{108e^3}$. It follows from Theorem 5.4 that problem (5.21)–(5.23) has at least one solution on $J$.

## 5.3 Existence Results for Impulsive NIFDE in Banach Space

### 5.3.1 Introduction

The purpose of this section is to establish existence and uniqueness results to the BVPs for NIFDEs

$$^cD^\nu_{t_k}y(t) = f(t, y, {}^cD^\nu_{t_k}y(t)), \quad \text{for each}, \quad t \in (t_k, t_{k+1}], \ k = 0, \dots, m, \ 0 < \nu \leq 1, \tag{5.24}$$

$$\Delta y|_{t=t_k} = I_k(y(t_k^-)), \qquad k = 1, \dots, m, \tag{5.25}$$

$$ay(0) + by(T) = c, \tag{5.26}$$

where $^cD^\nu_{t_k}$ is the Caputo fractional derivative, $(E, \|\cdot\|)$ is a real Banach space, $f: J \times E \times E \to E$ is a given function, $I_k: E \to E$, $a, b$ are real constants with $a + b \neq 0$ and $c \in E$, $0 = t_0 < t_1 < \dots < t_m < t_{m+1} = T$, $\Delta y|_{t=t_k} = y(t_k^+) - y(t_k^-)$, and $y(t_k^+) = \lim_{h\to 0^+} y(t_k + h)$ and $y(t_k^-) = \lim_{h\to 0^-} y(t_k + h)$ represent the right and left limits of $y(t)$ at $t = t_k$.

In this section, two results are discussed; the first is based on Darbo's fixed point theorem combined with the technique of measures of noncompactness, the second is based on Mönch's fixed point theorem. Finally, two examples are given to demonstrate the application of our main results.

### 5.3.2 Existence of Solutions

**Definition 5.13.** A function $y \in PC(J, E)$ whose $\nu$-derivative exists on $J_k$ is said to be a solution of (5.24)–(5.26) if $y$ satisfies the equation $^cD^\nu_{t_k}y(t) = f(t, y(t), {}^cD^\nu_{t_k}y(t))$ on $J_k$

and satisfies the conditions

$$\Delta y|_{t=t_k} = I_k(y(t_k^-)), \quad k = 1, \ldots, m,$$
$$ay(0) + by(T) = c.$$

To prove the existence of solutions to (5.24)–(5.26), we need the following auxiliary lemma.

**Lemma 5.14.** *Let $0 < \nu \le 1$, and let $\sigma\colon J \to E$ be continuous. A function $y$ is a solution of the fractional integral equation*

$$y(t) = \begin{cases} \frac{-1}{a+b}\left[b\sum_{i=1}^{m} I_i(y(t_i^-)) + \frac{b}{\Gamma(\nu)}\sum_{i=1}^{m}\int_{t_{i-1}}^{t_i}(t_i - s)^{\nu-1}\sigma(s)ds \right. \\ \left. + \frac{b}{\Gamma(\nu)}\int_{t_m}^{T}(T-s)^{\nu-1}\sigma(s)ds - c\right] + \frac{1}{\Gamma(\nu)}\int_0^t (t-s)^{\nu-1}\sigma(s)ds, & \text{if } t \in [0, t_1], \\ \frac{-1}{a+b}\left[b\sum_{i=1}^{m} I_i(y(t_i^-)) + \frac{b}{\Gamma(\nu)}\sum_{i=1}^{m}\int_{t_{i-1}}^{t_i}(t_i - s)^{\nu-1}\sigma(s)ds \right. \\ \left. + \frac{b}{\Gamma(\nu)}\int_{t_m}^{T}(T-s)^{\nu-1}\sigma(s)ds - c\right] + \sum_{i=1}^{k} I_i(y(t_i^-)) + \frac{1}{\Gamma(\nu)}\sum_{i=1}^{k}\int_{t_{i-1}}^{t_i}(t_i - s)^{\nu-1}\sigma(s)ds \\ + \frac{1}{\Gamma(\nu)}\int_{t_k}^{t}(t-s)^{\nu-1}\sigma(s)ds, & \text{if } t \in (t_k, t_{k+1}], \end{cases} \tag{5.27}$$

*where $k = 1, \ldots, m$, if and only if $y$ is a solution of the fractional BVP*

$${}^cD^\nu y(t) = \sigma(t), \quad t \in J_k,$$
$$\Delta y|_{t=t_k} = I_k(y(t_k^-)), \quad k = 1, \ldots, m,$$
$$ay(0) + by(T) = c.$$

We list the following conditions:

(5.15.1) The function $f\colon J \times E \times E \to E$ is continuous.

(5.15.2) There exist constants $K > 0$ and $0 < L < 1$ such that

$$\|f(t, u, v) - f(t, \bar{u}, \bar{v})\| \le K\|u - \bar{u}\| + L\|v - \bar{v}\|$$

for any $u, \bar{u}, v, \bar{v} \in E$ and $t \in J$.

(5.15.3) There exists a constant $\tilde{l} > 0$ such that

$$\|I_k(u) - I_k(\overline{u})\| \le \tilde{l}\|u - \overline{u}\|$$

for each $u, \overline{u} \in E$ and $k = 1, \ldots, m$.

We are now in a position to state and prove our existence result for problem (5.24)–(5.26) based on the concept of measures of noncompactness and Darbo's fixed point theorem.

**Remark 5.15** ([66]). Conditions (5.15.2) and (5.15.3) are respectively equivalent to the inequalities

$$\alpha\,(f(t, B_1, B_2)) \leq K\alpha(B_1) + L\alpha(B_2)$$
$$\alpha(I_k(B_1)) \leq \widetilde{l}\alpha(B_1)\,,$$

for any bounded sets $B_1, B_2 \subseteq E$, for each $t \in J$ and $k = 1, \ldots, m$.

**Theorem 5.16.** *Assume* (5.15.1)–(5.15.3) *hold. If*

$$\left(\frac{|b|}{|a+b|} + 1\right)\left(m\widetilde{l} + \frac{(m+1)KT^{\nu}}{(1-L)\Gamma(\nu+1)}\right) < 1\,, \tag{5.28}$$

*then BVP* (5.24)–(5.26) *has at least one solution on* $J$.

*Proof.* Transform problem (5.24)–(5.26) into a fixed point problem. Consider the operator $N\colon PC(J,E) \to PC(J,E)$ defined by

$$\begin{aligned} N(y)(t) = \frac{-1}{a+b}\Bigg[ & b\sum_{i=1}^{m} I_i(y(t_i^-)) + \frac{b}{\Gamma(\nu)}\sum_{i=1}^{m}\int_{t_{i-1}}^{t_i}(t_i - s)^{\nu-1}g(s)ds \\ & + \frac{b}{\Gamma(\nu)}\int_{t_m}^{T}(T-s)^{\nu-1}g(s)ds - c\Bigg] + \frac{1}{\Gamma(\nu)}\sum_{0<t_k<t}\int_{t_{k-1}}^{t_k}(t_k - s)^{\nu-1}g(s)ds \\ & + \frac{1}{\Gamma(\nu)}\int_{t_k}^{t}(t-s)^{\nu-1}g(s)ds + \sum_{0<t_k<t} I_k(y(t_k^-))\,, \end{aligned} \tag{5.29}$$

where $g \in C(J,E)$ is such that

$$g(t) = f(t, y(t), g(t))\,.$$

Clearly, the fixed points of operator $N$ are solutions of problem (5.24)–(5.26).

We will show that $N$ satisfies the assumptions of *Darbo's* fixed point theorem. The proof will be given in several claims.

*Claim 1: N is continuous.* Let $\{u_n\}$ be a sequence such that $u_n \to u$ in $PC(J,E)$. Then, for each $t \in J$,

$$\begin{aligned} \|N(u_n)(t) - N(u)(t)\| \leq \frac{|b|}{|a+b|}\Bigg[ & \sum_{i=1}^{m}\|I_k(u_n(t_k^-)) - I_k(u(t_k^-))\| \\ & + \frac{1}{\Gamma(\nu)}\sum_{i=1}^{m}\int_{t_{i-1}}^{t_i}(t_i - s)^{\nu-1}\|g_n(s) - g(s)\|ds \\ & + \frac{1}{\Gamma(\nu)}\int_{t_m}^{T}(T-s)^{\nu-1}\|g_n(s) - g(s)\|ds\Bigg] \end{aligned}$$

$$+ \frac{1}{\Gamma(\nu)} \sum_{0<t_k<t} \int_{t_{k-1}}^{t_k} (t_k - s)^{\nu-1} \|g_n(s) - g(s)\| ds$$
$$+ \frac{1}{\Gamma(\nu)} \int_{t_k}^{t} (t - s)^{\nu-1} \|g_n(s) - g(s)\| ds$$
$$+ \sum_{0<t_k<t} \|I_k(u_n(t_k^-)) - I_k(u(t_k^-))\| , \tag{5.30}$$

where $g_n, g \in C(J, E)$ such that

$$g_n(t) = f(t, u_n(t), g_n(t))$$

and

$$g(t) = f(t, u(t), g(t)) .$$

By (P2) we have

$$\begin{aligned} \|g_n(t) - g(t)\| &= \|f(t, u_n(t), g_n(t)) - f(t, u(t), g(t))\| \\ &\le K\|u_n(t) - u(t)\| + L\|g_n(t) - g(t)\| . \end{aligned}$$

Then

$$\|g_n(t) - g(t)\| \le \frac{K}{1 - L}\|u_n(t) - u(t)\| .$$

Since $u_n \to u$, we get $g_n(t) \to g(t)$ as $n \to \infty$ for each $t \in J$. Let $\eta > 0$ be such that, for each $t \in J$, we have $\|g_n(t)\| \le \eta$ and $\|g(t)\| \le \eta$. Then we have

$$\begin{aligned} (t - s)^{\nu-1}\|g_n(s) - g(s)\| &\le (t - s)^{\nu-1}[\|g_n(s)\| + \|g(s)\|] \\ &\le 2\eta(t - s)^{\nu-1} \end{aligned}$$

and

$$\begin{aligned} (t_k - s)^{\nu-1}\|g_n(s) - g(s)\| &\le (t_k - s)^{\nu-1}[\|g_n(s)\| + \|g(s)\|] \\ &\le 2\eta(t_k - s)^{\nu-1} . \end{aligned}$$

For each $t \in J$ the functions $s \to 2\eta(t - s)^{\nu-1}$ and $s \to 2\eta(t_k - s)^{\nu-1}$ are integrable on $[0, t]$; then the Lebesgue dominated convergence theorem and (5.30) imply that

$$\|N(u_n)(t) - N(u)(t)\| \to 0 \text{ as } n \to \infty .$$

Hence,

$$\|N(u_n) - N(u)\|_{PC} \to 0 \text{ as } n \to \infty .$$

Consequently, $N$ is continuous.

Let $R$ be the constant such that

$$R \geq \frac{\|c\|\Gamma(\nu+1)(1-L) + (|b| + |a+b|)\,[mc_1\Gamma(\nu+1)(1-L) + (m+1)T^\nu f^*]}{|a+b|\Gamma(\nu+1)(1-L) - (|b| + |a+b|)\left[m\tilde{l}\Gamma(\nu+1)(1-L) + (m+1)T^\nu K\right]}, \quad (5.31)$$

where $c_1 = \sup_{v\in E}\|I(v)\|$ and $f^* = \sup_{t\in J}\|f(t,0,0)\|$.

Define

$$D_R = \{u \in PC(J,E) : \|u\|_{PC} \leq R\}.$$

It is clear that $D_R$ is a bounded, closed, and convex subset of $PC(J,E)$.

*Claim 2:* $N(D_R) \subset D_R$. Let $u \in D_R$; we show that $Nu \in D_R$. For each $t \in J$ we have

$$\begin{aligned}\|N(y)(t)\| \leq{}& \frac{\|c\|}{|a+b|} + \frac{|b|}{|a+b|}\left[\sum_{i=1}^{m}\|I_i(y(t_i^-))\| + \frac{1}{\Gamma(\nu)}\sum_{i=1}^{m}\int_{t_{i-1}}^{t_i}(t_i-s)^{\nu-1}\|g(s)\|ds\right.\\ &\left.+\frac{1}{\Gamma(\nu)}\int_{t_m}^{T}(T-s)^{\nu-1}\|g(s)\|ds\right] + \frac{1}{\Gamma(\nu)}\sum_{0<t_k<t}\int_{t_{k-1}}^{t_k}(t_k-s)^{\nu-1}\|g(s)\|ds\\ &+\frac{1}{\Gamma(\nu)}\int_{t_k}^{t}(t-s)^{\nu-1}\|g(s)\|ds + \sum_{0<t_k<t}\|I_k(y(t_k^-))\|. \end{aligned} \quad (5.32)$$

By (P2) we have for each $t \in J$

$$\begin{aligned}\|g(t)\| &\leq \|f(t,u(t),g(t)) - f(t,0,0)\| + \|f(t,0,0)\|\\ &\leq K\|u(t)\| + L\|g(t)\| + f^*\\ &\leq K\|u(t)\|_{PC} + L\|g(t)\| + f^*\\ &\leq KR + L\|g(t)\| + f^*.\end{aligned}$$

Then

$$\|g(t)\| \leq \frac{f^* + KR}{1-L} := M.$$

Thus, (5.31), (5.32), and (5.15.3) imply that

$$\begin{aligned}\|Nu(t)\| \leq{}& \frac{\|c\|}{|a+b|} + \left(\frac{|b|}{|a+b|}+1\right)\left(\sum_{i=1}^{m}\|I_i(y(t_i^-)) - I_i(0)\| + \sum_{i=1}^{m}\|I_i(0)\|\right)\\ &+\left(\frac{|b|}{|a+b|}+1\right)\frac{(m+1)T^\nu M}{\Gamma(\nu+1)}\\ \leq{}& \frac{\|c\|}{|a+b|} + \left(\frac{|b|}{|a+b|}+1\right)\left[m(\tilde{l}R + c_1) + \frac{(m+1)T^\nu M}{\Gamma(\nu+1)}\right]\\ \leq{}& R.\end{aligned}$$

Thus, for each $t \in J$ we have $\|Nu(t)\| \leq R$. This implies that $\|Nu\|_{PC} \leq R$. Consequently,

$$N(D_R) \subset D_R.$$

*Claim 3: $N(D_R)$ is bounded and equicontinuous.* By Claim 2 we have $N(D_R) = \{N(u): u \in D_R\} \subset D_R$. Thus, for each $u \in D_R$ we have $\|N(u)\|_{PC} \le R$. Thus, $N(D_R)$ is bounded. Let $t_1, t_2 \in (0, T]$, $t_1 < t_2$, and let $u \in D_R$. Then

$$
\begin{aligned}
&\|N(u)(t_2) - N(u)(t_1)\| \\
&\le \frac{1}{\Gamma(\nu)} \int_0^{t_1} |(t_2 - s)^{\nu-1} - (t_1 - s)^{\nu-1}| \|g(s)\| ds + \frac{1}{\Gamma(\nu)} \int_{t_1}^{t_2} |(t_2 - s)^{\nu-1}| \|g(s)\| ds \\
&\quad + \sum_{0<t_k<t_2-t_1} \|I_k(u(t_k^-)) - I_k(0)\| + \sum_{0<t_k<t_2-t_1} \|I_k(0)\| \\
&\le \frac{M}{\Gamma(\nu+1)}[2(t_2 - t_1)^\nu + (t_2^\nu - t_1^\nu)] + (t_2 - t_1)(\tilde{l}\|u(t_k^-)\| + c_1) \\
&\le \frac{M}{\Gamma(\nu+1)}[2(t_2 - t_1)^\nu + (t_2^\nu - t_1^\nu)] + (t_2 - t_1)(\tilde{l}\|u\|_{PC} + c_1) \\
&\le \frac{M}{\Gamma(\nu+1)}[2(t_2 - t_1)^\nu + (t_2^\nu - t_1^\nu)] + (t_2 - t_1)(\tilde{l}R + c_1) .
\end{aligned}
$$

As $t_1 \to t_2$, the right-hand side of the preceding inequality tends to zero.

*Claim 4: The operator $N: D_R \to D_R$ is a strict set contraction.* Let $V \subset D_R$ and $t \in J$; then we have

$$
\begin{aligned}
\alpha(N(V)(t)) &= \alpha((Ny)(t), y \in V) \\
&\le \frac{|b|}{|a+b|} \left[ \sum_{i=1}^{m} \{\alpha(I_i(y(t_i^-))), y \in V\} \right. \\
&\quad + \frac{1}{\Gamma(\nu)} \sum_{i=1}^{m} \left\{ \int_{t_{i-1}}^{t_i} (t_i - s)^{\nu-1} \alpha(g(s)) ds, y \in V \right\} \\
&\quad \left. + \frac{1}{\Gamma(\nu)} \left\{ \int_{t_m}^{T} (T - s)^{\nu-1} \alpha(g(s)) ds, y \in V \right\} \right] \\
&\quad + \frac{1}{\Gamma(\nu)} \sum_{0<t_k<t} \left\{ \int_{t_{k-1}}^{t_k} (t_k - s)^{\nu-1} \alpha(g(s)) ds, y \in V \right\} \\
&\quad + \frac{1}{\Gamma(\nu)} \left\{ \int_{t_k}^{t} (t - s)^{\nu-1} \alpha(g(s)) ds, y \in V \right\} \\
&\quad + \sum_{0<t_k<t} \{\alpha(I_k(y(t_k^-))), y \in V\} .
\end{aligned}
$$

Then Remark 5.15 and Lemma 1.32 imply that, for each $s \in J$,

$$
\begin{aligned}
\alpha(\{g(s), y \in V\}) &= \alpha(\{f(s, y(s), g(s)), y \in V\}) \\
&\le K\alpha(\{y(s), y \in V\}) + L\alpha(\{g(s), y \in V\}) .
\end{aligned}
$$

Thus,

$$\alpha(\{g(s), y \in V\}) \leq \frac{K}{1-L}\alpha\{y(s), y \in V\}.$$

On the other hand, for each $t \in J$ and $k = 1, \ldots, m$ we have

$$\sum_{0<t_k<t} \alpha(\{I_k(y(t_k^-)), y \in V\}) \leq m\tilde{l}\alpha(\{y(t), y \in V\}).$$

Then

$$\begin{aligned}
\alpha(N(V)(t)) \leq \frac{|b|}{|a+b|}\Bigg[ & m\tilde{l}\alpha(\{y(t), y \in V\}) \\
& + \frac{mK}{\Gamma(\nu)(1-L)}\left\{\int_0^t (t-s)^{\nu-1}\{\alpha(y(s))\}ds, y \in V\right\} \\
& + \frac{K}{\Gamma(\nu)(1-L)}\left\{\int_0^T (T-s)^{\nu-1}\{\alpha(y(s))\}ds, y \in V\right\}\Bigg] \\
& + \frac{mK}{\Gamma(\nu)(1-L)}\left\{\int_0^t (t-s)^{\nu-1}\{\alpha(y(s))\}ds, y \in V\right\} \\
& + \frac{K}{\Gamma(\nu)(1-L)}\left\{\int_0^t (t-s)^{\nu-1}\{\alpha(y(s))\}ds, y \in V\right\} \\
& + m\tilde{l}\alpha(\{y(t), y \in V\}) \\
\leq & \left(\frac{|b|}{|a+b|}+1\right)\left(m\tilde{l} + \frac{(m+1)KT^{\nu}}{(1-L)\Gamma(\nu+1)}\right)\alpha_c(V).
\end{aligned}$$

Therefore,

$$\alpha_c(NV) \leq \left(\frac{|b|}{|a+b|}+1\right)\left(m\tilde{l} + \frac{(m+1)KT^{\nu}}{(1-L)\Gamma(\nu+1)}\right)\alpha_c(V).$$

Thus, operator $N$ is a set contraction. As a consequence of Theorem 1.45, the operator $N$ has a fixed point that is a solution of problem (5.24)–(5.26). □

Our next existence result for problem (5.24)–(5.26) is based on the concept of measures of noncompactness and *Mönch's* fixed point theorem.

**Theorem 5.17.** *Assume* (5.15.1)–(5.15.3) *and* (5.28) *hold. If*

$$m\tilde{l} < 1,$$

*then BVP* (5.24)–(5.26) *has at least one solution.*

*Proof.* Consider operator $N$ defined in (5.29). We will show that $N$ satisfies the assumptions of *Mönch's* fixed point theorem. We know that $N\colon D_R \to D_R$ is bounded and continuous, and we need to prove that the implication

$$[V = \overline{\operatorname{conv}} N(V) \text{ or } V = N(V) \cup \{0\}] \text{ implies } \alpha(V) = 0$$

holds for every subset $V$ of $D_R$. Now let $V$ be a subset of $D_R$ such that $V \subset \overline{\operatorname{conv}}(N(V) \cup \{0\})$. $V$ is bounded and equicontinuous, and therefore the function $t \to v(t) = \alpha(V(t))$ is continuous on $[0, T]$.

Using Lemma 5.14, we can write for each $t \in J$ and $k = 0, \dots, m$

$$N(y(t)) = y(0) + \sum_{i=1}^{k} I_i(y(t_i^-)) + \frac{1}{\Gamma(\alpha)} \sum_{i=1}^{k} \int_{t_{i-1}}^{t_i} (t_i - s)^{\alpha-1} g(s) ds$$
$$+ \frac{1}{\Gamma(\alpha)} \int_{t_k}^{t} (t - s)^{\alpha-1} g(s) ds\,,$$

where $g \in C(J, \mathbb{R})$ is such that

$$g(t) = f(t, y(t), g(t))\,.$$

By Remark 5.15, Lemma 1.33, and the properties of the measure $\alpha$, for each $t \in J$ we have

$$\begin{aligned}
v(t) &\le \alpha(N(V)(t) \cup \{0\}) \\
&\le \alpha(N(V)(t)) \\
&\le \alpha\{(Ny)(t), y \in V\} \\
&\le \alpha(y(0)) + \frac{mK}{\Gamma(\nu)(1-L)} \left\{ \int_0^t (t-s)^{\nu-1} \{\alpha(y(s))\} ds, y \in V \right\} \\
&\quad + \frac{K}{\Gamma(\nu)(1-L)} \left\{ \int_0^t (t-s)^{\nu-1} \{\alpha(y(s))\} ds, y \in V \right\} \\
&\quad + m\tilde{l}\alpha(\{y(t), y \in V\}) \\
&\le m\tilde{l}\alpha(\{y(t), y \in V\}) + \frac{(m+1)K}{(1-L)\Gamma(\nu)} \left\{ \int_0^t (t-s)^{\nu-1} \{\alpha(y(s))\} ds, y \in V \right\} \\
&= m\tilde{l}v(t) + \frac{(m+1)K}{(1-L)\Gamma(\nu)} \int_0^t (t-s)^{\nu-1} v(s) ds\,.
\end{aligned}$$

Then

$$v(t) \le \frac{(m+1)K}{(1-m\tilde{l})(1-L)\Gamma(\nu)} \int_0^t (t-s)^{\nu-1} v(s) ds\,.$$

Lemma 1.52 implies that $v(t) = 0$ for each $t \in J$. Then $V(t)$ is relatively compact in $E$. In view of the Ascoli–Arzelà theorem, $V$ is relatively compact in $D_R$. Applying now Theorem 1.46 we conclude that $N$ has a fixed point $y \in D_R$. Hence, $N$ has a fixed point that is a solution of problem (5.24)–(5.26). □

**Remark 5.18.** Our results for BVP (5.24)–(5.26) are appropriate for the following problems:

- Initial value problem: $a = 1, b = 0, c = 0$.
- Terminal value problem: $a = 0, b = 1, c$ arbitrary.
- Antiperiodic problem: $a = 1, b = 1, c = 0$.

However, our results are not applicable for the periodic problem, i.e., for $a = 1, b = -1, c = 0$.

### 5.3.3 Examples

*Example 1.* Consider the infinite system

$$ {}^cD^{\frac{1}{2}}_{t_k} y_n(t) = \frac{e^{-t}}{(11+e^t)}\left[\frac{y_n(t)}{1+y_n(t)} - \frac{{}^cD^{\frac{1}{2}}_{t_k} y_n(t)}{1+{}^cD^{\frac{1}{2}}_{t_k} y_n(t)}\right] \quad \text{for each } t \in J_0 \cup J_1\,, \tag{5.33} $$

$$ \Delta y_n|_{t=\frac{1}{2}} = \frac{y_n\left(\frac{1}{2}^-\right)}{10 + y_n\left(\frac{1}{2}^-\right)}\,, \tag{5.34} $$

$$ 2y_n(0) - y_n(1) = 3\,, \tag{5.35} $$

where $J_0 = [0, \frac{1}{2}]$, $J_1 = (\frac{1}{2}, 1]$, $t_0 = 0$, and $t_1 = \frac{1}{2}$.

Set

$$ E = l^1 = \left\{ y = (y_1, y_2, \ldots, y_n, \ldots)\,, \quad \sum_{n=1}^{\infty} |y_n| < \infty \right\}\,, $$

$$ f = (f_1, f_2, \ldots, f_n, \ldots)\,, $$

such that

$$ f(t,u,v) = \frac{e^{-t}}{(11+e^t)}\left[\frac{u}{1+u} - \frac{v}{1+v}\right]\,, \quad t \in [0,1],\ u, v \in E\,. $$

Clearly, the function $f$ is jointly continuous and $E$ is a Banach space with the norm $\|y\| = \sum_{n=1}^{\infty} |y_n|$.

For any $u, \bar{u}, v, \bar{v} \in E$, and $t \in [0, 1]$

$$ \|f(t,u,v) - f(t,\bar{u},\bar{v})\| \le \frac{1}{12}(\|u-\bar{u}\| + \|v-\bar{v}\|)\,. $$

Hence, condition (5.15.2) is satisfied by $K = L = \dfrac{1}{12}$. Let

$$ I_1(u) = \frac{u}{10+u}\,, \quad u \in E\,. $$

Let $u, v \in E$. Then we have

$$\|I_1(u) - I_1(v)\| = \left\| \frac{u}{10+u} - \frac{v}{10+v} \right\| \le \frac{1}{10}\|u - v\| .$$

Hence, condition (5.15.3) is satisfied by $\widetilde{l} = \frac{1}{10}$. The conditions

$$\begin{aligned}\left(\frac{|b|}{|a+b|} + 1\right)\left(m\widetilde{l} + \frac{(m+1)KT^{\nu}}{(1-L)\Gamma(\nu+1)}\right) &= \frac{1}{10} + \frac{\frac{2}{12}}{\left(1-\frac{1}{12}\right)\Gamma\left(\frac{3}{2}\right)} \\ &= \frac{8}{11\sqrt{\pi}} + \frac{1}{5} < 1\end{aligned}$$

are satisfied by $T = m = 1$, $a = 2$, $b = -1$, and $\nu = \frac{1}{2}$.

It follows from Theorem 5.16 that problem (5.33)–(5.35) has at least one solution on $J$.

*Example 2.* Consider the impulsive problem

$$ {}^cD^{\frac{1}{2}}_{t_k} y_n(t) = \frac{2 + \|y_n(t)\| + \|{}^cD^{\frac{1}{2}}_{t_k} y_n(t)\|}{108e^{t+3}\left(1 + \|y_n(t)\| + \|{}^cD^{\frac{1}{2}}_{t_k} y_n(t)\|\right)}, \quad \text{for each, } t \in J_0 \cup J_1 , \tag{5.36}$$

$$\Delta y_n|_{t=\frac{1}{3}} = \frac{\|y_n\left(\frac{1}{3}^-\right)\|}{6 + \|y_n\left(\frac{1}{3}^-\right)\|}, \tag{5.37}$$

$$y_n(0) = -y_n(1) , \tag{5.38}$$

where $J_0 = [0, \frac{1}{3}]$, $J_1 = (\frac{1}{3}, 1]$, $t_0 = 0$, and $t_1 = \frac{1}{3}$. Set

$$E = l^1 = \{y = (y_1, y_2, \dots, y_n, \dots), \quad \sum_{n=1}^{\infty} |y_n| < \infty\} ,$$

$$f = (f_1, f_2, \dots, f_n, \dots) ,$$

such that

$$f(t, u, v) = \frac{2 + \|u\| + \|v\|}{108e^{t+3}(1 + \|u\| + \|v\|)}, \quad t \in [0, 1], \ u, v \in E .$$

Clearly, the function $f$ is jointly continuous.

$E$ is a Banach space with the norm $\|y\| = \sum_{n=1}^{\infty} |y_n|$.

For any $u, \bar{u}, v, \bar{v} \in E$ and $t \in [0, 1]$

$$\|f(t, u, v) - f(t, \bar{u}, \bar{v})\| \le \frac{1}{108e^3}(\|u - \bar{u}\| + \|v - \bar{v}\|) .$$

Hence, condition (5.15.2) is satisfied by $K = L = \frac{1}{108e^3}$. Let

$$I_1(u) = \frac{\|u\|}{6 + \|u\|}, \quad u \in E .$$

Let $u, v \in E$. Then we have

$$\|I_1(u) - I_1(v)\| = \left\| \frac{u}{6+u} - \frac{v}{6+v} \right\| \le \frac{1}{6}\|u - v\| .$$

Hence, condition (5.15.3) is satisfied by $\widetilde{l} = \frac{1}{6}$.
The condition

$$\begin{aligned}\left(\frac{|b|}{|a+b|} + 1\right)\left(m\widetilde{l} + \frac{(m+1)KT^{\nu}}{(1-L)\Gamma(\nu+1)}\right) &= \frac{3}{2}\left(\frac{1}{6} + \frac{\frac{2}{12}}{\left(1 - \frac{1}{12}\right)\Gamma\left(\frac{3}{2}\right)}\right) \\ &= \frac{6}{11\sqrt{\pi}} + \frac{1}{4} < 1\end{aligned}$$

is satisfied by $T = m = 1$, $a = 1$, $b = 1$, and $\nu = \frac{1}{2}$. Also, we have

$$m\widetilde{l} = \frac{1}{6} < 1 .$$

It follows from Theorem 5.17 that problem (5.36)–(5.38) has at least one solution on $J$.

## 5.4 Notes and Remarks

The results of Chapter 5 are taken from Benchohra et al. [92]. Other results may be found in [57, 125, 249, 250].

# 6 Integrable Solutions for Implicit Fractional Differential Equations

## 6.1 Introduction

This chapter deals with the existence of integrable solutions for initial value problems (IVPs) for implicit fractional differential equations. We present results based on Schauder's fixed point theorem and the Banach contraction . Then we present other results with infinite delay. In the literature devoted to equations with finite delay, the state space is usually the space of all continuous functions on $[-r, 0]$, $r > 0$ and $\alpha = 1$ endowed with the uniform norm topology; see the book by Hale and Lunel [155]. When the delay is infinite, the selection of the state $\mathcal{B}$ (i.e., phase space) plays an important role in the study of both qualitative and quantitative theory for functional differential equations. A usual choice is a seminormed space satisfying suitable axioms introduced by Hale and Kato [154]. For a detailed discussion on this topic we refer the reader to the book by Hino et al. [162].

## 6.2 Integrable Solutions for NIFDE

### 6.2.1 Introduction

In this section we deal with the existence of integrable solutions for initial value problems (IVPs) for the fractional order implicit differential equation

$$^{c}D^{\alpha}y(t) = f(t, y(t), {}^{c}D^{\alpha}y(t)), \quad t \in J = [0, T] \tag{6.1}$$

$$y(0) = y_0, \tag{6.2}$$

where $f: J \times \mathbb{R} \times \mathbb{R} \to \mathbb{R}$ is a given function, $y_0 \in \mathbb{R}$.

### 6.2.2 Existence of solutions

**Definition 6.1.** A function $y \in L^1(J, \mathbb{R})$ is said to be a solution of IVP (6.1)–(6.2) if $y$ satisfies (6.1) and (6.2).

For the existence of solutions to problem (6.1)–(6.2), we need the following auxiliary lemma.

**Lemma 6.2.** *The solution of IVP* (6.1)–(6.2) *can be expressed by the integral equation*

$$y(t) = y_0 + I_0^{\alpha}x(t), \tag{6.3}$$

https://doi.org/10.1515/9783110553819-006

*where* $x \in L^1(J, \mathbb{R})$ *satisfies*

$$x(t) = f\left(t, y_0 + I_0^\alpha x(t), x(t)\right) . \tag{6.4}$$

*Proof.* Let ${}^cD^\alpha y(t) = x(t)$. Then

$$x(t) = f(t, y(t), x(t)) \tag{6.5}$$

and

$$\begin{aligned} y(t) &= y(0) + I^\alpha x(t) \\ &= y(0) + \frac{1}{\Gamma(\alpha)} \int_0^t (t-s)^{\alpha-1} x(s) ds . \end{aligned} \tag{6.6}$$

□

Let us introduce the following conditions:

**(6.2.1)** $f : J \times \mathbb{R}^2 \longrightarrow \mathbb{R}$ is measurable in $t \in J$ for any $(u_1, u_2) \in \mathbb{R}^2$ and continuous in $(u_1, u_2) \in \mathbb{R}^2$ for almost all $t \in J$.

**(6.2.2)** There exist a positive function $a \in L^1(J, \mathbb{R})$ and constants, $b_i > 0, i = 1, 2,$ such that

$$|f(t, u_1, u_2)| \leq |a(t)| + b_1|u_1| + b_2|u_2|, \forall (t, u_1, u_2) \in J \times \mathbb{R}^2 .$$

Our first result is based on Schauder's fixed point theorem.

**Theorem 6.3.** *Assume* (6.2.1) *and* (6.2.2) *hold. If*

$$\frac{b_1 T^{2\alpha}}{\Gamma(2\alpha + 1)} + \frac{b_2 T^\alpha}{\Gamma(\alpha + 1)} < 1 , \tag{6.7}$$

*then IVP* (6.1)–(6.2) *has at least one solution* $y \in L^1(J, \mathbb{R})$.

*Proof.* Transform problem (6.1)–(6.2) into a fixed point problem. Consider the operator

$$H : L^1(J, \mathbb{R}) \longrightarrow L^1(J, \mathbb{R})$$

defined by

$$(Hx)(t) = y_0 + I^\alpha x(t) , \tag{6.8}$$

where

$$x(t) = f\left(t, y_0 + I^\alpha x(t), x(t)\right) .$$

The operator $H$ is well defined, indeed, for each $x \in L^1(J, \mathbb{R})$, and from conditions (6.2.1) and (6.2.2) we obtain

$$
\begin{aligned}
\|Hx\|_{L_1} &= \int_0^T |Hx(t)|dt \\
&= \int_0^T |y_0 + I^\alpha x(t)|dt \\
&\leq T|y_0| + \int_0^T \left( \int_0^t \frac{(t-s)^{\alpha-1}}{\Gamma(\alpha)} |x(s)|ds \right) dt \\
&\leq T|y_0| + \int_0^T \left( \int_0^t \frac{(t-s)^{\alpha-1}}{\Gamma(\alpha)} |f(s, y_0 + I^\alpha x(s), x(s))|ds \right) dt \\
&\leq T|y_0| + \int_0^T \left( \int_0^t \frac{(t-s)^{\alpha-1}}{\Gamma(\alpha)} |a(s) + b_1(y_0 + I^\alpha x(s)) + b_2(x(s)|ds \right) dt \\
&\leq T|y_0| + \frac{T^\alpha}{\Gamma(\alpha+1)}\|a\|_{L_1} + \frac{b_1|y_0|T^{\alpha+1}}{\Gamma(\alpha+1)} + \frac{b_2 T^\alpha}{\Gamma(\alpha+1)}\|x\|_{L_1} \\
&\quad + b_1 \int_0^T \left( \int_0^t \frac{(t-s)^{\alpha-1}}{\Gamma(\alpha)} I^\alpha |x(s)|ds \right) dt \\
&\leq T|y_0| + \frac{T^\alpha}{\Gamma(\alpha+1)}\|a\|_{L_1} + \frac{b_1|y_0|T^{\alpha+1}}{\Gamma(\alpha+1)} + \frac{b_2 T^\alpha}{\Gamma(\alpha+1)}\|x\|_{L_1} \\
&\quad + \frac{b_1 T^{2\alpha}}{\Gamma(2\alpha+1)}\|x\|_{L_1} < +\infty\,.
\end{aligned}
\tag{6.9}
$$

Let

$$
r = \frac{T|y_0| + \left( \frac{T^\alpha \|a\|_{L_1} + b_1|y_0|T^{\alpha+1}}{\Gamma(\alpha+1)} \right)}{1 - \left( \frac{b_1 T^{2\alpha}}{\Gamma(2\alpha+1)} + \frac{b_2 T^\alpha}{\Gamma(\alpha+1)} \right)},
$$

and consider the set

$$
B_r = \{x \in L^1(J, \mathbb{R}) : \|x\|_{L_1} \leq r.\}\,.
$$

Clearly $B_r$ is nonempty, bounded, convex, and closed.

Now we will show that $HB_r \subset B_r$, indeed, for each $x \in B_r$, and from (6.7) and (6.9) we get

$$
\begin{aligned}
\|Hx\|_{L_1} &\leq T|y_0| + \left( \frac{T^\alpha \|a\|_{L_1} + b_1|y_0|T^{\alpha+1}}{\Gamma(\alpha+1)} \right) \\
&\quad + \left( \frac{b_1 T^{2\alpha}}{\Gamma(2\alpha+1)} + \frac{b_2 T^\alpha}{\Gamma(\alpha+1)} \right)\|x\|_{L_1} \\
&\leq r\,.
\end{aligned}
$$

Then $HB_r \subset B_r$. Assumption (6.2.1) implies that $H$ is continuous. Now we will show that $H$ is compact, that is, $HB_r$ is relatively compact. Clearly $HB_r$ is bounded in $L^1(J, \mathbb{R})$, i.e., condition (i) of the Kolmogorov compactness criterion is satisfied. It remains to show $(Hx)_h \longrightarrow (Hx)$ in $L^1(J, \mathbb{R})$ for each $x \in B_r$.

Let $x \in B_r$; then we have

$$
\begin{aligned}
&\|(Hx)_h - (Hx)\|_{L^1} \\
&= \int_0^T |(Hx)_h(t) - (Hx)(t)| dt \\
&= \int_0^T \left| \frac{1}{h} \int_t^{t+h} (Hx)(s) ds - (Hx)(t) \right| dt \\
&\le \int_0^T \left( \frac{1}{h} \int_t^{t+h} |(Hx)(s) - (Hx)(t)| ds \right) dt \\
&\le \int_0^T \left( \frac{1}{h} \int_t^{t+h} |I^\alpha x(s) - I^\alpha x(t)| ds \right) dt \\
&\le \int_0^T \frac{1}{h} \int_t^{t+h} |I^\alpha f(s, y_0 + I^\alpha x(s), x(s)) - I^\alpha f(t, y_0 + I^\alpha x(t), x(t))| ds dt .
\end{aligned}
$$

Since $x \in B_r \subset L^1(J, \mathbb{R})$, condition (6.2.2) implies that $f \in L^1(J, \mathbb{R})$. From Proposition 1.10 (v) it follows that $I^\alpha f \in L^1(J, \mathbb{R})$; thus, we have

$$
\frac{1}{h} \int_t^{t+h} |I^\alpha f(s, y_0 + I^\alpha x(s), x(s)) - I^\alpha f(t, y_0 + I^\alpha x(t), x(t))| ds \longrightarrow 0 \text{ as } h \longrightarrow 0, t \in J .
$$

Hence,

$$
(Hx)_h \longrightarrow (Hx) \ \text{ uniformly as } \ h \longrightarrow 0 .
$$

Then by the Kolmogorov compactness criterion, $HB_r$ is relatively compact. As a consequence of Schauder's fixed point theorem, IVP (6.1)–(6.2) has at least one solution in $B_r$. □

The following result is based on the Banach contraction principle.

**Theorem 6.4.** *Assume* (6.2.1) *holds and*

**(6.4.1)** *There exist constants $k_1, k_2 > 0$ such that*

$$
|f(t, x_1, y_1) - f(t, x_2, y_2)| \le k_1 |x_1 - x_2| + k_2 |y_1 - y_2| , \quad t \in J, \ x_1, x_2, y_1, y_2 \in \mathbb{R} .
$$

*If*

$$
\frac{k_1 T^{2\alpha}}{\Gamma(2\alpha + 1)} + \frac{k_2 T^\alpha}{\Gamma(\alpha + 1)} < 1 , \tag{6.10}
$$

*then IVP* (6.1)–(6.2) *has a unique solution $y \in L^1(J, \mathbb{R})$.*

*Proof.* We will use the Banach contraction principle to prove that $H$ defined by (6.8) has a fixed point. Let $x, y \in L^1(J, \mathbb{R})$ and $t \in J$. Then we have

$$\begin{aligned}|(Hx)(t) - (Hy)(t)| &= |I^\alpha [f(t, y_0 + I^\alpha x(t), x(t)) - f(t, y_0 + I^\alpha y(t), y(t))]| \\ &\leq k_1 I^{2\alpha}|x(t) - y(t)| + k_2 I^\alpha |x(t) - y(t)| \\ &\leq \frac{k_1}{\Gamma(2\alpha)} \int_0^t (t-s)^{2\alpha-1}|x(s) - y(s)|ds \\ &\quad + \frac{k_2}{\Gamma(\alpha)} \int_0^t (t-s)^{\alpha-1}|x(s) - y(s)|ds\,.\end{aligned}$$

Thus,

$$\begin{aligned}\|(Hx) - (Hy)\|_{L_1} &\leq \frac{k_1 T^{2\alpha}}{\Gamma(2\alpha+1)}\|x - y\|_{L_1} + \frac{k_2 T^\alpha}{\Gamma(\alpha+1)}\|x - y\|_{L_1} \\ &\leq \left(\frac{k_1 T^{2\alpha}}{\Gamma(2\alpha+1)} + \frac{k_2 T^\alpha}{\Gamma(\alpha+1)}\right)\|x - y\|_{L_1}\,.\end{aligned}$$

Consequently, by (6.10) $H$ is a contraction. As a consequence of the Banach contraction principle, we deduce that $H$ has a fixed point that is a solution of problem (6.1)–(6.2). □

### 6.2.3 Example

Let us consider the fractional IVP

$$^cD^\alpha y(t) = \frac{e^{-t}}{(e^t + 8)(1 + |y(t)| + |^cD^\alpha y(t)|)}\,, \quad t \in J := [0, 1],\ \alpha \in (0, 1]\,, \tag{6.11}$$

$$y(0) = 1\,. \tag{6.12}$$

Set

$$f(t, y, z) = \frac{e^{-t}}{(e^t + 8)(1 + y + z)}\,, \quad (t, y, z) \in J \times [0, +\infty) \times [0, +\infty)\,.$$

Let $y, z \in [0, +\infty)$ and $t \in J$. Then we have

$$\begin{aligned}|f(t, y_1, z_1) - f(t, y_2, z_2)| &= \left|\frac{e^{-t}}{e^t + 8}\left(\frac{1}{1 + y_1 + z_1} - \frac{1}{1 + y_2 + z_2}\right)\right| \\ &\leq \frac{e^{-t}(|y_1 - y_2| + |z_1 - z_2|)}{(e^t + 8)(1 + y_1 + z_1)(1 + y_2 + z_2)} \\ &\leq \frac{e^{-t}}{(e^t + 8)}(|y_1 - y_2| + |z_1 - z_2|) \\ &\leq \frac{1}{9}|y_1 - y_2| + \frac{1}{9}|z_1 - z_2|\,.\end{aligned}$$

Hence, condition (6.4.1) holds with $k_1 = k_2 = \frac{1}{9}$. Condition (6.10) is satisfied with $T = 1$. Indeed,

$$\frac{k_1 T^{2\alpha}}{\Gamma(2\alpha+1)} + \frac{k_2 T^{\alpha}}{\Gamma(\alpha+1)} = \frac{1}{9\Gamma(2\alpha+1)} + \frac{1}{9\Gamma(\alpha+1)} < 1 . \tag{6.13}$$

Then, by Theorem 6.4, problem (6.11)–(6.12) has a unique integrable solution on $[0, 1]$.

## 6.3 $L^1$-Solutions for NIFDEs with Nonlocal Conditions

### 6.3.1 Introduction

In this section, we deal with the existence of solutions of the nonlocal problem for fractional order implicit differential equation

$$^cD^\alpha y(t) = f(t, y(t), {}^cD^\alpha y(t)) , \quad a.e,\ t \in J =: (0, T] , \tag{6.14}$$

$$\sum_{k=1}^{m} a_k y(t_k) = y_0 , \tag{6.15}$$

where $f : J\times\mathbb{R}\times\mathbb{R} \to \mathbb{R}$ is a given function, $y_0 \in \mathbb{R}$, $a_k \in \mathbb{R}$, and $0 < t_1 < t_2 < \dots, t_m < T$, $k = 1, 2, \dots, m$.

Fractional differential equations with nonlocal conditions are discussed in [50] and references therein. Nonlocal conditions were initiated by Byszewski [118] when he proved the existence and uniqueness of mild and classical solutions of nonlocal Cauchy problems. As remarked by Byszewski ([116, 117]), the nonlocal condition can be more useful than the standard initial condition for describing some physical phenomena.

### 6.3.2 Existence of solutions

**Definition 6.5.** A function $y \in L^1([0, T], \mathbb{R})$ is said to be a solution of IVP (6.14)–(6.15) if $y$ satisfies (6.14) and (6.15).

Set

$$a = \frac{1}{\sum_{k=1}^m a_k} .$$

For the existence of solutions for nonlocal problem (6.14)–(6.15), we need the following auxiliary lemma.

**Lemma 6.6.** *Assume that $\sum_{k=1}^m a_k \neq 0$; then nonlocal problem (6.14)–(6.15) is equivalent to the integral equation*

$$y(t) = a y_0 - a \sum_{k=1}^{m} a_k \int_0^{t_k} \frac{(t_k - s)^{\alpha-1}}{\Gamma(\alpha)} x(s) ds + \int_0^t \frac{(t-s)^{\alpha-1}}{\Gamma(\alpha)} x(s) ds , \tag{6.16}$$

*where x is the solution of the functional integral equation*

$$x(t) = f\left(t, ay_0 - a\sum_{k=1}^{m} a_k \int_0^{t_k} \frac{(t_k - s)^{\alpha-1}}{\Gamma(\alpha)} x(s)ds) + \int_0^{t} \frac{(t - s)^{\alpha-1}}{\Gamma(\alpha)} x(s)ds, x(t)\right). \quad (6.17)$$

*Proof.* Let ${}^cD^\alpha y(t) = x(t))$ in equation (6.14); then

$$x(t) = f(t, y(t), x(t)) \quad (6.18)$$

and

$$\begin{aligned} y(t) &= y(0) + I^\alpha x(t) \\ &= y(0) + \int_0^t \frac{(t - s)^{\alpha-1}}{\Gamma(\alpha)} x(s)ds . \end{aligned} \quad (6.19)$$

Let $t = t_k$ in (6.19); we obtain

$$y(t_k) = y(0) + \int_0^{t_k} \frac{(t_k - s)^{\alpha-1}}{\Gamma(\alpha)} x(s)ds$$

and

$$\sum_{k=1}^{m} a_k y(t_k) = \sum_{k=1}^{m} a_k y(0) + \sum_{k=1}^{m} a_k \int_0^{t_k} \frac{(t_k - s)^{\alpha-1}}{\Gamma(\alpha)} x(s)ds . \quad (6.20)$$

Substituting (6.15) into (6.20), we get

$$y_0 = \sum_{k=1}^{m} a_k y(0) + \sum_{k=1}^{m} a_k \int_0^{t_k} \frac{(t_k - s)^{\alpha-1}}{\Gamma(\alpha)} x(s)ds$$

and

$$y(0) = a\left(y_0 - \sum_{k=1}^{m} a_k \int_0^{t_k} \frac{(t_k - s)^{\alpha-1}}{\Gamma(\alpha)} x(s)ds\right). \quad (6.21)$$

Substituting (6.21) into (6.18) and (6.19), we obtain (6.16) and (6.17).

To complete the proof, we prove that equation (6.16) satisfies nonlocal problem (6.14)–(6.15). Differentiating (6.16), we get

$${}^cD^\alpha y(t) = x(t) = f(t, y(t), {}^cD^\alpha y(t)) .$$

Letting $t = t_k$ in (6.16), we obtain

$$\begin{aligned} y(t_k) &= ay_0 - a\sum_{k=1}^{m} a_k \int_0^{t_k} \frac{(t_k - s)^{\alpha-1}}{\Gamma(\alpha)} x(s)ds) + \int_0^{t_k} \frac{(t_k - s)^{\alpha-1}}{\Gamma(\alpha)} x(s)ds \\ &= ay_0 + \left(1 - a\sum_{k=1}^{m} a_k\right) \int_0^{t_k} \frac{(t_k - s)^{\alpha-1}}{\Gamma(\alpha)} x(s)ds . \end{aligned}$$

Then

$$\sum_{k=1}^{m} a_k y(t_k) = \sum_{k=1}^{m} a_k a y_0 + \sum_{k=1}^{m} a_k \left(1 - a \sum_{k=1}^{m} a_k\right) \int_0^{t_k} \frac{(t_k - s)^{\alpha-1}}{\Gamma(\alpha)} x(s)ds = y_0 . \qquad \square$$

Let us introduce the following conditions:

**(6.6.1)** $f\colon [0, T] \times \mathbb{R}^2 \longrightarrow \mathbb{R}$ is measurable in $t \in [0, T]$ for any $(u_1, u_2) \in \mathbb{R}^2$ and continuous in $(u_1, u_2) \in \mathbb{R}^2$ for almost all $t \in [0, T]$.

**(6.6.2)** There exist a positive function $a \in L^1[0, T]$ and constants, $b_i > 0, i = 1, 2,$ such that

$$|f(t, u_1, u_2)| \le |a(t)| + b_1|u_1| + b_2|u_2| , \quad \forall (t, u_1, u_2) \in [0, T] \times \mathbb{R}^2 .$$

Our first result is based on Schauder's fixed point theorem.

**Theorem 6.7.** *Assume* (6.6.1)–(6.6.2). *If*

$$\frac{2b_1 T^\alpha}{\Gamma(\alpha + 1)} + b_2 < 1 , \tag{6.22}$$

*then IVP* (6.14)–(6.15) *has at least one solution* $y \in L^1([0, T], \mathbb{R})$.

*Proof.* Transform nonlocal problem (6.14)–(6.15) into a fixed point problem. Consider the operator

$$H\colon L^1([0, T], \mathbb{R}) \longrightarrow L^1([0, T], \mathbb{R})$$

defined by

$$(Hx)(t) = f\left( t, a y_0 - a \sum_{k=1}^{m} a_k \int_0^{t_k} \frac{(t_k - s)^{\alpha-1}}{\Gamma(\alpha)} x(s)ds) + \int_0^{t} \frac{(t - s)^{\alpha-1}}{\Gamma(\alpha)} x(s)ds, x(t) \right) . \tag{6.23}$$

Let

$$r = \frac{Tab_1|y_0| + \|a\|_{L_1}}{1 - \left(\frac{2b_1 T^\alpha}{\Gamma(\alpha+1)} + b_2\right)} ,$$

and consider the set

$$B_r = \{x \in L^1([0, T], \mathbb{R})\colon \|x\|_{L_1} \le r\} .$$

Clearly $B_r$ is nonempty, bounded, convex, and closed.

Now we will show that $HB_r \subset B_r$, indeed, for each $x \in B_r$, and from (6.22) and (6.23) we get

$$
\begin{aligned}
\|Hx\|_{L_1} &= \int_0^T |Hx(t)|dt \\
&= \int_0^T \left| f\left(t, ay_0 - a\sum_{k=1}^m a_k \int_0^{t_k} \frac{(t_k - s)^{\alpha-1}}{\Gamma(\alpha)} x(s)ds + \int_0^t \frac{(t-s)^{\alpha-1}}{\Gamma(\alpha)} x(s)ds, x(t)\right)\right| dt \\
&\leq \int_0^T \left[ |a(t)| + b_1 |ay_0 - a\sum_{k=1}^m a_k I^\alpha x(t)|_{t=t_k} + I^\alpha x(t)| + b_2|x(t)| \right] dt \\
&\leq Tab_1|y_0| + \|a\|_{L_1} + \frac{b_1 a \sum_{k=1}^m a_k t_k^\alpha}{\Gamma(\alpha+1)} \|x\|_{L_1} + \frac{b_1 T^\alpha}{\Gamma(\alpha+1)} \|x\|_{L_1} + b_2\|x\|_{L_1} \\
&\leq Tab_1|y_0| + \|a\|_{L_1} + \left(\frac{2b_1 T^\alpha}{\Gamma(\alpha+1)} + b_2\right)\|x\|_{L_1} \\
&\leq r\,.
\end{aligned}
$$

Then $HB_r \subset B_r$. Assumption (6.6.1) implies that $H$ is continuous. Now we will show that $H$ is compact, that is, $HB_r$ is relatively compact. Clearly $HB_r$ is bounded in $L^1([0,T],\mathbb{R})$, i.e., condition (i) of the Kolmogorov compactness criterion is satisfied. It remains to show $(Hx)_h \longrightarrow (Hx)$ in $L^1([0,T],\mathbb{R})$ for each $x \in B_r$.

Let $x \in B_r$; then we have

$$
\begin{aligned}
&\|(Hx)_h - (Hx)\|_{L^1} \\
&= \int_0^T |(Hx)_h(t) - (Hx)(t)|dt \\
&= \int_0^T \left| \frac{1}{h}\int_t^{t+h} (Hx)(s)ds - (Hx)(t)\right| dt \\
&\leq \int_0^T \left( \frac{1}{h}\int_t^{t+h} |(Hx)(s) - (Hx)(t)|ds \right) dt \\
&\leq \int_0^T \frac{1}{h}\int_t^{t+h} \left| f\left(t, ay_0 - a\sum_{k=1}^m a_k \int_0^{s_k} \frac{(s_k-\tau)^{\alpha-1}}{\Gamma(\alpha)} x(\tau)d\tau + \int_0^s \frac{(s-\tau)^{\alpha-1}}{\Gamma(\alpha)} x(\tau)d\tau, x(s)\right)\right. \\
&\quad \left. - f\left(t, ay_0 - a\sum_{k=1}^m a_k \int_0^{t_k} \frac{(t_k-s)^{\alpha-1}}{\Gamma(\alpha)} x(s)ds + \int_0^t \frac{(t-s)^{\alpha-1}}{\Gamma(\alpha)} x(s)ds, x(t)\right)\right| dsdt\,.
\end{aligned}
$$

Since $x \in B_r \subset L^1([0,T],\mathbb{R})$, condition (6.6.2) implies that $f \in L^1([0,T],\mathbb{R})$. Thus,

$$\frac{1}{h}\int_t^{t+h}\left|f\left(t, ay_0 - a\sum_{k=1}^{m} a_k \int_0^{s_k}\frac{(s_k-\tau)^{\alpha-1}}{\Gamma(\alpha)}x(\tau)d\tau + \int_0^{s}\frac{(s-\tau)^{\alpha-1}}{\Gamma(\alpha)}x(\tau)d\tau, x(s)\right)\right.$$
$$\left.-f\left(t, ay_0 - a\sum_{k=1}^{m} a_k \int_0^{t_k}\frac{(t_k-s)^{\alpha-1}}{\Gamma(\alpha)}x(s)ds + \int_0^{t}\frac{(t-s)^{\alpha-1}}{\Gamma(\alpha)}x(s)ds, x(t)\right)\right| ds \to 0$$

as $h \to 0$.

Hence,

$$(Hx)_h \to (Hx) \text{ uniformly as } h \to 0 .$$

Then, by the Kolmogorov compactness criterion, $HB_r$ is relatively compact. As a consequence of Schauder's fixed point theorem, nonlocal problem (6.14)–(6.15) has at least one solution in $B_r$. □

The following result is based on the Banach contraction principle.

**Theorem 6.8.** *Assume* (6.6.1) *holds and*

**(6.8.1)** *there exist constants* $k_1, k_2 > 0$ *such that*

$$|f(t,x_1,y_1)-f(t,x_2,y_2)| \le k_1|x_1-x_2|+k_2|y_1-y_2|, \quad t \in [0,T],\ x_1,x_2,y_1,y_2 \in \mathbb{R}.$$

*If*

$$\frac{2k_1T^\alpha}{\Gamma(\alpha+1)} + k_2 < 1 , \tag{6.24}$$

*then IVP* (6.14)–(6.15) *has a unique solution* $y \in L^1([0,T],\mathbb{R})$.

*Proof.* We will use the Banach contraction principle to prove that $H$ defined by (6.23) has a fixed point. Let $x, y \in L^1([0,T],\mathbb{R})$, and $t \in [0,T]$. Then we have

$$\begin{aligned}
&|(Hx)(t) - (Hy)(t)| \\
&= \left|f\left(t, ay_0 - a\sum_{k=1}^{m} a_k I^\alpha x(t)|_{t=t_k} + I^\alpha x(t), x(t)\right)\right. \\
&\quad \left.-f\left(t, ay_0 - a\sum_{k=1}^{m} a_k I^\alpha y(t)|_{t=t_k} + I^\alpha y(t), y(t)\right)\right| \\
&\le k_1 a \sum_{k=1}^{m} a_k \int_0^{t_k}\frac{(t_k-s)^{\alpha-1}}{\Gamma(\alpha)}|x(s)-y(s)|ds \\
&\quad + k_1\int_0^{t}\frac{(t-s)^{\alpha-1}}{\Gamma(\alpha)}|x(s)-y(s)|ds + k_2|x-y| .
\end{aligned}$$

Thus,

$$\begin{aligned}\|(Hx)-(Hy)\|_{L_1} &\le \frac{k_1 t_k^{\alpha} a \sum_{k=1}^{m} a_k}{\Gamma(\alpha+1)} \int_0^T |x(t)-y(t)|dt + \frac{k_1 T^{\alpha}}{\Gamma(\alpha+1)} \int_0^T |x(t)-y(t)|dt \\ &\quad + k_2 \int_0^T |x(t)-y(t)|dt \\ &\le \frac{2k_1 T^{\alpha}}{\Gamma(\alpha+1)} \|x-y\|_{L_1} + k_2 \|x-y\|_{L_1} \\ &\le \left( \frac{2k_1 T^{\alpha}}{\Gamma(\alpha+1)} + k_2 \right) \|x-y\|_{L_1} \,.\end{aligned}$$

Consequently, by (6.24), $H$ is a contraction. As a consequence of the Banach contraction principle, we deduce that $H$ has a fixed point that is a solution of nonlocal problem (6.14)–(6.15). □

### 6.3.3 Example

Let us consider the fractional nonlocal problem

$$^{c}D^{\alpha}y(t) = \frac{1}{(e^t+5)(1+|y(t)|+|^{c}D^{\alpha}y(t)|)}\,, \qquad t \in J := [0,1],\ \alpha \in (0,1]\,, \tag{6.25}$$

$$\sum_{k=1}^{m} a_k y(t_k) = 1\,, \tag{6.26}$$

where $a_k \in \mathbb{R}$, $0 < t_1 < t_2 < \cdots < 1$.

Set

$$f(t,y,z) = \frac{1}{(e^t+5)(1+y+z)}\,, \qquad (t,y,z) \in J \times [0,+\infty) \times [0,+\infty)\,.$$

Let $y, z \in [0,+\infty)$ and $t \in J$. Then we have

$$\begin{aligned}|f(t,y_1,z_1)-f(t,y_2,z_2)| &= \left| \frac{1}{e^t+5} \left( \frac{1}{1+y_1+z_1} - \frac{1}{1+y_2+z_2} \right) \right| \\ &\le \frac{|y_1-y_2|+|z_1-z_2|}{(e^t+5)(1+y_1+z_1)(1+y_2+z_2)} \\ &\le \frac{1}{(e^t+5)}(|y_1-y_2|+|z_1-z_2|) \\ &\le \frac{1}{6}|y_1-y_2| + \frac{1}{6}|z_1-z_2|\,.\end{aligned}$$

Hence condition (6.8.1) holds with $k_1 = k_2 = \frac{1}{6}$. Condition (6.24) is satisfied. Indeed,

$$\frac{2k_1}{\Gamma(\alpha+1)} + k_2 = \frac{1}{3\Gamma(\alpha+1)} + \frac{1}{6} < 1\,. \tag{6.27}$$

Then, by Theorem 3.2, nonlocal problem (6.25)–(6.26) has a unique integrable solution on $[0,1]$.

## 6.4 Integrable Solutions for NIFDEs with Infinite Delay

### 6.4.1 Introduction

In this section we deal with the existence of solutions for IVPs for implicit fractional order functional differential equations with infinite delay of the form

$$ {}^cD^\alpha y(t) = f(t, y_t, {}^cD^\alpha y_t)\,, \quad t \in J := [0, b] \tag{6.28}$$

$$ y(t) = \phi(t)\,, \quad t \in (-\infty, 0]\,, \tag{6.29}$$

where $f\colon J \times \mathcal{B} \times \mathcal{B} \to \mathbb{R}$ is a given function.

### 6.4.2 Existence of solutions

Set

$$\Omega = \{y\colon (-\infty, b] \to \mathbb{R}\colon y|_{(-\infty,0]} \in \mathcal{B} \text{ and } y|_J \in L^1(J)\}\,.$$

**Definition 6.9.** A function $y \in \Omega$ is said to be a solution of IVP (6.28)–(6.29) if $y$ satisfies (6.28) and (6.29).

For the existence of solutions to problem (6.28)–(6.29), we need the following auxiliary lemma.

**Lemma 6.10.** *The solution to IVP* (6.28)–(6.29) *can be expressed by*

$$ y(t) = \phi(0) + \frac{1}{\Gamma(\alpha)}\int_0^t (t-s)^{\alpha-1}x(s)ds, \qquad t \in J\,, \tag{6.30}$$

$$ y(t) = \phi(t), \qquad t \in (-\infty, 0]\,, \tag{6.31}$$

*where $x$ is the solution of the functional integral equation*

$$ x(t) = f\left(t, \phi(0) + \frac{1}{\Gamma(\alpha)}\int_0^t (t-s)^{\alpha-1}x_s ds, x_t\right)\,. \tag{6.32}$$

*Proof.* Let $y$ be a solution of (6.30)–(6.31); for $t \in J$ and $t \in (-\infty, 0]$ we have (6.28) and (6.29), respectively. □

To present the main result, let us introduce the following conditions:

**(6.10.1)** $f\colon J \times \mathcal{B}^2 \longrightarrow \mathbb{R}$ is measurable in $t \in J$ for any $(u_1, u_2) \in \mathcal{B}^2$ and continuous in $(u_1, u_2) \in \mathcal{B}^2$ for almost all $t \in J$.

**(6.10.2)** There exist constants $k_1, k_2 > 0$ such that

$$|f(t, x_1, y_1) - f(t, x_2, y_2)| \le k_1\|x_1 - x_2\|_{\mathcal{B}} + k_2\|y_1 - y_2\|_{\mathcal{B}}\,,$$

for $t \in J$ and every $x_1, x_2, y_1, y_2 \in \mathcal{B}$.

Our first existence result for IVP (6.28)–(6.29) is based on the Banach contraction principle. Set

$$K_b = \sup\{|K(t)| : t \in J\} .$$

**Theorem 6.11.** *Assume* (6.10.1)–(6.10.2). *If*

$$\frac{k_1 K_b b^{2\alpha}}{\Gamma(2\alpha+1)} + \frac{k_2 K_b b^{\alpha}}{\Gamma(\alpha+1)} < 1 , \tag{6.33}$$

*then IVP* (6.28)–(6.29) *has a unique solution on the interval* $(-\infty, b]$.

*Proof.* Transform problem (6.28)–(6.29) into a fixed point problem. Consider the operator $N : \Omega \longrightarrow \Omega$ defined by

$$(Ny)(t) = \begin{cases} \phi(t), & t \in (-\infty, 0] \\ \frac{1}{\Gamma(\alpha)} \int_0^t (t-s)^{\alpha-1} f(s, I^\alpha y_s, y_s) ds, & t \in J . \end{cases}$$

We use the Banach contraction principle to prove that $N$ has a fixed point.
Let $x(.) : (-\infty, b] \to \mathbb{R}$ be the function defined by

$$x(t) = \begin{cases} 0, & \text{if } t \in J \\ \phi(t), & \text{if } t \in (-\infty, 0] . \end{cases}$$

Then $x_0 = \phi$. For each $z \in L^1(J, \mathbb{R})$, with $z(0) = 0$, we denote by $\overline{z}$ the function defined by

$$\overline{z}(t) = \begin{cases} z(t), & \text{if } t \in J \\ 0, & \text{if } t \in (-\infty, 0] . \end{cases}$$

If $y(.)$ satisfies the integral equation

$$y(t) = \frac{1}{\Gamma(\alpha)} \int_0^t (t-s)^{\alpha-1} f(s, I^\alpha y_s, y_s) ds ,$$

we can decompose $y(.)$ as $y(t) = \overline{z}(t) + x(t), 0 \leq t \leq b$, then $y_t = \overline{z}_t + x_t$, for every $0 \leq t \leq b$, and the function $z(.)$ satisfies

$$z(t) = \frac{1}{\Gamma(\alpha)} \int_0^t (t-s)^{\alpha-1} f(s, I^\alpha(\overline{z}_s + x_s), \overline{z}_s + x_s) ds .$$

Set

$$L_0 = \{z \in L^1(J, \mathbb{R}) : z_0 = 0\} ,$$

and let $\|\cdot\|_b$ be the seminorm in $L_0$ defined by

$$\|z\|_b = \|z_0\|_{\mathcal{B}} + \int_0^b |z(t)| dt = \int_0^b |z(t)| dt , \quad z \in L_0 .$$

Then $L_0$ is a Banach space with norm $\|\cdot\|_b$. Let the operator $P\colon L_0 \to L_0$ be defined by

$$(Pz)(t) = \frac{1}{\Gamma(\alpha)} \int_0^t (t-s)^{\alpha-1} f(s, I^\alpha(\overline{z}_s + x_s), \overline{z}_s + x_s)ds\,, \quad t \in J\,. \tag{6.34}$$

The operator $N$ having a fixed point is equivalent to $P$ having a fixed point, and so we turn to proving that $P$ has a fixed point. We will show that $P\colon L_0 \to L_0$ is a contraction map. Indeed, consider $z, z^* \in L_0$. Then for each $t \in J$ we have

$$\begin{aligned}
&|(Pz)(t) - (Pz^*)(t)| \\
&\leq \frac{1}{\Gamma(\alpha)} \int_0^t (t-s)^{\alpha-1} |f(s, I^\alpha(\overline{z}_s + x_s), \overline{z}_s + x_s) - f(s, I^\alpha(\overline{z}^*_s + x_s), \overline{z}^*_s + x_s)| ds \\
&\leq \frac{1}{\Gamma(\alpha)} \int_0^t (t-s)^{\alpha-1} [k_1 \|I^\alpha(\overline{z}_s - \overline{z}^*_s)\|_{\mathcal{B}} + k_2 \|\overline{z}_s - \overline{z}^*_s\|_{\mathcal{B}}] ds \\
&\leq \frac{1}{\Gamma(\alpha)} \int_0^t (t-s)^{\alpha-1} K_b \left[k_1 \|I^\alpha(z(s) - z^*(s))\| + k_2 \|z(s) - z^*(s)\|\right] ds \\
&\leq \left( \frac{k_1 K_b b^{2\alpha}}{\Gamma(2\alpha+1)} + \frac{k_2 b^\alpha}{\Gamma(\alpha+1)} \right) \|z - z^*\|_b\,.
\end{aligned}$$

Therefore,

$$\|P(z) - P(z^*)\|_b \leq \left( \frac{k_1 K_b b^{2\alpha}}{\Gamma(2\alpha+1)} + \frac{k_2 K_b b^\alpha}{\Gamma(\alpha+1)} \right) \|z - z^*\|_b\,.$$

Consequently, by (6.33), $P$ is a contraction. The Banach contraction principle implies that $P$ has a unique fixed point that is the unique solution of problem (6.28)–(6.29). □

The next result is based on Schauder's fixed point theorem.

**Theorem 6.12.** *Assume that* (6.10.1) *holds and*

**(6.12.1)** *There exist a positive function $a \in L^1(J)$ and constants, $q_i > 0$, $i = 1, 2$, such that*

$$|f(t, u_1, u_2)| \leq |a(t)| + q_1 \|u_1\|_{\mathcal{B}} + q_2 \|u_2\|_{\mathcal{B}}\,, \quad \forall (t, u_1, u_2) \in J \times \mathbb{R}^2\,.$$

*If*

$$K_b \left( \frac{q_1 b^{2\alpha}}{\Gamma(2\alpha+1)} + \frac{q_2 b^\alpha}{\Gamma(\alpha+1)} \right) < 1\,, \tag{6.35}$$

*then IVP* (6.28)–(6.29) *has at least one solution $y \in L^1(J, \mathbb{R})$.*

*Proof.* Let $P\colon L_0 \to L_0$ be defined as in (6.34), and

$$r = \frac{\frac{b^\alpha \|a\|_{L^1}}{\Gamma(\alpha+1)} + M_b \|\phi\|_{\mathcal{B}} \left( \frac{q_1 b^{2\alpha}}{\Gamma(2\alpha+1)} + \frac{q_2 b^\alpha}{\Gamma(\alpha+1)} \right)}{1 - K_b \left( \frac{q_1 b^{2\alpha}}{\Gamma(2\alpha+1)} + \frac{q_2 b^\alpha}{\Gamma(\alpha+1)} \right)}\,,$$

where $M_b = \sup\{|M(t)| : t \in J\}$, and consider the set

$$B_r := \{z \in L_0, \|z\|_b \le r\} .$$

Clearly $B_r$ is nonempty, bounded, convex, and closed. We will show that operator $P$ satisfies the conditions of Schauder's fixed point theorem. The proof will be given in three steps.

*Step 1: P is continuous.* Let $z_n$ be a sequence such that $z_n \to z$ in $L_0$. Then

$$|(Pz_n)(t) - (Pz)(t)| \le \frac{1}{\Gamma(\alpha)} \int_0^t (t-s)^{\alpha-1} |f(s, I^\alpha(\overline{z}_{n_s} + x_s), \overline{z}_{n_s} + x_s) - f(s, I^\alpha(\overline{z}_s + x_s), \overline{z}_s + x_s)| ds .$$

Since $f$ is a continuous function, we have

$$\|P(z_n) - P(z)\|_b \le \frac{b^\alpha}{\Gamma(\alpha+1)} \|f(., I^\alpha(\overline{z}_{n_{(.)}} + x_{(.)}, \overline{z}_{n_{(.)}}) + x_{(.)}) - f(., I^\alpha(\overline{z}_{(.)} + x_{(.)}), \overline{z}_{(.)} + x_{(.)})\|_{L_1} \to 0$$

as $n \to \infty$.

*Step 2: P maps $B_r$ to itself.* Let $z \in B_r$. Since $f$ is a continuous function, we have for each $t \in [0, b]$

$$\begin{aligned} |(Pz)(t)| &\le \frac{1}{\Gamma(\alpha)} \int_0^t (t-s)^{\alpha-1} |f(s, I^\alpha(\overline{z}_s + x_s), \overline{z}_s + x_s)| ds \\ &\le \frac{1}{\Gamma(\alpha)} \int_0^t (t-s)^{\alpha-1} [a(t)| + q_1 \|I^\alpha(\overline{z}_s + x_s)\|_{\mathcal{B}} + q_2 \|\overline{z}_s + x_s\|_{\mathcal{B}}] ds \\ &\le \frac{b^\alpha \|a\|_{L_1}}{\Gamma(\alpha+1)} + \left( \frac{q_1 b^{2\alpha}}{\Gamma(2\alpha+1)} + \frac{q_2 b^\alpha}{\Gamma(\alpha+1)} \right) (K_b r + M_b \|\phi\|_{\mathcal{B}}) , \end{aligned}$$

where

$$\|\overline{z}_s + x_s\|_{\mathcal{B}} \le \|\overline{z}_s\|_{\mathcal{B}} + \|x_s\|_{\mathcal{B}} .$$

Hence, $\|P(z)\|_{L_1} \le r$. Then $P(B_r) \subset B_r$.

*Step 3: P is compact.* We will show that $P(B_r)$ is relatively compact. Clearly $P(B_r)$ is bounded in $L_0$, i.e., condition (i) of the Kolmogorov compactness criterion is satisfied. It remains to show $(Pz)_h \longrightarrow P(z)$, in $L_0$ for each $z \in B_r$.

Let $z \in B_r$; then we have

$$\begin{aligned}
&\|P(z)_h - P(z)\|_{L^1} \\
&= \int_0^b |(Pz)_h(t) - (Pz)(t)|dt \\
&= \int_0^b \left| \frac{1}{h} \int_t^{t+h} (Pz)(s)ds - (Pz)(t) \right| dt \\
&\le \int_0^b \left( \frac{1}{h} \int_t^{t+h} |(Pz)(s) - (Pz)(t)|ds \right) dt \\
&\le \int_0^b \frac{1}{h} \int_t^{t+h} |I^\alpha f\left(s, \overline{z}_s + x_s), \overline{z}_s + x_s\right) - I^\alpha f\left(t, I^\alpha(\overline{z}_t + x_t), \overline{z}_t + x_t\right)|dsdt\,.
\end{aligned}$$

Since $z \in B_r \subset L_0$, condition (6.12.1) implies that $f \in L_0$. From Proposition 1.10 it follows that $I^\alpha f \in L^1(J, \mathbb{R})$; then we have

$$\frac{1}{h} \int_t^{t+h} |I^\alpha f\left(\overline{z}_s + x_s), \overline{z}_s + x_s\right) - I^\alpha f\left(t, I^\alpha(\overline{z}_t + x_t), \overline{z}_t + x_t\right)|ds \longrightarrow 0 \text{ as } h \longrightarrow 0,\ t \in J\,.$$

Hence,

$$(Pz)_h \longrightarrow (Pz) \text{ uniformly as } h \longrightarrow 0\,.$$

Then by the Kolmogorov compactness criterion, $PB_r$ is relatively compact. As a consequence of Schauder's fixed point theorem, IVP (6.28)–(6.29) has at least one solution in $B_r$. □

### 6.4.3 Example

In this section we give an example to illustrate the usefulness of our main results. Let us consider the fractional initial value problem

$${}^cD^\alpha y(t) = \frac{ce^{-\gamma t+t}}{(e^t + e^{-t})(1 + \|y_t\| + \|{}^cD^\alpha y_t\|)}\,, \quad t \in J := [0, b],\ \alpha \in (0, 1]\,, \tag{6.36}$$

$$y(t) = \phi(t)\,, \quad t \in (-\infty, 0], \tag{6.37}$$

where $c > 1$ is fixed. Let $\gamma$ be a positive real constant and

$$B_\gamma = \{y \in L^1(-\infty, 0] : \lim_{\theta \to -\infty} e^{\gamma\theta} y(\theta), \text{ exists in } \mathbb{R}\}\,.$$

The norm of $B_\gamma$ is given by

$$\|y\|_\gamma = \int_{-\infty}^0 e^{\gamma\theta}|y(\theta)|d\theta\,.$$

Let $y\colon (-\infty, b] \to \mathbb{R}$ be such that $y_0 \in B_\gamma$. Then

$$\begin{aligned}\lim_{\theta\to-\infty} e^{\gamma\theta} y_t(\theta) &= \lim_{\theta\to-\infty} e^{\gamma\theta} y(t+\theta)\\ &= \lim_{\theta\to-\infty} e^{\gamma(\theta-t)} y(\theta)\\ &= e^{\gamma t} \lim_{\theta\to-\infty} e^{\gamma\theta} y_0(\theta) < \infty\,.\end{aligned}$$

Hence, $y_t \in B_\gamma$. Finally, we prove that

$$\|y_t\|_\gamma \le K(t)\int_0^t |y(s)|ds + M(t)\|y_0\|_\gamma\,,$$

where $K = M = 1$ and $H = 1$. We have

$$|y_t(\theta)| = |y(t+\theta)|\,.$$

If $\theta + t \le 0$, we get

$$|y_t(\theta)| \le \int_{-\infty}^{0} |y(s)|ds\,.$$

Then for $t + \theta \ge 0$ we have

$$|y_t(\theta)| \le \int_0^t |y(s)|ds\,.$$

Thus, for all $t + \theta \in J$ we get

$$|y_t(\theta)| \le \int_{-\infty}^{0} |y(s)|ds + \int_0^t |y(s)|ds\,.$$

Then

$$\|y_t\|_\gamma \le \|y_0\|_\gamma + \int_0^t |y(s)|ds\,.$$

It is clear that $(B_\gamma, \|\cdot\|)$ is a Banach space. We can conclude that $B_\gamma$ is a phase space. Set

$$f(t,y,z) = \frac{e^{-\gamma t+t}}{c(e^t + e^{-t})(1+y+z)}\,, \qquad (t,x,z) \in J \times B_\gamma \times B_\gamma\,.$$

For $t \in J$, $y_1, y_2, z_1, z_2 \in B_\gamma$ we have

$$\begin{aligned}|f(t,y_1,z_1) - f(t,y_2,z_2)| &= \frac{e^{-\gamma t+t}}{c(e^t+e^{-t})}\left|\frac{1}{1+y_1+z_1} - \frac{1}{1+y_2+z_2}\right|\\ &= \frac{e^{-\gamma t+t}(|y_1-y_2| + |z_1-z_2|)}{c(e^t+e^{-t})(1+y_1+z_1)(1+y_2+z_2)}\end{aligned}$$

$$
\begin{aligned}
&\le \frac{e^{-\gamma t} \times e^t(|y_1 - y_2| + |z_1 - z_2|)}{c(e^t + e^{-t})} \\
&\le \frac{e^{-\gamma t}(\|y_1 - y_2\|_\gamma + \|z_1 - z_2\|_\gamma)}{c} \\
&\le \frac{1}{c}\|y_1 - y_2\|_\gamma + \frac{1}{c}\|z_1 - z_2\|_\gamma .
\end{aligned}
$$

Hence condition (6.10.2) holds. We choose $b$ such that $\frac{K_b b^{2\alpha}}{c\Gamma(2\alpha+1)} + \frac{K_b b^{\alpha}}{c\Gamma(\alpha+1)} < 1$. Since $K_b = 1$, then

$$
\frac{b^{2\alpha}}{c\Gamma(2\alpha + 1)} + \frac{b^{\alpha}}{c\Gamma(\alpha + 1)} < 1 .
$$

Then, by Theorem 6.11, problem (6.36)–(6.37) has a unique integrable solution on $[-\infty, b]$.

## 6.5 An Existence Result of Integrable Solutions for NIFDEs

### 6.5.1 Introduction

This section deals with the existence of integrable solutions for an IVP for the implicit fractional order differential equation

$$
{}^cD^\alpha y(t) = f(t, y(t), {}^cD^\alpha y(t)) , \quad t \in J = [0, T] , \tag{6.38}
$$

$$
y(0) = y_0 , \tag{6.39}
$$

where $f : J \times \mathbb{R} \times \mathbb{R} \longrightarrow \mathbb{R}$ is a given function. We will use the technique of measures of noncompactness, which is often used in several branches of nonlinear analysis. In particular, that technique turns out to be a very useful tool in existence for several types of integral equations; details can be found in Akhmerov et al. [58] and Banaś et al. [81, 83].

The principal goal here is to prove the existence of integral solutions to problem (6.38)–(6.39) using Darbo's fixed point theorem.

### 6.5.2 Existence of solutions

Let us start by defining what we mean by a solution to problem (6.38)–(6.39).

**Definition 6.13.** A function $y \in L^1(J, \mathbb{R})$ is said to be a solution to IVP (6.38)–(6.39) if $y$ satisfies the equation ${}^cD^\alpha y(t) = f(t, y(t), {}^cD^\alpha y(t))$ on $J$ and the condition $y(0) = y_0$.

For the existence of solutions to problem (6.38)–(6.39), we need the following auxiliary lemma.

**Lemma 6.14.** *The solution to IVP (6.38)–(6.39) can be expressed by the integral equation*

$$y(t) = y_0 + \frac{1}{\Gamma(\alpha)} \int_0^t (t-s)^{\alpha-1} x(s) ds \,, \tag{6.40}$$

*where x is the solution of the functional integral equation*

$$x(t) = f\left( t, y_0 + \frac{1}{\Gamma(\alpha)} \int_0^t (t-s)^{\alpha-1} x(s) ds, x(t) \right) . \tag{6.41}$$

Let us introduce the following conditions:

**(6.14.1)** $f: J \times \mathbb{R} \times \mathbb{R} \longrightarrow \mathbb{R}$ satisfies the Carathéodory conditions.

**(6.14.2)** There exist a positive function $a \in L^1(J)$ and two constants, $q_1, q_2 > 0$, such that

$$|f(t, u_1, u_2)| \le |a(t)| + q_1|u_1| + q_2|u_2| \,, \quad \forall (t, u_1, u_2) \in J \times \mathbb{R} \times \mathbb{R} \,.$$

**(6.14.3)** We first consider two real numbers, $0 < |\rho| < \delta$. There exists a positive valued function $L_f(\cdot)$ that is continuous in a neighborhood of 0 with $L_f(0) = 0$ and two constants $k_1, k_2 > 0$ such that

$$|f(t+\rho, x_1, y_1) - f(t, x_2, y_2)| \le L_f(\rho) + k_1|x_1 - x_2| + k_2|y_1 - y_2| \,,$$
$$t \in [0, T], \ x_i, y_i \in \mathbb{R}, \ i = 1, 2 \,.$$

In this section, we study the existence of a solution to problem (6.38)–(6.39) by using the concept of measure of noncompactness in $L^1(J)$.

**Theorem 6.15.** *Assume* (6.14.1)–(6.14.3). *If*

$$\frac{k_1 T^{2\alpha}}{\Gamma(2\alpha+1)} + \frac{k_2 T^\alpha}{\Gamma(\alpha+1)} < 1 \,, \tag{6.42}$$

*then IVP* (6.38)–(6.39) *has at least one solution* $y \in L^1(J, \mathbb{R})$.

*Proof.* Consider the operator $N: L^1(J, \mathbb{R}) \to L^1(J, \mathbb{R})$ defined by

$$(Nx)(t) = y_0 + I^\alpha x(t) \,, \tag{6.43}$$

where $x(t) = f(t, y_0 + I^\alpha x(t), x(t))$. Clearly, the fixed points of the operator $N$ are solutions to problem (6.38)–(6.39). Let

$$r = \frac{T|y_0| + \left( \frac{T^\alpha \|a\|_{L^1} + q_1|y_0|T^{\alpha+1}}{\Gamma(\alpha+1)} \right)}{1 - \left( \frac{q_1 T^{2\alpha}}{\Gamma(2\alpha+1)} + \frac{q_2 T^\alpha}{\Gamma(\alpha+1)} \right)} \,,$$

where

$$\frac{q_1 T^{2\alpha}}{\Gamma(2\alpha+1)} + \frac{q_2 T^\alpha}{\Gamma(\alpha+1)} < 1 \,,$$

and consider the set

$$B_r = \{x \in L^1(J, \mathbb{R}) : \|x\|_{L^1} \le r,\ r > 0\} .$$

Clearly, the subset $B_r$ is closed, bounded, and convex. We will show that $N$ satisfies the assumptions of Darbo's fixed point theorem. The proof will be given in three steps.

*Step 1. $N$ is continuous.* Let $x_n$ be a sequence such that $x_n \to x$ in $B_r$. Then for each $t \in J$,

$$\begin{aligned}
&\|N(x_n) - N(x)\|_{L^1} \\
&= \|I^\alpha x_n(t) - I^\alpha x(t)\|_{L^1} \\
&= \left\| \frac{1}{\Gamma(\alpha)} \int_0^t (t-s)^{\alpha-1} (x_n(s) - x(s))\, ds \right\|_{L^1} \\
&\le \int_0^T \left( \frac{1}{\Gamma(\alpha)} \int_0^t (t-s)^{\alpha-1} |f(s, y_0 + I^\alpha x_n(s), x(s)) - f(s, y_0 + I^\alpha x(s), x(s))| ds \right) dt .
\end{aligned}$$

Since $f$ is of Carathéodory type, then by the Lebesgue dominated convergence theorem, we have

$$\|N(x_n) - N(x)\|_{L^1} \to 0 \text{ as } n \to \infty .$$

*Step 2. $N$ maps $B_r$ to itself.* Let $x$ be an arbitrary element in $B_r$. Then from (6.14.1)–(6.14.2) we obtain

$$\begin{aligned}
\|Nx\|_{L^1} &= \int_0^T |Nx(t)| dt \\
&= \int_0^T |y_0 + I^\alpha x(t)| dt \\
&\le T|y_0| + \int_0^T \left( \int_0^t \frac{(t-s)^{\alpha-1}}{\Gamma(\alpha)} |x(s)| ds \right) dt \\
&\le T|y_0| + \int_0^T \left( \int_0^t \frac{(t-s)^{\alpha-1}}{\Gamma(\alpha)} |f(s, y_0 + I^\alpha x(s), x(s))| ds \right) dt \\
&\le T|y_0| + \int_0^T \left( \int_0^t \frac{(t-s)^{\alpha-1}}{\Gamma(\alpha)} |a(s) + q_1 (y_0 + I^\alpha x(s)) + q_2(x(s))| ds \right) dt \\
&\le T|y_0| + \frac{T^\alpha}{\Gamma(\alpha+1)} \|a\|_{L^1} + \frac{q_1 |y_0| T^{\alpha+1}}{\Gamma(\alpha+1)} + \frac{q_2 T^\alpha}{\Gamma(\alpha+1)} \|x\|_{L^1} \\
&\quad + q_1 \int_0^T \left( \int_0^t \frac{(t-s)^{\alpha-1}}{\Gamma(\alpha)} I^\alpha |x(s)| ds \right) dt \\
&\le T|y_0| + \frac{T^\alpha}{\Gamma(\alpha+1)} \|a\|_{L^1} + \frac{q_1 |y_0| T^{\alpha+1}}{\Gamma(\alpha+1)} + \frac{q_2 T^\alpha}{\Gamma(\alpha+1)} \|x\|_{L^1} + \frac{q_1 T^{2\alpha}}{\Gamma(2\alpha+1)} \|x\|_{L^1} \\
&\le T|y_0| + \frac{T^\alpha \|a\|_{L^1} + q_1 |y_0| T^{\alpha+1}}{\Gamma(\alpha+1)} + \frac{q_1 T^{2\alpha} r}{\Gamma(2\alpha+1)} + \frac{q_2 T^\alpha r}{\Gamma(\alpha+1)} \le r .
\end{aligned}$$

Then $\|Nx\|_{L^1} \le r$. Thus, the operator $N$ maps $B_r$ to itself.

*Step 3. N is a contraction, i.e.,* $\mu(NX) \le k\mu(X)$, $k \in [0, 1)$. Now let us fix a nonempty subset $X$ of $B_r$. We first consider two real numbers, $0 < |\rho| < \delta$, and an arbitrary fixed $x \in X$. By (6.14.3) we have

$$\begin{aligned}
&\|Nx(t+\rho) - Nx(t)\|_{L^1} \\
&= \int_0^T |(Nx)(t+\rho) - (Nx)(t)|dt \\
&= \int_0^T |I^\alpha x(t+\rho) - I^\alpha x(t)|dt \\
&= \int_0^T |I^\alpha (x(t+\rho) - x(t))|dt \\
&= \int_0^T \left|I^\alpha \left(f(t+\rho, y_0 + I^\alpha x(t+\rho), x(t+\rho)) - f(t, y_0 + I^\alpha x(t), x(t))\right)\right| dt \\
&\le \int_0^T \left(\left|I^\alpha L_f(\rho)\right| + k_1 \left|I^{2\alpha}(x(t+\rho) - x(t))\right| + k_2 \left|I^\alpha (x(t+\rho) - x(t))\right|\right) dt \\
&\le \frac{T^\alpha}{\Gamma(\alpha+1)} \int_0^T |L_f(\rho)|dt + \frac{k_1 T^{2\alpha}}{\Gamma(2\alpha+1)} \int_0^T |x(t+\rho) - x(t)|dt \\
&\quad + \frac{k_2 T^\alpha}{\Gamma(\alpha+1)} \int_0^T |x(t+\rho) - x(t)|dt .
\end{aligned}$$

Hence, we have

$$\begin{aligned}
\|Nx(.+\rho) - Nx(.)\|_{L^1} \le\ & \frac{T^{\alpha+1}}{\Gamma(\alpha+1)} L_f(\rho) \\
& + \left(\frac{k_1 T^{2\alpha}}{\Gamma(2\alpha+1)} + \frac{k_2 T^\alpha}{\Gamma(\alpha+1)}\right) \|x(.+\rho) - x(.)\|_{L^1} .
\end{aligned}$$

Taking into account that

$$\lim_{\delta \to 0} \sup_{|\rho| \le \delta} L_f(\rho) = 0 ,$$

we get

$$\mu(NX) \le \left(\frac{k_1 T^{2\alpha}}{\Gamma(2\alpha+1)} + \frac{k_2 T^\alpha}{\Gamma(\alpha+1)}\right) \mu(X) .$$

Here, $\mu(.)$ is the measure of noncompactness in $L^1[0, T]$ by (4.3). This means that the operator $N$ is a contraction with respect to $\mu$. Since

$$\frac{k_1 T^{2\alpha}}{\Gamma(2\alpha+1)} + \frac{k_2 T^\alpha}{\Gamma(\alpha+1)} < 1 ,$$

by applying Darbo's fixed point theorem, we conclude that IVP (6.38)–(6.39) has at least one solution belonging to the set $B_r \subset L^1(J, \mathbb{R})$. □

### 6.5.3 Example

In this section we present an example to illustrate the usefulness of our main results. Let us consider the fractional initial value problem

$$^cD^\alpha y(t) = \frac{t(1 + |y(t)| + |^cD^\alpha y(t)|)}{(t+5)}, \quad t \in J := [0, 1], \ \alpha \in (0, 1], \tag{6.44}$$

$$y(0) = y_0 . \tag{6.45}$$

Set

$$f(t, y, z) = \frac{t(1 + y + z)}{(t+5)}, \quad (t, y, z) \in J \times \mathbb{R}_+ \times \mathbb{R}_+ .$$

Clearly, the function $f$ satisfies the Carathéodory conditions. For $y, z \in \mathbb{R}_+$ and $t \in J$, we have

$$\begin{aligned} |f(t, y, z)| &= \left|\frac{t}{t+5} + \frac{ty}{t+5} + \frac{tz}{t+5}\right| \\ &\le \left|\frac{t}{t+5}\right| + \left|\frac{ty}{t+5}\right| + \left|\frac{tz}{t+5}\right| \\ &\le \left|\frac{t}{t+5}\right| + \frac{1}{6}|y| + \frac{1}{6}|z| . \end{aligned}$$

We first show that $a \in L^1[0, 1]$, where $a(t) = \frac{t}{t+5}$; indeed, $a(t)$ is a measurable function and

$$\begin{aligned} \int_0^1 a(t)dt &= \int_0^1 \frac{t}{t+5}dt \\ &= [t - 5\ln|t+5|)]_0^1 \\ &= 1 - 5\ln 6 + 5\ln 5 < \infty . \end{aligned}$$

Then $a \in L^1[0, 1]$. Hence, (6.14.2) holds with $a(t) = \frac{t}{t+5}$ and $q_1 = q_2 = \frac{1}{6}$.

Moreover, for each $t \in [0, 1]$, $x_i, y_i \in \mathbb{R}_+$, $i = 1, 2$, we have

$$\begin{aligned} &|f(t+\rho, y_1, z_1) - f(t, y_2, z_2)| \\ &= \left|\frac{t+\rho}{t+\rho+5} - \frac{t}{t+5} + \frac{(t+\rho)y_1}{t+\rho+5} - \frac{ty_2}{t+5} + \frac{(t+\rho)z_1}{t+\rho+5} - \frac{tz_2}{t+5}\right| \\ &\le \frac{\rho}{\rho+5} + \left|\frac{(t+\rho)y_1}{t+\rho+5} - \frac{ty_2}{t+5}\right| + \left|\frac{(t+\rho)z_1}{t+\rho+5} - \frac{tz_2}{t+5}\right| \\ &\le L_f(\rho) + \left|\frac{(t+\rho)y_1}{t+\rho+5} - \frac{ty_2}{t+5}\right| + \left|\frac{(t+\rho)z_1}{t+\rho+5} - \frac{tz_2}{t+5}\right| , \end{aligned}$$

where $L_f(\rho) = \frac{\rho}{\rho+5}$.

As $\delta \to 0$, we have

$$|f(t+\rho, y_1, z_1) - f(t, y_2, z_2)| \le L_f(\rho) + \frac{1}{6}|y_1 - y_2| + \frac{1}{6}|z_1 - z_2| .$$

Then (6.14.3) holds with $L_f(\rho) = \frac{\rho}{\rho+5}$ and $k_1 = k_2 = \frac{1}{6}$. Condition (6.42) is satisfied for appropriate values of $\alpha \in (0, 1]$ with $T = 1$. Indeed,

$$\frac{k_1}{\Gamma(2\alpha+1)} + \frac{k_2}{\Gamma(\alpha+1)} < 1 \Leftrightarrow \frac{1}{\Gamma(\alpha+1)} + \frac{1}{\Gamma(2\alpha+1)} < 6 . \tag{6.46}$$

Then, by Theorem 6.15, problem (6.44)–(6.45) has at least one solution on $[0, 1]$ for values of $\alpha$ satisfying condition (6.46). For example,
If $\alpha = \frac{1}{2}$, then $\Gamma(\alpha+1) = \Gamma(\frac{3}{2}) \simeq 0.88$ and $\Gamma(2\alpha+1) = \Gamma(2) = 1$ and

$$\frac{k_1}{\Gamma(2\alpha+1)} + \frac{k_2}{\Gamma(\alpha+1)} = \frac{1}{6} + \frac{\frac{1}{6}}{0.88} \simeq 0.35659 < 1 .$$

## 6.6 Notes and Remarks

The results of Chapter 6 are taken from Benchohra et al. [107, 108]. Other results may be found in [179, 181].

# 7 Partial Hadamard Fractional Integral Equations and Inclusions

## 7.1 Introduction

The fractional calculus represents a powerful tool in applied mathematics for studying many problems in various fields of science and engineering, with many breakthrough results found in mathematical physics, finance, hydrology, biophysics, thermodynamics, control theory, statistical mechanics, astrophysics, cosmology, and bioengineering [242]. There have been significant developments in ordinary and partial fractional differential and integral equations in recent years; see the monographs of Abbas et al. [35, 35], Kilbas et al. [181], and Miller and Ross [200], the papers of Abbas et al. [24, 25, 43], Vityuk et al. [247], and the references therein.

In [119], Butzer et al. investigated the properties of the Hadamard fractional integral and derivative. In [120], they obtained the Mellin transform of the Hadamard fractional integral and differential operators, and in [220], Pooseh et al. obtained expansion formulas of the Hadamard operators in terms of integer order derivatives. Many other interesting properties of those operators and others are summarized in [239] and the references therein.

This chapter deals with the existence and uniqueness of solutions to several classes of Hadamard partial fractional integral equations. We present results based on Banach's contraction principle and others on the nonlinear alternative of Leray–Schauder type. This chapter initiates the study of Hadamard integral equations of two independent variables.

## 7.2 Functional Partial Hadamard Fractional Integral Equations

### 7.2.1 Introduction

This section deals with the existence and uniqueness of solutions to the Hadamard partial fractional integral equation of the form

$$u(x,y) = \mu(x,y) + \frac{1}{\Gamma(r_1)\Gamma(r_2)} \int_1^x \int_1^y \left(\log\frac{x}{s}\right)^{r_1-1} \left(\log\frac{y}{t}\right)^{r_2-1} \frac{f(s,t,u(s,t))}{st}\,dtds\,; \quad \text{if } (x,y) \in J\,, \tag{7.1}$$

where $J := [1,a] \times [1,b]$, $a, b > 1$, $r_1, r_2 > 0$, $\mu\colon J \to \mathbb{R}$, $f\colon J \times \mathbb{R} \to \mathbb{R}$ are given continuous functions.

https://doi.org/10.1515/9783110553819-007

We present two results for integral equation (7.1). The first one is based on Banach's contraction principle and the second one on the nonlinear alternative of Leray–Schauder type. This section initiates the study of Hadamard integral equations of two independent variables.

### 7.2.2 Main Results

**Definition 7.1.** A function $u \in C$ is said to be a solution of (7.1) if $u$ satisfies equation (7.1) on $J$.

Further, we present conditions for the existence and uniqueness of a solution to equation (7.1).

**Theorem 7.2.** *Make the following assumption:*
(7.2.1) *For any $u, v \in C$ and $(x, y) \in J$, there exists $k > 0$ such that*

$$|f(x, y, u) - f(x, y, v)| \leq k\|u - v\|_C .$$

*If*

$$L := \frac{k(\log a)^{r_1}(\log b)^{r_2}}{\Gamma(1 + r_1)\Gamma(1 + r_2)} < 1 , \tag{7.2}$$

*then there exists a unique solution of equation* (7.1) *on $J$.*

*Proof.* Transform integral equation (7.1) into a fixed point equation. Consider the operator $N\colon C \to C$ defined by

$$\begin{aligned}(Nu)(x, y) &= \mu(x, y)\\ &+ \frac{1}{\Gamma(r_1)\Gamma(r_2)} \int_1^x \int_1^y \left(\log \frac{x}{s}\right)^{r_1-1} \left(\log \frac{y}{t}\right)^{r_2-1} \frac{f(s, t, u(s, t))}{st} dtds .\end{aligned} \tag{7.3}$$

Let $v, w \in C$. Then for $(x, y) \in J$ we have

$$\begin{aligned}|(Nv)(x, y) - (Nw)(x, y)| &\leq \frac{1}{\Gamma(r_1)\Gamma(r_2)} \int_1^x \int_1^y \left|\log \frac{x}{s}\right|^{r_1-1} \left|\log \frac{y}{t}\right|^{r_2-1}\\ &\quad \times \frac{|f(s, t, u(s, t)) - f(s, t, v(s, t))|}{st} dtds\\ &\leq \frac{1}{\Gamma(r_1)\Gamma(r_2)} \int_1^x \int_1^y \left|\log \frac{x}{s}\right|^{r_1-1} \left|\log \frac{y}{t}\right|^{r_2-1}\\ &\quad \times \frac{k\|u - v\|_C}{st} dtds\\ &\leq \frac{k(\log a)^{r_1}(\log b)^{r_2}}{\Gamma(1 + r_1)\Gamma(1 + r_2)} \|v - w\|_C .\end{aligned}$$

Consequently,

$$\|N(v) - N(w)\|_C \leq L\|v - w\|_C .$$

From (7.2), $N$ is a contraction, so $N$ has a unique fixed point by Banach's contraction principle. □

**Theorem 7.3.** *Make the following assumption:*
(7.3.1) *There exist functions $p_1, p_2 \in C(J, \mathbb{R}_+)$ such that*

$$|f(x, y, u)| \leq p_1(x, y) + p_2(x, y)|u(x, y)| \quad \text{for any } u \in \mathbb{R} \text{ and } (x, y) \in J .$$

*Then integral equation (7.1) has at least one solution defined on $J$.*

*Proof.* Consider the operator $N$ defined in (7.3). We will show that the operator $N$ is continuous and completely continuous.

*Step 1. $N$ is continuous.* Let $\{u_n\}$ be a sequence such that $u_n \to u$ in $C$. Let $\eta > 0$ be such that $\|u_n\|_C \leq \eta$. Then

$$\begin{aligned}
|(Nu_n)(x, y) - (Nu)(x, y)| &\leq \frac{1}{\Gamma(r_1)\Gamma(r_2)} \int_1^x \int_1^y \left|\log \frac{x}{s}\right|^{r_1-1} \left|\log \frac{y}{t}\right|^{r_2-1} \\
&\quad \times \frac{|f(s, t, u_n(s, t)) - f(s, t, u(s, t))|}{st} dtds \\
&\leq \frac{1}{\Gamma(r_1)\Gamma(r_2)} \int_1^x \int_1^y \left|\log \frac{x}{s}\right|^{r_1-1} \left|\log \frac{y}{t}\right|^{r_2-1} \\
&\quad \times \frac{\sup_{(s,t)\in J} |f(s, t, u_n(s, t)) - f(s, t, u(s, t))|}{st} dtds \\
&\leq \frac{(\log a)^{r_1} (\log b)^{r_2}}{\Gamma(1 + r_1)\Gamma(1 + r_2)} \|f(., ., u_n(., .)) - f(., ., u(., .))\|_C .
\end{aligned}$$

From Lebesgue's dominated convergence theorem and the continuity of the function $f$ we get

$$|(Nu_n)(x, y) - (Nu)(x, y)| \to 0 \text{ as } n \to \infty .$$

*Step 2. $N$ maps bounded sets to bounded sets in $C$.* Indeed, it is enough to show that for any $\eta^* > 0$ there exists a positive constant $\tilde{\ell}$ such that, for each $u \in B_{\eta^*} = \{u \in C : \|u\|_C \leq \eta^*\}$, we have $\|N(u)\|_C \leq \tilde{\ell}$. Set

$$p_i^* = \sup_{(x,y)\in J} p_i(x, y) ; \quad i = 1, 2 .$$

From (7.3.1), for each $(x,y)\in J$ we have

$$\begin{aligned}|(Nu)(x,y)| &\le |\mu(x,y)| + \frac{1}{\Gamma(r_1)\Gamma(r_2)}\int_1^x\int_1^y \left|\log\frac{x}{s}\right|^{r_1-1}\left|\log\frac{y}{t}\right|^{r_2-1}\\ &\quad\times\frac{p_1(s,t)+p_2(s,t)\|u\|_C}{st}dtds\\ &\le \|\mu\|_\infty + \frac{(\log a)^{r_1}(\log b)^{r_2}}{\Gamma(1+r_1)\Gamma(1+r_2)}(p_1^*+p_2^*\eta^*)\\ &:= \tilde{\ell}\,.\end{aligned}$$

Hence,

$$\|N(u)\|_C \le \tilde{\ell}\,.$$

*Step 3: N maps bounded sets to equicontinuous sets in C.* Let $(x_1,y_1),(x_2,y_2)\in(1,a]\times(1,b]$, $x_1<x_2$, $y_1<y_2$, $B_{\eta^*}$ be a bounded set of $C$ as in Step 2, and let $u\in B_{\eta^*}$. Then

$$\begin{aligned}&|(Nu)(x_2,y_2)-(Nu)(x_1,y_1)| \le |\mu(x_1,y_1)-\mu(x_2,y_2)|\\ &+\frac{1}{\Gamma(r_1)\Gamma(r_2)}\int_1^{x_1}\int_1^{y_1}\left[\left|\log\frac{x_2}{s}\right|^{r_1-1}\left|\log\frac{y_2}{t}\right|^{r_2-1}-\left|\log\frac{x_1}{s}\right|^{r_1-1}\left|\log\frac{y_1}{t}\right|^{r_2-1}\right]\\ &\times\frac{|f(s,t,u(s,t))|}{st}dtds\\ &+\frac{1}{\Gamma(r_1)\Gamma(r_2)}\int_{x_1}^{x_2}\int_{y_1}^{y_2}\left|\log\frac{x_2}{s}\right|^{r_1-1}\left|\log\frac{y_2}{t}\right|^{r_2-1}\frac{|f(s,t,u(s,t))|}{st}dtds\\ &+\frac{1}{\Gamma(r_1)\Gamma(r_2)}\int_1^{x_1}\int_{y_1}^{y_2}\left|\log\frac{x_2}{s}\right|^{r_1-1}\left|\log\frac{y_2}{t}\right|^{r_2-1}\frac{|f(s,t,u(s,t))|}{st}dtds\\ &+\frac{1}{\Gamma(r_1)\Gamma(r_2)}\int_{x_1}^{x_2}\int_1^{y_1}\left|\log\frac{x_2}{s}\right|^{r_1-1}\left|\log\frac{y_2}{t}\right|^{r_2-1}\frac{|f(s,t,u(s,t))|}{st}dtds\,.\end{aligned}$$

Thus,

$$\begin{aligned}&|(Nu)(x_2,y_2)-(Nu)(x_1,y_1)| \le |\mu(x_1,y_1)-\mu(x_2,y_2)|\\ &+\frac{1}{\Gamma(r_1)\Gamma(r_2)}\int_1^{x_1}\int_1^{y_1}\left[\left|\log\frac{x_2}{s}\right|^{r_1-1}\left|\log\frac{y_2}{t}\right|^{r_2-1}-\left|\log\frac{x_1}{s}\right|^{r_1-1}\left|\log\frac{y_1}{t}\right|^{r_2-1}\right]\\ &\times\frac{p_1^*+p_2^*\eta^*}{st}dtds\\ &+\frac{1}{\Gamma(r_1)\Gamma(r_2)}\int_{x_1}^{x_2}\int_{y_1}^{y_2}\left|\log\frac{x_2}{s}\right|^{r_1-1}\left|\log\frac{y_2}{t}\right|^{r_2-1}\frac{p_1^*+p_2^*\eta^*}{st}dtds\end{aligned}$$

$$+\frac{1}{\Gamma(r_1)\Gamma(r_2)}\int_1^{x_1}\int_{y_1}^{y_2}\left|\log\frac{x_2}{s}\right|^{r_1-1}\left|\log\frac{y_2}{t}\right|^{r_2-1}\frac{p_1^*+p_2^*\eta^*}{st}dtds$$

$$+\frac{1}{\Gamma(r_1)\Gamma(r_2)}\int_{x_1}^{x_2}\int_1^{y_1}\left|\log\frac{x_2}{s}\right|^{r_1-1}\left|\log\frac{y_2}{t}\right|^{r_2-1}\frac{p_1^*+p_2^*\eta^*}{st}dtds$$

$$\le\frac{p_1^*+p_2^*\eta^*}{\Gamma(1+r_1)\Gamma(1+r_2)}$$

$$\times[2(\log y_2)^{r_2}(\log x_2-\log x_1)^{r_1}+2(\log x_2)^{r_1}(\log y_2-\log y_1)^{r_2}$$

$$+(\log x_1)^{r_1}(\log y_1)^{r_2}-(\log x_2)^{r_1}(\log y_2)^{r_2}$$

$$-2(\log x_2-\log x_1)^{r_1}(\log y_2-\log y_1)^{r_2}]\,.$$

As $x_1\to x_2$ and $y_1\to y_2$, the right-hand side of the preceding inequality tends to zero.

As a consequence of Steps 1–3, together with the Ascoli–Arzelà theorem, we can conclude that $N$ is continuous and completely continuous.

*Step 4. A priori bounds.* We now show that there exists an open set $U\subseteq C$ with $u\neq\lambda N(u)$ for $\lambda\in(0,1)$ and $u\in\partial U$. Let $u\in C$ be such that $u=\lambda N(u)$ for some $0<\lambda<1$. Thus, for each $(x,y)\in J$,

$$u(x,y)=\lambda\mu(x,y)+\frac{\lambda}{\Gamma(r_1)\Gamma(r_2)}\int_1^x\int_1^y\left(\log\frac{x}{s}\right)^{r_1-1}\left(\log\frac{y}{t}\right)^{r_2-1}\frac{f(s,t,u(s,t))}{st}dtds\,.$$

This implies that for each $(x,y)\in J$ we have

$$|u(x,y)|\le|\mu(x,y)|+\frac{1}{\Gamma(r_1)\Gamma(r_2)}\int_1^x\int_1^y\left|\log\frac{x}{s}\right|^{r_1-1}\left|\log\frac{y}{t}\right|^{r_2-1}$$

$$\times\frac{p_1(s,t)+p_2(s,t)|u(s,t)|}{st}dtds$$

$$\le\|\mu\|_\infty+\frac{p_1^*(\log a)^{r_1}(\log b)^{r_2}}{\Gamma(1+r_1)\Gamma(1+r_2)}$$

$$+\frac{p_2^*}{\Gamma(r_1)\Gamma(r_2)}\int_1^x\int_1^y\left|\log\frac{x}{s}\right|^{r_1-1}\left|\log\frac{y}{t}\right|^{r_2-1}\frac{|u(s,t)|}{st}dtds\,.$$

Thus, for each $(x,y)\in J$ we get

$$|u(x,y)|\le\|\mu\|_\infty+\frac{p_1^*(\log a)^{r_1}(\log b)^{r_2}}{\Gamma(1+r_1)\Gamma(1+r_2)}$$

$$+\frac{p_2^*}{\Gamma(r_1)\Gamma(r_2)}\int_1^x\int_1^y\left|\log\frac{x}{s}\right|^{r_1-1}\left|\log\frac{y}{t}\right|^{r_2-1}\frac{|u(s,t)|}{st}dtds$$

$$\le c+\int_1^x\int_1^y q(x,y,s,t)|u(s,t)|\,,$$

where

$$c := \|\mu\|_\infty + \frac{p_1^*(\log a)^{r_1}(\log b)^{r_2}}{\Gamma(1+r_1)\Gamma(1+r_2)}$$

and

$$q(x,y,s,t) := \frac{p_2^*}{st\Gamma(r_1)\Gamma(r_2)}\left|\log\frac{x}{s}\right|^{r_1-1}\left|\log\frac{y}{t}\right|^{r_2-1}.$$

From Lemma 1.56 we obtain

$$|u(x,y)| \le c\exp\left(\int_1^x\int_1^y B(s,t)dtds\right),$$

where

$$\begin{aligned} B(x,y) &= q(x,y,x,y) + \int_1^x D_1q(x,y,s,y)ds \\ &\quad + \int_1^y D_2q(x,y,x,t)dt + \int_1^x\int_1^y D_1D_2q(x,y,s,t)dtds \\ &\le \frac{p_2^*}{xy\Gamma(r_1)\Gamma(r_2)}(\log x)^{r_1-1}(\log y)^{r_2-1}. \end{aligned}$$

Hence,

$$\begin{aligned} |u(x,y)| &\le c\exp\left(\int_1^x\int_1^y \frac{p_2^*}{st\Gamma(r_1)\Gamma(r_2)}(\log s)^{r_1-1}(\log t)^{r_2-1}dtds\right) \\ &\le c\exp\left(\frac{p_2^*(\log a)^{r_1}(\log b)^{r_2}}{\Gamma(1+r_1)\Gamma(1+r_2)}\right) \\ &:= R. \end{aligned}$$

Set

$$U = \{u \in C : \|u\|_\infty < R+1\}.$$

By our choice of $U$, there is no $u \in \partial U$ such that $u = \lambda N(u)$ for $\lambda \in (0,1)$. As a consequence of the nonlinear alternative of the Leray–Schauder type [149], we deduce that $N$ has a fixed point $u$ in $\overline{U}$ that is a solution of our equation (7.1). □

### 7.2.3 An Example

Consider a partial Hadamard integral equation of the form

$$\begin{aligned} u(x,y) &= \mu(x,y) \\ &\quad + \int_1^x\int_1^y \left(\log\frac{x}{s}\right)^{r_1-1}\left(\log\frac{y}{t}\right)^{r_2-1}\frac{f(s,t,u(s,t))}{st\Gamma(r_1)\Gamma(r_2)}dtds\,; \quad (x,y)\in[1,e]\times[1,e], \end{aligned} \tag{7.4}$$

where

$$r_1, r_2 > 0\,, \quad \mu(x,y) = x + y^2; \quad (x,y) \in [1,e] \times [1,e]$$

and

$$f(x,y,u(x,y)) = \frac{cu(x,y)}{e^{x+y+2}}\,; \quad (x,y) \in [1,e] \times [1,e]\,,$$

with

$$c := \frac{e^4}{2}\Gamma(1+r_1)\Gamma(1+r_2)\,.$$

For each $u, \overline{u} \in \mathbb{R}$ and $(x,y) \in [1,e] \times [1,e]$ we have

$$|f(x,y,u(x,y)) - f(x,y,\overline{u}(x,y))| \le \frac{c}{e^4}\|u - \overline{u}\|_C\,.$$

Hence, condition (7.2.1) is satisfied by $k = \frac{c}{e^4}$. Condition (7.2) holds with $a = b = e$. Indeed,

$$\frac{k(\log a)^{r_1}(\log b)^{r_2}}{\Gamma(1+r_1)\Gamma(1+r_2)} = \frac{c}{e^4\Gamma(1+r_1)\Gamma(1+r_2)} = \frac{1}{2} < 1\,.$$

Consequently, Theorem 7.2 implies that integral equation (7.4) has a unique solution defined on $[1,e] \times [1,e]$.

## 7.3 Fredholm-Type Hadamard Fractional Integral Equations

### 7.3.1 Introduction

The qualitative properties and structure of the set of solutions of the Darboux problem for hyperbolic partial integer order differential equations have been studied by many authors, for instance, [43, 123, 146]. In [110], Bica et al. initiated the study of the Fredholm integral equation

$$x(t) = f(t) + \int_0^a g(t,s,x(s),x'(s))ds \tag{7.5}$$

in a Banach space setting. In [213], Pachpatte studied the qualitative behavior of solutions of equation (7.5) and its further generalization. Inspired by the results in [110, 212, 213], Pachpatte in [214] studied the Fredholm-type integral equation

$$u(x,y) = f(x,y) + \int_0^a\int_0^b g(x,y,s,t,u(s,t),D_1u(s,t),D_2u(s,t))dtds\,, \tag{7.6}$$

where $u$ is an unknown function. Recently, in [24], Abbas and Benchohra studied some uniqueness results for the Fredholm-type Riemann–Liouville integral equation

$$\begin{aligned} u(x,y) = \mu(x,y) + \frac{1}{\Gamma(r_1)\Gamma(r_2)}\int_0^a\int_0^b (a-s)^{r_1-1}(b-t)^{r_2-1} \\ \times f(x,y,s,t,u(s,t),({}^cD_\theta^r u)(s,t))dtds\,; \quad \text{if } (x,y) \in J := [0,a] \times [0,b]\,, \end{aligned} \tag{7.7}$$

where $a, b \in (0, \infty)$, $\theta = (0, 0)$, ${}^{c}D^{r}_{\theta}$ is the standard Caputo's fractional derivative of order $r = (r_1, r_2) \in (0, 1] \times (0, 1]$, $\mu : J \to \mathbb{R}^n$, and $f : J \times J \times \mathbb{R}^n \times \mathbb{R}^n \to \mathbb{R}^n$ are given continuous functions.

This section deals with the existence and uniqueness of solutions to the Fredholm-type Hadamard partial integral equation

$$u(x,y) = \mu(x,y) + \frac{1}{\Gamma(r_1)\Gamma(r_2)} \int_1^a \int_1^b \left(\ln \frac{a}{s}\right)^{r_1-1} \left(\ln \frac{b}{t}\right)^{r_2-1} \times \frac{f(x,y,s,t,u(s,t),({}^{H}D^{r}_{\sigma}u)(s,t))}{st} dtds \, ; \quad \text{if } (x,y) \in J := [1,a] \times [1,b] \, , \tag{7.8}$$

where $a, b \in (1, \infty)$, $\sigma = (1, 1)$, ${}^{H}D^{r}_{\theta}$ is the standard Hadamard fractional derivative of order $r = (r_1, r_2) \in (0, 1] \times (0, 1]$, $\mu : J \to \mathbb{R}^n$, and $f : J \times J \times \mathbb{R}^n \times \mathbb{R}^n \to \mathbb{R}^n$ are given continuous functions.

### 7.3.2 Main Results

Define the space $E = E(J, \mathbb{R}^n)$ by

$$E := \{w \in C(J) : \ {}^{H}D^{r}_{\sigma}w \ \text{ exists and } \ {}^{H}D^{r}_{\sigma}w \in C(J)\} \, .$$

For $w \in E$, use the notation

$$\|w(x,y)\|_1 = \|w(x,y)\| + \|{}^{H}D^{r}_{\sigma}w(x,y)\| \, .$$

In the space $E$ we define the norm

$$\|w\|_E = \sup_{(x,y)\in J} \|w(x,y)\|_1 \, .$$

**Lemma 7.4.** *$(E, \|\cdot\|_E)$ is a Banach space.*

*Proof.* Let $\{u_n\}_{n=0}^{\infty}$ be a Cauchy sequence in the space $(E, \|\cdot\|_E)$. Then

$$\forall \epsilon > 0, \exists N > 0 \text{ such that for all } n, m > N \text{ we have } \|u_n - u_m\|_E < \epsilon \, .$$

Thus, $\{u_n(x,y)\}_{n=0}^{\infty}$ and $\{({}^{H}D^{r}_{\sigma}u_n)(x,y)\}_{n=0}^{\infty}$ are Cauchy sequences in $\mathbb{R}^n$. Then $\{u_n(x,y)\}_{n=0}^{\infty}$ converges to some $u(x,y)$ in $\mathbb{R}^n$, and $\{{}^{H}D^{r}_{\sigma}u_n\}_{n=0}^{\infty}$ converges uniformly to some $v(x,y) \in E$. Next, we need to prove that $u \in E$ and $v = {}^{H}D^{r}_{\sigma}u$. According to the uniform convergence of $\{({}^{H}D^{r}_{\sigma}u_n)(x,y)\}_{n=0}^{\infty}$ and the dominated convergence theorem, we obtain

$$v(x,y) = \lim_{n\to\infty} ({}^{H}D^{r}_{\sigma}u_n)(x,y) \, .$$

Thus, $\{{}^{H}D^{r}_{\sigma}u_n\}_{n=0}^{\infty}$ converges uniformly to ${}^{H}D^{r}_{\sigma}u$ in $E$. Hence, $u \in E$ and

$$v(x,y) = ({}^{H}D^{r}_{\sigma}u)(x,y) \, .$$

□

**Definition 7.5.** By a solution to equation (7.8), we mean every function $w \in E$ such that $w$ satisfies (7.8) on $J$.

Next we present conditions for the existence of solutions of integral equation (7.8).

**Theorem 7.6.** *Make the following assumptions:*

(7.6.1) *There exist* $0 < r_3 < \min\{r_1, r_2\}$, *functions* $\rho_1 : J \times J \to \mathbb{R}^+$, $\varphi : J \to \mathbb{R}^+$, *with* $\rho_1(x, y, \cdot, \cdot), \varphi \in L^{\frac{1}{r_3}}(J)$, *and a nondecreasing function* $\psi : [0, \infty) \to (0, \infty)$ *such that*

$$\|f(x, y, s, t, u, v)\| \leq \rho_1(x, y, s, t)(\|u\| + \|v\|) \tag{7.9}$$

*and*

$$\begin{aligned} &\|f(x_1, y_1, s, t, u, v) - f(x_2, y_2, s, t, u, v)\| \\ &\leq \varphi(s, t)(|x_1 - x_2| + |y_1 - y_2|)\psi(\|u\| + \|v\|) \\ &\text{for each } (x, y), (s, t), (x_1, y_1), (x_2, y_2) \in J \text{ and } u, v \in \mathbb{R}^n . \end{aligned} \tag{7.10}$$

(7.6.2) *There exist nonnegative constants* $\alpha, \beta_1, \beta_2$ *such that for* $(x, y) \in J$ *we have*

$$\begin{cases} \|\mu(x, y)\|_1 & \leq \alpha , \\ \int_1^a \int_1^b \rho_1^{\frac{1}{r_3}}(x, y, s, t)dtds & \leq \beta_1^{\frac{1}{r_3}} , \\ \int_1^a \int_1^b \rho_2^{\frac{1}{r_3}}(x, y, s, t)dtds & \leq \beta_2^{\frac{1}{r_3}} , \end{cases} \tag{7.11}$$

*where*

$$\rho_2(x, y, \cdot, \cdot) \in L^{\frac{1}{r_3}}(J) \text{ and } \rho_2(x, y, s, t) = ({}^H D_\sigma^r \rho_1)(x, y, s, t) .$$

*If*

$$\ell := \frac{(\beta_1 + \beta_2)(\ln a)^{(\omega_1+1)(1-r_3)}(\ln b)^{(\omega_2+1)(1-r_3)}}{(\omega_1 + 1)^{(1-r_3)}(\omega_2 + 1)^{(1-r_3)}\Gamma(r_1)\Gamma(r_2)} < 1 , \tag{7.12}$$

*where* $\omega_1 = \frac{r_1-1}{1-r_3}$, $\omega_2 = \frac{r_2-1}{1-r_3}$, *then the Fredholm–Hadamard integral equation* (7.8) *has at least one solution on* $J$.

**Remark 7.7.** It is clear that condition (7.9) implies

$$\|({}^H D_\sigma^r f)(x, y, s, t, u, v)\| \leq \rho_2(x, y, s, t)(\|u\| + \|v\|) . \tag{7.13}$$

*Proof.* Let $u \in E$, and define the operator $N : E \to E$ by

$$\begin{aligned} (Nu)(x, y) =& \mu(x, y) + \frac{1}{\Gamma(r_1)\Gamma(r_2)} \int_1^a \int_1^b \left(\ln \frac{a}{s}\right)^{r_1-1} \left(\ln \frac{b}{t}\right)^{r_2-1} \\ &\times \frac{f(x, y, s, t, u(s, t), ({}^H D_\sigma^r u)(s, t))}{st} dtds . \end{aligned} \tag{7.14}$$

Differentiating both sides of (7.14) by applying the Hadamard fractional derivative, we get

$$ {}^{H}D^{r}_{\sigma}(Nu)(x,y) = {}^{H}D^{r}_{\sigma}\mu(x,y) + \frac{1}{\Gamma(r_1)\Gamma(r_2)} \int_1^a\int_1^b \left(\ln\frac{a}{s}\right)^{r_1-1}\left(\ln\frac{b}{t}\right)^{r_2-1} \times \frac{{}^{H}D^{r}_{\sigma}f(x,y,s,t,u(s,t),({}^{H}D^{r}_{\sigma}u)(s,t))}{st}dtds\,. \tag{7.15}$$

Set

$$M = \frac{\alpha}{1-\ell} \quad \text{and} \quad D = \{u \in E \colon \|u\|_E \le M\}\,.$$

Clearly, $D$ is a closed convex subset of $E$. Now we show that $N$ maps $D$ to itself. Evidently, $N(u)$, ${}^{H}D^{r}_{\theta}(Nu)$ are continuous on $J$. From (7.11) and using (7.6.1) and (7.6.2), for each $(x,y) \in J$ we have

$$\begin{aligned}
\|(Nu)(x,y)\|_1 &\le \|\mu(x,y)\|_1 \\
&\quad + \frac{1}{\Gamma(r_1)\Gamma(r_2)} \int_1^a\int_1^b \left|\ln\frac{a}{s}\right|^{r_1-1}\left|\ln\frac{b}{t}\right|^{r_2-1} \\
&\quad \times \left\|\frac{f(x,y,s,t,u(s,t),({}^{H}D^{r}_{\sigma}u)(s,t))}{st}\right\| dtds \\
&\quad + \frac{1}{\Gamma(r_1)\Gamma(r_2)} \int_1^a\int_1^b \left(\ln\frac{a}{s}\right)^{r_1-1}\left(\ln\frac{b}{t}\right)^{r_2-1} \\
&\quad \times \left\|\frac{{}^{H}D^{r}_{\sigma}f(x,y,s,t,u(s,t),({}^{H}D^{r}_{\sigma}u)(s,t))}{st}\right\| dtds \\
&\le \|\mu(x,y)\|_1 \\
&\quad + \frac{1}{\Gamma(r_1)\Gamma(r_2)} \left(\int_1^a\int_1^b \frac{1}{st}\left|\ln\frac{a}{s}\right|^{\frac{r_1-1}{1-r_3}}\left|\ln\frac{b}{t}\right|^{\frac{r_2-1}{1-r_3}} dtds\right)^{1-r_3} \\
&\quad \times \left(\int_1^a\int_1^b \|f(x,y,s,t,u(s,t),({}^{H}D^{r}_{\sigma}u)(s,t))\|^{\frac{1}{r_3}} dtds\right)^{r_3} \\
&\quad + \frac{1}{\Gamma(r_1)\Gamma(r_2)} \left(\int_1^a\int_1^b \frac{1}{st}\left|\ln\frac{a}{s}\right|^{\frac{r_1-1}{1-r_3}}\left|\ln\frac{b}{t}\right|^{\frac{r_2-1}{1-r_3}} dtds\right)^{1-r_3} \\
&\quad \times \left(\int_1^a\int_1^b \|{}^{H}D^{r}_{\sigma}f(x,y,s,t,u(s,t),({}^{H}D^{r}_{\sigma}u)(s,t))\|^{\frac{1}{r_3}} dtds\right)^{r_3}.
\end{aligned}$$

Then for each $(x, y) \in J$ we obtain

$$\begin{aligned}
\|(Nu)(x,y)\|_1 &\leq \|\mu(x,y)\|_1 + \frac{(\ln a)^{(\omega_1+1)(1-r_3)}(\ln b)^{(\omega_2+1)(1-r_3)}}{(\omega_1+1)^{(1-r_3)}(\omega_2+1)^{(1-r_3)}\Gamma(r_1)\Gamma(r_2)} \\
&\times \left[ \left( \int_1^a \int_1^b \rho_1^{\frac{1}{r_3}}(x,y,s,t)\|u(s,t)\|_1^{\frac{1}{r_3}} dtds \right)^{r_3} + \left( \int_1^a \int_1^b \rho_2^{\frac{1}{r_3}}(x,y,s,t)\|u(s,t)\|_1^{\frac{1}{r_3}} dtds \right)^{r_3} \right] \\
&\leq \alpha + \frac{(\ln a)^{(\omega_1+1)(1-r_3)}(\ln b)^{(\omega_2+1)(1-r_3)}}{(\omega_1+1)^{(1-r_3)}(\omega_2+1)^{(1-r_3)}\Gamma(r_1)\Gamma(r_2)} \\
&\times \left[ \|u\|_E \left( \int_1^a \int_1^b \rho_1^{\frac{1}{r_3}}(x,y,s,t) dtds \right)^{r_3} + \|u|_E \left( \int_0^a \int_0^b \rho_2^{\frac{1}{r_3}}(x,y,s,t) dtds \right)^{r_3} \right] \\
&\leq \alpha + \frac{(M\beta_1 + M\beta_2)(\ln a)^{(\omega_1+1)(1-r_3)}(\ln b)^{(\omega_2+1)(1-r_3)}}{(\omega_1+1)^{(1-r_3)}(\omega_2+1)^{(1-r_3)}\Gamma(r_1)\Gamma(r_2)} \\
&= \alpha + M\ell\,.
\end{aligned}$$

From (7.12) and the definition of $M$ we get

$$\|N(u)\|_E \leq M\,.$$

Hence, $N(u) \in D$. This proves that the operator $N$ maps $D$ to itself. Next we verify that the operator $N$ satisfies the assumptions of Schauder's fixed point theorem. The proof will be given in several steps.

*Step 1. $N$ is continuous.* Let $\{u_n\}$ be a sequence such that $u_n \to u$ in $D$. Then

$$\begin{aligned}
\|(Nu_n)(x,y) - (Nu)(x,y)\|_1 \leq{}& \frac{1}{\Gamma(r_1)\Gamma(r_2)} \int_1^a \int_1^b \frac{1}{st} \left|\ln \frac{a}{s}\right|^{r_1-1} \left|\ln \frac{b}{t}\right|^{r_2-1} \\
&\times \|f(x,y,s,t,u_n(s,t),({}^H D_\sigma^r u_n)(s,t)) \\
&- f(x,y,s,t,u(s,t),({}^H D_\sigma^r u)(s,t))\| dtds \\
&+ \frac{1}{\Gamma(r_1)\Gamma(r_2)} \int_1^a \int_1^b \frac{1}{st} \left|\ln \frac{a}{s}\right|^{r_1-1} \left|\ln \frac{b}{t}\right|^{r_2-1} \\
&\times \|{}^H D_\sigma^r f(x,y,s,t,u_n(s,t),({}^H D_\sigma^r u_n)(s,t)) \\
&- {}^H D_\sigma^r f(x,y,s,t,u(s,t),({}^H D_\sigma^r u)(s,t))\| dtds\,.
\end{aligned}$$

Thus,

$$\begin{aligned}
\|(Nu_n)(x,y)-(Nu)(x,y)\|_1 \le{}& \frac{1}{\Gamma(r_1)\Gamma(r_2)}\int_1^a\int_1^b \frac{1}{st}\left|\ln\frac{a}{s}\right|^{r_1-1}\left|\ln\frac{b}{t}\right|^{r_2-1}\\
&\times \sup_{(s,t)\in J}\|f\left(x,y,s,t,u_n(s,t),({}^HD_\sigma^r u_n)(s,t)\right)\\
&-f\left(x,y,s,t,u(s,t),({}^HD_\sigma^r u)(s,t)\right)\|dtds\\
&+\frac{1}{\Gamma(r_1)\Gamma(r_2)}\int_1^a\int_1^b \frac{1}{st}\left|\ln\frac{a}{s}\right|^{r_1-1}\left|\ln\frac{b}{t}\right|^{r_2-1}\\
&\times \sup_{(s,t)\in J}\|{}^HD_\sigma^r f\left(x,y,s,t,u_n(s,t),({}^HD_\sigma^r u_n)(s,t)\right)\\
&-{}^HD_\sigma^r f\left(x,y,s,t,u(s,t),({}^HD_\sigma^r u)(s,t)\right)\|dtds\\
\le{}& \frac{(\ln a)^{r_1}(\ln b)^{r_2}}{\Gamma(1+r_1)\Gamma(1+r_2)}\left(\|f\left(x,y,\cdot,\cdot,u_n(\cdot,\cdot),({}^HD_\sigma^r u_n)(\cdot,\cdot)\right)\right.\\
&-f\left(x,y,\cdot,\cdot,u(\cdot,\cdot),({}^HD_\sigma^r u)(\cdot,\cdot)\right)\|\\
&+\|{}^HD_\sigma^r f\left(x,y,\cdot,\cdot,u_n(\cdot,\cdot),({}^HD_\sigma^r u_n)(\cdot,\cdot)\right)\\
&\left.-{}^HD_\sigma^r f\left(x,y,\cdot,\cdot,u(\cdot,\cdot),({}^HD_\sigma^r u)(\cdot,\cdot)\right)\|\right)\\
\le{}& \frac{(\ln a)^{r_1}(\ln b)^{r_2}}{\Gamma(1+r_1)\Gamma(1+r_2)}\|f\left(x,y,\cdot,\cdot,u_n(\cdot,\cdot),({}^HD_\sigma^r u_n)(\cdot,\cdot)\right)\\
&-f\left(x,y,\cdot,\cdot,u(\cdot,\cdot),({}^HD_\sigma^r u)(\cdot,\cdot)\right)\|_1\,.
\end{aligned}$$

Hence, from Lebesgue's dominated convergence theorem and the continuity of the function $f$ we get

$$\|(N(u_n)-N(u)\|_E \to 0 \text{ as } n\to\infty\,.$$

*Step 2. $N(D)$ is bounded.* This is clear since $N(D)\subset D$ and $D$ is bounded.

*Step 3. $N(D)$ is equicontinuous.* Let $(x_1,y_1),(x_2,y_2)\in(1,a]\times(1,b]$, $x_1<x_2$, $y_1<y_2$, and let $u\in D$. Then

$$\begin{aligned}
\|(Nu)(x_2,y_2)-(Nu)(x_1,y_1)\|_1 \le{}& \|\mu(x_1,y_1)-\mu(x_2,y_2)\|_1\\
&+\frac{1}{\Gamma(r_1)\Gamma(r_2)}\int_1^a\int_1^b \frac{1}{st}\left|\ln\frac{a}{s}\right|^{r_1-1}\left|\ln\frac{b}{t}\right|^{r_2-1}\\
&\times\|f(x_2,y_2,s,t,u(s,t),({}^HD_\sigma^r u)(s,t))\\
&-f(x_1,y_1,s,t,u(s,t),({}^HD_\sigma^r u)(s,t))\|dtds\,,\\
&+\frac{1}{\Gamma(r_1)\Gamma(r_2)}\int_1^a\int_1^b \frac{1}{st}\left|\ln\frac{a}{s}\right|^{r_1-1}\left|\ln\frac{b}{t}\right|^{r_2-1}\\
&\times\|{}^HD_\sigma^r f(x_2,y_2,s,t,u(s,t),({}^HD_\sigma^r u)(s,t))\\
&-{}^HD_\sigma^r f(x_1,y_1,s,t,u(s,t),({}^HD_\sigma^r u)(s,t))\|dtds\,.
\end{aligned}$$

Thus,

$$\begin{aligned}\|(Nu)(x_2,y_2)-(Nu)(x_1,y_1)\|_1 &\le \|\mu(x_1,y_1)-\mu(x_2,y_2)\|_1\\&+\frac{(\ln a)^{(\omega_1+1)(1-r_3)}(\ln b)^{(\omega_2+1)(1-r_3)}}{(\omega_1+1)^{(1-r_3)}(\omega_2+1)^{(1-r_3)}\Gamma(r_1)\Gamma(r_2)}\\&\times\Bigg[\Bigg(\int_1^a\int_1^b \|f(x_2,y_2,s,t,u(s,t),({}^HD_\sigma^r u)(s,t))\\&-f(x_1,y_1,s,t,u(s,t),({}^HD_\sigma^r u)(s,t))\|^{\frac{1}{r_3}}dtds\Bigg)^{r_3}\\&+\Bigg(\int_1^a\int_1^b \|{}^HD_\sigma^r f(x_2,y_2,s,t,u(s,t),({}^HD_\sigma^r u)(s,t))\\&-{}^HD_\sigma^r f(x_1,y_1,s,t,u(s,t),({}^HD_\sigma^r u)(s,t))\|^{\frac{1}{r_3}}dtds\Bigg)^{r_3}\Bigg].\end{aligned}$$

Hence,

$$\begin{aligned}\|(Nu)(x_2,y_2)-(Nu)(x_1,y_1)\|_1 &\le \|\mu(x_1,y_1)-\mu(x_2,y_2)\|_1\\&+\frac{(\ln a)^{(\omega_1+1)(1-r_3)}(\ln b)^{(\omega_2+1)(1-r_3)}}{(\omega_1+1)^{(1-r_3)}(\omega_2+1)^{(1-r_3)}\Gamma(r_1)\Gamma(r_2)}\\&\times(|x_1-x_2|+|y_1-y_2|)\psi(\|u\|_1)\\&\times\Bigg[\Bigg(\int_1^a\int_1^b\|\varphi(s,t)\|^{\frac{1}{r_3}}dtds\Bigg)^{r_3}\\&+\Bigg(\int_1^a\int_1^b\|({}^HD_\sigma^r\varphi)(s,t)\|^{\frac{1}{r_3}}dtds\Bigg)^{r_3}\Bigg]\\&\le \|\mu(x_1,y_1)-\mu(x_2,y_2)\|_1\\&+\frac{(\ln a)^{(\omega_1+1)(1-r_3)}(\ln b)^{(\omega_2+1)(1-r_3)}}{(\omega_1+1)^{(1-r_3)}(\omega_2+1)^{(1-r_3)}\Gamma(r_1)\Gamma(r_2)}\\&\times[(\|\varphi\|_{L^{\frac{1}{r_3}}})^{r_3}+\left(\|{}^HD_\sigma^r\varphi\|_{L^{\frac{1}{r_3}}}\right)^{r_3}]\psi(M)\\&\times(|x_1-x_2|+|y_1-y_2|)\,.\end{aligned}$$

As $x_1\to x_2$ and $y_1\to y_2$, the right-hand side of the preceding inequality tends to zero.

As a consequence of Steps 1–3, together with the Ascoli–Arzelà theorem, we can conclude that $N$ is continuous and completely continuous. From an application of Schauder's theorem [149], we deduce that $N$ has a fixed point $u$ that is a solution of integral equation (7.8). □

Now we define the Banach space

$$X := \{w \in C(J):\ {}^H D^{r_1}_{1,x} w,\ {}^H D^{r_2}_{1,y} w \text{ exist and } {}^H D^{r_1}_{1,x} w,\ {}^H D^{r_2}_{1,y} w \in C(J)\},$$

with the norm

$$\|w\|_X = \sup_{(x,y)\in J} \|w(x,y)\|_1 ,$$

where

$$\|w(x,y)\|_1 = \|w(x,y)\| + \|{}^H D^{r_1}_{1,x} w(x,y)\| + \|{}^H D^{r_2}_{1,y} w(x,y)\| .$$

**Corollary 7.8.** *Consider the Fredholm-type Hadamard integral equation*

$$\begin{aligned} u(x,y) = \mu(x,y) + \frac{1}{\Gamma(r_1)\Gamma(r_2)} \int_1^a \int_1^b \left(\ln \frac{x}{s}\right)^{r_1-1} \left(\ln \frac{y}{t}\right)^{r_2-1} & \\ \times f(x,y,s,t,u(s,t),({}^H D^{r_1}_{1,s} u)(s,t),({}^H D^{r_2}_{1,t} u)(s,t))dtds ; & \qquad (7.16) \\ \text{if } (x,y) \in J := [1,a]\times[1,b] . & \end{aligned}$$

*Make the following assumptions:*

(7.8.1) *There exist* $0 < r_3 < \min\{r_1, r_2\}$, *functions* $\rho_1 : J \times J \to \mathbb{R}^+$, $\varphi : J \to \mathbb{R}^+$, *with* $\rho_1(x,y,\cdot,\cdot), \varphi \in L^{\frac{1}{r_3}}(J)$, *and a nondecreasing function* $\psi : [0,\infty) \to (0,\infty)$ *such that*

$$\|f(x,y,s,t,u,v,w)\| \le \rho_1(x,y,s,t)(\|u\| + \|v\| + \|w\|) \qquad (7.17)$$

*and*

$$\begin{aligned} &\|f(x_1,y_1,s,t,u,v,w) - f(x_2,y_2,s,t,u,v,w)\| \\ &\le \varphi(s,t)(|x_1-x_2| + |y_1-y_2|)\psi(\|u\| + \|v\| + \|w\|) \qquad (7.18) \\ &\text{for each } (x,y),(s,t),(x_1,y_1),(x_2,y_2) \in J \text{ and } u,v,w \in \mathbb{R}^n . \end{aligned}$$

(7.8.2) *There exist nonnegative constants* $\alpha, \beta_1, \beta_2, \beta_3$ *such that for* $(x,y) \in J$ *we have*

$$\begin{cases} \|\mu(x,y)\|_1 & \le \alpha , \\ \int_1^a \int_1^b \rho_1^{\frac{1}{r_3}}(x,y,s,t)dtds & \le \beta_1^{\frac{1}{r_3}} , \\ \int_1^a \int_1^b \rho_2^{\frac{1}{r_3}}(x,y,s,t)dtds & \le \beta_2^{\frac{1}{r_3}} , \\ \int_1^a \int_1^b \rho_3^{\frac{1}{r_3}}(x,y,s,t)dtds & \le \beta_3^{\frac{1}{r_3}} , \end{cases} \qquad (7.19)$$

*where*

$$\rho_2(x,y,s,t) = ({}^H D^{r}_{1,x} \rho_1)(x,y,s,t), \text{ and } \rho_3(x,y,s,t) = ({}^H D^{r}_{1,y} \rho_1)(x,y,s,t) .$$

*If*

$$\frac{(\beta_1+\beta_2+\beta_3)a^{(\omega_1+1)(1-r_3)} b^{(\omega_2+1)(1-r_3)}}{(\omega_1+1)^{(1-r_3)}(\omega_2+1)^{(1-r_3)}\Gamma(r_1)\Gamma(r_2)} < 1 , \qquad (7.20)$$

*where* $\omega_1 = \frac{r_1-1}{1-r_3}$, $\omega_2 = \frac{r_2-1}{1-r_3}$, *then equation* (7.16) *has at least one solution on* $J$ *in* $X$.

### 7.3.3 An Example

As an application of our results we consider the Fredholm partial Hadamard integral equation

$$u(x,y)=\mu(x,y)+\int_1^e\int_1^e(1-\ln s)^{r_1-1}(1-\ln t)^{r_2-1}\frac{f(x,y,s,t,u(s,t),{}^H(D_\sigma^r u)(s,t))}{st\Gamma(r_1)\Gamma(r_2)}dtds \tag{7.21}$$

for $(x,y)\in[1,e]\times[1,e]$, where $r_1,r_2>0$, $\mu(x,y)=x+y^2$; $(x,y)\in[1,e]\times[1,e]$, and

$$f(x,y,s,t,u(x,y),v(x,y))=c(x+y)st^2\frac{u(x,y)+v(x,y)}{e^{x+y+5}}\,,\qquad (x,y)\in[1,e]\times[1,e]\,,$$

with

$$c:=\frac{(\omega_1+1)^{(1-r_3)}(\omega_2+1)^{(1-r_3)}\Gamma(r_1)\Gamma(r_2)}{\frac{2e^{-3}}{(\Gamma(1+r_1))^{r_3}(\Gamma(1+r_2))^{r_3}}\left(1+\frac{1}{\Gamma(1-r_1)\Gamma(1-r_2)}\right)}\,,$$

$$0<r_3<\min\{r_1,r_2\}\,,\quad \omega_1=\frac{r_1-1}{1-r_3}\,,\quad \text{and } \omega_2=\frac{r_2-1}{1-r_3}\,.$$

For each $u,v\in\mathbb{R}$ and $(x,y)\in[1,e]\times[1,e]$ we have

$$|f(x,y,u,v)|\le 2ce^{-3}(|u|+|v|)\,,$$

and for each $(x,y),(s,t),(x_1,y_1),(x_2,y_2)\in[1,e]\times[1,e]$, and $u,v\in\mathbb{R}$ we have

$$|f(x_1,y_1,s,t,u,v)-f(x_2,y_2,s,t,u,v)|\le 2ce^{-3}(|x_1-x_2|+|y_1-y_2|)(|u|+|v|)\,.$$

Hence, condition (7.6.1) is satisfied by

$$\rho_1=ce^{-3}\,,\quad \rho_2=\frac{ce^{-3}}{\Gamma(1-r_1)\Gamma(1-r_2)}\,,\quad \varphi(s,t)=2ce^{-3}\,,\quad \psi(x)=1\,.$$

Also, (7.6.2) is satisfied by

$$\alpha=(e+e^2)\left(1+\frac{1}{\Gamma(1-r_1)\Gamma(1-r_2)}\right),\quad \beta_1=c\frac{e^{-3}}{(\Gamma(1+r_1))^{r_3}(\Gamma(1+r_2))^{r_3}}\,,$$

and

$$\beta_2=c\frac{e^{-3}}{\Gamma(1-r_1)\Gamma(1-r_2)\Gamma(1+r_1))^{r_3}(\Gamma(1+r_2))^{r_3}}\,.$$

Condition (7.12) holds with $a=b=e$. Indeed,

$$\begin{aligned}\ell&=\frac{(\beta_1+\beta_2)(\ln a)^{(\omega_1+1)(1-r_3)}(\ln b)^{(\omega_2+1)(1-r_3)}}{(\omega_1+1)^{(1-r_3)}(\omega_2+1)^{(1-r_3)}\Gamma(r_1)\Gamma(r_2)}\\&=\frac{c\frac{e^{-3}}{(\Gamma(1+r_1))^{r_3}(\Gamma(1+r_2))^{r_3}}\left(1+\frac{1}{\Gamma(1-r_1)\Gamma(1-r_2)}\right)}{(\omega_1+1)^{(1-r_3)}(\omega_2+1)^{(1-r_3)}\Gamma(r_1)\Gamma(r_2)}\\&=\frac{1}{2}<1\,.\end{aligned}$$

Consequently, Theorem 7.6 implies the Fredholm–Hadamard integral equation (7.21) has at least one solution on $[1,e]\times[1,e]$.

## 7.4 Upper and Lower Solutions Method for Partial Hadamard Fractional Integral Equations and Inclusions

### 7.4.1 Introduction

In this section, we use the upper and lower solutions method combined with Schauder's fixed point theorem and a fixed point theorem for condensing multivalued maps to Martelli to investigate the existence of solutions for some classes of partial Hadamard fractional integral equations and inclusions.

The method of upper and lower solutions has been successfully applied to study the existence of solutions for ordinary and partial differential equations and inclusions. See the monographs by Benchohra et al. [101], the papers of Abbas et al. [25, 20, 17, 15, 29], Pachpatte [211], and the references therein.

In this section, we use the method of upper and lower solutions for the existence of solutions to the Hadamard partial fractional integral equation

$$\begin{aligned} u(x,y) &= \mu(x,y) \\ &\quad + \frac{1}{\Gamma(r_1)\Gamma(r_2)} \int_1^x \int_1^y \left(\log\frac{x}{s}\right)^{r_1-1} \left(\log\frac{y}{t}\right)^{r_2-1} \frac{f(s,t,u(s,t))}{st}\, dt\, ds\,; \quad \text{if } (x,y) \in J\,, \end{aligned} \tag{7.22}$$

where $J := [1,a] \times [1,b]$, $a, b > 1$, $r_1, r_2 > 0$, $\mu : J \to \mathbb{R}$, $f : J \times \mathbb{R} \to \mathbb{R}$ are given continuous functions. Next we discuss the existence of solutions to the Hadamard partial fractional integral inclusion

$$u(x,y) - \mu(x,y) \in ({}^H I_\sigma^r F)(x,y,u(x,y));\ (x,y) \in J\,, \tag{7.23}$$

where $\sigma = (1,1)$, $F : J \times \mathbb{R} \to \mathcal{P}(\mathbb{R})$ is a compact-valued multivalued map, ${}^H I_\sigma^r F$ is the definite Hadamard integral for the set-valued function $F$ of order $r = (r_1, r_2) \in (0,\infty) \times (0,\infty)$, and $\mu : J \to \mathbb{R}$ is a given continuous function; additionally, $\mathcal{P}(\mathbb{R})$ is the family of all nonempty subsets of $\mathbb{R}$.

This section initiates the application of the upper and lower solutions method to these new classes of problems.

### 7.4.2 Existence Results for Partial Hadamard Fractional Integral Equations

Let us start by defining what we mean by a solution of integral equation (7.1).

**Definition 7.9.** A function $u \in C$ is said to be a solution of (7.1) if $u$ satisfies equation (7.1) on $J$.

**Definition 7.10.** A function $z \in C$ is said to be a lower solution of integral equation (7.1) if $z$ satisfies

$$u(x,y) \le \mu(x,y) + \int_1^x \int_1^y \left(\log \frac{x}{s}\right)^{r_1-1} \left(\log \frac{y}{t}\right)^{r_2-1} \frac{f(s,t,u(s,t))}{st\Gamma(r_1)\Gamma(r_2)} dtds\,; \quad (x,y) \in J\,.$$

The function $z$ is said to be an upper solution of (7.22) if the reverse inequality holds.

Further, we present our main result for equation (7.1).

**Theorem 7.11.** *Assume*
(7.11.1) *There exist $v$ and $w \in C$, lower and upper solutions to equation* (7.1) *such that $v \le w$.*
*Then integral equation* (7.22) *has at least one solution $u$ such that*

$$v(x,y) \le u(x,y) \le w(x,y) \text{ for all } (x,y) \in J\,.$$

*Proof.* Consider the modified integral equation

$$u(x,y) = \mu(x,y) + \frac{1}{\Gamma(r_1)\Gamma(r_2)} \int_1^x \int_1^y \left(\log \frac{x}{s}\right)^{r_1-1} \left(\log \frac{y}{t}\right)^{r_2-1} \frac{g(s,t,u(s,t))}{st} dtds\,, \quad (7.24)$$

where

$$g(x,y,u(x,y)) = f(x,y,h(x,y,u(x,y)))\,,$$
$$h(x,y,u(x,y)) = \max\{v(x,y), \min\{u(x,y), w(x,y)\}\}$$

for each $(x,y) \in J$.

A solution of (7.24) is a fixed point of the operator $N: C \to C$ defined by

$$(Nu)(x,y) = \mu(x,y) + \frac{1}{\Gamma(r_1)\Gamma(r_2)} \int_1^x \int_1^y \left(\log \frac{x}{s}\right)^{r_1-1} \left(\log \frac{y}{t}\right)^{r_2-1} \frac{g(s,t,u(s,t))}{st} dtds\,.$$

Notice that $g$ is a continuous function, and from (7.11.1) there exists $M > 0$ such that

$$|g(x,y,u)| \le M\,, \quad \text{for each } (x,y) \in J, \text{ and } u \in \mathbb{R}\,. \quad (7.25)$$

Set

$$\eta = \|\mu\|_C + \frac{M(\log a)^{r_1}(\log b)^{r_2}}{\Gamma(1+r_1)\Gamma(1+r_2)}$$

and

$$D = \{u \in C : \|u\|_C \le \eta\}\,.$$

Clearly, $D$ is a closed convex subset of $C$ and $N$ maps $D$ to itself. We will show that $N$ satisfies the assumptions of Theorem 1.42. The proof will be given in several steps.

*Step 1. N is continuous.* Let $\{u_n\}$ be a sequence such that $u_n \to u$ in $D$. Then

$$|(Nu_n)(x,y) - (Nu)(x,y)| \le \frac{1}{\Gamma(r_1)\Gamma(r_2)} \int_1^x \int_1^y \left|\log\frac{x}{s}\right|^{r_1-1} \left|\log\frac{y}{t}\right|^{r_2-1} \times \frac{|g(s,t,u_n(s,t)) - g(s,t,u(s,t))|}{st} dtds$$
$$\le \frac{1}{\Gamma(r_1)\Gamma(r_2)} \int_1^x \int_1^y \left|\log\frac{x}{s}\right|^{r_1-1} \left|\log\frac{y}{t}\right|^{r_2-1} \times \frac{\sup_{(s,t)\in J} |g(s,t,u_n(s,t)) - g(s,t,u(s,t))|}{st} dtds$$
$$\le \frac{(\log a)^{r_1}(\log b)^{r_2}}{\Gamma(1+r_1)\Gamma(1+r_2)} \|g(\cdot,\cdot,u_n(\cdot,\cdot)) - g(\cdot,\cdot,u(\cdot,\cdot))\|_C .$$

From Lebesgue's dominated convergence theorem and the continuity of the function $g$ we get

$$|(Nu_n)(x,y) - (Nu)(x,y)| \to 0 \quad \text{as} \quad n \to \infty .$$

*Step 2. $N(D)$ is bounded.* This is clear since $N(D) \subset D$ and $D$ is bounded.

*Step 3. $N(D)$ is equicontinuous.* Let $(x_1,y_1), (x_2,y_2) \in (1,a] \times (1,b]$, $x_1 < x_2$, $y_1 < y_2$, and let $u \in D$. Then

$$|(Nu)(x_2,y_2) - (Nu)(x_1,y_1)| \le |\mu(x_1,y_1) - \mu(x_2,y_2)|$$
$$+ \int_1^{x_1}\int_1^{y_1} \left| \left(\log\frac{x_2}{s}\right)^{r_1-1} \left(\log\frac{y_2}{t}\right)^{r_2-1} - \left(\log\frac{x_1}{s}\right)^{r_1-1} \left(\log\frac{y_1}{t}\right)^{r_2-1} \right| \times \frac{|g(s,t,u(s,t))|}{st\Gamma(r_1)\Gamma(r_2)} dtds$$
$$+ \frac{1}{\Gamma(r_1)\Gamma(r_2)} \int_{x_1}^{x_2}\int_{y_1}^{y_2} \left|\log\frac{x_2}{s}\right|^{r_1-1} \left|\log\frac{y_2}{t}\right|^{r_2-1} \frac{|g(s,t,u(s,t))|}{st} dtds$$
$$+ \frac{1}{\Gamma(r_1)\Gamma(r_2)} \int_{1}^{x_1}\int_{y_1}^{y_2} \left|\log\frac{x_2}{s}\right|^{r_1-1} \left|\log\frac{y_2}{t}\right|^{r_2-1} \frac{|g(s,t,u(s,t))|}{st} dtds$$
$$+ \frac{1}{\Gamma(r_1)\Gamma(r_2)} \int_{x_1}^{x_2}\int_{1}^{y_1} \left|\log\frac{x_2}{s}\right|^{r_1-1} \left|\log\frac{y_2}{t}\right|^{r_2-1} \frac{|g(s,t,u(s,t))|}{st} dtds .$$

Thus,

$$|(Nu)(x_2,y_2) - (Nu)(x_1,y_1)| \le |\mu(x_1,y_1) - \mu(x_2,y_2)|$$
$$+ \int_1^{x_1}\int_1^{y_1} \left| \left(\log\frac{x_2}{s}\right)^{r_1-1} \left(\log\frac{y_2}{t}\right)^{r_2-1} - \left(\log\frac{x_1}{s}\right)^{r_1-1} \left(\log\frac{y_1}{t}\right)^{r_2-1} \right| \times \frac{M}{st\Gamma(r_1)\Gamma(r_2)} dtds$$

$$+\frac{1}{\Gamma(r_1)\Gamma(r_2)}\int_{x_1}^{x_2}\int_{y_1}^{y_2}\left|\log\frac{x_2}{s}\right|^{r_1-1}\left|\log\frac{y_2}{t}\right|^{r_2-1}\frac{M}{st}dtds$$

$$+\frac{1}{\Gamma(r_1)\Gamma(r_2)}\int_{1}^{x_1}\int_{y_1}^{y_2}\left|\log\frac{x_2}{s}\right|^{r_1-1}\left|\log\frac{y_2}{t}\right|^{r_2-1}\frac{M}{st}dtds$$

$$+\frac{1}{\Gamma(r_1)\Gamma(r_2)}\int_{x_1}^{x_2}\int_{1}^{y_1}\left|\log\frac{x_2}{s}\right|^{r_1-1}\left|\log\frac{y_2}{t}\right|^{r_2-1}\frac{M}{st}dtds$$

$$\le|\mu(x_1,y_1)-\mu(x_2,y_2)|$$
$$+\frac{M}{\Gamma(1+r_1)\Gamma(1+r_2)}$$
$$\times\,[2(\log y_2)^{r_2}(\log x_2-\log x_1)^{r_1}+2(\log x_2)^{r_1}(\log y_2-\log y_1)^{r_2}$$
$$+(\log x_1)^{r_1}(\log y_1)^{r_2}-(\log x_2)^{r_1}(\log y_2)^{r_2}$$
$$-\,2(\log x_2-\log x_1)^{r_1}(\log y_2-\log y_1)^{r_2}]\,.$$

As $x_1 \to x_2$ and $y_1 \to y_2$, the right-hand side of the preceding inequality tends to zero.

As a consequence of Steps 1–3, together with the Ascoli–Arzelà theorem, we can conclude that $N$ is continuous and completely continuous. From an application of Theorem 1.42 we deduce that $N$ has a fixed point $u$ that is a solution of equation (8.11).

*Step 4. The solution $u$ of* (7.24) *satisfies*

$$v(x,y)\le u(x,y)\le w(x,y)\ \text{ for all }\ (x,y)\in J\,.$$

Let $u$ be the preceding solution of (7.24). We prove that

$$u(x,y)\le w(x,y)\ \text{ for all }\ (x,y)\in J\,.$$

Assume that $u - w$ attains a positive maximum on $J$ at $(\overline{x},\overline{y}) \in J$; then

$$(u-w)(\overline{x},\overline{y})=\max\{u(x,y)-w(x,y):\ (x,y)\in J\}>0\,.$$

We distinguish the following cases.

*Case 1.* If $(\overline{x},\overline{y}) \in (1,a)\times[1,b]$, then there exists $(x^*,y^*) \in (1,a)\times[1,b]$ such that

$$[u(x,y^*)-w(x,y^*)]+[u(x^*,y)-w(x^*,y)]-[u(x^*,y^*)-w(x^*,y^*)]\le 0\,;$$
$$\text{for all }(x,y)\in([x^*,\overline{x}]\times\{y^*\})\cup(\{x^*\}\times[y^*,b])\,, \tag{7.26}$$

and

$$u(x,y)-w(x,y)>0;\ \text{ for all }\ (x,y)\in(x^*,\overline{x}]\times(y^*,b]\,. \tag{7.27}$$

By the definition of $h$ we have

$$u(x,y)=\mu(x,y)+\frac{1}{\Gamma(r_1)\Gamma(r_2)}\int_1^x\int_1^y\left(\log\frac{x}{s}\right)^{r_1-1}\left(\log\frac{y}{t}\right)^{r_2-1}\frac{g(s,t,u(s,t))}{st}dtds \tag{7.28}$$

for all $(x, y) \in [x^*, \overline{x}] \times [y^*, b]$, where

$$g(x, y, u(x, y)) = f(x, y, w(x, y)), \quad (x, y) \in [x^*, \overline{x}] \times [y^*, b].$$

Thus equation (7.28) gives

$$\begin{aligned} &u(x, y) + u(x^*, y^*) - u(x, y^*) - u(x^*, y) \\ &= \int_{x^*}^{x}\int_{y^*}^{y} \left(\log \frac{x}{s}\right)^{r_1-1} \left(\log \frac{y}{t}\right)^{r_2-1} \frac{g(s, t, u(s, t))}{st\Gamma(r_1)\Gamma(r_2)} dtds. \end{aligned}$$

Using the fact that $w$ is an upper solution of (7.1) we get

$$u(x, y) + u(x^*, y^*) - u(x, y^*) - u(x^*, y) \le w(x, y) + w(x^*, y^*) - w(x, y^*) - w(x^*, y).$$

Then

$$[u(x, y) - w(x, y)] \le [u(x, y^*) - w(x, y^*)] + [u(x^*, y) - w(x^*, y)] - [u(x^*, y^*) - w(x^*, y^*)]. \tag{7.29}$$

Thus, from (7.26), (7.27), and (7.29) we obtain the contradiction

$$\begin{aligned} &0 < [u(x, y) - w(x, y)] \le [u(x, y^*) - w(x, y^*)] \\ &+ [u(x^*, y) - w(x^*, y)] - [u(x^*, y^*) - w(x^*, y^*)] \le 0 \quad \text{for all } (x, y) \in [x^*, \overline{x}] \times [y^*, b]. \end{aligned}$$

*Case 2.* If $\overline{x} = 1$, then

$$w(1, \overline{y}) < u(1, \overline{y}) \le w(1, \overline{y}),$$

which is a contradiction. Thus,

$$u(x, y) \le w(x, y) \quad \text{for all } (x, y) \in J.$$

Analogously, we can prove that

$$u(x, y) \ge v(x, y), \quad \text{for all } (x, y) \in J.$$

This shows that integral equation (7.24) has a solution $u$ that satisfies $v \le u \le w$, which is a solution of (7.22). □

### 7.4.3 Existence Results for Partial Hadamard Fractional Integral Inclusions

**Definition 7.12.** A function $z \in C$ is said to be a lower solution of (7.23) if there exists a function $f \in S_{F \circ z}$ such that $z$ satisfies

$$z(x, y) \le \mu(x, y) + \int_1^x\int_1^y \left(\log \frac{x}{s}\right)^{r_1-1} \left(\log \frac{y}{t}\right)^{r_2-1} \frac{f(s, t)}{st\Gamma(r_1)\Gamma(r_2)} dtds; \quad (x, y) \in J.$$

The function $z$ is said to be an upper solution of (7.23) if the reverse inequality holds.

**Theorem 7.13.** *Make the following assumptions:*
(7.13.1) *The multifunction $F\colon J\times\mathbb{R}\longrightarrow\mathcal{P}_{cp,cv}(\mathbb{R})$ is $L^1$-Carathéodory.*
(7.13.2) *There exist $v$ and $w\in C$, lower and upper solutions for the integral inclusion (7.23), such that $v\le w$.*
*Then the Hadamard integral inclusion (7.23) has at least one solution $u$ such that*

$$v(x,y)\le u(x,y)\le w(x,y)\ \text{for all}\ (x,y)\in J\,.$$

**Remark 7.14.** Solutions of inclusion (7.23) are solutions of the Hadamard integral inclusion

$$u(x,y)\in\left\{\mu(x,y)+({}^HI_\sigma^r f)(x,y)\colon f\in S_{F\circ u}\right\}\,;\quad (x,y)\in J\,.$$

*Proof.* Consider the modified integral inclusion

$$u(x,y)-\mu(x,y)\in({}^HI_\sigma^r F)(x,y,(gu)(x,y))\,;\quad (x,y)\in J\,, \tag{7.30}$$

where $g\colon C\longrightarrow C$ is the truncation operator defined by

$$(gu)(x,y)=\begin{cases}v(x,y); & u(x,y)<v(x,y),\\ u(x,y); & v(x,y)\le u(x,y)\le w(x,y),\\ w(x,y); & w(x,y)<u(x,y).\end{cases}$$

A solution of (7.30) is a fixed point of the operator $N\colon C\to\mathcal{P}(C)$ defined by

$$(Nu)(x,y)=\left\{h\in C\colon\begin{cases}h(x,y)=\mu(x,y)\\ +\int_1^x\int_1^y\left(\log\frac{x}{s}\right)^{r_1-1}\left(\log\frac{y}{t}\right)^{r_2-1}\frac{f(s,t)}{st\Gamma(r_1)\Gamma(r_2)}dtds;\ (x,y)\in J\,,\end{cases}\right\}$$

where

$$\begin{aligned}&f\in\tilde S^1_{F\circ g(u)}=\{f\in S^1_{F\circ g(u)}\colon f(x,y)\ge f_1(x,y)\text{ on }A_1\text{ and }f(x,y)\le f_2(x,y)\text{ on }A_2\}\,,\\ &A_1=\{(x,y)\in J\colon u(x,y)<v(x,y)\le w(x,y)\}\,,\\ &A_2=\{(x,y)\in J\colon u(x,y)\le w(x,y)<u(x,y)\}\,,\end{aligned}$$

and

$$S^1_{F\circ g(u)}=\{f\in L^1(J)\colon f(x,y)\in F(t,x,(gu)(x,y));\ \text{for }(x,y)\in J\}\,.$$

**Remark 7.15.** (A) For each $u\in C$ the set $\tilde S_{F\circ g(u)}$ is nonempty. In fact, (7.13.1) implies the existence of $f_3\in S_{F\circ g(u)}$, so we set

$$f=f_1\chi_{A_1}+f_2\chi_{A_2}+f_3\chi_{A_3}\,,$$

where $\chi_{A_i}$ is a characteristic function of $A_i$; $i=1,2,3$ and

$$A_3=\{(x,y)\in J\colon v(x,y)\le u(x,y)\le w(x,y)\}\,.$$

Then, by decomposability, $f\in\tilde S_{F\circ g(u)}$.

(B) By the definition of $g$, it is clear that $F(.,.,(gu)(.,.))$ is an $L^1$-Carathéodory multivalued map with compact convex values, and there exists $\phi \in C(J, \mathbb{R}_+)$ such that

$$\|F(t, x, (gu)(x, y))\|_{\mathcal{P}} \le \phi(x, y)\,; \quad \text{for each } u \in \mathbb{R} \text{ and } (x, y) \in J\,.$$

Set

$$\phi^* := \sup_{(x,y)\in J} \phi(x, y)\,.$$

From Remark 7.14 and the fact that $g(u) = u$ for all $v \le u \le w$, the problem of finding the solutions of integral inclusion (7.23) is reduced to finding the solutions of the operator inclusion $u \in N(u)$. We will show that $N$ is a completely continuous multivalued map, u.s.c. with convex closed values. The proof will be given in several steps.

*Step 1: $N(u)$ is convex for each $u \in C$.* Indeed, if $h_1, h_2$ belong to $N(u)$, then there exist $f_1, f_2 \in \tilde{S}^1_{F\circ g(u)}$ such that for each $(x, y) \in J$ we have

$$h_i(x, y) = \mu(x, y) + \int_1^x \int_1^y \left(\log \frac{x}{s}\right)^{r_1-1} \left(\log \frac{y}{t}\right)^{r_2-1} \frac{f_i(s, t)}{st\Gamma(r_1)\Gamma(r_2)} dtds\,; \quad i = 1, 2\,.$$

Let $0 \le \xi \le 1$. Then for each $(x, y) \in J$ we have

$$(\xi h_1 + (1-\xi)h_2)(x, y) = \mu(x, y) + \int_1^x \int_1^y \left(\log \frac{x}{s}\right)^{r_1-1} \left(\log \frac{y}{t}\right)^{r_2-1} \frac{((\xi f_1 + (1-\xi)f_2))(s, t)}{st\Gamma(r_1)\Gamma(r_2)} dtds\,.$$

Since $\tilde{S}^1_{F\circ g(u)}$ is convex (because $F$ has convex values), we have

$$\xi h_1 + (1-\xi)h_2 \in N(u)\,.$$

*Step 2: $N$ sends bounded sets of $C$ to bounded sets.* Indeed, we can prove that $N(C)$ is bounded. It is enough to show that there exists a positive constant $\ell$ such that for each $h \in N(u)$, $u \in C$ one has $\|h\|_C \le \ell$.
If $h \in N(u)$, then there exists $f \in \tilde{S}^1_{F\circ g(u)}$ such that for each $(x, y) \in J$ we have

$$h(x, y) = \mu(x, y) + \int_1^x \int_1^y \left(\log \frac{x}{s}\right)^{r_1-1} \left(\log \frac{y}{t}\right)^{r_2-1} \frac{f(s, t)}{st\Gamma(r_1)\Gamma(r_2)} dtds\,.$$

Then we get

$$|h(x, y)| \le |\mu(x, y)| + \int_1^x \int_1^y \left|\log \frac{x}{s}\right|^{r_1-1} \left|\log \frac{y}{t}\right|^{r_2-1} \frac{\phi(s, t)}{st\Gamma(r_1)\Gamma(r_2)} dtds\,.$$

Thus, we obtain

$$
\begin{aligned}
|h(x,y)| &\le \|\mu\|_C \\
&\quad + \int_{\phi^*}^{x}\int_{1}^{y} \left|\log\frac{x}{s}\right|^{r_1-1}\left|\log\frac{y}{t}\right|^{r_2-1}\frac{\phi^*}{st\Gamma(r_1)\Gamma(r_2)}dtds \\
&\le \|\mu\|_C + \frac{(\log a)^{r_1}(\log b)^{r_2}\phi^*}{\Gamma(1+r_1)\Gamma(1+r_2)} := \ell .
\end{aligned}
$$

Hence,

$$
\|h\|_C \le \ell .
$$

*Step 3: N sends bounded sets of C to equicontinuous sets.* Let $(x_1,y_1),(x_2,y_2)\in J$, $x_1<x_2, y_1<y_2$ and $B_\rho=\{u\in C:\ \|u\|_C\le\rho\}$ be a bounded set of $C$. For each $u\in B_\rho$ and $h\in N(u)$ there exists $f\in\tilde{S}^1_{F\circ g(u)}$ such that for each $(x,y)\in J$ we get

$$
\begin{aligned}
|h(x_2,y_2)-h(x_1,y_1)| &\le |\mu(x_1,y_1)-\mu(x_2,y_2)| \\
&\quad + \int_{1}^{x_1}\int_{1}^{y_1}\left|\left(\log\frac{x_2}{s}\right)^{r_1-1}\left(\log\frac{y_2}{t}\right)^{r_2-1}-\left(\log\frac{x_1}{s}\right)^{r_1-1}\left(\log\frac{y_1}{t}\right)^{r_2-1}\right| \\
&\quad \times \frac{|f(s,t)|}{st\Gamma(r_1)\Gamma(r_2)}dtds \\
&\quad + \frac{1}{\Gamma(r_1)\Gamma(r_2)}\int_{x_1}^{x_2}\int_{y_1}^{y_2}\left|\log\frac{x_2}{s}\right|^{r_1-1}\left|\log\frac{y_2}{t}\right|^{r_2-1}\frac{|f(s,t)|}{st}dtds \\
&\quad + \frac{1}{\Gamma(r_1)\Gamma(r_2)}\int_{1}^{x_1}\int_{y_1}^{y_2}\left|\log\frac{x_2}{s}\right|^{r_1-1}\left|\log\frac{y_2}{t}\right|^{r_2-1}\frac{|f(s,t)|}{st}dtds \\
&\quad + \frac{1}{\Gamma(r_1)\Gamma(r_2)}\int_{x_1}^{x_2}\int_{1}^{y_1}\left|\log\frac{x_2}{s}\right|^{r_1-1}\left|\log\frac{y_2}{t}\right|^{r_2-1}\frac{|f(s,t)|}{st}dtds .
\end{aligned}
$$

Hence,

$$
\begin{aligned}
|h(x_2,y_2)-h(x_1,y_1)| &\le |\mu(x_1,y_1)-\mu(x_2,y_2)| \\
&\quad + \frac{\phi^*}{\Gamma(1+r_1)\Gamma(1+r_2)} \\
&\quad \times\big[2(\log y_2)^{r_2}(\log x_2-\log x_1)^{r_1}+2(\log x_2)^{r_1}(\log y_2-\log y_1)^{r_2} \\
&\quad +(\log x_1)^{r_1}(\log y_1)^{r_2}-(\log x_2)^{r_1}(\log y_2)^{r_2} \\
&\quad -2(\log x_2-\log x_1)^{r_1}(\log y_2-\log y_1)^{r_2}\big] .
\end{aligned}
$$

As $x_1\to x_2$ and $y_1\to y_2$, the right-hand side of the preceding inequality tends to zero.

As a consequence of Steps 1–3, together with the Ascoli–Arzelà theorem, we can conclude that $N$ is completely continuous and, therefore, a condensing multivalued map.

*Step 4: N has a closed graph.* Let $u_n \to u_*$, $h_n \in N(u_n)$, and $h_n \to h_*$. We need to show that $h_* \in N(u_*)$.
$h_n \in N(u_n)$ means that there exists $f_n \in \tilde{S}^1_{F\circ g(u_n)}$ such that for each $(x,y) \in J$ we have

$$h_n(x,y) = \mu(x,y) + \int_1^x \int_1^y \left(\log\frac{x}{s}\right)^{r_1-1} \left(\log\frac{y}{t}\right)^{r_2-1} \frac{f_n(s,t)}{st\Gamma(r_1)\Gamma(r_2)} dtds\,.$$

We must show that there exists $f_* \in \tilde{S}^1_{F\circ g(u_*)}$ such that, for each $(x,y) \in J$,

$$h_*(x,y) = \mu(x,y) + \int_1^x \int_1^y \left(\log\frac{x}{s}\right)^{r_1-1} \left(\log\frac{y}{t}\right)^{r_2-1} \frac{f_*(s,t)}{st\Gamma(r_1)\Gamma(r_2)} dtds\,.$$

Now we consider the linear continuous operator

$$\begin{aligned} \Lambda : L^1(J) &\longrightarrow C(J)\,, \\ f &\longmapsto \Lambda f \end{aligned}$$

defined by

$$(\Lambda f)(x,y) = \mu(x,y) + \int_1^x \int_1^y \left(\log\frac{x}{s}\right)^{r_1-1} \left(\log\frac{y}{t}\right)^{r_2-1} \frac{f(s,t)}{st\Gamma(r_1)\Gamma(r_2)} dtds\,.$$

**Remark 7.16.** Remark 7.15 (B) implies that the operator $\Lambda$ is well defined.

From Lemma 1.25 it follows that $\Lambda \circ \tilde{S}^1_F$ is a closed graph operator. Clearly we have

$$|h_n(x,y) - h_*(x,y)| = \mu(x,y) + \int_1^x \int_1^y \left(\log\frac{x}{s}\right)^{r_1-1} \left(\log\frac{y}{t}\right)^{r_2-1} \frac{|f_n(s,t) - f_*(s,t)|}{st\Gamma(r_1)\Gamma(r_2)} dtds \to 0$$

as $n \to \infty$.

Moreover, from the definition of $\Lambda$ we have

$$|h_n(x,y) - h_*(x,y)| \in \Lambda(\tilde{S}^1_{F\circ g(u_n)})\,; \quad \text{if } (x,y) \in J\,.$$

Since $u_n \to u_*$, it follows from Lemma 1.25 that for some $f_* \in \Lambda(\tilde{S}^1_{F\circ g(u_*)})$ we have

$$h_*(x,y) = \mu(x,y) + \int_1^x \int_1^y \left(\log\frac{x}{s}\right)^{r_1-1} \left(\log\frac{y}{t}\right)^{r_2-1} \frac{f_*(s,t)}{st\Gamma(r_1)\Gamma(r_2)} dtds\,.$$

From Lemma 1.24 we can conclude that $N$ is u.s.c.

*Step 5:* The set $\Omega = \{u \in C\colon \lambda u \in N(u) \text{ for some } \lambda > 1\}$ is bounded. Let $u \in \Omega$. Then there exists $f \in \Lambda(\tilde{S}^1_{F\circ g(u)})$ such that

$$\lambda u(x,y) = \mu(x,y) + \int_1^x \int_1^y \left(\log\frac{x}{s}\right)^{r_1-1} \left(\log\frac{y}{t}\right)^{r_2-1} \frac{f(s,t)}{st\Gamma(r_1)\Gamma(r_2)} dtds\,.$$

As in Step 2, this implies that for each $(x,y) \in J$ we have

$$\|u\|_C \le \frac{\ell}{\lambda} < \ell\,.$$

This shows that $\Omega$ is bounded. As a consequence of Theorem 1.48, we deduce that $N$ has a fixed point that is a solution of (7.30) on $J$.

*Step 6: The solution $u$ of* (7.30) *satisfies*

$$v(x,y) \le u(x,y) \le w(x,y)\,; \quad \text{for all } (x,y) \in J\,.$$

First we prove that

$$u(x,y) \le w(x,y)\,; \quad \text{for all } (x,y) \in J\,.$$

Assume that $u - w$ attains a positive maximum on $J$ at $(\overline{x},\overline{y}) \in J$; then

$$(u-w)(\overline{x},\overline{y}) = \max\{u(x,y) - w(x,y);\ (x,y) \in J\} > 0\,.$$

We distinguish the following cases.

*Case 1.* If $(\overline{x},\overline{y}) \in (1,a) \times [1,b]$, then there exists $(x^*,y^*) \in (1,a) \times [1,b]$ such that

$$\begin{aligned} &[u(x,y^*) - w(x,y^*)] + [u(x^*,y) - w(x^*,y)] \\ &- [u(x^*,y^*) - w(x^*,y^*)] \le 0; \quad \text{for all } (x,y) \in ([x^*,\overline{x}] \times \{y^*\}) \cup (\{x^*\} \times [y^*,b]) \end{aligned} \tag{7.31}$$

and

$$u(x,y) - w(x,y) > 0\,; \quad \text{for all } (x,y) \in (x^*,\overline{x}] \times (y^*,b]\,. \tag{7.32}$$

For all $(x,y) \in [x^*,\overline{x}] \times [y^*,b]$ we have

$$u(x,y) = \mu(x,y) + \frac{1}{\Gamma(r_1)\Gamma(r_2)} \int_1^x \int_1^y \left(\log\frac{x}{s}\right)^{r_1-1} \left(\log\frac{y}{t}\right)^{r_2-1} \frac{f(s,t)}{st} dtds\,, \tag{7.33}$$

where $f \in S_{F\circ u}$. Thus, equation (7.33) gives

$$\begin{aligned} &u(x,y) + u(x^*,y^*) - u(x,y^*) - u(x^*,y) \\ &= \int_{x^*}^x \int_{y^*}^y \left(\log\frac{x}{s}\right)^{r_1-1} \left(\log\frac{y}{t}\right)^{r_2-1} \frac{f(s,t)}{st\Gamma(r_1)\Gamma(r_2)} dtds\,. \end{aligned} \tag{7.34}$$

From (7.34) and using the fact that $w$ is an upper solution of (7.23) we get

$$u(x,y)+u(x^*,y^*)-u(x,y^*)-u(x^*,y)\le w(x,y)+w(x^*,y^*)-w(x,y^*)-w(x^*,y).$$

Then

$$[u(x,y)-w(x,y)]\le[u(x,y^*)-w(x,y^*)]+[u(x^*,y)-w(x^*,y)]-[u(x^*,y^*)-w(x^*,y^*)]. \tag{7.35}$$

Thus, from (7.31), (7.32), and (7.35) we obtain the contradiction

$$0<[u(x,y)-w(x,y)]\le[u(x,y^*)-w(x,y^*)]$$
$$+[u(x^*,y)-w(x^*,y)]-[u(x^*,y^*)-w(x^*,y^*)]\le 0;\quad \text{for all } (x,y)\in[x^*,\overline{x}]\times[y^*,b].$$

*Case 2.* If $\overline{x}=1$, then

$$w(1,\overline{y})<u(1,\overline{y})\le w(1,\overline{y}),$$

which is a contradiction. Thus,

$$u(x,y)\le w(x,y);\quad \text{for all } (x,y)\in J.$$

Analogously, we can prove that

$$u(x,y)\ge v(x,y)\ \text{ for all } (x,y)\in J.$$

This shows that problem (7.30) has a solution $u$ that satisfies $v\le u\le w$, which is a solution of integral inclusion (7.23). □

## 7.5 Notes and Remarks

The results of Chapter 7 are taken from Abbas et al. [1, 12, 32]. Other results may be found in [29, 42, 119, 120].

# 8 Stability Results for Partial Hadamard Fractional Integral Equations and Inclusions

## 8.1 Introduction

This chapter deals with some existence and Ulam stability results for several classes of partial integral equations via Hadamard's fractional integral by applying some fixed point theorems.

## 8.2 Ulam Stabilities for Partial Hadamard Fractional Integral Equations

### 8.2.1 Introduction

This section deals with the existence the Ulam stability of solutions to the Hadamard partial fractional integral equation

$$u(x,y) = \mu(x,y) + \frac{1}{\Gamma(r_1)\Gamma(r_2)} \int_1^x \int_1^y \left(\log\frac{x}{s}\right)^{r_1-1} \left(\log\frac{y}{t}\right)^{r_2-1} \frac{f(s,t,u(s,t))}{st} dtds \; ; \text{ if } (x,y) \in J\,, \tag{8.1}$$

where $J := [1,a] \times [1,b]$, $a, b > 1, r_1, r_2 > 0, \mu : J \to \mathbb{R}, f : J \times \mathbb{R} \to \mathbb{R}$ are given continuous functions.

We present two results for integral equation (8.1). The first one is based on Banach's contraction principle and the second one on the nonlinear alternative of the Leray–Schauder type.

### 8.2.2 Existence and Ulam Stabilities Results

In this section, we discuss the existence of solutions and present conditions for the Ulam stability for the Hadamard integral equation (8.1).

The following conditions will be used in the sequel.

(8.1.1) There exist functions $p_1, p_2 \in C(J, \mathbb{R}_+)$ such that for any $u \in \mathbb{R}$ and $(x,y) \in J$,

$$|f(x,y,u)| \le p_1(x,y) + \frac{p_2(x,y)}{1+|u(x,y)|}|u(x,y)|\,,$$

with

$$p_i^* = \sup_{(x,y)\in J} p_i(x,y)\,; \quad i = 1,2\,.$$

https://doi.org/10.1515/9783110553819-008

(8.1.2) There exists $\lambda_\Phi > 0$ such that for each $(x, y) \in J$ we have

$$({}^H I_\sigma^r \Phi)(x, y) \le \lambda_\Phi \Phi(x, y) .$$

**Theorem 8.1.** *Assume* (8.1.1). *If*

$$\frac{(\log a)^{r_1} (\log b)^{r_2}}{\Gamma(1 + r_1)\Gamma(1 + r_2)} p_2^* < 1 , \tag{8.2}$$

*then integral equation* (8.1) *has a solution defined on* $J$.

*Proof.* Let $\rho > 0$ be a constant such that

$$\rho > \frac{\|\mu\|_\infty + \frac{(\log a)^{r_1} (\log b)^{r_2}}{\Gamma(1+r_1)\Gamma(1+r_2)} p_1^*}{1 - \frac{(\log a)^{r_1} (\log b)^{r_2}}{\Gamma(1+r_1)\Gamma(1+r_2)} p_2^*} .$$

We use Schauder's fixed point theorem [149] to prove that the operator $N\colon C \to C$ defined by

$$(Nu)(x, y) = \mu(x, y) + \frac{1}{\Gamma(r_1)\Gamma(r_2)} \int_1^x \int_1^y \left(\log \frac{x}{s}\right)^{r_1 - 1} \left(\log \frac{y}{t}\right)^{r_2 - 1} \frac{f(s, t, u(s, t))}{st} dt ds \tag{8.3}$$

has a fixed point. The proof will be given in four steps.

*Step 1: $N$ transforms the ball $B_\rho := \{u \in \mathcal{C}\colon \|u\|_C \le \rho\}$ into itself.* For any $u \in B_\rho$ and each $(x, y) \in J$ we have

$$\begin{aligned} |(Nu)(x, y)| &\le |\mu(x, y)| + \frac{1}{\Gamma(r_1)\Gamma(r_2)} \int_1^x \int_1^y \left|\log \frac{x}{s}\right|^{r_1 - 1} \left|\log \frac{y}{t}\right|^{r_2 - 1} \\ &\quad \times \frac{p_1(s, t) + p_2(s, t)\|u\|_C}{st} dt ds \\ &\le \|\mu\|_\infty + \frac{(\log a)^{r_1} (\log b)^{r_2}}{\Gamma(1 + r_1)\Gamma(1 + r_2)} (p_1^* + p_2^* \rho) . \end{aligned}$$

Thus, by (8.2) and the definition of $\rho$ we get $\|(Nu)\|_C \le \rho$. This implies that $N$ transforms the ball $B_\rho$ into itself.

*Step 2: $N\colon B_\rho \to B_\rho$ is continuous.*
Let $\{u_n\}_{n \in \mathbb{N}}$ be a sequence such that $u_n \to u$ in $B_\rho$. Then

$$\begin{aligned} |(Nu_n)(x, y) - (Nu)(x, y)| &\le \frac{1}{\Gamma(r_1)\Gamma(r_2)} \int_1^x \int_1^y \left|\log \frac{x}{s}\right|^{r_1 - 1} \left|\log \frac{y}{t}\right|^{r_2 - 1} \\ &\quad \times \frac{|f(s, t, u_n(s, t)) - f(s, t, u(s, t))|}{st} dt ds \end{aligned}$$

$$\leq \frac{1}{\Gamma(r_1)\Gamma(r_2)} \int_1^x \int_1^y \left|\log \frac{x}{s}\right|^{r_1-1} \left|\log \frac{y}{t}\right|^{r_2-1}$$
$$\times \frac{\sup_{(s,t)\in J} |f(s,t,u_n(s,t)) - f(s,t,u(s,t))|}{st} dtds$$
$$\leq \frac{(\log a)^{r_1} (\log b)^{r_2}}{\Gamma(1+r_1)\Gamma(1+r_2)} \|f(\cdot,\cdot,u_n(\cdot,\cdot)) - f(\cdot,\cdot,u(\cdot,\cdot))\|_C .$$

From Lebesgue's dominated convergence theorem and the continuity of the function $f$ we get

$$|(Nu_n)(x,y) - (Nu)(x,y)| \to 0 \text{ as } n \to \infty .$$

*Step 3*: *$N(B_\rho)$ is bounded.* This is clear since $N(B_\rho) \subset B_\rho$ and $B_\rho$ is bounded.

*Step 4*: *$N(B_\rho)$ is equicontinuous.*

Let $(x_1,y_1), (x_2,y_2) \in J$, $x_1 < x_2$, $y_1 < y_2$. Then

$$|(Nu)(x_2,y_2) - (Nu)(x_1,y_1)| \leq |\mu(x_1,y_1) - \mu(x_2,y_2)|$$
$$+ \frac{1}{\Gamma(r_1)\Gamma(r_2)} \int_1^{x_1} \int_1^{y_1} \left[\left|\log \frac{x_2}{s}\right|^{r_1-1} \left|\log \frac{y_2}{t}\right|^{r_2-1} - \left|\log \frac{x_1}{s}\right|^{r_1-1} \left|\log \frac{y_1}{t}\right|^{r_2-1}\right]$$
$$\times \frac{|f(s,t,u(s,t))|}{st} dtds$$
$$+ \frac{1}{\Gamma(r_1)\Gamma(r_2)} \int_{x_1}^{x_2} \int_{y_1}^{y_2} \left|\log \frac{x_2}{s}\right|^{r_1-1} \left|\log \frac{y_2}{t}\right|^{r_2-1} \frac{|f(s,t,u(s,t))|}{st} dtds$$
$$+ \frac{1}{\Gamma(r_1)\Gamma(r_2)} \int_1^{x_1} \int_{y_1}^{y_2} \left|\log \frac{x_2}{s}\right|^{r_1-1} \left|\log \frac{y_2}{t}\right|^{r_2-1} \frac{|f(s,t,u(s,t))|}{st} dtds$$
$$+ \frac{1}{\Gamma(r_1)\Gamma(r_2)} \int_{x_1}^{x_2} \int_1^{y_1} \left|\log \frac{x_2}{s}\right|^{r_1-1} \left|\log \frac{y_2}{t}\right|^{r_2-1} \frac{|f(s,t,u(s,t))|}{st} dtds .$$

Thus,

$$|(Nu)(x_2,y_2) - (Nu)(x_1,y_1)| \leq |\mu(x_1,y_1) - \mu(x_2,y_2)|$$
$$+ \frac{1}{\Gamma(r_1)\Gamma(r_2)} \int_1^{x_1} \int_1^{y_1} \left[\left|\log \frac{x_2}{s}\right|^{r_1-1} \left|\log \frac{y_2}{t}\right|^{r_2-1} - \left|\log \frac{x_1}{s}\right|^{r_1-1} \left|\log \frac{y_1}{t}\right|^{r_2-1}\right]$$
$$\times \frac{p_1^* + p_2^*\rho}{st} dtds$$
$$+ \frac{1}{\Gamma(r_1)\Gamma(r_2)} \int_{x_1}^{x_2} \int_{y_1}^{y_2} \left|\log \frac{x_2}{s}\right|^{r_1-1} \left|\log \frac{y_2}{t}\right|^{r_2-1} \frac{p_1^* + p_2^*\rho}{st} dtds$$
$$+ \frac{1}{\Gamma(r_1)\Gamma(r_2)} \int_1^{x_1} \int_{y_1}^{y_2} \left|\log \frac{x_2}{s}\right|^{r_1-1} \left|\log \frac{y_2}{t}\right|^{r_2-1} \frac{p_1^* + p_2^*\rho}{st} dtds$$

$$+\frac{1}{\Gamma(r_1)\Gamma(r_2)}\int_{x_1}^{x_2}\int_{1}^{y_1}\left|\log\frac{x_2}{s}\right|^{r_1-1}\left|\log\frac{y_2}{t}\right|^{r_2-1}\frac{p_1^*+p_2^*\rho}{st}dtds$$
$$\leq\frac{p_1^*+p_2^*\rho}{\Gamma(1+r_1)\Gamma(1+r_2)}$$
$$\times[2(\log y_2)^{r_2}(\log x_2-\log x_1)^{r_1}+2(\log x_2)^{r_1}(\log y_2-\log y_1)^{r_2}$$
$$+(\log x_1)^{r_1}(\log y_1)^{r_2}-(\log x_2)^{r_1}(\log y_2)^{r_2}-2(\log x_2-\log x_1)^{r_1}(\log y_2-\log y_1)^{r_2}]\,.$$

As $x_1 \to x_2$ and $y_1 \to y_2$, the right-hand side of the preceding inequality tends to zero.

As a consequence of Steps 1–4, together with the Ascoli–Arzelà theorem, we can conclude that $N$ is continuous and compact. From an application of Schauder's theorem [149], we deduce that $N$ has a fixed point $u$ that is a solution of integral equation (8.1). □

Now we are concerned with the stability of solutions for integral equation (8.1).

Recall $N: C \to C$ as defined in 8.3. Let $\epsilon > 0$, and let $\Phi: J \to [0,\infty)$ be a continuous function. We consider the inequalities

$$|u(x,y)-(Nu)(x,y)| \leq \epsilon\,; \quad (x,y) \in J\,, \tag{8.4}$$
$$|u(x,y)-(Nu)(x,y)| \leq \Phi(x,y)\,; \quad (x,y) \in J\,, \tag{8.5}$$
$$|u(x,y)-(Nu)(x,y)| \leq \epsilon\Phi(x,y)\,; \quad (x,y) \in J\,. \tag{8.6}$$

**Definition 8.2** ([35, 233]). Equation (8.1) is Ulam–Hyers stable if there exists a real number $c_N > 0$ such that for each $\epsilon > 0$ and for each solution $u \in C$ of inequality (8.4) there exists a solution $v \in C$ of equation (8.1) with

$$|u(x,y)-v(x,y)| \leq \epsilon c_N\,; \quad (x,y) \in J\,.$$

**Definition 8.3** ([35, 233]). Equation (8.1) is generalized Ulam–Hyers stable if there exists $c_N: C([0,\infty),[0,\infty))$, with $c_N(0) = 0$, such that for each $\epsilon > 0$ and for each solution $u \in \mathcal{C}$ of (8.4) there exists a solution $v \in C$ of equation (8.1) with

$$|u(x,y)-v(x,y)| \leq c_N(\epsilon)\,; \quad (x,y) \in J\,.$$

**Definition 8.4** ([35, 233]). Equation (8.1) is Ulam–Hyers–Rassias stable with respect to $\Phi$ if there exists a real number $c_{N,\Phi} > 0$ such that for each $\epsilon > 0$ and for each solution $u \in C$ of (8.6) there exists a solution $v \in C$ of equation (8.1) with

$$|u(x,y)-v(x,y)| \leq \epsilon c_{N,\Phi}\Phi(x,y)\,; \quad (x,y) \in J\,.$$

**Definition 8.5** ([35, 233]). Equation (8.1) is generalized Ulam–Hyers–Rassias stable with respect to $\Phi$ if there exists a real number $c_{N,\Phi} > 0$ such that for each solution $u \in C$ of (8.5) there exists a solution $v \in \mathcal{C}$ of equation (8.1) with $|u(x,y)-v(x,y)| \leq c_{N,\Phi}\Phi(x,y)$; $(x,y) \in J$.

**Remark 8.6.** It is clear that (i) Definition 8.2 implies Definition 8.3, (ii) Definition 8.4 implies Definition 8.5, and (iii) Definition 8.4 for $\Phi(.,.) = 1$ implies Definition 8.2.

One could make similar remarks for inequalities (8.4) and (8.6).

**Theorem 8.7.** *Assume* (8.1.1), (8.1.2), *and* (8.2) *hold. Furthermore, suppose that there exist* $q_i \in C(J, \mathbb{R}_+)$, $i = 1, 2$, *such that for each* $(x, y) \in J$ *we have*

$$p_i(x, y) \leq q_i(x, y)\Phi(x, y) .$$

*Then integral equation* (8.1) *is generalized Ulam–Hyers–Rassias stable.*

*Proof.* Let $u$ be a solution of inequality (8.5). By Theorem 8.1 there exists $v$ that is a solution of integral equation (8.1). Hence,

$$v(x, y) = \mu(x, y) + \int_1^x \int_1^y \left(\log \frac{x}{s}\right)^{r_1-1} \left(\log \frac{y}{t}\right)^{r_2-1} \frac{f(s, t, v(s, t))}{st\Gamma(r_1)\Gamma(r_2)} dtds .$$

By inequality (8.5), for each $(x, y) \in J$ we have

$$\left| u(x, y) - \mu(x, y) - \int_1^x \int_1^y \left(\log \frac{x}{s}\right)^{r_1-1} \left(\log \frac{y}{t}\right)^{r_2-1} \frac{f(s, t, u(s, t))}{st\Gamma(r_1)\Gamma(r_2)} dtds \right| \leq \Phi(x, y) .$$

Set

$$q_i^* = \sup_{(x,y)\in J} q_i(x, y) ; \quad i = 1, 2 .$$

For each $(x, y) \in J$ we have

$$\begin{aligned}
|u(x, y) - v(x, y)| &\leq \left| u(x, y) - \mu(x, y) - \int_1^x \int_1^y \left(\log \frac{x}{s}\right)^{r_1-1} \left(\log \frac{y}{t}\right)^{r_2-1} \frac{f(s, t, u(s, t))}{st\Gamma(r_1)\Gamma(r_2)} dtds \right| \\
&\quad + \int_1^x \int_1^y \left|\log \frac{x}{s}\right|^{r_1-1} \left|\log \frac{y}{t}\right|^{r_2-1} \frac{|f(s, t, u(s, t)) - f(s, t, v(s, t))|}{st\Gamma(r_1)\Gamma(r_2)} dtds \\
&\leq \Phi(x, y) + \frac{1}{\Gamma(r_1)\Gamma(r_2)} \int_1^x \int_1^y \left|\log \frac{x}{s}\right|^{r_1-1} \left|\log \frac{y}{t}\right|^{r_2-1} \\
&\quad \times \left(2q_1^* + \frac{q_2^*|u(s, t)|}{1 + |u|} + \frac{q_2^*|v(s, t)|}{1 + |v|}\right) \frac{\Phi(s, t)}{st} dtds \\
&\leq \Phi(x, y) + 2(q_1^* + q_2^*)({}^H I_\sigma^r \Phi)(x, y) \\
&\leq [1 + 2(q_1^* + q_2^*)\lambda_\phi]\Phi(x, y) \\
&:= c_{N,\Phi}\Phi(x, y) .
\end{aligned}$$

Hence, integral equation (8.1) is generalized Ulam–Hyers–Rassias stable. □

### 8.2.3 An Example

As an application of our results we consider the partial Hadamard integral equation

$$u(x,y) = \mu(x,y) + \int_1^x \int_1^y \left(\log \frac{x}{s}\right)^{r_1-1} \left(\log \frac{y}{t}\right)^{r_2-1} \frac{f(s,t,u(s,t))}{st\Gamma(r_1)\Gamma(r_2)} dtds\,, \quad (x,y) \in [1,e] \times [1,e]\,, \tag{8.7}$$

where

$$r_1, r_2 > 0\,, \quad \mu(x,y) = x + y^2\,, \quad (x,y) \in [1,e] \times [1,e]\,,$$

and

$$f(x,y,u(x,y)) = cxy^2 \left(e^{-4} + \frac{u(x,y)}{e^{x+y+5}}\right)\,, \quad (x,y) \in [1,e] \times [1,e]\,,$$

with

$$c := \frac{e^4}{2}\Gamma(1+r_1)\Gamma(1+r_2)\,.$$

For each $u \in \mathbb{R}$ and $(x,y) \in [1,e] \times [1,e]$ we have

$$|f(x,y,u(x,y))| \leq ce^{-4}(1+|u|)\,.$$

Hence, condition (8.1.1) is satisfied by $p_1^* = p_2^* = ce^{-4}$. Condition (8.2) holds with $a = b = e$. Indeed,

$$\frac{(\log a)^{r_1}(\log b)^{r_2} p_2^*}{\Gamma(1+r_1)\Gamma(1+r_2)} = \frac{c}{e^4\Gamma(1+r_1)\Gamma(1+r_2)} = \frac{1}{2} < 1\,.$$

Consequently, Theorem 8.1 implies that Hadamard integral equation (8.7) has a solution defined on $[1,e] \times [1,e]$. Also, condition (8.1.2) is satisfied by

$$\Phi(x,y) = e^3, \text{ and } \lambda_\Phi = \frac{1}{\Gamma(1+r_1)\Gamma(1+r_2)}\,.$$

Indeed, for each $(x,y) \in [1,e] \times [1,e]$ we get

$$\begin{aligned}({}^H I_\sigma^r \Phi)(x,y) &\leq \frac{e^3}{\Gamma(1+r_1)\Gamma(1+r_2)} \\ &= \lambda_\Phi \Phi(x,y)\,.\end{aligned}$$

Consequently, Theorem 8.7 implies that equation (8.7) is generalized Ulam–Hyers–Rassias stable.

## 8.3 Global Stability Results for Volterra-Type Partial Hadamard Fractional Integral Equations

### 8.3.1 Introduction

In [47], Abbas et al. studied some existence and stability results for the nonlinear quadratic Volterra integral equation of Riemann–Liouville fractional order

$$u(t,x) = f(t,x,u(t,x),u(\alpha(t),x)) + \frac{1}{\Gamma(r)}\int_0^{\beta(t)} (\beta(t)-s)^{r-1} \times g(t,x,s,u(s,x),u(\gamma(s),x))ds\,, \quad (t,x) \in \mathbb{R}_+ \times [0,b]\,, \tag{8.8}$$

where $b > 0$, $\mathbb{R}_+ = [0,\infty)$, $r \in (0,\infty)$, $\alpha,\beta,\gamma\colon \mathbb{R}_+ \to \mathbb{R}_+$, and $f\colon \mathbb{R}_+\times[0,b]\times\mathbb{R}\times\mathbb{R} \to \mathbb{R}$ and $g\colon \mathbb{R}_+ \times [0,b] \times \mathbb{R}_+ \times \mathbb{R} \times \mathbb{R} \to \mathbb{R}$ are given continuous functions.

This section deals with the global existence and stability of solutions to the nonlinear quadratic Volterra partial integral equation of Hadamard fractional order

$$u(t,x) = f(t,x,u(t,x),u(\alpha(t),x)) + \frac{1}{\Gamma(r_1)\Gamma(r_2)}\int_1^{\beta(t)}\int_1^{x} \left(\log\frac{\beta(t)}{s}\right)^{r_1-1}\left(\log\frac{x}{\xi}\right)^{r_2-1} \times g(t,x,s,\xi,u(s,\xi),u(\gamma(s),\xi))\frac{d\xi ds}{s\xi}\,, \quad (t,x) \in J := [1,\infty)\times[1,b]\,, \tag{8.9}$$

where $b > 1$, $r_1, r_2 \in (0,\infty)$, $\alpha,\beta,\ \gamma\colon [1,\infty) \to [1,\infty)$, and $f\colon J \times \mathbb{R} \times \mathbb{R} \to \mathbb{R}$ and $g\colon J\times J\times\mathbb{R}\times\mathbb{R} \to \mathbb{R}$ are given continuous functions. Our existence results are based upon Schauder's fixed point theorem. Also, we obtain some results about the local asymptotic stability of solutions of the equation in question. Finally, we present an example illustrating the applicability of the imposed conditions.

### 8.3.2 Existence and Global Stability Results

In this section, we are concerned with the existence and the asymptotic stability of solutions for Hadamard partial integral equation (8.9).

In the sequel, we will use the following conditions.

(8.3.1) The function $\alpha\colon [1,\infty) \to [1,\infty)$ satisfies $\lim_{t\to\infty}\alpha(t) = \infty$.

(8.3.2) There exist constants $M, L > 0$ and a nondecreasing function $\psi_1\colon [0,\infty) \to (0,\infty)$ such that $M < \frac{L}{2}$,

$$|f(t,x,u_1,v_1) - f(t,x,u_2,v_2)| \le \frac{M(|u_1-u_2|+|v_1-v_2|)}{(1+\alpha(t))(L+|u_1-u_2|+|v_1-v_2|)}\,,$$

and

$$|f(t_1,x_1,u,v) - f(t_2,x_2,u,v)| \le (|t_1-t_2|+|x_1-x_2|)\psi_1(|u|+|v|)$$

for each $(t,x), (t_1,x_1), (t_2,x_2) \in J$ and $u, v, u_1, v_1, u_2, v_2 \in \mathbb{R}$.

(8.3.3) The function $t \to f(t, x, 0, 0)$ is bounded on $J$ with

$$f^* = \sup_{(t,x)\in[1,\infty)\times[1,b]} f(t, x, 0, 0)$$

and

$$\lim_{t\to\infty} |f(t, x, 0, 0)| = 0 \,; \quad x \in [1, b] \,.$$

(8.3.4) There exist continuous functions $p, q, \varphi : J \to \mathbb{R}_+$ and a nondecreasing function $\psi_2 : [0, \infty) \to (0, \infty)$ such that

$$|g(t_1, x_1, s, \xi, u, v) - g(t_2, x_2, s, \xi, u, v)| \le \varphi(s, \xi)(|t_1 - t_2| + |x_1 - x_2|)\psi_2(|u| + |v|)$$

and

$$|g(t, x, s, \xi, u, v)| \le \frac{p(t, x)q(s, \xi)}{1 + \alpha(t) + |u| + |v|}$$

for each $(t, x), (s, \xi), (t_1, x_1), (t_2, x_2) \in J$ and $u, v \in \mathbb{R}$. Moreover, assume that

$$\lim_{t\to\infty} p(t, x) \int_1^{\beta(t)} \int_1^x \left|\log \frac{\beta(t)}{s}\right|^{r_1-1} \left|\log \frac{x}{\xi}\right|^{r_2-1} q(s, \xi)d\xi ds = 0 \,.$$

**Theorem 8.8.** *Assume* (8.3.1)–(8.3.4). *Then integral equation* (7.1) *has at least one solution in the space* $BC$. *Moreover, solutions of equation* (7.1) *are globally asymptotically stable.*

*Proof.* Set $d^* := \sup_{(t,x)\in J} d(t, x)$, where

$$d(t, x) = \frac{p(t, x)}{\Gamma(r_1)\Gamma(r_2)} \int_1^{\beta(t)} \int_1^x \left|\log \frac{\beta(t)}{s}\right|^{r_1-1} \left|\log \frac{x}{\xi}\right|^{r_2-1} q(s, \xi)d\xi ds \,.$$

From condition (8.3.4) we infer that $d^*$ is finite. Let us define the operator $N$ such that, for any $u \in BC$,

$$(Nu)(t, x) = f(t, x, u(t, x), u(\alpha(t), x)) + \frac{1}{\Gamma(r_1)\Gamma(r_2)} \int_1^{\beta(t)} \int_1^x \left(\log \frac{\beta(t)}{s}\right)^{r_1-1} \left(\log \frac{x}{\xi}\right)^{r_2-1}$$
$$\times g(t, x, s, \xi, u(s, \xi), u(\gamma(s), \xi))\frac{d\xi ds}{s\xi} \,, \quad (t, x) \in J \,. \tag{8.10}$$

By considering the conditions of this theorem, we infer that $N(u)$ is continuous on $J$. Now we prove that $N(u) \in BC$ for any $u \in BC$. For arbitrarily fixed $(t, x) \in J$ we have

$$|(Nu)(t, x)| \le |f(t, x, u(t, x), u(\alpha(t), x)) - f(t, x, 0, 0)| + |f(t, x, 0, 0)|$$
$$+ \frac{1}{\Gamma(r_1)\Gamma(r_2)} \int_1^{\beta(t)} \int_1^x \left|\log \frac{\beta(t)}{s}\right|^{r_1-1} \left|\log \frac{x}{\xi}\right|^{r_2-1}$$
$$\times |g(t, x, s, \xi, u(s, \xi), u(\gamma(s), \xi))| \frac{d\xi ds}{s\xi}|$$

$$\leq \frac{M(|u(t,x)|+|u(\alpha(t),x)|)}{(1+\alpha(t))(L+|u(t,x)|+|u(\alpha(t),x)|)}+|f(t,x,0,0)|$$
$$+\frac{p(t,x)}{\Gamma(r_1)\Gamma(r_2)}\int_1^{\beta(t)}\int_1^{x}\left|\log\frac{\beta(t)}{s}\right|^{r_1-1}\left|\log\frac{x}{\xi}\right|^{r_2-1}$$
$$\times\frac{q(s,\xi)}{1+\alpha(t)+|u(s,\xi))|+|u(\gamma(s),\xi))|}\frac{d\xi ds}{s\xi}$$
$$\leq \frac{M(|u(t,x)|+|u(\alpha(t),x)|)}{|u(t,x)|+|u(\alpha(t),x)|}+f^*+d^*\,.$$

Thus,

$$\|N(u)\|_{BC}\leq M+f^*+d^*\,. \tag{8.11}$$

Hence, $N(u)\in BC$. Equation (8.11) yields that $N$ transforms the ball $B_\eta:=B(0,\eta)$ into itself, where $\eta=M+f^*+d^*$. We will show that $N\colon B_\eta\to B_\eta$ satisfies the assumptions of Schauder's fixed point theorem [149]. The proof will be given in several steps and cases.

*Step 1: $N$ is continuous.* Let $\{u_n\}_{n\in\mathbb{N}}$ be a sequence such that $u_n\to u$ in $B_\eta$. Then for each $(t,x)\in J$ we have

$$|(Nu_n)(t,x)-(Nu)(t,x)|\leq|f(t,x,u_n(t,x),u_n(\alpha(t),x))-f(t,x,u(t,x),u(\alpha(t),x))|$$
$$+\frac{1}{\Gamma(r_1)\Gamma(r_2)}\int_1^{\beta(t)}\int_1^{x}\left|\log\frac{\beta(t)}{s}\right|^{r_1-1}\left|\log\frac{x}{\xi}\right|^{r_2-1}$$
$$\times\sup_{(s,\xi)\in J}|g(t,x,s,\xi,u_n(s,\xi),u_n(\gamma(s),\xi))$$
$$-g(t,x,s,\xi,u(s,\xi),u(\gamma(s),\xi))|\frac{d\xi ds}{s\xi}$$
$$\leq\frac{2M}{L}\|u_n-u\|_{BC}$$
$$+\frac{1}{\Gamma(r_1)\Gamma(r_2)}\int_1^{\beta(t)}\int_1^{x}\left|\log\frac{\beta(t)}{s}\right|^{r_1-1}\left|\log\frac{x}{\xi}\right|^{r_2-1}$$
$$\times\|g(t,x,.,.,u_n(.,.),u_n(\gamma(.),.))$$
$$-g(t,x,.,.,u(.,.),u(\gamma(.),.))\|_{BC}d\xi ds\,. \tag{8.12}$$

*Case 1.* If $(t,x)\in[1,T]\times[1,b]$, $T>1$, then, since $u_n\to u$ as $n\to\infty$ and $g,\gamma$ are continuous, then (8.12) gives

$$\|N(u_n)-N(u)\|_{BC}\to 0\ \text{ as }\ n\to\infty\,.$$

*Case 2.* If $(t,x) \in (T,\infty) \times [1,b]$, $T > 1$, then from (8.3.4) and (8.12), for each $(t,x) \in J$, we have

$$\begin{aligned}|(Nu_n)(t,x) - (Nu)(t,x)| &\leq \frac{2M}{L}\|u_n - u\|_{BC} \\ &\quad + \frac{2p(t,x)}{\Gamma(r_1)\Gamma(r_2)} \int_1^{\beta(t)} \int_1^{x} \left|\log\frac{\beta(t)}{s}\right|^{r_1-1} \left|\log\frac{x}{\xi}\right|^{r_2-1} \frac{q(s,\xi)}{s\xi} d\xi ds \\ &\leq \frac{2M}{L}\|u_n - u\|_{BC} + d(t,x)\,.\end{aligned}$$

Thus, we get

$$|(Nu_n)(t,x) - (Nu)(t,x)| \leq \frac{2M}{L}\|u_n - u\|_{BC} + d(t,x)\,. \tag{8.13}$$

Since $u_n \to u$ as $n \to \infty$ and $t \to \infty$, then (8.13) gives

$$\|N(u_n) - N(u)\|_{BC} \to 0 \text{ as } n \to \infty\,.$$

*Step 2: $N(B_\eta)$ is uniformly bounded.* This is clear since $N(B_\eta) \subset B_\eta$ and $B_\eta$ is bounded.

*Step 3: $N(B_\eta)$ is equicontinuous on every compact subset* $[1,a] \times [1,b]$ *of* $J$, $a > 0$. Let $(t_1,x_1), (t_2,x_2) \in [1,a] \times [1,b]$, $t_1 < t_2$, $x_1 < x_2$, and let $u \in B_\eta$. Also, without loss of generality, suppose that $\beta(t_1) \leq \beta(t_2)$. Then we have

$$\begin{aligned}&|(Nu)(t_2,x_2) - (Nu)(t_1,x_1)| \\ &\leq |f(t_2,x_2,u(t_2,x_2),u(\alpha(t_2),x_2)) - f(t_2,x_2,u(t_1,x_1),u(\alpha(t_1),x_1))| \\ &\quad + |f(t_2,x_2,u(t_1,x_1),u(\alpha(t_1),x_1)) - f(t_1,x_1,u(t_1,x_1),u(\alpha(t_1),x_1))| \\ &\quad + \frac{1}{\Gamma(r_1)\Gamma(r_2)} \int_1^{\beta(t_2)} \int_1^{x_2} \left|\log\frac{\beta(t_2)}{s}\right|^{r_1-1} \left|\log\frac{x_2}{\xi}\right|^{r_2-1} \\ &\quad \times |g(t_2,x_2,s,\xi,u(s,\xi),u(\gamma(s),\xi)) - g(t_1,x_1,s,\xi,u(s,\xi),u(\gamma(s),\xi))| d\xi ds \\ &\quad + \left| \frac{1}{\Gamma(r_1)\Gamma(r_2)} \int_1^{\beta(t_2)} \int_1^{x_2} \left(\log\frac{\beta(t_2)}{s}\right)^{r_1-1} \left(\log\frac{x_2}{\xi}\right)^{r_2-1} \right. \\ &\quad \times g(t_1,x_1,s,\xi,u(s,\xi),u(\gamma(s),\xi)) d\xi ds \\ &\quad - \frac{1}{\Gamma(r_1)\Gamma(r_2)} \int_1^{\beta(t_1)} \int_1^{x_1} \left(\log\frac{\beta(t_2)}{s}\right)^{r_1-1} \left(\log\frac{x_2}{\xi}\right)^{r_2-1} \\ &\quad \left. \times g(t_1,x_1,s,\xi,u(s,\xi),u(\gamma(s),\xi)) d\xi ds \right| \\ &\quad + \frac{1}{\Gamma(r_1)\Gamma(r_2)} \int_1^{\beta(t_1)} \int_1^{x_1} \left| \left(\log\frac{\beta(t_2)}{s}\right)^{r_1-1} \left(\log\frac{x_2}{\xi}\right)^{r_2-1} \right. \\ &\quad \left. - \left(\log\frac{\beta(t_1)}{s}\right)^{r_1-1} \left(\log\frac{x_1}{\xi}\right)^{r_2-1} \right| |g(t_1,x_1,s,\xi,u(s,\xi),u(\gamma(s),\xi))| d\xi ds\,.\end{aligned}$$

Thus, we obtain

$$\begin{aligned}
&|(Nu)(t_2,x_2)-(Nu)(t_1,x_1)| \\
&\leq \frac{M}{L}(|u(t_2,x_2)-u(t_1,x_1)|+|u(\alpha(t_2),x_2)-u(\alpha(t_1),x_1)|) \\
&\quad + (|t_2-t_1|+|x_2-x_1|)\psi_1(2\|u\|_{BC}) \\
&\quad + \frac{1}{\Gamma(r_1)\Gamma(r_2)}\int_1^{\beta(t_2)}\int_1^{x_2}\left|\log\frac{\beta(t_2)}{s}\right|^{r_1-1}\left|\log\frac{x_2}{\xi}\right|^{r_2-1} \\
&\quad\quad \times \varphi(s,\xi)(|t_2-t_1|+|x_2-x_1|)\psi_2(2\|u\|_{BC})d\xi ds \\
&\quad + \frac{1}{\Gamma(r_1)\Gamma(r_2)}\int_{\beta(t_1)}^{\beta(t_2)}\int_1^{x_2}\left|\log\frac{\beta(t_2)}{s}\right|^{r_1-1}\left|\log\frac{x_2}{\xi}\right|^{r_2-1} \\
&\quad\quad \times |g(t_1,x_1,s,\xi,u(s,\xi),u(\gamma(s),\xi))|d\xi ds \\
&\quad + \frac{1}{\Gamma(r_1)\Gamma(r_2)}\int_1^{\beta(t_2)}\int_{x_1}^{x_2}\left|\log\frac{\beta(t_2)}{s}\right|^{r_1-1}\left|\log\frac{x_2}{\xi}\right|^{r_2-1} \\
&\quad\quad \times |g(t_1,x_1,s,\xi,u(s,\xi),u(\gamma(s),\xi))|d\xi ds \\
&\quad + \frac{1}{\Gamma(r_1)\Gamma(r_2)}\int_{\beta(t_1)}^{\beta(t_2)}\int_{x_1}^{x_2}\left|\log\frac{\beta(t_2)}{s}\right|^{r_1-1}\left|\log\frac{x_2}{\xi}\right|^{r_2-1} \\
&\quad\quad \times |g(t_1,x_1,s,\xi,u(s,\xi),u(\gamma(s),\xi))|d\xi ds \\
&\quad + \frac{1}{\Gamma(r_1)\Gamma(r_2)}\int_1^{\beta(t_1)}\int_1^{x_1}\left|\left(\log\frac{\beta(t_2)}{s}\right)^{r_1-1}\left(\log\frac{x_2}{\xi}\right)^{r_2-1}\right. \\
&\quad - \left.\left(\log\frac{\beta(t_1)}{s}\right)^{r_1-1}\left(\log\frac{x_1}{\xi}\right)^{r_2-1}\right| |g(t_1,x_1,s,\xi,u(s,\xi),u(\gamma(s),\xi))|d\xi ds\,.
\end{aligned}$$

Hence, we get

$$\begin{aligned}
&|(Nu)(t_2,x_2)-(Nu)(t_1,x_1)| \\
&\leq \frac{M}{L}(|u(t_2,x_2)-u(t_1,x_1)|+|u(\alpha(t_2),x_2)-u(\alpha(t_1),x_1)|) \\
&\quad + (|t_2-t_1|+|x_2-x_1|)\psi_1(2\eta) \\
&\quad + \frac{(|t_2-t_1|+|x_2-x_1|)\psi_2(2\eta)}{\Gamma(r_1)\Gamma(r_2)} \\
&\quad\quad \times \int_1^{\beta(t_2)}\int_1^{x_2}\left|\log\frac{\beta(t_2)}{s}\right|^{r_1-1}\left|\log\frac{x_2}{\xi}\right|^{r_2-1}\varphi(s,\xi)d\xi ds \\
&\quad + \frac{p(t_1,x_1)}{\Gamma(r_1)\Gamma(r_2)}\int_{\beta(t_1)}^{\beta(t_2)}\int_1^{x_2}\left|\log\frac{\beta(t_2)}{s}\right|^{r_1-1}\left|\log\frac{x_2}{\xi}\right|^{r_2-1} q(s,\xi)|d\xi ds
\end{aligned}$$

$$+\frac{p(t_1,x_1)}{\Gamma(r_1)\Gamma(r_2)}\int_1^{\beta(t_2)}\int_{x_1}^{x_2}\left|\log\frac{\beta(t_2)}{s}\right|^{r_1-1}\left|\log\frac{x_2}{\xi}\right|^{r_2-1}q(s,\xi)|d\xi ds$$

$$+\frac{p(t_1,x_1)}{\Gamma(r_1)\Gamma(r_2)}\int_{\beta(t_1)}^{\beta(t_2)}\int_{x_1}^{x_2}\left|\log\frac{\beta(t_2)}{s}\right|^{r_1-1}\left|\log\frac{x_2}{\xi}\right|^{r_2-1}q(s,\xi)|d\xi ds$$

$$+\frac{p(t_1,x_1)}{\Gamma(r_1)\Gamma(r_2)}\int_1^{\beta(t_1)}\int_1^{x_1}\left|\left(\log\frac{\beta(t_2)}{s}\right)^{r_1-1}\left(\log\frac{x_2}{\xi}\right)^{r_2-1}\right.$$

$$\left.-\left(\log\frac{\beta(t_1)}{s}\right)^{r_1-1}\left(\log\frac{x_1}{\xi}\right)^{r_2-1}\right|q(s,\xi)|d\xi ds\,.$$

From the continuity of $\alpha, \beta, f, g$ and as $t_1 \to t_2$ and $x_1 \to x_2$, the right-hand side of the preceding inequality tends to zero.

*Step 4: $N(B_\eta)$ is equiconvergent.* Let $(t,x) \in J$ and $u \in B_\eta$; then we have

$$\begin{aligned}
|u(t,x)| &\le |f(t,x,u(t,x),u(\alpha(t),x)) - f(t,x,0,0) + f(t,x,0,0)| \\
&\quad + \left|\frac{1}{\Gamma(r_1)\Gamma(r_2)}\int_1^{\beta(t)}\int_1^{x}\left(\log\frac{\beta(t)}{s}\right)^{r_1-1}\left(\log\frac{x}{\xi}\right)^{r_2-1}\right. \\
&\quad \left.\times g(t,x,s,\xi,u(s,\xi),u(\gamma(s),\xi))\frac{d\xi ds}{s\xi}\right| \\
&\le \frac{M(|u(t,x)| + |u(\alpha(t),x)|)}{(1+\alpha(t))(L + |u(t,x)| + |u(\alpha(t),x)|)} + |f(t,x,0,0)| \\
&\quad + \frac{p(t,x)}{\Gamma(r_1)\Gamma(r_2)}\int_1^{\beta(t)}\int_1^{x}\left(\log\frac{\beta(t)}{s}\right)^{r_1-1}\left(\log\frac{x}{\xi}\right)^{r_2-1} \\
&\quad \times \frac{q(s,\xi)}{1+\alpha(t)+|u(s,\xi)|+|u(\gamma(s),\xi)|}d\xi ds \\
&\le \frac{M}{1+\alpha(t)} + |f(t,x,0,0)| \\
&\quad + \frac{p(t,x)}{\Gamma(r_1)\Gamma(r_2)(1+\alpha(t))}\int_1^{\beta(t)}\int_1^{x}\left(\log\frac{\beta(t)}{s}\right)^{r_1-1}\left(\log\frac{x}{\xi}\right)^{r_2-1}q(s,\xi)d\xi ds \\
&\le \frac{M}{1+\alpha(t)} + |f(t,x,0,0)| + \frac{d^*}{1+\alpha(t)}\,.
\end{aligned}$$

Thus, for each $x \in [1,b]$ we get

$$|u(t,x)| \to 0, \quad \text{as } t \to +\infty\,.$$

Hence,

$$|u(t,x) - u(+\infty,x)| \to 0, \quad \text{as } t \to +\infty\,.$$

As a consequence of Steps 1–4, together with Lemma 1.57, we can conclude that $N\colon B_\eta \to B_\eta$ is continuous and compact. From an application of Schauder's fixed point theorem [149] we deduce that $N$ has a fixed point $u$ that is a solution of Hadamard integral equation (8.9).

*Step 5: the uniform global attractivity.* Let us assume that $u_0$ is a solution of integral equation (8.9) with the conditions of this theorem. Consider the ball $B(u_0, \eta)$ with $\eta^* = \frac{LM^*}{L-2M}$, where

$$M^* := \frac{1}{\Gamma(r_1)\Gamma(r_2)} \sup_{(t,x)\in J} \left\{ \int_1^{\beta(t)} \int_1^x \left(\log \frac{\beta(t)}{s}\right)^{r_1-1} \left(\log \frac{x}{\xi}\right)^{r_2-1} \times |g(t,x,s,\xi,u(s,\xi),u(\gamma(s),\xi)) - g(t,x,s,\xi,u_0(s,\xi),u_0(\gamma(s),\xi))| d\xi ds;\ u \in BC \right\}.$$

Taking $u \in B(u_0, \eta^*)$, we then have

$$\begin{aligned} |(Nu)(t,x) - u_0(t,x)| &= |(Nu)(t,x) - (Nu_0)(t,x)| \\ &\le |f(t,x,u(t,x),u(\alpha(t),x)) - f(t,x,u_0(t,x),u_0(\alpha(t),x))| \\ &\quad + \frac{1}{\Gamma(r_1)\Gamma(r_2)} \int_1^{\beta(t)} \int_1^x \left(\log \frac{\beta(t)}{s}\right)^{r_1-1} \left(\log \frac{x}{\xi}\right)^{r_2-1} \\ &\quad \times |g(t,x,s,\xi,u(s,\xi),u(\gamma(s),\xi)) \\ &\quad - g(t,x,s,\xi,u_0(s,\xi),u_0(\gamma(s),\xi))| \frac{d\xi ds}{s\xi} \\ &\le \frac{2M}{L\|}\|u - u_0\|_{BC} + M^* \\ &\le \frac{2M}{L}\eta^* + M^* = \eta^*. \end{aligned}$$

Thus, we observe that $N$ is a continuous function such that $N(B(u_0, \eta^*)) \subset B(u_0, \eta^*)$. Moreover, if $u$ is a solution of equation (8.9), then

$$\begin{aligned} |u(t,x) - u_0(t,x)| &= |(Nu)(t,x) - (Nu_0)(t,x)| \\ &\le |f(t,x,u(t,x),u(\alpha(t),x)) - f(t,x,u_0(t,x),u_0(\alpha(t),x))| \\ &\quad + \frac{1}{\Gamma(r_1)\Gamma(r_2)} \int_1^{\beta(t)} \int_1^x \left(\log \frac{\beta(t)}{s}\right)^{r_1-1} \left(\log \frac{x}{\xi}\right)^{r_2-1} \\ &\quad \times |g(t,x,s,\xi,u(s,\xi),u(\gamma(s),\xi)) \\ &\quad - g(t,x,s,\xi,u_0(s,\xi),u_0(\gamma(s),\xi))| d\xi ds. \end{aligned}$$

Thus,

$$|u(t,x)-u_0(t,x)| \leq \frac{M}{L}(|u(t,x)-u_0(t,x)|+|u(\alpha(t),x)-u_0(\alpha(t),x)|) + \frac{p(t,x)}{\Gamma(r_1)\Gamma(r_2)}\int_1^{\beta(t)}\int_1^{x}\left(\log\frac{\beta(t)}{s}\right)^{r_1-1}\left(\log\frac{x}{\xi}\right)^{r_2-1} q(s,\xi)d\xi ds\,. \tag{8.14}$$

By using (8.14) and the fact that $\alpha(t)\to\infty$ as $t\to\infty$, we get

$$\lim_{t\to\infty}|u(t,x)-u_0(t,x)| \leq \lim_{t\to\infty}\frac{L.p(t,x)}{\Gamma(r_1)\Gamma(r_2)(L-2M)}\int_1^{\beta(t)}\int_1^{x}\left(\log\frac{\beta(t)}{s}\right)^{r_1-1}\left(\log\frac{x}{\xi}\right)^{r_2-1} \times q(s,\xi)d\xi ds = 0\,.$$

Consequently, all solutions of integral equation (7.1) are globally asymptotically stable. □

### 8.3.3 An Example

As an application of our results we consider the partial Hadamard integral equation of fractional order

$$u(t,x) = \frac{tx}{10(1+t+t^2+t^3)}(1+2\sin(u(t,x))) + \frac{1}{\Gamma^2\left(\frac{1}{3}\right)}\int_1^{t}\int_1^{x}\left(\log\frac{t}{s}\right)^{\frac{-2}{3}}\left(\log\frac{x}{\xi}\right)^{\frac{-2}{3}} \times \frac{\ln(1+2x(s\xi)^{-1}|u(s,\xi)|)}{(1+t+2|u(s,\xi)|)^2(1+x^2+t^4)}d\xi ds\,; \quad (t,x)\in[1,\infty)\times[1,e]\,, \tag{8.15}$$

where $r_1 = r_2 = \frac{1}{3}$, $\alpha(t) = \beta(t) = \gamma(t) = t$,

$$f(t,x,u,v) = \frac{tx(1+\sin(u)+\sin(v))}{10(1+t)(1+t^2)}\,,$$

and

$$g(t,x,s,\xi,u,v) = \frac{\ln(1+x(s\xi)^{-1}(|u|+|v|))}{(1+t+|u|+|v|)^2(1+x^2+t^4)}$$

for $(t,x),(s,\xi)\in[1,\infty)\times[1,e]$, and $u,v\in\mathbb{R}$.

We can easily check that the conditions of Theorem 8.8 are satisfied. In fact, we have that the function $f$ is continuous and satisfies (8.3.2), where $M=\frac{1}{10}$, $L=1$. Also, $f$ satisfies (8.3.3), with $f^* = \frac{e}{10}$. Next, let us note that the function $g$ satisfies (8.3.4), where $p(t,x) = \frac{1}{1+x^2+t^4}$ and $q(s,\xi) = (s\xi)^{-1}$.

Additionally,

$$\lim_{t\to\infty} p(t,x)\int_1^t\int_1^x \left|\log\frac{t}{s}\right|^{\frac{-2}{3}}\left|\log\frac{x}{\xi}\right|^{\frac{-2}{3}} q(s,\xi)d\xi ds$$

$$= \lim_{t\to\infty}\frac{x}{1+x^2+t^4}\int_1^t\int_1^x \left|\log\frac{t}{s}\right|^{\frac{-2}{3}}\left|\log\frac{x}{\xi}\right|^{\frac{-2}{3}}\frac{d\xi ds}{s\xi}$$

$$= \lim_{t\to\infty}\frac{9x(\log t)^{\frac{1}{3}}}{1+x^2+t^4} = 0\,.$$

Hence by Theorem 8.8, equation (8.15) has a solution defined on $[1,\infty)\times[1,e]$, and solutions of this equation are globally asymptotically stable.

## 8.4 Ulam Stabilities for Hadamard Fractional Integral Equations in Fréchet Spaces

### 8.4.1 Introduction

In this section, we present some results concerning the existence and Ulam stabilities of solutions for some functional integral equations of Hadamard fractional order. We use an extension of the Burton–Kirk fixed point theorem in Fréchet spaces.

Recently some interesting results on the existence and Ulam stabilities of the solutions of some classes of differential equations were obtained by Abbas et al. [5, 24, 25, 28]. This section deals with the existence and Ulam stabilities of solutions of the following Hadamard fractional integral equations:

$$\begin{aligned} u(t,x) &= \mu(t,x) + f(t,x,({}^H I_\sigma^r u)(t,x), u(t,x)) \\ &\quad + \frac{1}{\Gamma(r_1)\Gamma(r_2)}\int_1^t\int_1^x \left(\log\frac{t}{s}\right)^{r_1-1}\left(\log\frac{x}{y}\right)^{r_2-1} \\ &\quad \times g(t,x,s,y,u(s,y))\frac{dyds}{sy}\,, \quad (t,x)\in J := [1,+\infty)\times[1,b]\,, \end{aligned} \tag{8.16}$$

where $b>1$, $\sigma=(1,1)$, $r=(r_1,r_2)$, $r_1,r_2\in(0,\infty)$, $\mu\colon J\to\mathbb{R}$, $f\colon J\times\mathbb{R}\times\mathbb{R}\to\mathbb{R}$ and $g\colon J'\times\mathbb{R}\to\mathbb{R}$ are given continuous functions, and $J'=\{(t,x,s,y)\in J^2 : s\le t, y\le x\}$. Our investigations are conducted in Fréchet spaces with an application of the fixed point theorem of Burton–Kirk to the existence of solutions of integral equation (8.16), and we prove that all solutions are generalized Ulam–Hyers–Rassias stable.

### 8.4.2 Existence and Ulam Stabilities Results

Here we are concerned with the existence and the Ulam stability of solutions for integral equation (8.16). Set

$$J'_p = \{(t,x,s,y):\ 1 \le s \le t \le p,\ 1 \le y \le x \le b\}\ ; \quad p \in \mathbb{N}\backslash\{0,1\}\,.$$

The following conditions will be used in the sequel:

(8.4.1) There exist continuous functions $l, k\colon J_p \to \mathbb{R}_+$ such that

$$|f(t,x,u_1,v_1) - f(t,x,u_2,v_2)| \le \frac{l(t,x)|u_1-u_2| + k(t,x)|v_1-v_2|}{1+|u_1-u_2|+|v_1-v_2|}$$

for each $(t,x) \in J_p$ and each $u_1,u_2,v_1,v_2 \in \mathbb{R}$.

(8.4.2) There exist continuous functions $P, Q, \varphi\colon J'_p \to \mathbb{R}_+$ and a nondecreasing function $\psi\colon [0,\infty) \to (0,\infty)$ such that

$$|g(t,x,s,y,u)| \le \frac{P(t,x,s,y) + Q(t,x,s,y)|u|}{1+|u|}$$

for $(t,x,s,y) \in J'$, $u \in \mathbb{R}$, and

$$\begin{aligned}&|g(t_1,x_1,s,y,u) - g(t_2,x_2,s,y,u)| \le \varphi(s,y)(|t_1-t_2| + |x_1-x_2|)\\ &\times \psi(|u|)\,; \quad (t_1,x_1,s,y), (t_2,x_2,s,y) \in J'_p,\ u \in \mathbb{R}\,.\end{aligned}$$

(8.4.3) There exist continuous functions $P_1, Q_1\colon J_p \to [0,\infty)$ such that for each $(t,s), (t,x) \in J_p$ we have

$$P(t,x,s,y,w) \le \phi(t,x)P_1(s,y), \quad \text{and} \quad Q(t,x,s,y,w) \le \phi(t,x)Q_1(s,y)\,.$$

**Theorem 8.9.** *Assume* (8.4.1) *and* (8.4.2). *If*

$$\ell := k_p + \frac{l_p(\log p)^{r_1}(\log b)^{r_2}}{\Gamma(1+r_1)\Gamma(1+r_2)} < 1\,, \tag{8.17}$$

*where*

$$k_p = \sup_{(t,x)\in J_p} k(t,x), \ l_p = \sup_{(t,x)\in J_p} l(t,x);\ p \in \mathbb{N}\backslash\{0,1\}\,,$$

*then Hadamard integral equation* (8.16) *has at least one solution in the space C. Furthermore, if condition* (8.4.3) *holds, then equation* (8.16) *is generalized Ulam–Hyers–Rassias stable.*

*Proof.* Let us define the operators $A, B\colon C \to C$ defined by

$$(Au)(t,x) = \int_1^t\int_1^x \left(\log\frac{t}{s}\right)^{r_1-1}\left(\log\frac{x}{y}\right)^{r_2-1} \frac{g(t,x,s,y,u(s,y))}{sy\Gamma(r_1)\Gamma(r_2)}dyds\ ; \quad (t,x) \in J\,, \tag{8.18}$$

$$(Bu)(t,x) = \mu(t,x) + f(t,x,({}^HI^r_\sigma u)(t,x), u(t,x))\ ; \quad (t,x) \in J\,. \tag{8.19}$$

We will show that operators $A$ and $B$ satisfy all the conditions of Theorem 1.43. The proof will be given in several steps.

*Step 1. $A$ is compact.* To this end, we must prove that $A$ is continuous and it transforms every bounded set into a relatively compact set. Let $M \subset C$ be a bounded set of $C$. The proof will be given in several claims.

*Claim 1. $A$ is continuous.* Let $\{u_n\}_{n\in\mathbb{N}\setminus\{0,1\}}$ be a sequence in $M$ such that $u_n \to u$ in $M$. Then, for each $(t,x) \in J_p;\ p \in \mathbb{N}\setminus\{0,1\}$, we have

$$\begin{aligned}
&|(Au_n)(t,x) - (Au)(t,x)| \\
&\le \frac{1}{\Gamma(r_1)\Gamma(r_2)} \int_1^t \int_1^x \left|\log\frac{t}{s}\right|^{r_1-1} \left|\log\frac{x}{y}\right|^{r_2-1} \\
&\quad \times |g(t,x,s,y,u_n(s,y)) - g(t,x,s,y,u(s,y))|dyds \\
&\le \frac{1}{\Gamma(r_1)\Gamma(r_2)} \int_1^t \int_1^x \left|\log\frac{t}{s}\right|^{r_1-1} \left|\log\frac{x}{y}\right|^{r_2-1} \\
&\quad \times |g(t,x,s,y,u_n(s,y)) - g(t,x,s,y,u(s,y))|dyds\,.
\end{aligned} \tag{8.20}$$

Since $u_n \to u$ as $n \to \infty$ and $g$ is continuous, (8.20) gives

$$\|A(u_n) - A(u)\|_p \to 0 \quad \text{as} \quad n \to \infty\,.$$

*Claim 2. $A$ maps bounded sets to bounded sets in $C$.* For arbitrarily fixed $(t,x) \in J_p$ and $u \in M$, we have

$$\begin{aligned}
|(Au)(t,x)| &\le \frac{1}{\Gamma(r_1)\Gamma(r_2)} \int_1^t \int_1^x \left|\log\frac{t}{s}\right|^{r_1-1} \left|\log\frac{x}{y}\right|^{r_2-1} \\
&\quad \times |g(t,x,s,y,u(s,y))|dyds \\
&\le \frac{1}{\Gamma(r_1)\Gamma(r_2)} \int_1^t \int_1^x \left|\log\frac{t}{s}\right|^{r_1-1} \left|\log\frac{x}{y}\right|^{r_2-1} \\
&\quad \times \frac{P(t,x,s,y) + Q(t,x,s,y)|u(s,y)|}{1+|u(s,y)|} dyds \\
&\le \frac{1}{\Gamma(r_1)\Gamma(r_2)} \int_1^t \int_1^x \left|\log\frac{t}{s}\right|^{r_1-1} \left|\log\frac{x}{y}\right|^{r_2-1} \\
&\quad \times (P(t,x,s,y) + Q(t,x,s,y))dyds \\
&\le P_p + Q_p\,,
\end{aligned}$$

where

$$P_p = \sup_{(t,x)\in J_p} \int_1^t \int_1^x \left|\log\frac{t}{s}\right|^{r_1-1} \left|\log\frac{x}{y}\right|^{r_2-1} \frac{P(t,x,s,y)}{\Gamma(r_1)\Gamma(r_2)} dyds$$

and

$$Q_p = \sup_{(t,x)\in J_p} \int_1^t \int_1^x \left|\log \frac{t}{s}\right|^{r_1-1} \left|\log \frac{x}{y}\right|^{r_2-1} \frac{Q(t,x,s,y)}{\Gamma(r_1)\Gamma(r_2)} dyds \,.$$

Thus,

$$\|A(u)\|_p \le P_p + Q_p := \ell'_p \,.$$

*Claim 3. A maps bounded sets to equicontinuous sets in C.* Let $(t_1, x_1), (t_2, x_2) \in J_p$, $t_1 < t_2, x_1 < x_2$, and let $u \in M$; thus, we have

$$\begin{aligned}
&|(Au)(t_2,x_2) - (Au)(t_1,x_1)| \\
&\le \frac{1}{\Gamma(r_1)\Gamma(r_2)} \left| \int_1^{t_2}\int_1^{x_2} \left|\log \frac{t_2}{s}\right|^{r_1-1} \left|\log \frac{x_2}{y}\right|^{r_2-1} \right. \\
&\quad \times [g(t_2,x_2,s,y,u(s,y)) - g(t_1,x_1,s,y,u(s,y))]dyds| \\
&\quad + \frac{1}{\Gamma(r_1)\Gamma(r_2)} \left| \int_1^{t_2}\int_1^{x_2} \left|\log \frac{t_2}{s}\right|^{r_1-1} \left|\log \frac{x_2}{y}\right|^{r_2-1} g(t_1,x_1,s,y,u(s,y))dyds \right. \\
&\quad \left. - \int_1^{t_2}\int_1^{x_2} \left|\log \frac{t_1}{s}\right|^{r_1-1} \left|\log \frac{x_1}{y}\right|^{r_2-1} g(t_1,x_1,s,y,u(s,y))dyds \right| \\
&\quad + \frac{1}{\Gamma(r_1)\Gamma(r_2)} \left| \int_1^{t_2}\int_1^{x_2} \left|\log \frac{t_1}{s}\right|^{r_1-1} \left|\log \frac{x_1}{y}\right|^{r_2-1} g(t_1,x_1,s,y,u(s,y))dyds \right. \\
&\quad \left. - \int_1^{t_1}\int_1^{x_1} \left|\log \frac{t_1}{s}\right|^{r_1-1} \left|\log \frac{x_1}{y}\right|^{r_2-1} g(t_1,x_1,s,y,u(s,y))dyds \right| .
\end{aligned}$$

Thus,

$$\begin{aligned}
&|(Au)(t_2,x_2) - (Au)(t_1,x_1)| \\
&\le \frac{1}{\Gamma(r_1)\Gamma(r_2)} \int_1^{t_2}\int_1^{x_2} \left|\log \frac{t_2}{s}\right|^{r_1-1} \left|\log \frac{x_2}{y}\right|^{r_2-1} \\
&\quad \times |g(t_2,x_2,s,y,u(s,y)) - g(t_1,x_1,s,y,u(s,y))|\, dyds \\
&\quad + \frac{1}{\Gamma(r_1)\Gamma(r_2)} \int_1^{t_1}\int_1^{x_1} \left| \left(\log \frac{t_2}{s}\right)^{r_1-1} \left(\log \frac{x_2}{y}\right)^{r_2-1} - \left(\log \frac{t_1}{s}\right)^{r_1-1} \left(\log \frac{x_1}{y}\right)^{r_2-1} \right| \\
&\quad \times |g(t_1,x_1,s,y,u(s,y))|\, dyds \\
&\quad + \frac{1}{\Gamma(r_1)\Gamma(r_2)} \int_1^{t_1}\int_{x_1}^{x_2} \left|\log \frac{t_2}{s}\right|^{r_1-1} \left|\log \frac{x_2}{y}\right|^{r_2-1} |g(t_1,x_1,s,y,u(s,y))|dyds
\end{aligned}$$

$$+\frac{1}{\Gamma(r_1)\Gamma(r_2)}\int_{t_1}^{t_2}\int_{1}^{x_1}\left|\log\frac{t_2}{s}\right|^{r_1-1}\left|\log\frac{x_2}{y}\right|^{r_2-1}|g(t_1,x_1,s,y,u(s,y))|dyds$$

$$+\frac{1}{\Gamma(r_1)\Gamma(r_2)}\int_{t_1}^{t_2}\int_{x_1}^{x_2}\left|\log\frac{t_2}{s}\right|^{r_1-1}\left|\log\frac{x_2}{y}\right|^{r_2-1}|g(t_1,x_1,s,y,u(s,y))|dyds\,.$$

Hence,

$$\begin{aligned}
&|(Au)(t_2,x_2)-(Au)(t_1,x_1)|\\
&\le\frac{1}{\Gamma(r_1)\Gamma(r_2)}\int_{1}^{t_2}\int_{1}^{x_2}\left|\log\frac{t_2}{s}\right|^{r_1-1}\left|\log\frac{x_2}{y}\right|^{r_2-1}\\
&\quad\times\varphi(s,y)(|t_1-t_2|+|x_1-x_2|)\psi(\ell_p)dyds\\
&\quad+\frac{1}{\Gamma(r_1)\Gamma(r_2)}\int_{1}^{t_1}\int_{1}^{x_1}\left|\left(\log\frac{t_2}{s}\right)^{r_1-1}\left(\log\frac{x_2}{y}\right)^{r_2-1}-\left(\log\frac{t_1}{s}\right)^{r_1-1}\left(\log\frac{x_1}{y}\right)^{r_2-1}\right|\\
&\quad\times(P(t_1,x_1,s,y)+Q(t_1,x_1,s,y))dyds\\
&\quad+\frac{1}{\Gamma(r_1)\Gamma(r_2)}\int_{t_2}^{t_2}\int_{1}^{x_2}\left|\log\frac{t_2}{s}\right|^{r_1-1}\left|\log\frac{x_2}{y}\right|^{r_2-1}(P(t_1,x_1,s,y)+Q(t_1,x_1,s,y))dyds\\
&\quad+\frac{1}{\Gamma(r_1)\Gamma(r_2)}\int_{1}^{t_1}\int_{x_1}^{x_2}\left|\log\frac{t_2}{s}\right|^{r_1-1}\left|\log\frac{x_2}{y}\right|^{r_2-1}(P(t_1,x_1,s,y)+Q(t_1,x_1,s,y))dyds\\
&\quad+\frac{1}{\Gamma(r_1)\Gamma(r_2)}\int_{t_1}^{t_2}\int_{x_1}^{x_2}\left|\log\frac{t_2}{s}\right|^{r_1-1}\left|\log\frac{x_2}{y}\right|^{r_2-1}(P(t_1,x_1,s,y)+Q(t_1,x_1,s,y))dyds\,.
\end{aligned}$$

From the continuity of functions $P$, $Q$, $\varphi$ and as $t_1\longrightarrow t_2$ and $x_1\longrightarrow x_2$, the right-hand side of the preceding inequality tends to zero. As a consequence of Claims 1–3 and from the Ascoli–Arzelà theorem, we can conclude that $A$ is continuous and compact.

*Step 2. $B$ is a contraction.* Consider $v, w\in C$. Then, by (8.4.1), for any $p\in\mathbb{N}\backslash\{0,1\}$ and each $(t,x)\in J_p$, we have

$$\begin{aligned}
|(Bv)(t,x)-(Bw)(t,x)|&\le l(t,x)|^H I_\sigma^r(v-w)(t,x)|+k(t,x)|(v-w)(t,x)|\\
&\le\left(k(t,x)+\frac{l(t,x)(\log p)^{r_1}(\log b)^{r_2}}{\Gamma(1+r_1)\Gamma(1+r_2)}\right)|v-w|\,.
\end{aligned}$$

Thus,

$$\|(B(v)-B(w)\|_p\le\left(k_p+\frac{l_p(\log p)^{r_1}(\log b)^{r_2}}{\Gamma(1+r_1)\Gamma(1+r_2)}\right)\|v-w\|_p\,.$$

By (8.17) we conclude that $B$ is a contraction.

*Step 3.* The set $\mathcal{E} := \{u \in C(J) : u = \lambda A(u) + \lambda B(\frac{u}{\lambda}), \lambda \in (0,1)\}$ is bounded. Let $u \in C$ such that $u = \lambda A(u) + \lambda B(\frac{u}{\lambda})$ for some $\lambda \in (0,1)$. Then for any $p \in \mathbb{N}\backslash\{0,1\}$ and each $(t,x) \in J_p$ we have

$$
\begin{aligned}
|u(t,x)| &\le \lambda|A(u)| + \lambda|B\left(\frac{u}{\lambda}\right)| \\
&\le |\mu(t,x)| + |f(t,x,0,0)| + k(t,x) + l(t,x) \\
&\quad + \frac{1}{\Gamma(r_1)\Gamma(r_2)} \int_1^t \int_1^x \left|\log\frac{t}{s}\right|^{r_1-1} \left|\log\frac{x}{y}\right|^{r_2-1} \\
&\quad \times \frac{P(t,x,s,y) + Q(t,x,s,y)}{sy} dyds \\
&\le \mu_p + f_p + k_p + l_p + P_p + Q_p ,
\end{aligned}
$$

where

$$
\mu_p = \sup_{(t,x)\in[1,p]\times[1,b]} \mu(t,x), \quad f_p = \sup_{(t,x)\in[1,p]\times[1,b]} |f(t,x,0,0)|; \; p \in \mathbb{N}\backslash\{0,1\} .
$$

Thus,

$$
\|u\|_p \le \mu_p + f_p + k_p + l_p + P_p + Q_p =: \ell_p^* .
$$

Hence, the set $\mathcal{E}$ is bounded.

As a consequence of Steps 1–3 and from an application of Theorem 1.43, we deduce that $N$ has a fixed point $u$ that is a solution of integral equation (8.16).

*Step 4. The generalized Ulam–Hyers–Rassias stability.* Set

$$
P_{1p} = \sup_{(s,y)\in J_p} P_1(s,y), \text{ and } Q_{1p} = \sup_{(s,y)\in J_p} Q_1(s,y) .
$$

Let $u$ be a solution of inequality (8.18) and $v$ be a solution of equation (8.16). Then

$$
\begin{aligned}
v(t,x) &= \mu(t,x) + f(t,x,{}^H I_\sigma^r v(t,x), v(t,x)) \\
&\quad + \frac{1}{\Gamma(r_1)\Gamma(r_2)} \int_1^t \int_1^x \left(\log\frac{t}{s}\right)^{r_1-1} \left(\log\frac{x}{y}\right)^{r_2-1} \\
&\quad \times g(t,x,s,y,v(s,y)) \frac{dyds}{sy} , \quad (t,x) \in J := [1,+\infty) \times [1,b] .
\end{aligned}
$$

From inequality (8.18) and condition (8.4.3), for each $(t,x) \in J_p$, we have

$$
\begin{aligned}
|u(t,x) - v(x,y)| &\le |u(t,x) - (Nu)(t,x)| + |(Nu)(t,x) - (Nv)(t,x)| \\
&\le \phi(x,y) + |f(t,x,({}^H I_\sigma^r u)(t,x,),u(t,x)) - f(t,x,{}^H I_\sigma^r v(t,x),v(t,x))| \\
&\quad + \frac{1}{\Gamma(r_1)\Gamma(r_2)} \int_1^t \int_1^x \left|\log \frac{t}{s}\right|^{r_1-1} \left|\log \frac{x}{y}\right|^{r_2-1} \\
&\qquad \times |g(t,x,s,y,u(s,y)) - g(t,x,s,y,v(s,y))| \frac{dyds}{sy} \\
&\le \phi(x,y) + l(t,x)|({}^H I_\sigma^r u)(t,x) - {}^H I_\sigma^r v(t,x)| + k(t,x)|u(t,x) - v(t,x)| \\
&\quad + \frac{2\phi(t,x)}{\Gamma(r_1)\Gamma(r_2)} \int_1^t \int_1^x \left|\log \frac{t}{s}\right|^{r_1-1} \left|\log \frac{x}{y}\right|^{r_2-1} \phi(t,x) \\
&\qquad \times (P_1(s,y) + Q_1(s,t)) dyds \\
&\le \phi(x,y) + \ell_p |u(t,x) - v(t,x)| \\
&\quad + \frac{2\phi(t,x)}{\Gamma(r_1)\Gamma(r_2)} \int_1^t \int_1^x \left|\log \frac{t}{s}\right|^{r_1-1} \left|\log \frac{x}{y}\right|^{r_2-1} (P_{1p} + Q_{1p}) dyds \\
&\le \phi(t,x) + \ell_p |u(t,x) - v(t,x)| \\
&\quad + \frac{2(P_{1p} + Q_{1p})\phi(t,x)}{\Gamma(r_1)\Gamma(r_2)} \int_1^t \int_1^x \left|\log \frac{t}{s}\right|^{r_1-1} \left|\log \frac{x}{y}\right|^{r_2-1} dyds\,.
\end{aligned}
$$

Thus, for each $(t,x) \in J_p$ we obtain

$$
\begin{aligned}
|u(t,x) - v(x,y)| &\le \frac{\phi(t,x)}{1-\ell_p} \left( 1 + \frac{2(P_{1p} + Q_{1p})}{\Gamma(r_1)\Gamma(r_2)} \int_1^t \int_1^x \left|\log \frac{t}{s}\right|^{r_1-1} \left|\log \frac{x}{y}\right|^{r_2-1} dyds \right) \\
&\le \frac{1}{1-\ell_p} \left( 1 + \frac{2(P_{1p} + Q_{1p})(\log p)^{r_1} (\log b)^{r_2}}{\Gamma(1+r_1)\Gamma(1+r_2)} \right) \phi(t,x) \\
&:= c_{N,\phi} \phi(t,x)\,.
\end{aligned}
$$

Hence, for each $(t,x) \in J_p$ we get

$$
|u(t,x) - v(x,y)| \le c_{N,\phi} \phi(x,y)\,.
$$

Consequently, equation (8.16) is generalized Ulam–Hyers–Rassias stable. □

### 8.4.3 An Example

Consider the Hadamard fractional order integral equation

$$u(t,x)=\frac{xe^{3-2t}}{1+t+x^2}+\frac{xe^{-t-2}}{c_p(1+e^{-2p}|({}^HI^r_\sigma u)(t,x)|+e^{-p}|u(t,x)|)}$$
$$+\int_1^t\int_1^x\left(\log\frac{t}{s}\right)^{r_1-1}\left(\log\frac{x}{y}\right)^{r_2-1}\frac{g(t,x,s,y,u(s,y))}{\Gamma(r_1)\Gamma(r_2)}dyds\,,$$
$$(t,x)\in[1,+\infty)\times[1,e]\,, \tag{8.21}$$

where $c_p=e^{-p}+\frac{e^{-2p}p^{r_1}}{\Gamma(1+r_1)\Gamma(1+r_2)}$; $p\in\mathbb{N}\backslash\{0,1\}$, $r=(r_1,r_2)\in(0,\infty)\times(0,\infty)$,

$$g(t,x,s,y,u)=\frac{xs^{\frac{-3}{4}}(1+|u|)\sin\sqrt{t}\sin s}{(1+x^2+t^2)(1+|u|)}\quad\text{if } (t,x,s,y)\in J' \text{ and } u\in\mathbb{R}\,,$$

and

$$J'=\{(t,x,s,y):\ 1\le s\le t\ \text{ and }\ 1\le x\le y\le e\}\,.$$

Set

$$\mu(t,x)=\frac{xe^{3-2t}}{1+t+x^2},\ f(t,x,u,v)=\frac{xe^{-t-2}}{c_p(1+e^{-2p}|u|+e^{-p}|v|)};\quad p\in\mathbb{N}\backslash\{0,1\}\,.$$

The function $f$ is continuous and satisfies (8.4.1), with $k(t,x)=\frac{xe^{-t-2-p}}{c_p}$, $l(t,x)=\frac{xe^{-t-2-2p}}{c_p}$, $k_p=\frac{e^{-2-p}}{c_p}$, and $l_p=\frac{e^{-2-2p}}{c_p}$. Additionally, the function $g$ is continuous and satisfies (8.4.2), with

$$P(t,x,s,y)=Q(t,x,s,y)=\frac{xs^{\frac{-3}{4}}\sin\sqrt{t}\sin s}{1+x^2+t^2};\quad (t,x,s,y)\in J'\,.$$

Further, the function $g$ is continuous and satisfies (8.4.3), with

$$P_1(s,y)=Q(s,y)=s^{\frac{-3}{4}}\sin s\,,\quad P_{1p}=Q_{1p}=p^{\frac{-3}{4}}$$

and

$$\phi(t,x)=\frac{x\sin\sqrt{t}}{1+x^2+t^2}\,.$$

Finally, we will show that condition (8.17) holds with $b=e$. Indeed, for each $p\in\mathbb{N}\backslash\{0,1\}$ we get

$$k_p+\frac{l_p(\log p)^{r_1}(\log b)^{r_2}}{\Gamma(1+r_1)\Gamma(1+r_2)}=\frac{1}{c_p}\left(e^{-2-p}+\frac{e^{-2-2p}p^{r_1}}{\Gamma(1+r_1)\Gamma(1+r_2)}\right)=e^{-2}<1\,.$$

Hence, by Theorem 8.9, equation (8.21) has a solution defined on $[1,+\infty)\times[1,e]$ and (8.21) is generalized Ulam–Hyers–Rassias stable.

## 8.5 Ulam Stability Results for Hadamard Partial Fractional Integral Inclusions via Picard Operators

### 8.5.1 Introduction

In this section, using weakly Picard operators theory, we investigate some existence results and Ulam-type stability concepts of fixed point inclusions due to Rus for a class of partial Hadamard fractional integral inclusions.

In [47, 37, 39], Abbas et al. studied some Ulam stabilities for functional fractional partial differential and integral inclusions via Picard operators. In this section, we discuss the Ulam–Hyers and the Ulam–Hyers–Rassias stability for the new class of fractional partial integral inclusions

$$u(x,y)-\mu(x,y)\in({}^{H}I_{\sigma}^{r}F)(x,y,u(x,y))\,;\quad (x,y)\in J:=[1,a]\times[1,b]\,, \tag{8.22}$$

where $a,b>1$, $\sigma=(1,1)$, $F\colon J\times E\to\mathcal{P}(E)$ is a set-valued function with nonempty values in a (real or complex) separable Banach space $E$, $\mathcal{P}(E)$ is the family of all nonempty subsets of $E$, and $\mu\colon J\to E$ is a given continuous function.

### 8.5.2 Picard Operators Theory

In what follows we will give some basic definitions and results on Picard operators [228, 229]. Let $(X,d)$ be a metric space and $A\colon X\to X$ an operator. We denote by $F_A$ the set of the fixed points of $A$. We denote by $A^0:=1_X,\ A^1:=A,\dots,A^{n+1}:=A^n\circ A;\ n\in\mathbb{N}$ the iterate operators of the operator $A$.

**Definition 8.10.** The operator $A\colon X\to X$ is a Picard operator (PO) if there exists $x^*\in X$ such that
(i) $F_A=\{x^*\}$,
(ii) The sequence $(A^n(x_0))_{n\in\mathbb{N}}$ converges to $x^*$ for all $x_0\in X$.

**Definition 8.11.** The operator $A\colon X\to X$ is a weakly Picard operator (WPO) if the sequence $(A^n(x))_{n\in\mathbb{N}}$ converges for all $x\in X$ and its limit (which may depend on $x$) is a fixed point of $A$.

**Definition 8.12.** If $A$ is a WPO, then we consider the operator $A^\infty$ defined by

$$A^\infty\colon X\to X\,;\quad A^\infty(x)=\lim_{n\to\infty}A^n(x)\,.$$

**Remark 8.13.** It is clear that $A^\infty(X)=F_A$.

**Definition 8.14.** Let $A$ be a WPO and $c>0$. The operator $A$ is a c-weakly Picard operator if

$$d(x,A^\infty(x))\le c\,d(x,A(x))\,;\quad x\in X\,.$$

In the multivalued case we have the following concepts (see [218, 235]).

**Definition 8.15.** Let $(X, d)$ be a metric space and $F: X \to \mathcal{P}_{cl}(X)$ a multivalued operator. By definition, $F$ is a multivalued weakly Picard operator (MWPO) if for each $u \in X$ and each $v \in F(x)$ there exists a sequence $(u_n)_{n\in\mathbb{N}}$ such that

(i) $u_0 = u, u_1 = v$,
(ii) $u_{n+1} \in F(u_n)$ for each $n \in \mathbb{N}$,
(iii) the sequence $(u_n)_{n\in\mathbb{N}}$ is convergent and its limit is a fixed point of $F$.

**Remark 8.16.** A sequence $(u_n)_{n\in\mathbb{N}}$ satisfying conditions (i) and (ii) in the preceding definition is called a sequence of successive approximations of $F$ starting from $(x, y) \in$ Graph$(F)$.

If $F: X \to \mathcal{P}_{cl}(X)$ is a MWPO, then we define $F_1$: Graph$(F) \to \mathcal{P}(Fix(F))$ by the formula $F_1(x, y) :=$ {$u \in$ Fix$(F)$: there exists a sequence of successive approximations of $F$ starting from $(x, y)$ that converges to $u$}.

**Definition 8.17.** Let $(X, d)$ be a metric space, and let $\Psi: [0, \infty) \to [0, \infty)$ be an increasing function that is continuous at 0 and $\Psi(0) = 0$. Then $F: X \to \mathcal{P}_{cl}(X)$ is said to be a multivalued $\Psi$-weakly Picard operator ($\Psi$-MWPO) if it is a MWPO and there exists a selection $A^\infty$: Graph$(F) \to$ Fix$(F)$ of $F^\infty$ such that

$$d(u, A^\infty(u, v)) \le \Psi(d(u, v))\,; \quad \text{for all } (u, v) \in \mathrm{Graph}(F)\,.$$

If there exists $c > 0$ such that $\Psi(t) = ct$ for each $t \in [0, \infty)$, then $F$ is called a multivalued c-weakly Picard operator ($c$-MWPO).

Let us recall the notion of comparison.

**Definition 8.18.** A function $\varphi: [0, \infty) \to [0, \infty)$ is said to be a comparison function (see [228]) if it is increasing and $\varphi^n \to 0$ as $n \to \infty$.

As a consequence, we have $\varphi(t) < t$ for each $t > 0$, $\varphi(0) = 0$, and $\varphi$ is continuous at 0.

**Definition 8.19.** A function $\varphi: [0, \infty) \to [0, \infty)$ is said to be a strict comparison function (see [228]) if it is strictly increasing and $\sum_{n=1}^{\infty} \varphi^n(t) < \infty$ for each $t > 0$.

**Example 8.20.** *The mappings $\varphi_1, \varphi_2: [0, \infty) \to [0, \infty)$ given by $\varphi_1(t) = ct$, $c \in [0, 1)$, and $\varphi_2(t) = \frac{t}{1+t}$, $t \in [0, \infty)$, are strict comparison functions.*

**Definition 8.21.** A multivalued operator $N: X \to \mathcal{P}_{cl}(X)$ is called

(a) $\gamma$-Lipschitz if and only if there exists $\gamma \ge 0$ such that

$$H_d(N(u), N(v)) \le \gamma d(u, v) \quad \text{for each } u, v \in X\,,$$

(b) a multivalued $\gamma$-contraction if and only if it is $\gamma$-Lipschitz with $\gamma \in [0, 1)$,
(c) a multivalued $\varphi$-contraction if and only if there exists a strict comparison function $\varphi: [0, \infty) \to [0, \infty)$ such that

$$H_d(N(u), N(v)) \le \varphi(d(u, v)) \quad \text{for each } u, v \in X\,.$$

The following result, a generalization of the Covitz–Nadler fixed point principle (see [130]), is known in the literature as Węgrzyk's fixed point theorem.

**Lemma 8.22** ([255]). *Let $(X, d)$ be a complete metric space. If $A: X \to \mathcal{P}_{cl}(X)$ is a $\varphi$-contraction, then $Fix(A)$ is nonempty, and for any $u_0 \in X$ there exists a sequence of successive approximations of $A$ starting from $u_0$, which converges to a fixed point of $A$.*

The next result is known as Węgrzyk's theorem.

**Lemma 8.23** ([255]). *Let $(X, d)$ be a Banach space. If an operator $A: X \to \mathcal{P}_{cl}(X)$ is a $\varphi$-contraction, then $A$ is a MWPO.*

Now we present an important characterization lemma from the point of view of Ulam–Hyers stability.

**Lemma 8.24** ([217]). *Let $(X, d)$ be a metric space. If $A: X \to \mathcal{P}_{cp}(X)$ is a $\Psi$-MWPO, then the fixed point inclusion $u \in A(u)$ is generalized Ulam–Hyers stable. In particular, if $A$ is $c$-MWPO, then the fixed point inclusion $u \in A(u)$ is Ulam–Hyers stable.*

Another Ulam–Hyers stability result, more efficient for applications, was proved in [193].

**Theorem 8.25** ([193]). *Let $(X, d)$ be a complete metric space and $A: X \to \mathcal{P}_{cp}(X)$ a multivalued $\varphi$-contraction. Then:*

(i) Existence of fixed point: *$A$ is a MWPO;*
(ii) Ulam–Hyers stability for fixed point inclusion: *If additionally $\varphi(ct) \le c\varphi(t)$ for every $t \in [0, \infty)$ (where $c > 1$) and $t = 0$ is a point of uniform convergence for the series $\sum_{n=1}^{\infty} \varphi^n(t)$, then $A$ is a $\Psi$-MWPO, with $\Psi(t) := t + \sum_{n=1}^{\infty} \varphi^n(t)$, for each $t \in [0, \infty)$;*
(iii) Data dependence of fixed point set: *Let $S: X \to \mathcal{P}_{cl}(X)$ be a multivalued $\varphi$-contraction and $\eta > 0$ be such that $H_d(S(x), A(x)) \le \eta$ for each $x \in X$. Suppose that $\varphi(ct) \le c\varphi(t)$ for every $t \in [0, \infty)$ (where $c > 1$) and $t = 0$ is a point of uniform convergence for the series $\sum_{n=1}^{\infty} \varphi^n(t)$. Then $H_d(Fix(S), Fix(F)) \le \Psi(\eta)$.*

### 8.5.3 Existence and Ulam Stability Results

In this section, we present conditions for the existence and Ulam stability of Hadamard integral inclusion (8.22).

**Theorem 8.26.** *Make the following assumptions:*

(8.21.1) *$(x, y) \longmapsto F(x, y, u)$ is jointly measurable for each $u \in E$.*
(8.21.2) *$u \longmapsto F(x, y, u)$ is lower semicontinuous for almost all $(x, y) \in J$.*
(8.21.3) *There exist $p \in L^{\infty}(J, [0, \infty))$ and a strict comparison function $\varphi: [0, \infty) \to [0, \infty)$ such that for each $(x, y) \in J$ and each $u, v \in E$ we have*

$$H_d(F(x, y, u), F(x, y, \overline{u})) \le p(x, y)\varphi(\|u - \overline{u}\|_E) \tag{8.23}$$

*and*

$$\frac{(\log a)^{r_1}(\log b)^{r_2}\|p\|_{L^\infty}}{\Gamma(1+r_1)\Gamma(1+r_2)} \leq 1\,. \tag{8.24}$$

(8.21.4) *There exists an integrable function $q\colon [1,b] \to [0,\infty)$ such that for each $x \in [1,a]$ and $u \in E$ we have $F(x,y,u) \subset q(y)B(0,1)$, a.e. $y \in [1,b]$, where $B(0,1) = \{u \in E\colon \|u\|_E < 1\}$.*

*Then we have that:*

(a) *The integral inclusion (7.1) has at least one solution and N is a MWPO.*

(b) *If additionally $\varphi(ct) \leq c\varphi(t)$ for every $t \in [0,\infty)$ (where $c > 1$) and $t = 0$ is a point of uniform convergence for the series $\sum_{n=1}^{\infty}\varphi^n(t)$, then integral inclusion (7.1) is generalized Ulam–Hyers stable, and N is a $\Psi$-MWPO, with the function $\Psi$ defined by $\Psi(t) := t + \sum_{n=1}^{\infty}\varphi^n(t)$, for each $t \in [0,\infty)$. Moreover, in this case the continuous data dependence of the solution set of integral inclusion (8.23) holds.*

**Remark 8.27.** For each $u \in \mathcal{C}$, the set $S_{F\circ u}$ is nonempty since, by (8.21.1), $F$ has a measurable selection (see [121] Theorem III.6).

*Proof.* The proof will be given in two steps.

*Step 1. $N(u) \in P_{cp}(\mathcal{C})$ for each $u \in \mathcal{C}$.* From the continuity of $\mu$ and Theorem 2 in Rybiński [236] we have that for each $u \in \mathcal{C}$ there exists $f \in S_{F\circ u}$, for all $(x,y) \in J$, such that $f(x,y)$ is integrable with respect to $y$ and continuous with respect to $x$. Then the function $v(x,y) = \mu(x,y) + {}^H I_\sigma^r f(x,y)$ has the property $v \in N(u)$. Moreover, from (8.21.1) and (8.21.4), via Theorem 8.6.3. in Aubin and Frankowska [69], we get that $N(u)$ is a compact set for each $u \in \mathcal{C}$.

*Step 2. $H_d(N(u), N(\overline{u})) \leq \varphi(\|u-\overline{u}\|_\infty)$ for each $u, \overline{u} \in \mathcal{C}$.* Let $u, \overline{u} \in \mathcal{C}$ and $h \in N(u)$. Then there exists $f(x,y) \in F(x,y,u(x,y))$ such that for each $(x,y) \in J$ we have

$$h(x,y) = \mu(x,y) + {}^H I_\sigma^r f(x,y)\,.$$

From (8.21.3) it follows that

$$H_d(F(x,y,u(x,y)), F(x,y,\overline{u}(x,y))) \leq p(x,y)\varphi(\|u(x,y)-\overline{u}(x,y)\|_E)\,.$$

Hence, there exists $w(x,y) \in F(x,y,\overline{u}(x,y)$ such that

$$\|f(x,y) - w(x,y)\|_E \leq p(x,y)\varphi(\|u(x,y)-\overline{u}(x,y)\|_E)\,; \quad (x,y) \in J\,.$$

Consider $U\colon J \to \mathcal{P}(E)$ given by

$$U(x,y) = \{w \in E\colon \|f(x,y) - w(x,y)\|_E \leq p(x,y)\varphi(\|u(x,y)-\overline{u}(x,y)\|_E)\}\,.$$

Since the multivalued operator $u(x,y) = U(x,y) \cap F(x,y,\overline{u}(x,y))$ is measurable (see Proposition III.4 in [121]), there exists a function $\overline{f}(x,y)$ that is a measurable selection for $u$. Thus, $\overline{f}(x,y) \in F(x,y,\overline{u}(x,y))$, and for each $(x,y) \in J$,

$$\|f(x,y) - \overline{f}(x,y)\|_E \leq p(x,y)\varphi(\|u(x,y)-\overline{u}(x,y)\|_E)\,.$$

Let us define for each $(x, y) \in J$

$$\overline{h}(x, y) = \mu(x, y) + {}^H I_\sigma^r \overline{f}(x, y) .$$

Then for each $(x, y) \in J$ we have

$$\begin{aligned}\|h(x, y) - \overline{h}(x, y)\|_E &\leq {}^H I_\sigma^r \|f(x, y) - \overline{f}(x, y)\|_E \\ &\leq {}^H I_\sigma^r (p(x, y)\varphi(\|u(x, y) - \overline{u}(x, y)\|_E)) \\ &\leq \|p\|_{L^\infty}\varphi(\|u - \overline{u}\|_\infty)\left(\int_1^x\int_1^y \frac{|\log \frac{x}{s}|^{r_1-1} |\log \frac{y}{t}|^{r_2-1}}{st\Gamma(r_1)\Gamma(r_2)} dtds\right) \\ &\leq \frac{(\log a)^{r_1}(\log b)^{r_2}\|p\|_{L^\infty}}{\Gamma(1 + r_1)\Gamma(1 + r_2)}\varphi(\|u - \overline{u}\|_\infty) .\end{aligned}$$

Thus, by (10.2), we get

$$\|h - \overline{h}\|_\infty \leq \varphi(\|u - \overline{u}\|_\infty) .$$

By an analogous relation, obtained by interchanging the roles of $u$ and $\overline{u}$, it follows that

$$H_d(N(u), N(\overline{u})) \leq \varphi(\|u - \overline{u}\|_\infty) .$$

Hence, $N$ is a $\varphi$-contraction.

(a) By Lemma 8.22, $N$ has a fixed point that is a solution of inclusion (7.1) on $J$, and by [Theorem 8.25 (i)], $N$ is a MWPO.

(b) We will prove that the fixed point inclusion problem (7.1) is generalized Ulam–Hyers stable. Indeed, let $\epsilon > 0$ and $v \in \mathcal{C}$ for which there exists $u \in \mathcal{C}$ such that

$$u(x, y) \in \mu(x, y) + ({}^H I_\sigma^r F)(x, y, v(x, y)) , \quad \text{if } (x, y) \in J ,$$

and

$$\|u - v\|_\infty \leq \epsilon .$$

Then $H_d(v, N(v)) \leq \epsilon$. Moreover, by the preceding proof we have that $N$ is a multivalued $\varphi$-contraction, and using [Theorem 8.25 (i)-(ii)], we obtain that $N$ is a $\Psi$-MWPO. Then, by Lemma 8.24, we obtain that the fixed point problem $u \in N(u)$ is generalized Ulam–Hyers stable. Thus, integral inclusion (8.22) is generalized Ulam–Hyers stable.

Concerning the conclusion of the theorem, we apply [Theorem 8.25 (iii)]. □

### 8.5.4 An Example

Let

$$E = l^1 = \left\{w = (w_1, w_2, \ldots, w_n, \ldots) : \sum_{n=1}^{\infty} |w_n| < \infty\right\}$$

be a Banach space with norm

$$\|w\|_E = \sum_{n=1}^{\infty} |w_n| ,$$

and consider the partial functional fractional order integral inclusion of the form

$$u(x,y) \in \mu(x,y) + ({}^H I_\sigma^r F)(x,y,u(x,y)) , \quad \text{a.e. } (x,y) \in [1,e] \times [1,e] , \tag{8.25}$$

where $r = (r_1, r_2), r_1, r_2 \in (0, \infty)$,

$$u = (u_1, u_2, \ldots, u_n, \ldots) , \quad \mu(x,y) = (x + e^{-y}, 0, \ldots, 0, \ldots) ,$$

and

$$\begin{aligned} &F(x,y,u(x,y)) \\ &= \{v \in C([1,e] \times [1,e], \mathbb{R}) : \|f_1(x,y,u(x,y))\|_E \le \|v\|_E \le \|f_2(x,y,u(x,y))\|_E\} , \end{aligned}$$

$(x,y) \in [1,e] \times [1,e]$, where $f_1, f_2 : [1,e] \times [1,e] \times E \to E$,

$$f_k = (f_{k,1}, f_{k,2}, \ldots, f_{k,n}, \ldots) ; \quad k \in \{1,2\}, \ n \in \mathbb{N} ,$$

$$f_{1,n}(x,y,u_n(x,y)) = \frac{xy^2 u_n}{(1 + \|u_n\|_E) e^{10+x+y}} , \quad n \in \mathbb{N} ,$$

and

$$f_{2,n}(x,y,u_n(x,y)) = \frac{xy^2 u_n}{e^{10+x+y}} ; \quad n \in \mathbb{N} .$$

We assume that $F$ is closed and convex valued. We can see that the solutions of the inclusion (7.4) are solutions of the fixed point inclusion $u \in A(u)$, where $A : C([1,e] \times [1,e], \mathbb{R}) \to \mathcal{P}(C([1,e] \times [1,e], \mathbb{R}))$ is the multifunction operator defined by

$$(Au)(x,y) = \left\{\mu(x,y) + ({}^H I_\sigma^r f)(x,y);\ f \in S_{F \circ u}\right\} ; \quad (x,y) \in [1,e] \times [1,e] .$$

For each $(x,y) \in [1,e] \times [1,e]$ and all $z_1, z_2 \in E$ we have

$$\|f_2(x,y,z_2) - f_1(x,y,z_1)\|_E \le xy^2 e^{-10-x-y} \|z_2 - z_1\|_E .$$

Thus, conditions (8.21.1)–(8.21.3) are satisfied by $p(x,y) = xy^2 e^{-10-x-y}$. Condition (10.2) holds with $a = b = e$. Indeed, $\|p\|_{L^\infty} = e^{-9}$, $\Gamma(1+r_i) > \frac{1}{2}$; $i = 1, 2$. A simple computation shows that

$$\zeta := \frac{(\log a)^{r_1} (\log b)^{r_2} \|p\|_{L^\infty}}{\Gamma(1+r_1)\Gamma(1+r_2)} < 4e^{-9} < 1 .$$

Condition (8.21.4) is satisfied by $q(y) = \frac{y^2 e^{-10-y}}{\|F\|_{\mathcal{P}}}$; $y \in [1,e]$, where

$$\|F\|_{\mathcal{P}} = \sup\{\|f\|_{\mathcal{C}} : f \in S_{F \circ u}\} ; \quad \text{for all } u \in \mathcal{C} .$$

Consequently, by Theorem 8.26, we draw the following conclusions:
(a) Integral inclusion (8.25) has least one solution and $A$ is a (MWPO).

(b) The function $\varphi\colon [0,\infty) \to [0,\infty)$ defined by $\varphi(t) = \zeta t$ satisfies $\varphi(\zeta t) \le \zeta\varphi(t)$ for every $t \in [0,\infty)$ and $t = 0$ is a point of uniform convergence for the series $\sum_{n=1}^{\infty}(\zeta t)^n$. Then the integral inclusion (7.4) is generalized Ulam–Hyers stable, and $A$ is a $\Psi$-MWPO, with the function $\Psi$ defined by $\Psi(t) := t + (1 - \zeta t)^{-1}$ for each $t \in [0, \zeta^{-1})$. Moreover, the continuous data dependence of the solution set of integral inclusion (8.23) holds.

## 8.6 Notes and Remarks

The results of Chapter 8 are taken from Abbas et al. [2, 10, 9, 11]. Other results may be found in [24, 25, 22, 31, 153].

# 9 Hadamard–Stieltjes Fractional Integral Equations

## 9.1 Introduction

If $u$ is a real function defined on the interval $[a, b]$, then the symbol $\bigvee_a^b u$ denotes the variation of $u$ on $[a, b]$. We say that $u$ is of bounded variation on the interval $[a, b]$ whenever $\bigvee_a^b u$ is finite. If $w\colon [a, b] \times [c, b] \to \mathbb{R}$, then the symbol $\bigvee_{t=p}^q w(t, s)$ indicates the variation of the function $t \to w(t, s)$ on the interval $[p, q] \subset [a, b]$, where $s$ is arbitrarily fixed in $[c, d]$. In the same way we define $\bigvee_{s=p}^q w(t, s)$. For the properties of functions of bounded variation we refer the reader to [204].

If $u$ and $\varphi$ are two real functions defined on the interval $[a, b]$, then under some conditions (see [204]) we can define the Stieltjes integral (in the Riemann–Stieltjes sense)

$$\int_a^b u(t)d\varphi(t)$$

of the function $u$ with respect to $\varphi$. In this case we say that $u$ is Stieltjes integrable on $[a, b]$ with respect to $\varphi$. Several conditions are known guaranteeing Stieltjes integrability [204]. One of the most frequently used requires that $u$ be continuous and $\varphi$ be of bounded variation on $[a, b]$.

If $u$ is Stieltjes integrable on the interval $[a, b]$ with respect to a function $\varphi$ of bounded variation, then

$$\left|\int_a^b u(t)d\varphi(t)\right| \le \int_a^b |u(t)|d\left(\bigvee_a^t \varphi\right).$$

If $u$ and $v$ are Stieltjes integrable functions on the interval $[a, b]$ with respect to a nondecreasing function $\varphi$ such that $u(t) \le v(t)$ for $t \in [a, b]$. Then

$$\int_a^b u(t)d\varphi(t) \le \int_a^b v(t)d\varphi(t).$$

In the sequel we consider Stieltjes integrals of the form

$$\int_a^b u(t)d_s g(t, s)$$

and Hadamard–Stieltjes integrals of fractional order

$$\frac{1}{\Gamma(r)}\int_1^t \left(\log\frac{t}{s}\right)^{q-1} u(s)d_s g(t, s),$$

where $g\colon [1, \infty) \times [1, \infty) \to \mathbb{R}$, $q \in (0, \infty)$, and the symbol $d_s$ indicates the integration with respect to $s$.

https://doi.org/10.1515/9783110553819-009

**Definition 9.1.** Let $r_1, r_2 \geq 0$, $\sigma = (1,1)$ and $r = (r_1, r_2)$. For $w \in L^1(J, \mathbb{R})$ define the Hadamard–Stieltjes partial fractional integral of order $r$ by the expression

$$({}^{HS}I_\sigma^r w)(x,y) = \frac{1}{\Gamma(r_1)\Gamma(r_2)} \int_1^x \int_1^y \left(\log\frac{x}{s}\right)^{r_1-1} \left(\log\frac{y}{t}\right)^{r_2-1} \frac{w(s,t)}{st} d_t g_2(y,t) d_s g_1(x,s) ,$$

where $g_1, g_2 : [1,\infty) \times [1,\infty) \to \mathbb{R}$.

## 9.2 Existence and Stability of Solutions for Hadamard–Stieltjes Fractional Integral Equations

### 9.2.1 Introduction

We give some existence results and Ulam stability results for a class of Hadamard–Stieltjes integral equations. We present two results, the first one an existence result based on Schauder's fixed point theorem, and the second one about the generalized Ulam–Hyers–Rassias stability.

This section deals with the existence of the Ulam stability of solutions to the Hadamard–Stieltjes fractional integral equation

$$\begin{aligned} u(x,y) &= \mu(x,y) \\ &+ \int_1^x \int_1^y \left(\log\frac{x}{s}\right)^{r_1-1} \left(\log\frac{y}{t}\right)^{r_2-1} \frac{f(s,t,u(s,t))}{st\Gamma(r_1)\Gamma(r_2)} d_t g_2(y,t) d_s g_1(x,s) ; \quad \text{if } (x,y) \in J , \end{aligned} \tag{9.1}$$

where $J := [1,a] \times [1,b]$, $a, b > 1$, $r_1, r_2 > 0$, and $\mu : J \to \mathbb{R}$, $f : J \times \mathbb{R} \to \mathbb{R}$, $g_1 : [1,a]^2 \to \mathbb{R}$, and $g_2 : [1,b]^2 \to \mathbb{R}$ are given continuous functions.

Our investigations are conducted with an application of Schauder's fixed point theorem for the existence of solutions of integral equation (9.1). Also, we obtain some results on the generalized Ulam–Hyers–Rassias stability of solutions of (9.1). Finally, we present an example illustrating the applicability of the imposed conditions.

### 9.2.2 Existence and Ulam Stabilities Results

In this section, we discuss the existence of solutions and present conditions for the Ulam stability for the Hadamard integral equation (7.1).

The following conditions will be used in the sequel:

(9.2.1) There exist functions $p_1, p_2 \in C(J, \mathbb{R}_+)$ such that for any $u \in \mathbb{R}$ and $(x,y) \in J$

$$|f(x,y,u)| \leq p_1(x,y) + \frac{p_2(x,y)}{1+|u(x,y)|}|u(x,y)| ,$$

with

$$p_i^* = \sup_{(x,y)\in J} \sup_{(s,t)\in[1,x]\times[1,y]} \left|\log \frac{x}{s}\right|^{r_1-1} \left|\log \frac{y}{t}\right|^{r_2-1} \frac{p_i(s,t)}{st\Gamma(r_1)\Gamma(r_2)}; \quad i = 1, 2 .$$

(9.2.2) For all $x_1, x_2 \in [1, a]$ such that $x_1 < x_2$ the function $s \mapsto g(x_2, s) - g(x_1, s)$ is nondecreasing on $[1, a]$. Also, for all $y_1, y_2 \in [1, b]$ such that $y_1 < y_2$ the function $s \mapsto g(y_2, t) - g(y_1, t)$ is nondecreasing on $[1, b]$.

(9.2.3) The functions $s \mapsto g_1(0, s)$ and $t \mapsto g_2(0, t)$ are nondecreasing on $[1, a]$ and $[1, b]$, respectively.

(9.2.4) The functions $s \mapsto g_1(x, s)$ and $x \mapsto g_1(x, s)$ are continuous on $[1, a]$ for each fixed $x \in [1, a]$ and $s \in [1, a]$, respectively. Also, the functions $t \mapsto g_2(y, t)$ and $y \mapsto g_2(y, t)$ are continuous on $[1, b]$ for each fixed $y \in [1, b]$ or $t \in [1, b]$, respectively.

(9.2.5) There exists $\lambda_\Phi > 0$ such that for each $(x, y) \in J$ we have

$$({}^{HS}I_\sigma^r \Phi)(x, y) \le \lambda_\Phi \Phi(x, y) .$$

Set

$$g^* = \sup_{(x,y)\in J} \bigvee_{k_2=1}^{y} g_2(y, k_2) \bigvee_{k_1=1}^{x} g_1(x, k_1) .$$

**Theorem 9.2.** *Assume* (9.2.1)–(9.2.4)*; then integral equation* (9.1) *has a solution defined on* $J$.

*Proof.* Let $\rho > 0$ be a constant such that

$$\rho > \|\mu\|_\infty + g^*(p_1^* + p_2^*) .$$

We will use Schauder's fixed point theorem [149] to prove that the operator $N\colon C \to C$ defined by

$$(Nu)(x, y) = \mu(x, y) + \int_1^x \int_1^y \left(\log \frac{x}{s}\right)^{r_1-1} \left(\log \frac{y}{t}\right)^{r_2-1} \frac{f(s, t, u(s, t))}{st\Gamma(r_1)\Gamma(r_2)} d_t g_2(y, t) d_s g_1(x, s) \tag{9.2}$$

has a fixed point. The proof will be given in four steps.

*Step 1: N transforms the ball* $B_\rho := \{u \in \mathcal{C}\colon \|u\|_C \le \rho\}$ *into itself.* For any $u \in B_\rho$ and each $(x, y) \in J$ we have

$$\begin{aligned} |(Nu)(x, y)| \le |\mu(x, y)| &+ \frac{1}{\Gamma(r_1)\Gamma(r_2)} \int_1^x \int_1^y \left|\log \frac{x}{s}\right|^{r_1-1} \left|\log \frac{y}{t}\right|^{r_2-1} \\ &\times \frac{p_1(s, t)}{st} |d_t g_2(y, t) d_s g_1(x, s)| \\ &+ \frac{1}{\Gamma(r_1)\Gamma(r_2)} \int_1^x \int_1^y \left|\log \frac{x}{s}\right|^{r_1-1} \left|\log \frac{y}{t}\right|^{r_2-1} \end{aligned}$$

$$\begin{aligned}
&\times \frac{p_2(s,t)|u(s,t)|}{st(1+|u(s,t)|)}|d_t g_2(y,t) d_s g_1(x,s)| \\
&\leq \|\mu\|_C + \frac{1}{\Gamma(r_1)\Gamma(r_2)} \int_1^x \int_1^y \left|\log \frac{x}{s}\right|^{r_1-1} \left|\log \frac{y}{t}\right|^{r_2-1} \\
&\times \frac{p_1(s,t)+p_2(s,t)\rho}{st} d_t \bigvee_{k_2=1}^{t} g_2(y,k_2) d_s \bigvee_{k_1=1}^{s} g_1(x,k_1) \\
&\leq \|\mu\|_C + (p_1^* + p_2^*) \int_1^x \int_1^y d_t \bigvee_{k_2=1}^{t} g_2(y,k_2) d_s \bigvee_{k_1=1}^{s} g_1(x,k_1) \\
&\leq \|\mu\|_C + g^*(p_1^* + p_2^*) \\
&\leq \rho\,.
\end{aligned}$$

Thus, $\|(Nu)\|_C \leq \rho$. This implies that $N$ transforms the ball $B_\rho$ into itself.

*Step 2: $N\colon B_\rho \to B_\rho$ is continuous.* Let $\{u_n\}_{n\in\mathbb{N}}$ be a sequence such that $u_n \to u$ in $B_\rho$. Then

$$\begin{aligned}
|(Nu_n)(x,y) - (Nu)(x,y)| &\leq \frac{1}{\Gamma(r_1)\Gamma(r_2)} \int_1^x \int_1^y \left|\log \frac{x}{s}\right|^{r_1-1} \left|\log \frac{y}{t}\right|^{r_2-1} \\
&\quad \times \frac{|f(s,t,u_n(s,t)) - f(s,t,u(s,t))|}{st} d_t g_2(y,t) d_s g_1(x,s) \\
&\leq \frac{\sup_{(s,t)\in J} |f(s,t,u_n(s,t)) - f(s,t,u(s,t))|}{\Gamma(r_1)\Gamma(r_2)} \\
&\quad \times \int_1^x \int_1^y \left|\log \frac{x}{s}\right|^{r_1-1} \left|\log \frac{y}{t}\right|^{r_2-1} d_t \bigvee_{k_2=1}^{t} g_2(y,k_2) d_s \bigvee_{k_1=1}^{s} g_1(x,k_1) \\
&\leq g^* \|f(.,.,u_n(.,.)) - f(.,.,u(.,.))\|_C\,.
\end{aligned}$$

From Lebesgue's dominated convergence theorem and the continuity of the function $f$ we get

$$|(Nu_n)(x,y) - (Nu)(x,y)| \to 0 \text{ as } n \to \infty\,.$$

*Step 3: $N(B_\rho)$ is bounded.* This is clear since $N(B_\rho) \subset B_\rho$ and $B_\rho$ is bounded.

*Step 4: $N(B_\rho)$ is equicontinuous.* Let $(x_1,y_1), (x_2,y_2) \in J$, $x_1 < x_2$, $y_1 < y_2$. Then

$$\begin{aligned}
&|(Nu)(x_2,y_2) - (Nu)(x_1,y_1)| \leq |\mu(x_1,y_1) - \mu(x_2,y_2)| \\
&+ \left| \frac{1}{\Gamma(r_1)\Gamma(r_2)} \int_1^{x_1} \int_1^{y_1} \left|\log \frac{x_2}{s}\right|^{r_1-1} \left|\log \frac{y_2}{t}\right|^{r_2-1} \times \frac{f(s,t,u(s,t))}{st} d_t g_2(y_2,t) d_s g_1(x_2,s) \right. \\
&\left. - \frac{1}{\Gamma(r_1)\Gamma(r_2)} \int_1^{x_1} \int_1^{y_1} \left|\log \frac{x_1}{s}\right|^{r_1-1} \left|\log \frac{y_1}{t}\right|^{r_2-1} \times \frac{f(s,t,u(s,t))}{st} d_t g_2(y_1,t) d_s g_1(x_1,s) \right|
\end{aligned}$$

$$+\frac{1}{\Gamma(r_1)\Gamma(r_2)}\int_{x_1}^{x_2}\int_{y_1}^{y_2}\left|\log\frac{x_2}{s}\right|^{r_1-1}\left|\log\frac{y_2}{t}\right|^{r_2-1}\frac{|f(s,t,u(s,t))|}{st}|d_t g_2(y_2,t)d_s g_1(x_2,s)|$$
$$+\frac{1}{\Gamma(r_1)\Gamma(r_2)}\int_{1}^{x_1}\int_{y_1}^{y_2}\left|\log\frac{x_2}{s}\right|^{r_1-1}\left|\log\frac{y_2}{t}\right|^{r_2-1}\frac{|f(s,t,u(s,t))|}{st}|d_t g_2(y_2,t)d_s g_1(x_2,s)|$$
$$+\frac{1}{\Gamma(r_1)\Gamma(r_2)}\int_{x_1}^{x_2}\int_{1}^{y_1}\left|\log\frac{x_2}{s}\right|^{r_1-1}\left|\log\frac{y_2}{t}\right|^{r_2-1}\frac{|f(s,t,u(s,t))|}{st}|d_t g_2(y_2,t)d_s g_1(x_2,s)|\,.$$

Thus, we obtain

$$\begin{aligned}|(Nu)(x_2,y_2)-(Nu)(x_1,y_1)| \le{}& |\mu(x_1,y_1)-\mu(x_2,y_2)|\\ &+(p_1^*+p_2^*)\int_1^{x_1}\int_1^{y_1}\left|d_t\bigvee_{k_2=1}^{t} g_2(y_2,k_2)d_s\bigvee_{k_1=1}^{s} g_1(x_2,k_1)\right.\\ &\left.-d_t\bigvee_{k_2=1}^{t} g_2(y_1,k_2)d_s\bigvee_{k_1=1}^{s} g_1(x_1,k_1)\right|\\ &+(p_1^*+p_2^*)\int_{x_1}^{x_2}\int_{y_1}^{y_2} d_t\bigvee_{k_2=1}^{t} g_2(y_2,k_2)d_s\bigvee_{k_1=1}^{s} g_1(x_2,k_1)\\ &+(p_1^*+p_2^*)\int_{1}^{x_1}\int_{y_1}^{y_2} d_t\bigvee_{k_2=1}^{t} g_2(y_2,k_2)d_s\bigvee_{k_1=1}^{s} g_1(x_2,k_1)\\ &+(p_1^*+p_2^*)\int_{x_1}^{x_2}\int_{1}^{y_1} d_t\bigvee_{k_2=1}^{t} g_2(y_2,k_2)d_s\bigvee_{k_1=1}^{s} g_1(x_2,k_1)\,.\end{aligned}$$

Hence, we get

$$\begin{aligned}&|(Nu)(x_2,y_2)-(Nu)(x_1,y_1)| \le |\mu(x_1,y_1)-\mu(x_2,y_2)|\\ &+(p_1^*+p_2^*)\left|\bigvee_{k_2=1}^{y_1} g_2(y_2,k_2)\bigvee_{k_1=1}^{x_1} g_1(x_2,k_1)-\bigvee_{k_2=1}^{y_1} g_2(y_1,k_2)\bigvee_{k_1=1}^{x_1} g_1(x_1,k_1)\right|\\ &+(p_1^*+p_2^*)\bigvee_{k_2=y_1}^{y_2} g_2(y_2,k_2)\bigvee_{k_1=x_1}^{x_2} g_1(x_2,k_1)\\ &+(p_1^*+p_2^*)\bigvee_{k_2=y_1}^{y_2} g_2(y_2,k_2)\bigvee_{k_1=1}^{x_2} g_1(x_2,k_1)\\ &+(p_1^*+p_2^*)\bigvee_{k_2=1}^{y_2} g_2(y_2,k_2)\bigvee_{k_1=x_1}^{x_2} g_1(x_2,k_1)\,.\end{aligned}$$

As $x_1 \to x_2$ and $y_1 \to y_2$, the right-hand side of the preceding inequality tends to zero.

As a consequence of Steps 1–4, together with the Ascoli–Arzelà theorem, we can conclude that $N$ is continuous and compact. From an application of Schauder's theorem [149] we deduce that $N$ has a fixed point $u$ that is a solution of integral equation (9.1). □

Now we are concerned with the stability of solutions for integral equation (9.1). Let $\epsilon > 0$ and $\Phi\colon J \to [0, \infty)$ be a continuous function. We consider the inequalities

$$|u(x,y) - (Nu)(x,y)| \le \epsilon; \quad (x,y) \in J\,, \tag{9.3}$$

$$|u(x,y) - (Nu)(x,y)| \le \Phi(x,y); \quad (x,y) \in J\,, \tag{9.4}$$

$$|u(x,y) - (Nu)(x,y)| \le \epsilon\Phi(x,y); \quad (x,y) \in J\,. \tag{9.5}$$

**Theorem 9.3.** *Assume (9.2.1)–(9.2.5). If there exist $q_i \in C(J, \mathbb{R}_+)$; $i = 1, 2$ such that for each $(x,y) \in J$ we have*

$$p_i(x,y) \le q_i(x,y)\Phi(x,y)\,,$$

*then integral equation (9.1) is generalized Ulam–Hyers–Rassias stable.*

*Proof.* Let $u$ be a solution of inequality (9.4). By Theorem 9.2, there exists $v$ that is a solution of integral equation (9.1). Hence,

$$v(x,y) = \mu(x,y) + \int_1^x\int_1^y \left(\log\frac{x}{s}\right)^{r_1-1}\left(\log\frac{y}{t}\right)^{r_2-1}\frac{f(s,t,v(s,t))}{st\Gamma(r_1)\Gamma(r_2)}d_tg_2(y,t)d_sg_1(x,s)\,.$$

By inequality (9.4) for each $(x,y) \in J$ we have

$$\begin{aligned}&\Bigg|u(x,y) - \mu(x,y) - \int_1^x\int_1^y \left(\log\frac{x}{s}\right)^{r_1-1}\left(\log\frac{y}{t}\right)^{r_2-1}\\&\times\frac{f(s,t,u(s,t))}{st\Gamma(r_1)\Gamma(r_2)}d_tg_2(y,t)d_sg_1(x,s)\Bigg| \le \Phi(x,y)\,.\end{aligned}$$

Set

$$q_i^* = \sup_{(x,y)\in J} q_i(x,y)\,; \quad i = 1, 2\,.$$

For each $(x,y) \in J$ we have

$$\begin{aligned}&|u(x,y) - v(x,y)|\\&\le \Bigg|u(x,y) - \mu(x,y) - \int_1^x\int_1^y \left(\log\frac{x}{s}\right)^{r_1-1}\left(\log\frac{y}{t}\right)^{r_2-1}\\&\quad\times\frac{f(s,t,u(s,t))}{st\Gamma(r_1)\Gamma(r_2)}d_tg_2(y,t)d_sg_1(x,s)\Bigg|\\&\quad+\int_1^x\int_1^y \left|\log\frac{x}{s}\right|^{r_1-1}\left|\log\frac{y}{t}\right|^{r_2-1}\\&\quad\times\frac{|f(s,t,u(s,t)) - f(s,t,v(s,t))|}{st\Gamma(r_1)\Gamma(r_2)}d_tg_2(y,t)d_sg_1(x,s)\\&\le \Phi(x,y) + \frac{1}{\Gamma(r_1)\Gamma(r_2)}\int_1^x\int_1^y \left|\log\frac{x}{s}\right|^{r_1-1}\left|\log\frac{y}{t}\right|^{r_2-1}\\&\quad\times\left(2q_1^* + \frac{q_2^*|u(s,t)|}{1+|u|} + \frac{q_2^*|v(s,t)|}{1+|v|}\right)\frac{\Phi(s,t)}{st}d_tg_2(y,t)d_sg_1(x,s)\end{aligned}$$

$$\begin{aligned}&\le \Phi(x,y) + 2(q_1^* + q_2^*)({}^{HS}I_\sigma^r \Phi)(x,y)\\&\le [1 + 2(q_1^* + q_2^*)\lambda_\phi]\Phi(x,y)\\&:= c_{N,\Phi}\Phi(x,y)\,.\end{aligned}$$

Hence, integral equation (9.1) is generalized Ulam–Hyers–Rassias stable. □

### 9.2.3 An Example

As an application of our results we consider the Hadamard–Stieltjes integral equation

$$u(x,y) = \mu(x,y) + \int_1^x\int_1^y \left(\log\frac{x}{s}\right)^{r_1-1}\left(\log\frac{y}{t}\right)^{r_2-1}\frac{f(s,t,u(s,t))}{st\Gamma(r_1)\Gamma(r_2)}d_tg_2(y,t)d_sg_1(x,s)\,;$$
$$(x,y)\in[1,e]\times[1,e]\,, \tag{9.6}$$

where

$$\begin{aligned}&r_1, r_2 > 0, && \mu(x,y) = x + y^2; && (x,y)\in[1,e]\times[1,e],\\&g_1(x,s) = s, && g_2(y,t) = t, && s,t\in[1,e]\,,\end{aligned}$$

and

$$f(x,y,u(x,y)) = xy^2\left(e^{-7} + \frac{u(x,y)}{e^{x+y+5}}\right), \quad (x,y)\in[1,e]\times[1,e]\,.$$

Condition (9.2.1) is satisfied with $p_1(x,y) = xy^2e^{-7}$ and $p_2^* = \frac{xy^2}{e^{x+y+5}}$. We can see that the functions $g_1$ and $g_2$ satisfy (9.2.2)–(9.2.4). Consequently, Theorem 9.2 implies that Hadamard integral equation (9.6) has a solution defined on $[1,e]\times[1,e]$.
Condition (9.2.5) is satisfied by

$$\Phi(x,y) = e^3, \text{ and } \lambda_\Phi = \frac{1}{\Gamma(1+r_1)\Gamma(1+r_2)}\,.$$

Indeed, for each $(x,y)\in[1,e]\times[1,e]$ we get

$$\begin{aligned}({}^{HS}I_\sigma^r\Phi)(x,y) &\le \frac{e^3}{\Gamma(1+r_1)\Gamma(1+r_2)}\\&= \lambda_\Phi\Phi(x,y)\,.\end{aligned}$$

Consequently, Theorem 9.3 implies that equation (9.6) is generalized Ulam–Hyers–Rassias stable.

## 9.3 Global Stability Results for Volterra-Type Fractional Hadamard–Stieltjes Partial Integral Equations

### 9.3.1 Introduction

This section deals with the existence and global stability of solutions of a new class of Volterra partial integral equations of Hadamard–Stieltjes fractional order.

Recently, Abbas et al. [47, 33, 38] studied some existence and stability results for some classes of nonlinear quadratic Volterra integral equations of Riemann–Liouville fractional order. This section deals with the global existence and stability of solutions to the nonlinear quadratic Volterra partial integral equation of Hadamard fractional order

$$u(t,x) = f(t,x,u(t,x),u(\alpha(t),x)) + \frac{1}{\Gamma(r_1)\Gamma(r_2)} \int_1^{\beta(t)} \int_1^{x} \left(\log\frac{\beta(t)}{s}\right)^{r_1-1} \left(\log\frac{x}{\xi}\right)^{r_2-1} \times h(t,x,s,\xi,u(s,\xi),u(\gamma(s),\xi)) \frac{d_\xi g_2(x,\xi) d_s g_1(t,s)}{s\xi}, \quad (t,x) \in J, \tag{9.7}$$

where $J := [1,\infty) \times [1,b]$, $b > 1$, $r_1, r_2 \in (0,\infty)$, $\alpha, \beta, \gamma\colon [1,\infty) \to [1,\infty)$, $f\colon J \times \mathbb{R} \times \mathbb{R} \to \mathbb{R}$, $g_1\colon \Delta_1 \to \mathbb{R}$, $g_2\colon \Delta_2 \to \mathbb{R}$, and $h\colon J_1 \times \mathbb{R} \times \mathbb{R} \to \mathbb{R}$ are given continuous functions, $\Delta_1 = \{(t,s)\colon 1 \le s \le t\}$, $\Delta_2 = \{(x,\xi)\colon 1 \le \xi \le x \le b\}$, $J_1 = \{(t,x,s,\xi)\colon (t,s) \in \Delta_1,$ and $(x,\xi) \in \Delta_2\}$.

In this section we provide some existence and asymptotic stability of such a new class of fractional integral equations. Finally, we present an example illustrating the applicability of the imposed conditions.

### 9.3.2 Existence and Asymptotic Stability Results

In this section, we are concerned with the existence and asymptotic stability of solutions to the Hadamard partial integral equation (9.7).

Let us introduce the following conditions:

(9.4.1) The function $\alpha\colon [1,\infty) \to [1,\infty)$ satisfies $\lim_{t\to\infty} \alpha(t) = \infty$.

(9.4.2) There exist constants $M, L > 0$ and a nondecreasing function $\psi_1\colon [0,\infty) \to (0,\infty)$ such that $M < \frac{L}{2}$,

$$|f(t,x,u_1,v_1) - f(t,x,u_2,v_2)| \le \frac{M(|u_1-u_2| + |v_1-v_2|)}{(1+\alpha(t))(L + |u_1-u_2| + |v_1-v_2|)},$$

and

$$|f(t_1,x_1,u,v) - f(t_2,x_2,u,v)| \le (|t_1-t_2| + |x_1-x_2|)\psi_1(|u|+|v|)$$

for each $(t,x), (t_1,x_1), (t_2,x_2) \in J$ and $u, v, u_1, v_1, u_2, v_2 \in \mathbb{R}$.

(9.4.3) The function $t \to f(t, x, 0, 0)$ is bounded on $J$ with

$$f^* = \sup_{(t,x)\in[1,\infty)\times[1,b]} f(t, x, 0, 0)$$

and

$$\lim_{t\to\infty} |f(t, x, 0, 0)| = 0; \ x \in [1, b] .$$

(9.4.4) There exist continuous functions $\varphi : J \to \mathbb{R}_+, p : J_1 \to \mathbb{R}_+$ and a nondecreasing function $\psi_2 : [0, \infty) \to (0, \infty)$ such that

$$|h(t_1, x_1, s, \xi, u, v) - h(t_2, x_2, s, \xi, u, v)| \le \varphi(s, \xi)(|t_1 - t_2| + |x_1 - x_2|)\psi_2(|u| + |v|)$$

and

$$|h(t, x, s, \xi, u, v)| \le \frac{p(t, x, s, \xi)}{1 + \alpha(t) + |u| + |v|}$$

for each $(t, s), (t_1, s), (t_2, s) \in \Delta_1, (x, \xi), (x_1, \xi), (x_2, \xi) \in \Delta_2$ and $u, v \in \mathbb{R}$. Moreover, assume that

$$\lim_{t\to\infty} \int_1^{\beta(t)} \int_1^x \left|\log \frac{\beta(t)}{s}\right|^{r_1-1} \left|\log \frac{x}{\xi}\right|^{r_2-1} p(t, x, s, \xi) d_\xi \bigvee_{k_2=1}^{\xi} g_2(x, k_2) d_s \bigvee_{k_1=1}^{s} g_1(t, k_1) = 0$$

for each $x \in [1, b]$.

(9.4.5) The functions $s \mapsto g_1(t, s)$ and $\xi \mapsto g_2(x, \xi)$ have bounded variations for each fixed $t \in [1, \infty)$ or $x \in [1, b]$, respectively. Moreover, the functions $s \mapsto g_1(1, s)$ and $\xi \mapsto g_2(1, \xi)$ are nondecreasing on $[1, \infty)$ or $[1, b]$, respectively,

(9.4.6) For each $(t, s), (t_1, s), (t_2, s) \in \Delta_1, (x, \xi), (x_1, \xi), (x_2, \xi) \in \Delta_2$ we have

$$\left| \bigvee_{k_2=1}^{x_2} g_2(x_2, k_2) \bigvee_{k_1=1}^{t_2} g_1(t_2, k_1) - \bigvee_{k_2=1}^{x_1} g_2(x_1, k_2) d_s \bigvee_{k_1=1}^{t_1} g_1(t_1, k_1) \right| \to 0$$

as $t_1 \to t_2$ and $x_1 \to x_2$.

(9.4.7) $g_1(t, 1) = g_2(x, 1) = 0$ for any $t \in [1, \infty)$ and any $x \in [1, b]$.

**Theorem 9.4.** *Assume (9.4.1)–(9.4.7). Then integral equation (9.7) has at least one solution in the space BC. Moreover, solutions of equation (9.7) are globally asymptotically stable.*

*Proof.* Set $d^* := \sup_{(t,x)\in J} d(t, x)$, where

$$d(t, x) = \int_1^{\beta(t)} \int_1^x \left|\log \frac{\beta(t)}{s}\right|^{r_1-1} \left|\log \frac{x}{\xi}\right|^{r_2-1} \frac{p(t, x, s, \xi)}{\Gamma(r_1)\Gamma(r_2)} d_\xi \bigvee_{k_2=1}^{\xi} g_2(x, k_2) d_s \bigvee_{k_1=1}^{s} g_1(t, k_1) .$$

From condition (9.4.4), we infer that $d^*$ is finite. Let us define the operator $N$ such that, for any $u \in BC$,

$$(Nu)(t, x) = f(t, x, u(t, x), u(\alpha(t), x)) + \int_1^{\beta(t)} \int_1^x \left(\log \frac{\beta(t)}{s}\right)^{r_1-1} \left(\log \frac{x}{\xi}\right)^{r_2-1}$$
$$\times h(t, x, s, \xi, u(s, \xi), u(y(s), \xi)) \frac{d_\xi g_2(x, \xi) d_s g_1(t, s)}{s\xi\Gamma(r_1)\Gamma(r_2)} , \quad (t, x) \in J . \quad (9.8)$$

By considering the conditions of this theorem, we infer that $N(u)$ is continuous on $J$. Now we prove that $N(u) \in BC$ for any $u \in BC$. For arbitrarily fixed $(t, x) \in J$ we have

$$
\begin{aligned}
|(Nu)(t,x)| &\le |f(t,x,u(t,x),u(\alpha(t),x)) - f(t,x,0,0)| + |f(t,x,0,0)| \\
&\quad + \frac{1}{\Gamma(r_1)\Gamma(r_2)} \int_1^{\beta(t)} \int_1^{x} \left|\log\frac{\beta(t)}{s}\right|^{r_1-1} \left|\log\frac{x}{\xi}\right|^{r_2-1} \\
&\quad \times |h(t,x,s,\xi,u(s,\xi),u(\gamma(s),\xi))| \frac{|d_\xi g_2(x,\xi) d_s g_1(t,s)|}{s\xi} \\
&\le \frac{M(|u(t,x)| + |u(\alpha(t),x)|)}{(1+\alpha(t))(L + |u(t,x)| + |u(\alpha(t),x)|)} + |f(t,x,0,0)| \\
&\quad + \frac{1}{\Gamma(r_1)\Gamma(r_2)} \int_1^{\beta(t)} \int_1^{x} \left|\log\frac{\beta(t)}{s}\right|^{r_1-1} \left|\log\frac{x}{\xi}\right|^{r_2-1} \\
&\quad \times \frac{p(t,x,s,\xi)}{1+\alpha(t) + |u(s,\xi))| + |u(\gamma(s),\xi))|} d_\xi \bigvee_{k_2=1}^{\xi} g_2(x,k_2) d_s \bigvee_{k_1=1}^{s} g_1(t,k_1) \\
&\le \frac{M(|u(t,x)| + |u(\alpha(t),x)|)}{|u(t,x)| + |u(\alpha(t),x)|} + f^* + d^* .
\end{aligned}
$$

Thus,

$$
\|N(u)\|_{BC} \le M + f^* + d^* . \tag{9.9}
$$

Hence, $N(u) \in BC$. Equation (9.9) yields that $N$ transforms the ball $B_\eta := B(0,\eta)$ into itself, where $\eta = M + f^* + d^*$. We will show that $N\colon B_\eta \to B_\eta$ satisfies the conditions of Theorem 1.42. The proof will be given in several steps and cases.

*Step 1: N is continuous.* Let $\{u_n\}_{n\in\mathbb{N}}$ be a sequence such that $u_n \to u$ in $B_\eta$. Then for each $(t, x) \in J$ we have

$$
\begin{aligned}
&|(Nu_n)(t,x) - (Nu)(t,x)| \le |f(t,x,u_n(t,x),u_n(\alpha(t),x)) - f(t,x,u(t,x),u(\alpha(t),x))| \\
&\quad + \frac{1}{\Gamma(r_1)\Gamma(r_2)} \int_1^{\beta(t)} \int_1^{x} \left|\log\frac{\beta(t)}{s}\right|^{r_1-1} \left|\log\frac{x}{\xi}\right|^{r_2-1} \\
&\quad \times \sup_{(s,\xi)\in J} |h(t,x,s,\xi,u_n(s,\xi),u_n(\gamma(s),\xi)) - h(t,x,s,\xi,u(s,\xi),u(\gamma(s),\xi))| \\
&\quad \times \frac{|d_\xi g_2(x,\xi) d_s g_1(t,s)|}{s\xi} \\
&\le \frac{2M}{L} \|u_n - u\|_{BC} \\
&\quad + \frac{1}{\Gamma(r_1)\Gamma(r_2)} \int_1^{\beta(t)} \int_1^{x} \left|\log\frac{\beta(t)}{s}\right|^{r_1-1} \left|\log\frac{x}{\xi}\right|^{r_2-1} \\
&\quad \times \|h(t,x,.,.,u_n(.,.),u_n(\gamma(.),.)) - h(t,x,.,.,u(.,.),u(\gamma(.),.))\|_{BC} \\
&\quad \times d_\xi \bigvee_{k_2=1}^{\xi} g_2(x,k_2) d_s \bigvee_{k_1=1}^{s} g_1(t,k_1) .
\end{aligned} \tag{9.10}
$$

*Case 1.* If $(t,x) \in [1,T] \times [1,b]$, $T > 1$, then, since $u_n \to u$ as $n \to \infty$ and $h, \gamma$ are continuous, (9.10) gives

$$\|N(u_n) - N(u)\|_{BC} \to 0 \text{ as } n \to \infty .$$

*Case 2.* If $(t,x) \in (T,\infty) \times [1,b]$, $T > 1$, then from (9.4.4) and (9.10), for each $(t,x) \in J$, we have

$$\begin{aligned} |(Nu_n)(t,x) - (Nu)(t,x)| &\le \frac{2M}{L}\|u_n - u\|_{BC} \\ &\quad + \frac{2}{\Gamma(r_1)\Gamma(r_2)} \int_1^{\beta(t)}\int_1^{x} \left|\log\frac{\beta(t)}{s}\right|^{r_1-1} \left|\log\frac{x}{\xi}\right|^{r_2-1} \\ &\quad \times p(t,x,s,\xi)d_\xi \bigvee_{k_2=1}^{\xi} g_2(x,k_2)d_s \bigvee_{k_1=1}^{s} g_1(t,k_1) . \\ &\le \frac{2M}{L}\|u_n - u\|_{BC} + 2d(t,x) . \end{aligned}$$

Thus, we get

$$|(Nu_n)(t,x) - (Nu)(t,x)| \le \frac{2M}{L}\|u_n - u\|_{BC} + 2d(t,x) . \tag{9.11}$$

Since $u_n \to u$ as $n \to \infty$ and $t \to \infty$, then (6.6) gives

$$\|N(u_n) - N(u)\|_{BC} \to 0 \text{ as } n \to \infty .$$

*Step 2: $N(B_\eta)$ is uniformly bounded.* This is clear since $N(B_\eta) \subset B_\eta$ and $B_\eta$ is bounded.

*Step 3: $N(B_\eta)$ is equicontinuous on every compact subset* $[1,a] \times [1,b]$ *of $J$, $a > 0$.* Let $(t_1,x_1), (t_2,x_2) \in [1,a] \times [1,b]$, $t_1 < t_2$, $x_1 < x_2$, and let $u \in B_\eta$. Also, without loss of generality, suppose that $\beta(t_1) \le \beta(t_2)$. Then we have

$$\begin{aligned} &|(Nu)(t_2,x_2) - (Nu)(t_1,x_1)| \\ &\quad \le |f(t_2,x_2,u(t_2,x_2),u(\alpha(t_2),x_2)) - f(t_2,x_2,u(t_1,x_1),u(\alpha(t_1),x_1))| \\ &\qquad + |f(t_2,x_2,u(t_1,x_1),u(\alpha(t_1),x_1)) - f(t_1,x_1,u(t_1,x_1),u(\alpha(t_1),x_1))| \\ &\qquad + \frac{1}{\Gamma(r_1)\Gamma(r_2)} \int_1^{\beta(t_2)}\int_1^{x_2} \left|\log\frac{\beta(t_2)}{s}\right|^{r_1-1} \left|\log\frac{x_2}{\xi}\right|^{r_2-1} \\ &\qquad \times |h(t_2,x_2,s,\xi,u(s,\xi),u(\gamma(s),\xi)) - h(t_1,x_1,s,\xi,u(s,\xi),u(\gamma(s),\xi))| \\ &\qquad \times |d_\xi g_2(x_2,\xi)d_s g_1(t_2,s)| \\ &\qquad + \left| \frac{1}{\Gamma(r_1)\Gamma(r_2)} \int_1^{\beta(t_2)}\int_1^{x_2} \left(\log\frac{\beta(t_2)}{s}\right)^{r_1-1} \left(\log\frac{x_2}{\xi}\right)^{r_2-1} \right. \\ &\qquad \times h(t_1,x_1,s,\xi,u(s,\xi),u(\gamma(s),\xi))d_\xi g_2(x_2,\xi)d_s g_1(t_2,s) \end{aligned}$$

$$
\begin{aligned}
&- \frac{1}{\Gamma(r_1)\Gamma(r_2)} \int_1^{\beta(t_1)} \int_1^{x_1} \left(\log \frac{\beta(t_2)}{s}\right)^{r_1-1} \left(\log \frac{x_2}{\xi}\right)^{r_2-1} \\
&\times h(t_1, x_1, s, \xi, u(s,\xi), u(\gamma(s),\xi)) d_\xi g_2(x_2,\xi) d_s g_1(t_2, s) \Big| \\
&+ \left| \frac{1}{\Gamma(r_1)\Gamma(r_2)} \int_1^{\beta(t_1)} \int_1^{x_1} \left(\log \frac{\beta(t_2)}{s}\right)^{r_1-1} \left(\log \frac{x_2}{\xi}\right)^{r_2-1} \right. \\
&\times h(t_1, x_1, s, \xi, u(s,\xi), u(\gamma(s),\xi)) (d_\xi g_2(x_2,\xi) d_s g_1(t_2, s) - d_\xi g_2(x_1,\xi) d_s g_1(t_1, s)) \Big| \\
&+ \frac{1}{\Gamma(r_1)\Gamma(r_2)} \int_1^{\beta(t_1)} \int_1^{x_1} \left| \left(\log \frac{\beta(t_2)}{s}\right)^{r_1-1} \left(\log \frac{x_2}{\xi}\right)^{r_2-1} \right. \\
&\left. - \left(\log \frac{\beta(t_1)}{s}\right)^{r_1-1} \left(\log \frac{x_1}{\xi}\right)^{r_2-1} \right| |h(t_1, x_1, s, \xi, u(s,\xi), u(\gamma(s),\xi))| \\
&\times |d_\xi g_2(x_1,\xi) d_s g_1(t_1, s)| \,.
\end{aligned}
$$

Thus, we obtain

$$
\begin{aligned}
&|(Nu)(t_2, x_2) - (Nu)(t_1, x_1)| \\
&\quad \leq \frac{M}{L}(|u(t_2, x_2) - u(t_1, x_1)| + |u(\alpha(t_2), x_2) - u(\alpha(t_1), x_1)|) \\
&\quad + (|t_2 - t_1| + |x_2 - x_1|)\psi_1(2\|u\|_{BC}) \\
&\quad + \frac{1}{\Gamma(r_1)\Gamma(r_2)} \int_1^{\beta(t_2)} \int_1^{x_2} \left|\log \frac{\beta(t_2)}{s}\right|^{r_1-1} \left|\log \frac{x_2}{\xi}\right|^{r_2-1} \\
&\quad \times \varphi(s,\xi)(|t_2 - t_1| + |x_2 - x_1|)\psi_2(2\|u\|_{BC}) d_\xi \bigvee_{k_2=1}^{\xi} g_2(x_2, k_2) d_s \bigvee_{k_1=1}^{s} g_1(t_2, k_1) \\
&\quad + \frac{1}{\Gamma(r_1)\Gamma(r_2)} \int_{\beta(t_1)}^{\beta(t_2)} \int_1^{x_2} \left|\log \frac{\beta(t_2)}{s}\right|^{r_1-1} \left|\log \frac{x_2}{\xi}\right|^{r_2-1} \\
&\quad \times |h(t_1, x_1, s, \xi, u(s,\xi), u(\gamma(s),\xi))| d_\xi \bigvee_{k_2=1}^{\xi} g_2(x_2, k_2) d_s \bigvee_{k_1=1}^{s} g_1(t_2, k_1) \\
&\quad + \frac{1}{\Gamma(r_1)\Gamma(r_2)} \int_1^{\beta(t_2)} \int_{x_1}^{x_2} \left|\log \frac{\beta(t_2)}{s}\right|^{r_1-1} \left|\log \frac{x_2}{\xi}\right|^{r_2-1} \\
&\quad \times |h(t_1, x_1, s, \xi, u(s,\xi), u(\gamma(s),\xi))| d_\xi \bigvee_{k_2=1}^{\xi} g_2(x_2, k_2) d_s \bigvee_{k_1=1}^{s} g_1(t_2, k_1) \\
&\quad + \frac{1}{\Gamma(r_1)\Gamma(r_2)} \int_{\beta(t_1)}^{\beta(t_2)} \int_{x_1}^{x_2} \left|\log \frac{\beta(t_2)}{s}\right|^{r_1-1} \left|\log \frac{x_2}{\xi}\right|^{r_2-1}
\end{aligned}
$$

$$\times |h(t_1, x_1, s, \xi, u(s,\xi), u(\gamma(s),\xi))| d_\xi \bigvee_{k_2=1}^{\xi} g_2(x_2, k_2) d_s \bigvee_{k_1=1}^{s} g_1(t_2, k_1)$$

$$+ \frac{1}{\Gamma(r_1)\Gamma(r_2)} \int_1^{\beta(t_1)} \int_1^{x_1} \left|\log \frac{\beta(t_2)}{s}\right|^{r_1-1} \left|\log \frac{x_2}{\xi}\right|^{r_2-1}$$

$$\times |h(t_1, x_1, s, \xi, u(s,\xi), u(\gamma(s),\xi))| \| d_\xi g_2(x_2,\xi) d_s g_1(t_2, s) - d_\xi g_2(x_1,\xi) d_s g_1(t_1, s))|$$

$$+ \frac{1}{\Gamma(r_1)\Gamma(r_2)} \int_1^{\beta(t_1)} \int_1^{x_1} \left| \left(\log \frac{\beta(t_2)}{s}\right)^{r_1-1} \left(\log \frac{x_2}{\xi}\right)^{r_2-1}\right.$$

$$\left. - \left(\log \frac{\beta(t_1)}{s}\right)^{r_1-1} \left(\log \frac{x_1}{\xi}\right)^{r_2-1} \right| |h(t_1, x_1, s, \xi, u(s,\xi), u(\gamma(s),\xi))|$$

$$\times d_\xi \bigvee_{k_2=1}^{\xi} g_2(x_1, k_2) d_s \bigvee_{k_1=1}^{s} g_1(t_1, k_1)\,.$$

Hence, we get

$$\begin{aligned}
&|(Nu)(t_2, x_2) - (Nu)(t_1, x_1)| \\
&\le \frac{M}{L}(|u(t_2, x_2) - u(t_1, x_1)| + |u(\alpha(t_2), x_2) - u(\alpha(t_1), x_1)|) \\
&\quad + (|t_2 - t_1| + |x_2 - x_1|)\psi_1(2\eta) \\
&\quad + \frac{(|t_2 - t_1| + |x_2 - x_1|)\psi_2(2\eta)}{\Gamma(r_1)\Gamma(r_2)} \\
&\quad \times \int_1^{\beta(t_2)} \int_1^{x_2} \left|\log \frac{\beta(t_2)}{s}\right|^{r_1-1} \left|\log \frac{x_2}{\xi}\right|^{r_2-1} \varphi(s,\xi) d_\xi \bigvee_{k_2=1}^{\xi} g_2(x_2, k_2) d_s \bigvee_{k_1=1}^{s} g_1(t_2, k_1) \\
&\quad + \frac{1}{\Gamma(r_1)\Gamma(r_2)} \int_{\beta(t_1)}^{\beta(t_2)} \int_1^{x_2} \left|\log \frac{\beta(t_2)}{s}\right|^{r_1-1} \left|\log \frac{x_2}{\xi}\right|^{r_2-1} \\
&\quad \times p(t_1, x_1, s, \xi) d_\xi \bigvee_{k_2=1}^{\xi} g_2(x_2, k_2) d_s \bigvee_{k_1=1}^{s} g_1(t_2, k_1) \\
&\quad + \frac{1}{\Gamma(r_1)\Gamma(r_2)} \int_1^{\beta(t_2)} \int_{x_1}^{x_2} \left|\log \frac{\beta(t_2)}{s}\right|^{r_1-1} \left|\log \frac{x_2}{\xi}\right|^{r_2-1} \\
&\quad \times p(t_1, x_1, s, \xi) d_\xi \bigvee_{k_2=1}^{\xi} g_2(x_2, k_2) d_s \bigvee_{k_1=1}^{s} g_1(t_2, k_1) \\
&\quad + \frac{1}{\Gamma(r_1)\Gamma(r_2)} \int_{\beta(t_1)}^{\beta(t_2)} \int_{x_1}^{x_2} \left|\log \frac{\beta(t_2)}{s}\right|^{r_1-1} \left|\log \frac{x_2}{\xi}\right|^{r_2-1} \\
&\quad \times p(t_1, x_1, s, \xi) d_\xi \bigvee_{k_2=1}^{\xi} g_2(x_2, k_2) d_s \bigvee_{k_1=1}^{s} g_1(t_2, k_1)
\end{aligned}$$

$$+\frac{1}{\Gamma(r_1)\Gamma(r_2)}\int_1^{\beta(t_1)}\int_1^{x_1}\left|\log\frac{\beta(t_2)}{s}\right|^{r_1-1}\left|\log\frac{x_2}{\xi}\right|^{r_2-1}p(t_1,x_1,s,\xi)$$
$$\times\left|d_\xi\bigvee_{k_2=1}^{\xi}g_2(x_2,k_2)d_s\bigvee_{k_1=1}^{s}g_1(t_2,k_1)-d_\xi\bigvee_{k_2=1}^{\xi}g_2(x_1,k_2)d_s\bigvee_{k_1=1}^{s}g_1(t_1,k_1)\right|$$
$$+\frac{1}{\Gamma(r_1)\Gamma(r_2)}\int_1^{\beta(t_1)}\int_1^{x_1}\left|\left(\log\frac{\beta(t_2)}{s}\right)^{r_1-1}\left(\log\frac{x_2}{\xi}\right)^{r_2-1}\right.$$
$$\left.-\left(\log\frac{\beta(t_1)}{s}\right)^{r_1-1}\left(\log\frac{x_1}{\xi}\right)^{r_2-1}\right|p(t_1,x_1,s,\xi)d_\xi\bigvee_{k_2=1}^{\xi}g_2(x_1,k_2)d_s\bigvee_{k_1=1}^{s}g_1(t_1,k_1)\,.$$

From the continuity of $\alpha,\beta,\varphi,p$ and as $t_1\to t_2$ and $x_1\to x_2$, the right-hand side of the preceding inequality tends to zero.

*Step 4: $N(B_\eta)$ is equiconvergent.* Let $(t,x)\in J$ and $u\in B_\eta$; then we have

$$\begin{aligned}|u(t,x)|&\le|f(t,x,u(t,x),u(\alpha(t),x))-f(t,x,0,0)+f(t,x,0,0)|\\&\quad+\left|\frac{1}{\Gamma(r_1)\Gamma(r_2)}\int_1^{\beta(t)}\int_1^{x}\left(\log\frac{\beta(t)}{s}\right)^{r_1-1}\left(\log\frac{x}{\xi}\right)^{r_2-1}\right.\\&\quad\left.\times h(t,x,s,\xi,u(s,\xi),u(\gamma(s),\xi))\frac{d_\xi g_2(x,\xi)d_s g_1(t,s)}{s\xi}\right|\\&\le\frac{M(|u(t,x)|+|u(\alpha(t),x)|)}{(1+\alpha(t))(L+|u(t,x)|+|u(\alpha(t),x)|)}+|f(t,x,0,0)|\\&\quad+\frac{1}{\Gamma(r_1)\Gamma(r_2)}\int_1^{\beta(t)}\int_1^{x}\left(\log\frac{\beta(t)}{s}\right)^{r_1-1}\left(\log\frac{x}{\xi}\right)^{r_2-1}\\&\quad\times\frac{p(t,x,s,\xi)}{1+\alpha(t)+|u(s,\xi)|+|u(\gamma(s),\xi)|}d_\xi g_2(x,\xi)d_s g_1(t,s)\\&\le\frac{M}{1+\alpha(t)}+|f(t,x,0,0)|\\&\quad+\frac{1}{\Gamma(r_1)\Gamma(r_2)(1+\alpha(t))}\int_1^{\beta(t)}\int_1^{x}\left(\log\frac{\beta(t)}{s}\right)^{r_1-1}\left(\log\frac{x}{\xi}\right)^{r_2-1}\\&\quad\times p(t,x,s,\xi)d_\xi\bigvee_{k_2=1}^{\xi}g_2(x,k_2)d_s\bigvee_{k_1=1}^{s}g_1(t,k_1)\\&\le\frac{M}{1+\alpha(t)}+|f(t,x,0,0)|+\frac{d^*}{1+\alpha(t)}\,.\end{aligned}$$

Thus, for each $x\in[1,b]$ we get

$$|u(t,x)|\to 0,\quad\text{as }t\to+\infty\,.$$

Hence,

$$|u(t,x)-u(+\infty,x)|\to 0,\quad\text{as }t\to+\infty\,.$$

As a consequence of Steps 1–4, together with Lemma 1.57, we can conclude that $N\colon B_\eta \to B_\eta$ is continuous and compact. From an application of Theorem 1.42, we deduce that $N$ has a fixed point $u$ that is a solution of the Hadamard integral equation (9.7).

*Step 5: Uniform global attractivity.* Let us assume that $u_0$ is a solution of integral equation (9.7) with the conditions of this theorem. Consider the ball $B(u_0, \eta)$ with $\eta^* = \frac{LM^*}{L-2M}$, where

$$\begin{aligned} M^* := \frac{1}{\Gamma(r_1)\Gamma(r_2)} \sup_{(t,x)\in J} \Bigg\{ & \int_1^{\beta(t)} \int_1^{x} \left|\log \frac{\beta(t)}{s}\right|^{r_1-1} \left|\log \frac{x}{\xi}\right|^{r_2-1} \\ & \times |h(t,x,s,\xi,u(s,\xi),u(\gamma(s),\xi)) - h(t,x,s,\xi,u_0(s,\xi),u_0(\gamma(s),\xi))| \\ & \times d_\xi \bigvee_{k_2=1}^{\xi} g_2(x,k_2) d_s \bigvee_{k_1=1}^{s} g_1(t,k_1);\ u \in BC \Bigg\} . \end{aligned}$$

Taking $u \in B(u_0, \eta^*)$, we then have

$$\begin{aligned} & |(Nu)(t,x) - u_0(t,x)| = |(Nu)(t,x) - (Nu_0)(t,x)| \\ & \le |f(t,x,u(t,x),u(\alpha(t),x)) - f(t,x,u_0(t,x),u_0(\alpha(t),x))| \\ & \quad + \frac{1}{\Gamma(r_1)\Gamma(r_2)} \int_1^{\beta(t)} \int_1^{x} \left|\log \frac{\beta(t)}{s}\right|^{r_1-1} \left|\log \frac{x}{\xi}\right|^{r_2-1} \\ & \quad \times |h(t,x,s,\xi,u(s,\xi),u(\gamma(s),\xi)) - h(t,x,s,\xi,u_0(s,\xi),u_0(\gamma(s),\xi))| \\ & \quad \times d_\xi \bigvee_{k_2=1}^{\xi} g_2(x,k_2) d_s \bigvee_{k_1=1}^{s} g_1(t,k_1) \\ & \le \frac{2M}{L} \|u - u_0\|_{BC} + M^* \\ & \le \frac{2M}{L} \eta^* + M^* = \eta^* . \end{aligned}$$

Thus we observe that $N$ is a continuous function such that $N(B(u_0, \eta^*)) \subset B(u_0, \eta^*)$. Moreover, if $u$ is a solution of equation (7.1), then

$$\begin{aligned} & |u(t,x) - u_0(t,x)| = |(Nu)(t,x) - (Nu_0)(t,x)| \\ & \le |f(t,x,u(t,x),u(\alpha(t),x)) - f(t,x,u_0(t,x),u_0(\alpha(t),x))| \\ & \quad + \frac{1}{\Gamma(r_1)\Gamma(r_2)} \int_1^{\beta(t)} \int_1^{x} \left|\log \frac{\beta(t)}{s}\right|^{r_1-1} \left|\log \frac{x}{\xi}\right|^{r_2-1} \\ & \quad \times |h(t,x,s,\xi,u(s,\xi),u(\gamma(s),\xi)) - h(t,x,s,\xi,u_0(s,\xi),u_0(\gamma(s),\xi))| \\ & \quad \times d_\xi \bigvee_{k_2=1}^{\xi} g_2(x,k_2) d_s \bigvee_{k_1=1}^{s} g_1(t,k_1) . \end{aligned}$$

Thus,

$$|u(t,x)-u_0(t,x)| \leq \frac{M}{L}(|u(t,x)-u_0(t,x)|+|u(\alpha(t),x)-u_0(\alpha(t),x)|)$$
$$+\frac{1}{\Gamma(r_1)\Gamma(r_2)}\int_2^{\beta(t)}\int_1^{x}\left|\log\frac{\beta(t)}{s}\right|^{r_1-1}\left|\log\frac{x}{\xi}\right|^{r_2-1}$$
$$\times p(t,x,s,\xi)d_\xi \bigvee_{k_2=1}^{\xi} g_2(x,k_2)d_s \bigvee_{k_1=1}^{s} g_1(t,k_1)\,. \tag{9.12}$$

By using (9.12) and the fact that $\alpha(t) \to \infty$ as $t \to \infty$, we get

$$\lim_{t\to\infty}|u(t,x)-u_0(t,x)| \leq \lim_{t\to\infty}\frac{2L}{\Gamma(r_1)\Gamma(r_2)(L-2M)}\int_1^{\beta(t)}\int_1^{x}\left|\log\frac{\beta(t)}{s}\right|^{r_1-1}\left|\log\frac{x}{\xi}\right|^{r_2-1}$$
$$\times p(t,x,s,\xi)d_\xi \bigvee_{k_2=1}^{\xi} g_2(x,k_2)d_s \bigvee_{k_1=1}^{s} g_1(t,k_1) = 0\,.$$

Consequently, all solutions of the Hadamard–Volterra–Stieltjes integral equation (7.1) are globally asymptotically stable. □

### 9.3.3 An Example

As an application of our results we consider the partial Hadamard Volterra–Stieltjes integral equation of fractional order

$$u(t,x) = \frac{tx}{10(1+t+t^2+t^3)}(1+2\sin(u(t,x))) + \frac{1}{\Gamma^2\left(\frac{1}{3}\right)}\int_1^{t}\int_1^{x}\left(\log\frac{t}{s}\right)^{\frac{-2}{3}}\left(\log\frac{x}{\xi}\right)^{\frac{-2}{3}}$$
$$\times \frac{\ln(1+2x(s\xi)^{-1}|u(s,\xi)|)}{(1+t+2|u(s,\xi)|)^2(1+x^2+t^4)}d_\xi g_2(x,\xi)d_s g_1(t,s)\,, \tag{9.13}$$
$$(t,x) \in [1,\infty)\times[1,e]\,,$$

where $r_1 = r_2 = \frac{1}{3}$, $\alpha(t) = \beta(t) = \gamma(t) = t$, $g_1(t,s) = s$, $g_2(x,\xi) = \xi$; $s,\xi \in [1,e]$,

$$f(t,x,u,v) = \frac{tx(1+\sin(u)+\sin(v))}{10(1+t)(1+t^2)}\,,$$

and

$$h(t,x,s,\xi,u,v) = \frac{\ln(1+x(s\xi)^{-1}(|u|+|v|))}{(1+t+|u|+|v|)^2(1+x^2+t^4)}$$

for $(t,x),(s,\xi) \in [1,\infty)\times[1,e]$, and $u,v \in \mathbb{R}$.

We can easily check that the conditions of Theorem 9.4 are satisfied. In fact, we have that the function $f$ is continuous and satisfies (9.4.2), where $M = \frac{1}{10}$, $L = 1$. Also, $f$

satisfies (9.4.3), with $f^* = \frac{e}{10}$. Next, let us notice that the function $h$ satisfies (9.4.4), where $p(t,x,s,\xi) = \frac{1}{s\xi(1+x^2+t^4)}$. Also,

$$\begin{aligned}
&\lim_{t\to\infty}\int_1^t\int_1^x \left|\log\frac{t}{s}\right|^{\frac{-2}{3}}\left|\log\frac{x}{\xi}\right|^{\frac{-2}{3}} p(t,x,s,\xi)d_\xi \bigvee_{k_2=1}^{\xi} g_2(x,k_2)d_s \bigvee_{k_1=1}^{s} g_1(t,k_1)\\
&= \lim_{t\to\infty}\frac{x}{1+x^2+t^4}\int_1^t\int_1^x \left|\log\frac{t}{s}\right|^{\frac{-2}{3}}\left|\log\frac{x}{\xi}\right|^{\frac{-2}{3}}\frac{1}{s\xi}d_\xi \bigvee_{k_2=1}^{\xi} g_2(x,k_2)d_s \bigvee_{k_1=1}^{s} g_1(t,k_1)\\
&= \lim_{t\to\infty}\frac{x}{1+x^2+t^4}\int_1^t\int_1^x \left|\log\frac{t}{s}\right|^{\frac{-2}{3}}\left|\log\frac{x}{\xi}\right|^{\frac{-2}{3}}\frac{d\xi ds}{s\xi}\\
&= \lim_{t\to\infty}\frac{9x(\log t)^{\frac{1}{3}}}{1+x^2+t^4} = 0\,.
\end{aligned}$$

Hence, by Theorem 9.4, the Volterra–Stieltjes equation (9.13) has a solution defined on $[1,\infty)\times[1,e]$, and solutions of this equation are globally asymptotically stable.

## 9.4 Volterra-Type Nonlinear Multidelay Hadamard–Stieltjes Fractional Integral Equations

### 9.4.1 Introduction

This section deals with the existence and global attractivity of solutions for Volterra–Stieltjes quadratic delay integral equations of Hadamard fractional order.

In [24, 25, 32, 30], the authors studied the existence and stability of solutions for some integral equations. Motivated by the aforementioned papers, in this section we establish some sufficient conditions for the existence and attractivity of solutions of the class of Volterra-type delay fractional order Hadamard–Stieltjes quadratic integral equations,

$$\begin{aligned}
u(t,x) &= \mu(t,x) + \frac{f(t,x,u(t,x))}{\Gamma(r_1)\Gamma(r_2)}\int_1^t\int_1^x \left(\log\frac{t}{s}\right)^{r_1-1}\left(\log\frac{x}{y}\right)^{r_2-1}\\
&\quad\times h(t,x,s,y,u(s-\tau_1,y-\xi_1),\dots,u(s-\tau_m,y-\xi_m))\\
&\quad\times\frac{1}{sy}d_y g_2(x,y)d_s g_1(t,s)\,;\quad \text{if } (t,x)\in J := [1,+\infty)\times[1,b]\,, \qquad (9.14)\\
u(t,x) &= \Phi(t,x) \quad \text{if } (t,x)\in\tilde{J} := [-T,\infty)\times[-\xi,b]\backslash(1,\infty)\times(1,b]\,, \qquad (9.15)
\end{aligned}$$

where $b>1$, $r_1,r_2\in(0,\infty)$, $\tau_i,\xi_i\geq -1;\ i=1\dots,m$, $T=\max_{i=1\dots,m}\{\tau_i\}$, $\xi=\max_{i=1\dots,m}\{\xi_i\}$, $\mu\colon J\to\mathbb{R}$, $f\colon J\times\mathbb{R}\to\mathbb{R}$, $g_1\colon\Delta_1\to\mathbb{R}$, $g_2\colon\Delta_2\to\mathbb{R}$, $h\colon J'\times\mathbb{R}^m\to\mathbb{R}$ are given continuous functions, $\Delta_1=\{(t,s)\colon 1\leq s\leq t\}$, $\Delta_2=\{(x,y)\colon 1\leq y\leq x\leq b\}$,

$J_1 = \{(t, x, s, y) : (t, s) \in \Delta_1 \text{ and } (x, y) \in \Delta_2\}$, and $\Phi : \tilde{J} \to \mathbb{R}$ is continuous with $\mu(t, 1) = \Phi(t, 1)$ for each $t \in [1, +\infty)$ and $\mu(1, x) = \Phi(1, x)$ for each $x \in [1, b]$.

We use the Schauder fixed point theorem for the existence of solutions of problem (9.14)–(9.15), and we prove that all solutions are uniformly globally attractive.

Let $\emptyset \neq \Omega \subset BC$, let $G : \Omega \to \Omega$, and consider the solutions of the equation

$$(Gu)(t, x) = u(t, x) . \tag{9.16}$$

**Definition 9.5** ([35]). The solutions of equation (9.16) are locally attractive if there exists a ball $B(u_0, \eta)$ in the space $BC$ such that, for arbitrary solutions $v = v(t, x)$ and $w = w(t, x)$ of equations (9.16) belonging to $B(u_0, \eta) \bigcap \Omega$, we have that, for each $x \in [1, b]$,

$$\lim_{t\to\infty} (v(t, x) - w(t, x)) = 0 . \tag{9.17}$$

When the limit (9.17) is uniform with respect to $B(u_0, \eta)$, the solutions of equation (9.16) are said to be uniformly locally attractive (or, equivalently, that solutions of (9.16) are locally asymptotically stable).

**Definition 9.6** ([35]). The solution $v = v(t, x)$ of equation (9.16) is said to be globally attractive if (9.17) holds for each solution $w = w(t, x)$ of (9.16). If condition (9.17) is satisfied uniformly with respect to the set $\Omega$, then solutions of equation (9.16) are said to be globally asymptotically stable (or uniformly globally attractive).

### 9.4.2 Existence and Attractivity Results

Let us start in this section by defining what we mean by a solution of problem (9.14)–(9.15).

**Definition 9.7.** By a solution of problem (9.14)–(9.15) we mean every function $u \in BC$ such that $u$ satisfies equation (9.14) on $J$ and equation (9.15) on $\tilde{J}$.

The following conditions will be used in the sequel.

(9.9.1) The functions $\mu$ and $\Phi$ are in $BC$. Moreover, assume that

$$\lim_{t\to\infty} \mu(t, x) = 0 ; \quad x \in [1, b] .$$

(9.9.2) There exist a positive function $P \in BC$ and a nondecreasing function $\psi_1 : [0, \infty) \to (0, \infty)$ such that

$$|f(t, x, u) - f(t, x, v)| \leq P(t, x)|u - v| ; \quad (t, x) \in J, \ u, v \in \mathbb{R}$$

and

$$|f(t_1, x_1, u) - f(t_2, x_2, u)| \leq (|t_1 - t_2| + |x_1 - x_2|)\psi_1(|u|); \quad (t_1, x_1), (t_2, x_2) \in J, \ u \in \mathbb{R}.$$

(9.9.3) The function $t \to f(t,x,0)$ is bounded on $J$, and $\lim_{t\to\infty} f(t,x,0) = 0$; $x \in [1,b]$.

(9.9.4) The functions $s \mapsto g_1(t,s)$ and $y \mapsto g_2(x,y)$ have bounded variations for each fixed $t \in [1,\infty)$ or $x \in [1,b]$, respectively. Moreover, the functions $s \mapsto g_1(1,s)$ and $y \mapsto g_2(1,y)$ are nondecreasing on $[1,\infty)$ or $[1,b]$, respectively.

(9.9.5) For each $(t_1,s),(t_2,s) \in \triangle_1, (x_1,y_1),(x_2,y_2) \in \triangle_2$ we have

$$\left| \bigvee_{k_2=1}^{x_1} g_2(x_2,k_2) \bigvee_{k_1=1}^{t_1} g_1(t_2,k_1) - \bigvee_{k_2=1}^{x_1} g_2(x_1,k_2) \bigvee_{k_1=1}^{t_1} g_1(t_1,k_1)) \right| \to 0$$

as $t_1 \to t_2$ and $x_1 \to x_2$.

(9.9.6) $g_1(t,1) = g_2(x,1) = 0$ for any $t \in [1,\infty)$ and any $x \in [1,b]$.

(9.9.7) There exists a continuous function $Q\colon J' \to \mathbb{R}_+$ and a nondecreasing function $\psi_2\colon [0,\infty) \to (0,\infty)$ such that

$$|h(t,x,s,y,u_1,\dots,u_m)| \le \frac{Q(t,x,s,y)}{1+\sum_{i=1}^m |u_i|}\,; \quad (t,x,s,y) \in J',\ u_i \in \mathbb{R},\ i = 1\dots m\,,$$

and

$$\begin{aligned} &|h(t_1,x_1,s,y,u_1,\dots,u_m) - h(t_2,x_2,s,y,u_1,\dots,u_m)| \\ &\le \varphi(s,y)(|t_1-t_2| + |x_1-x_2|)\psi_2\left(\sum_{i=1}^m |u_i|\right)\,; \\ &(t_1,s),(t_2,s) \in \triangle_1,\ (x_1,y),(x_2,y) \in \triangle_2,\ u_i \in \mathbb{R},\ i = 1\dots m\,. \end{aligned}$$

Moreover, assume that

$$\lim_{t\to\infty} \int_1^t \left|\log \frac{t}{s}\right|^{r_1-1} \frac{Q(t,x,s,y)}{s} d_s G_1(t,s) = 0$$

for each $(x,y) \in \triangle_2$, where $G_1(t,s) = \bigvee_{k_1=1}^{s} g_1(t,k_1)$.

**Remark 9.8.** Set

$$\mu^* := \sup_{(t,x)\in J} \mu(t,x), \quad \Phi^* := \sup_{(t,x)\in \tilde{J}} \Phi(t,x), \quad f^* := \sup_{(t,x)\in J} f(t,x,0), \quad p^* := \sup_{(t,x)\in J} P(t,x)$$

and

$$q^* := \sup_{(t,x)\in J} \int_1^t \int_1^x \left|\log \frac{t}{s}\right|^{r_1-1} \left|\log \frac{x}{y}\right|^{r_2-1} \frac{Q(t,x,s,y)}{sy\Gamma(r_1)\Gamma(r_2)} d_y G_2(x,y) d_s G_1(t,s)\,,$$

where $G_2(x,y) = \bigvee_{k_2=1}^{y} g_2(x,k_2)$. From our conditions we infer that $\mu^*, \Phi^*, f^*, p^*$, and $q^*$ are finite.

Now we prove the following theorem concerning the existence and attractivity of solutions of problem (9.14)–(9.15).

**Theorem 9.9.** *Assume* (9.9.1)–(9.9.7). *If*

$$p^* q^* < 1\,, \tag{9.18}$$

*then problem* (9.14)–(9.15) *has at least one solution in the space BC. Moreover, the solutions of problem* (9.14)–(9.15) *are uniformly globally attractive.*

*Proof.* Define the operator $N$ such that, for any $u \in BC$,

$$(Nu)(t,x) = \begin{cases} \Phi(t,x), & (t,x) \in \tilde{J}, \\ \mu(t,x) + \dfrac{f(t,x,u(t,x))}{\Gamma(r_1)\Gamma(r_2)} \displaystyle\int_1^t \int_1^x \left(\log \frac{t}{s}\right)^{r_1-1} \left(\log \frac{x}{y}\right)^{r_2-1} & \\ \times h(t,x,s,y,u(s-\tau_1, y-\xi_1), \ldots, u(s-\tau_m, y-\xi_m)) & \\ \times \dfrac{1}{sy} d_y g_2(x,y) d_s g_1(t,s), & (t,x) \in J. \end{cases} \tag{9.19}$$

The operator $N$ maps $BC$ to $BC$. Indeed, the map $N(u)$ is continuous on $J$ for any $u \in BC$, and for each $(t,x) \in J$ we have

$$\begin{aligned}
|(Nu)(t,x)| &\le |\mu(t,x)| + \frac{|f(t,x,0)|}{\Gamma(r_1)\Gamma(r_2)} \int_1^t \int_1^x \left|\log \frac{t}{s}\right|^{r_1-1} \left|\log \frac{x}{y}\right|^{r_2-1} \\
&\quad \times |h(t,x,s,y,u(s-\tau_1,y-\xi_1),\ldots,u(s-\tau_m,y-\xi_m))| \frac{d_y g_2(x,y) d_s g_1(t,s)}{sy} \\
&\quad + \frac{|f(t,x,u(t,x)) - f(t,x,0)|}{\Gamma(r_1)\Gamma(r_2)} \int_1^t \int_1^x \left|\log \frac{t}{s}\right|^{r_1-1} \left|\log \frac{x}{y}\right|^{r_2-1} \\
&\quad \times |h(t,x,s,y,u(s-\tau_1,y-\xi_1),\ldots,u(s-\tau_m,y-\xi_m))| d_y G_2(x,y) d_s G_1(t,s) \\
&\le \mu^* + \frac{f^*}{\Gamma(r_1)\Gamma(r_2)} \int_1^t \int_1^x \left|\log \frac{t}{s}\right|^{r_1-1} \left|\log \frac{x}{y}\right|^{r_2-1} \\
&\quad \times \frac{Q(t,x,s,y)}{1 + \sum_{i=1}^m |u_i(s-\tau_i, y-\xi_i)|} d_y G_2(x,y) d_s G_1(t,s) \\
&\quad + \frac{P(t,x)|u(t,x)|}{\Gamma(r_1)\Gamma(r_2)} \int_1^t \int_1^x \left|\log \frac{t}{s}\right|^{r_1-1} \left|\log \frac{x}{y}\right|^{r_2-1} \\
&\quad \times Q(t,x,s,y) d_y G_2(x,y) d_s G_1(t,s) \\
&\le \mu^* + \frac{f^*}{\Gamma(r_1)\Gamma(r_2)} \int_1^t \int_1^x \left|\log \frac{t}{s}\right|^{r_1-1} \left|\log \frac{x}{y}\right|^{r_2-1} \\
&\quad \times Q(t,x,s,y) d_y G_2(x,y) d_s G_1(t,s)
\end{aligned}$$

$$+\frac{p^*\|u\|_{BC}}{\Gamma(r_1)\Gamma(r_2)}\int_1^t\int_1^x\left|\log\frac{t}{s}\right|^{r_1-1}\left|\log\frac{x}{y}\right|^{r_2-1}$$
$$\times Q(t,x,s,y)d_yG_2(x,y)d_sG_1(t,s)$$
$$\le \mu^*+q^*(f^*+p^*\|u\|_{BC}).$$

Thus, for each $(t,x)\in J$ we have

$$|(Nu)(t,x)|\le \mu^*+q^*(f^*+p^*\|u\|_{BC}). \tag{9.20}$$

Also, for $(t,x)\in\tilde{J}$ we have

$$|(Nu)(t,x)|=|\Phi(t,x)|\le\Phi^*.$$

Thus,

$$\|Nu\|_{BC}\le\max\{\Phi^*,\mu^*+q^*(f^*+p^*\|u\|_{BC})\}.$$

Hence, $N(u)\subset BC$. This proves that the operator $N$ maps $BC$ to itself.

The issue of finding the solutions of problem (9.14)–(9.15) is reduced to finding the solutions of the operator equation $N(u)=u$. We can show that $N$ transforms the ball $B_\eta:=B(0,\eta)$ into itself, where $\eta=\max\{\Phi^*,\eta^*\}$ and $\eta^*>\frac{\mu^*+q^*f^*}{1-p^*q^*}$. We will show that $N:B_\eta\to B_\eta$ satisfies the conditions of Theorem 1.42. The proof will be given in several steps and cases.

*Step 1: N is continuous.* Let $\{u_n\}_{n\in\mathbb{N}}$ be a sequence such that $u_n\to u$ in $B_\eta$. Then for each $(t,x)\in[-T,\infty)\times[-\xi,b]$ we have

$$|(Nu_n)(t,x)-(Nu)(t,x)|$$
$$\le|f(t,x,u_n(t,x))-f(t,x,u(t,x))|\int_1^t\int_1^x\left|\log\frac{t}{s}\right|^{r_1-1}\left|\log\frac{x}{y}\right|^{r_2-1}$$
$$\times|h(t,x,s,y,u_n(s-\tau_1,y-\xi_1),\dots,u_n(s-\tau_m,y-\xi_m))|\frac{d_yg_2(x,y)d_sg_1(t,s)}{sy\Gamma(r_1)\Gamma(r_2)}$$
$$+|f(t,x,u(t,x))|\int_1^t\int_1^x\left|\log\frac{t}{s}\right|^{r_1-1}\left|\log\frac{x}{y}\right|^{r_2-1}$$
$$\times|h(t,x,s,y,u_n(s-\tau_1,y-\xi_1),\dots,u_n(s-\tau_m,y-\xi_m))$$
$$-h(t,x,s,y,u(s-\tau_1,y-\xi_1),\dots,u(s-\tau_m,y-\xi_m))|\frac{d_yg_2(x,y)d_sg_1(t,s)}{sy\Gamma(r_1)\Gamma(r_2)}$$
$$\le p^*q^*|u_n(t,x)-u(t,x)|+\frac{f^*+p^*\|u\|_{BC}}{\Gamma(r_1)\Gamma(r_2)}\int_1^t\int_1^x\left|\log\frac{t}{s}\right|^{r_1-1}\left|\log\frac{x}{y}\right|^{r_2-1}$$
$$\times|h(t,x,s,y,u_n(s-\tau_1,y-\xi_1),\dots,u_n(s-\tau_m,y-\xi_m))$$
$$-h(t,x,s,y,u(s-\tau_1,y-\xi_1),\dots,u(s-\tau_m,y-\xi_m))|d_yG_2(x,y)d_sG_1(t,s)$$

$$\le p^* q^* |u_n(t,x) - u(t,x)| + \frac{f^* + p^*\eta}{\Gamma(r_1)\Gamma(r_2)} \int_1^t \int_1^x \left|\log \frac{t}{s}\right|^{r_1-1} \left|\log \frac{x}{y}\right|^{r_2-1}$$
$$\times |h(t,x,s,y,u_n(s-\tau_1, y-\xi_1), \ldots, u_n(s-\tau_m, y-\xi_m))$$
$$- h(t,x,s,y,u(s-\tau_1, y-\xi_1), \ldots, u(s-\tau_m, y-\xi_m))| d_y G_2(x,y) d_s G_1(t,s) . \tag{9.21}$$

*Case 1.* If $(t,x) \in \tilde{J} \cup [1,a] \times [1,b]$, $a > 1$, then, since $u_n \to u$ as $n \to \infty$ and $h$ is continuous, (9.21) gives

$$\|N(u_n) - N(u)\|_{BC} \to 0 \text{ as } n \to \infty .$$

*Case 2.* If $(t,x) \in (a,\infty) \times [1,b]$, $a > 1$, then from (9.9.6) and (9.21) we get

$$|(Nu_n)(t,x) - (Nu)(t,x)|$$
$$\le p^* q^* |u_n(t,x) - u(t,x)| + \frac{f^* + p^*\eta}{\Gamma(r_1)\Gamma(r_2)} \int_1^t \int_1^x \left|\log \frac{t}{s}\right|^{r_1-1} \left|\log \frac{x}{y}\right|^{r_2-1}$$
$$\times 2Q(t,x,s,y) d_y G_2(x,y) d_s G_1(t,s)$$
$$\le p^* q^* |u_n(t,x) - u(t,x)| + \frac{2(f^* + p^*\eta)}{\Gamma(r_1)\Gamma(r_2)} \int_1^x \left|\log \frac{x}{y}\right|^{r_2-1} \left[ \int_1^t \left|\log \frac{t}{s}\right|^{r_1-1} \right.$$
$$\left. \times\ Q(t,x,s,y) d_s G_1(t,s) \right] d_y G_2(x,y) . \tag{9.22}$$

Thus, (9.22) gives

$$\|N(u_n) - N(u)\|_{BC} \to 0 \text{ as } n \to \infty .$$

*Step 2: $N(B_\eta)$ is uniformly bounded.* This is clear since $N(B_\eta) \subset B_\eta$ and $B_\eta$ is bounded.

*Step 3: $N(B_\eta)$* is equicontinuous on every compact subset $[-T,a] \times [-\xi,b]$ of $[-T,\infty) \times [-\xi,b]$; $a > 1$. Let $(t_1,x_1), (t_2,x_2) \in [1,a] \times [1,b]$, $t_1 < t_2$, $x_1 < x_2$, and let $u \in B_\eta$. Then we have

$$|(Nu)(t_2,x_2) - (Nu)(t_1,x_1)| \le |\mu(t_2,x_2) - \mu(t_1,x_1)|$$
$$+ \frac{|f(t_2,x_2,u(t_2,x_2)) - f(t_1,x_1,u(t_1,x_1))|}{\Gamma(r_1)\Gamma(r_2)} \int_1^{t_2} \int_1^{x_2} \left|\log \frac{t_2}{s}\right|^{r_1-1} \left|\log \frac{x_2}{y}\right|^{r_2-1}$$
$$\times |h(t_2,x_2,s,y,u(s-\tau_1,y-\xi_1), \ldots, u(s-\tau_m,y-\xi_m))\| d_y g_2(x_2,y) d_s g_1(t_2,s)|$$
$$+ \frac{|f(t_1,x_1,u(t_1,x_1))|}{\Gamma(r_1)\Gamma(r_2)} \int_1^{t_2} \int_1^{x_2} \left|\log \frac{t_2}{s}\right|^{r_1-1} \left|\log \frac{x_2}{y}\right|^{r_2-1}$$
$$\times |h(t_2,x_2,s,y,u(s-\tau_1,y-\xi_1), \ldots, u(s-\tau_m,y-\xi_m))$$

$$
\begin{aligned}
& - h(t_1, x_1, s, y, u(s-\tau_1, y-\xi_1), \dots, u(s-\tau_m, y-\xi_m))| \\
& \quad \times |d_\xi g_2(x_2, \xi) d_s g_1(t_2, s)| \\
& + \left| \frac{|f(t_1, x_1, u(t_1, x_1))|}{\Gamma(r_1)\Gamma(r_2)} \int_1^{t_2}\int_1^{x_2} \left|\log \frac{t_2}{s}\right|^{r_1-1} \left|\log \frac{x_2}{y}\right|^{r_2-1} \right. \\
& \quad \times h(t_1, x_1, s, y, u(s-\tau_1, y-\xi_1), \dots, u(s-\tau_m, y-\xi_m)) d_\xi g_2(x_2, \xi) d_s g_1(t_2, s) \\
& - \frac{|f(t_1, x_1, u(t_1, x_1))|}{\Gamma(r_1)\Gamma(r_2)} \int_1^{t_1}\int_1^{x_1} \left(\log \frac{t_2}{s}\right)^{r_1-1} \left(\log \frac{x_2}{\xi}\right)^{r_2-1} \\
& \quad \left. \times h(t_1, x_1, s, y, u(s-\tau_1, y-\xi_1), \dots, u(s-\tau_m, y-\xi_m)) d_y g_2(x_2, y) d_s g_1(t_2, s) \right| \\
& + \left| \frac{|f(t_1, x_1, u(t_1, x_1))|}{\Gamma(r_1)\Gamma(r_2)} \int_1^{t_1}\int_1^{x_1} \left(\log \frac{t_2}{s}\right)^{r_1-1} \left(\log \frac{x_2}{y}\right)^{r_2-1} \right. \\
& \quad \times h(t_1, x_1, s, y, u(s-\tau_1, y-\xi_1), \dots, u(s-\tau_m, y-\xi_m)) \\
& \quad \left. \times (d_y g_2(x_2, y) d_s g_1(t_2, s) - d_y g_2(x_1, y) d_s g_1(t_1, s)) \right| \\
& + \frac{|f(t_1, x_1, u(t_1, x_1))|}{\Gamma(r_1)\Gamma(r_2)} \int_1^{t_1}\int_1^{x_1} \left| \left(\log \frac{t_2}{s}\right)^{r_1-1} \left(\log \frac{x_2}{y}\right)^{r_2-1} - \left(\log \frac{t_1}{s}\right)^{r_1-1} \left(\log \frac{x_1}{y}\right)^{r_2-1} \right| \\
& \quad \times |h(t_1, x_1, s, y, u(s-\tau_1, y-\xi_1), \dots, u(s-\tau_m, y-\xi_m))| |d_y g_2(x_1, y) d_s g_1(t_1, s)| .
\end{aligned}
$$

Thus, we obtain

$$
\begin{aligned}
& |(Nu)(t_2, x_2) - (Nu)(t_1, x_1)| \le |\mu(t_2, x_2) - \mu(t_1, x_1)| \\
& + \frac{(|t_1 - t_2| + |x_1 - x_2|)\psi_1(|u|)}{\Gamma(r_1)\Gamma(r_2)} \int_1^{t_2}\int_1^{x_2} \left|\log \frac{t_2}{s}\right|^{r_1-1} \left|\log \frac{x_2}{y}\right|^{r_2-1} \\
& \times Q(t_2, x_2, s, y) |d_y g_2(x_2, y) d_s g_1(t_2, s)| \\
& + \frac{|P(t_1, x_1)|u| + |f(t_1, x_1, 0)|}{\Gamma(r_1)\Gamma(r_2)} \int_1^{t_2}\int_1^{x_2} \left|\log \frac{t_2}{s}\right|^{r_1-1} \left|\log \frac{x_2}{y}\right|^{r_2-1} \\
& \times \varphi(s, y)(|t_1 - t_2| + |x_1 - x_2|)\psi_2\left(\sum_{i=1}^{m} |u|\right) |d_\xi g_2(x_2, \xi) d_s g_1(t_2, s)| \\
& + \frac{|P(t_1, x_1)|u| + |f(t_1, x_1, 0)|}{\Gamma(r_1)\Gamma(r_2)} \int_1^{t_2}\int_{x_1}^{x_2} \left|\log \frac{t_2}{s}\right|^{r_1-1} \left|\log \frac{x_2}{y}\right|^{r_2-1} \\
& \times Q(t_1, x_1, s, y) |d_\xi g_2(x_2, \xi) d_s g_1(t_2, s)| \\
& + \frac{|P(t_1, x_1)|u| + |f(t_1, x_1, 0)|}{\Gamma(r_1)\Gamma(r_2)} \int_{t_1}^{t_2}\int_1^{x_2} \left|\log \frac{t_2}{s}\right|^{r_1-1} \left|\log \frac{x_2}{y}\right|^{r_2-1} \\
& \times Q(t_1, x_1, s, y) |d_y g_2(x_2, y) d_s g_1(t_2, s)|
\end{aligned}
$$

$$
\begin{aligned}
&+\frac{|P(t_1,x_1)|u|+|f(t_1,x_1,0)|}{\Gamma(r_1)\Gamma(r_2)}\int_{t_1}^{t_2}\int_{x_1}^{x_2}\left|\log\frac{t_2}{s}\right|^{r_1-1}\left|\log\frac{x_2}{y}\right|^{r_2-1}\\
&\quad\times Q(t_1,x_1,s,y)|d_yg_2(x_2,y)d_sg_1(t_2,s)|\\
&+\frac{|P(t_1,x_1)|u|+|f(t_1,x_1,0)|}{\Gamma(r_1)\Gamma(r_2)}\int_{1}^{t_1}\int_{1}^{x_1}\left(\log\frac{t_2}{s}\right)^{r_1-1}\left(\log\frac{x_2}{y}\right)^{r_2-1}\\
&\quad\times Q(t_1,x_1,s,y)|d_yg_2(x_2,y)d_sg_1(t_2,s)-d_yg_2(x_1,y)d_sg_1(t_1,s)|\\
&+\frac{|P(t_1,x_1)|u|+|f(t_1,x_1,0)|}{\Gamma(r_1)\Gamma(r_2)}\int_{1}^{t_1}\int_{1}^{x_1}Q(t_1,x_1,s,y)\\
&\quad\times\left|\left(\log\frac{t_2}{s}\right)^{r_1-1}\left(\log\frac{x_2}{y}\right)^{r_2-1}-\left(\log\frac{t_1}{s}\right)^{r_1-1}\left(\log\frac{x_1}{y}\right)^{r_2-1}\right|\\
&\quad\times|d_yg_2(x_1,y)d_sg_1(t_1,s)|\,.
\end{aligned}
$$

Hence, we get

$$
\begin{aligned}
&|(Nu)(t_2,x_2)-(Nu)(t_1,x_1)|\le|\mu(t_2,x_2)-\mu(t_1,x_1)|\\
&+\frac{(|t_1-t_2|+|x_1-x_2|)\psi_1(\eta)}{\Gamma(r_1)\Gamma(r_2)}\int_{1}^{t_2}\int_{1}^{x_2}\left|\log\frac{t_2}{s}\right|^{r_1-1}\left|\log\frac{x_2}{y}\right|^{r_2-1}\\
&\times Q(t_2,x_2,s,y)d_yG_2(x_2,y)d_sG_1(t_2,s)\\
&+\frac{(p^*\eta+f^*)(|t_1-t_2|+|x_1-x_2|)\psi_2(m\eta)}{\Gamma(r_1)\Gamma(r_2)}\int_{1}^{t_2}\int_{1}^{x_2}\left|\log\frac{t_2}{s}\right|^{r_1-1}\left|\log\frac{x_2}{y}\right|^{r_2-1}\\
&\times\varphi(s,y)d_yG_2(x_2,y)d_sG_1(t_2,s)\\
&+\frac{p^*\eta+f^*}{\Gamma(r_1)\Gamma(r_2)}\int_{1}^{t_2}\int_{x_1}^{x_2}\left|\log\frac{t_2}{s}\right|^{r_1-1}\left|\log\frac{x_2}{y}\right|^{r_2-1}Q(t_1,x_1,s,y)d_yG_2(x_2,y)d_sG_1(t_2,s)\\
&+\frac{p^*\eta+f^*}{\Gamma(r_1)\Gamma(r_2)}\int_{t_1}^{t_2}\int_{1}^{x_2}\left|\log\frac{t_2}{s}\right|^{r_1-1}\left|\log\frac{x_2}{y}\right|^{r_2-1}Q(t_1,x_1,s,y)d_yG_2(x_2,y)d_sG_1(t_2,s)\\
&+\frac{p^*\eta+f^*}{\Gamma(r_1)\Gamma(r_2)}\int_{t_1}^{t_2}\int_{x_1}^{x_2}\left|\log\frac{t_2}{s}\right|^{r_1-1}\left|\log\frac{x_2}{y}\right|^{r_2-1}Q(t_1,x_1,s,y)|d_yG_2(x_2,y)d_sG_1(t_2,s)\\
&+\frac{p^*\eta+f^*}{\Gamma(r_1)\Gamma(r_2)}\sup_{(s,y)\in[1,t_1]\times[1,x_1]}\left[\left(\log\frac{t_2}{s}\right)^{r_1-1}\left(\log\frac{x_2}{y}\right)^{r_2-1}Q(t_1,x_1,s,y)\right]\\
&\times\left|\bigvee_{k_2=1}^{x_1}g_2(x_2,k_2)\bigvee_{k_1=1}^{t_1}g_1(t_2,k_1)-\bigvee_{k_2=1}^{x_1}g_2(x_1,k_2)d_s\bigvee_{k_1=1}^{t_1}g_1(t_1,k_1)\right|
\end{aligned}
$$

$$+\frac{p^*\eta+f^*}{\Gamma(r_1)\Gamma(r_2)}\int_1^{t_1}\int_1^{x_1}Q(t_1,x_1,s,y)$$
$$\times\left|\left(\log\frac{t_2}{s}\right)^{r_1-1}\left(\log\frac{x_2}{y}\right)^{r_2-1}-\left(\log\frac{t_1}{s}\right)^{r_1-1}\left(\log\frac{x_1}{y}\right)^{r_2-1}\right|$$
$$\times d_yG_2(x_1,y)d_sG_1(t_1,s)\,.$$

From the continuity of $\mu, Q, g_1, g_2$ and as $t_1 \to t_2$ and $x_1 \to x_2$, the right-hand side of the preceding inequality tends to zero. The equicontinuity for the cases $t_1 < t_2 < 0$, $x_1 < x_2 < 0$ and $t_1 \le 0 \le t_2$, $x_1 \le 0 \le x_2$ is obvious.

*Step 4: $N(B_\eta)$ is equiconvergent.* Let $(t,x) \in J$ and $u \in B_\eta$; then we have

$$|(Nu)(t,x)| \le |\mu(t,x)| + \frac{|f(t,x,0)|}{\Gamma(r_1)\Gamma(r_2)}\int_1^{t}\int_1^{x}\left|\log\frac{t}{s}\right|^{r_1-1}\left|\log\frac{x}{y}\right|^{r_2-1}$$
$$\times|h(t,x,s,y,u(s-\tau_1,y-\xi_1),\dots,u(s-\tau_m,y-\xi_m))||d_yg_2(x,y)d_sg_1(t,s)|$$
$$+\frac{|f(t,x,u(t,x))-f(t,x,0)|}{\Gamma(r_1)\Gamma(r_2)}\int_1^{t}\int_1^{x}\left|\log\frac{t}{s}\right|^{r_1-1}\left|\log\frac{x}{y}\right|^{r_2-1}$$
$$\times|h(t,x,s,y,u(s-\tau_1,y-\xi_1),\dots,u(s-\tau_m,y-\xi_m))||d_yg_2(x,y)d_sg_1(t,s)|$$
$$\le |\mu(t,x)| + \frac{f^*}{\Gamma(r_1)\Gamma(r_2)}\int_1^{t}\int_1^{x}\left|\log\frac{t}{s}\right|^{r_1-1}\left|\log\frac{x}{y}\right|^{r_2-1}$$
$$\times\frac{Q(t,x,s,y)}{1+\sum_{i=1}^m|u_i(s-\tau_i,y-\xi_i)|}d_yG_2(x,y)d_sG_1(t,s)$$
$$+\frac{P(t,x)|u(t,x)|}{\Gamma(r_1)\Gamma(r_2)}\int_1^{t}\int_1^{x}\left|\log\frac{t}{s}\right|^{r_1-1}\left|\log\frac{x}{y}\right|^{r_2-1}$$
$$\times Q(t,x,s,y)d_yG_2(x,y)d_sG_1(t,s)$$
$$\le |\mu(t,x)| + \frac{f^*+p^*\eta}{\Gamma(r_1)\Gamma(r_2)}\int_1^{t}\int_1^{x}\left|\log\frac{t}{s}\right|^{r_1-1}\left|\log\frac{x}{y}\right|^{r_2-1}$$
$$\times Q(t,x,s,y)d_yG_2(x,y)d_sG_1(t,s)\,.$$

Thus, for each $x \in [0,b]$ we have

$$|(Nu)(t,x)| \to 0, \quad \text{as } t \to +\infty\,.$$

Hence, for each $x \in [0,b]$ we get

$$|(Nu)(t,x)-(Nu)(+\infty,x)| \to 0, \quad \text{as } t \to +\infty\,.$$

As a consequence of Steps 1–4, together with Lemma 1.57, we can conclude that $N\colon B_\eta \to B_\eta$ is continuous and compact. From an application of Schauder's theorem, we deduce that $N$ has a fixed point $u$ that is a solution of problem (9.14)–(9.15).

*Step 5: Uniform global attractivity of solutions.* Let us assume that $u$ and $v$ are two solutions of problem (9.14)–(9.15). Then for each $(t,x) \in J$ we have

$$
\begin{aligned}
|u(t,x) - v(t,x)| &= |(Nu)(t,x) - (Nv)(t,x)| \\
&\leq p^* q^* |u(t,x) - v(t,x)| + \frac{f^* + p^* \|u\|_{BC}}{\Gamma(r_1)\Gamma(r_2)} \int_1^t \int_1^x \left|\log \frac{t}{s}\right|^{r_1-1} \left|\log \frac{x}{y}\right|^{r_2-1} \\
&\quad \times 2Q(t,x,s,y) d_y G_2(x,y) d_s G_1(t,s) . \\
&\leq p^* q^* |u(t,x) - v(t,x)| + \frac{2(f^* + p^* \|u\|_{BC})}{\Gamma(r_1)\Gamma(r_2)} \\
&\quad \times \int_1^x \left|\log \frac{x}{y}\right|^{r_2-1} \left[ \int_1^t \left|\log \frac{t}{s}\right|^{r_1-1} Q(t,x,s,y) d_s G(t,s) \right] d_y G_2(x,y) .
\end{aligned}
$$

Thus, for each $(t,x) \in J$ we get

$$
\begin{aligned}
|u(t,x) - v(t,x)| \leq{}& \frac{2(f^* + p^* \|u\|_{BC})}{(1 - p^* q^*)\Gamma(r_1)\Gamma(r_2)} \\
&\times \int_1^x \left|\log \frac{x}{y}\right|^{r_2-1} \left[ \int_1^t \left|\log \frac{t}{s}\right|^{r_1-1} Q(t,x,s,y) d_s G(t,s) \right] d_y G_2(x,y) .
\end{aligned}
\tag{9.23}
$$

Using (9.23), we deduce that for each $x \in [0,b]$ we get

$$
\lim_{t \to \infty} |u(t,x) - v(t,x)| = 0 .
$$

Consequently, all solutions of problem (9.14)–(9.15) are uniformly globally attractive. □

### 9.4.3 An Example

As an application and to illustrate our results, we consider the problem of fractional order Volterra–Stieltjes quadratic multidelay Hadamard integral equations

$$
\begin{aligned}
u(t,x) ={}& \mu(t,x) + \frac{f(t,x,u(t,x))}{\Gamma(r_1)\Gamma(r_2)} \int_1^t \int_1^x \left(\log \frac{t}{s}\right)^{r_1-1} \left(\log \frac{x}{y}\right)^{r_2-1} \\
&\times h\left(t,x,s,y,u(s-1,y-2), u\left(s - \frac{1}{2}, y - \frac{2}{5}\right)\right) \\
&\times \frac{1}{sy} d_y g_2(x,y) d_s g_1(t,s) \quad \text{if } (t,x) \in J := [1,+\infty) \times [1,e] ,
\end{aligned}
\tag{9.24}
$$

$$
u(t,x) = \frac{2}{(2+t^2)(2+x^2)} \quad \text{if } (t,x) \in \tilde{J} := [-1,\infty) \times [-2,e] \backslash (1,\infty) \times (1,e] , \tag{9.25}
$$

where

$$r_1 = \frac{1}{4}, \quad r_2 = \frac{1}{2}, \quad \mu(t,x) = \frac{1}{2+t^2+x^2}; \quad (t,x) \in J,$$
$$f(t,x,u) = 1 + \frac{e^{x-t}|u|}{1+|u|}; \quad (t,x) \in J, \quad u \in \mathbb{R},$$
$$g_1(t,s) = s; \quad (t,s) \in [1,+\infty)^2, \quad g_2(x,y) = y; \quad (x,y) \in [1,e]^2,$$
$$h(t,x,s,y,u_1,u_2) = \frac{cxs^{\frac{-3}{4}}|u| \sin\sqrt{t} \sin s}{(1+t^2+y^2)(2+|u_1|+|u_2|)}; \quad (t,x,s,y) \in J', \quad u_1,u_2 \in \mathbb{R},$$
$$c = \frac{\pi}{16e\Gamma\left(\frac{1}{4}\right)} \text{ and } J' = \{(t,x,s,y) : 1 \le s \le t, \ 1 \le y \le x \le e\}.$$

Set

$$\tau_1 = 1, \tau_2 = \frac{1}{2}, \quad \xi_1 = 2, \quad \xi_2 = \frac{2}{5} \text{ and } \Phi(t,x) = \frac{2}{(2+t^2)(2+x^2)}; \quad (t,x) \in \tilde{J}.$$

Then $T = e$ and $\xi = 2$.
First, we can see that for each $x \in [1,e]$ we have $\lim\limits_{t\to+\infty} \frac{1}{2+t^2+x^2} = 0$. Then (9.9.1) is satisfied by $\mu^* = \frac{1}{4}$, and $\Phi^* = \frac{2}{9}$.
Next, the function $f$ is continuous, $f^* = 1$, and

$$|f(t,x,u) - f(t,x,v)| \le e^{x-t}|u-v|; \quad (t,x) \in J, \ u,v \in \mathbb{R}.$$

Then (9.9.2) is satisfied by $P(t,x) = e^{x-t}$; $(t,x) \in J$, and then $p^* = e^2$. Also, we can easily see that the function $g$ satisfies conditions (9.9.3)–(9.9.5). Also, the function $h$ satisfies condition (9.9.6). Indeed, $h$ is continuous and

$$|h(t,x,s,y,u_1,u_2)| \le \frac{Q(t,x,s,y)}{1+|u_1|+|u_2|}, \quad (t,x,s,y) \in J', \ u_1,u_2 \in \mathbb{R},$$

where

$$Q(t,x,s,y) = \frac{cxs^{\frac{-3}{4}} \sin\sqrt{t} \sin s}{1+t^2+y^2}, \quad (t,x,s,y) \in J'.$$

Then

$$\begin{aligned}
\left| \int_1^t \left|\log\frac{t}{s}\right|^{r_1-1} \frac{Q(t,x,s,y)}{s} d_s g_1(t,s) \right| &\le \int_1^t \left|\log\frac{t}{s}\right|^{\frac{-3}{4}} cxs^{\frac{-7}{4}} |\sin\sqrt{t}\sin s| d_s G_1(t,s) \\
&\le cx|\sin\sqrt{t}| \int_1^t \left|\log\frac{t}{s}\right|^{\frac{-3}{4}} s^{\frac{-7}{4}} ds \\
&\le \frac{cx\Gamma^2\left(\frac{1}{4}\right)}{\sqrt{\pi}} \left|\frac{\sin\sqrt{t}}{\sqrt{\log t}}\right| \\
&\le \frac{cx\Gamma^2\left(\frac{1}{4}\right)}{\sqrt{\pi \log t}} \longrightarrow 0 \text{ as } t \to \infty
\end{aligned}$$

and

$$q^* := \sup_{(t,x)\in J} \frac{1}{\Gamma(r_1)\Gamma(r_2)} \int_1^t \int_1^x \left|\log \frac{t}{s}\right|^{r_1-1} \left|\log \frac{x}{y}\right|^{r_2-1} \frac{Q(t,x,s,y)}{sy} dyds$$

$$\leq \sup_{(t,x)\in J} \frac{1}{\Gamma\left(\frac{3}{2}\right)} \frac{cx\Gamma\left(\frac{1}{4}\right)}{\sqrt{\pi}} \left|\frac{\sin\sqrt{t}}{\sqrt{\log t}}\right| \leq \frac{2ec\Gamma\left(\frac{1}{4}\right)}{\pi} = \frac{1}{8}.$$

Finally, we can see that $p^* q^* \leq \frac{e^2}{8} < 1$. Consequently, by Theorem 9.9, problem (7.4)–(9.25) has at least one solution in the space $BC([-1,\infty) \times [-2,e])$, and all solutions of (9.24)–(9.25) are uniformly globally attractive.

## 9.5 Notes and Remarks

The results of Chapter 9 are taken from Abbas et al. [4, 3]. Other results may be found in [19, 18, 16, 30, 33, 38].

# 10 Ulam Stabilities for Random Hadamard Fractional Integral Equations

## 10.1 Introduction

Let $\beta_E$ be the $\sigma$-algebra of Borel subsets of $E$. A mapping $v\colon \Omega \to E$ is said to be measurable if for any $B \in \beta_E$ one has

$$v^{-1}(B) = \{w \in \Omega\colon v(w) \in B\} \subset \mathcal{A}\,.$$

To define integrals of sample paths of a random process, it is necessary to define a jointly measurable map.

**Definition 10.1.** A mapping $T\colon \Omega \times E \to E$ is called jointly measurable if for any $B \in \beta_E$ one has

$$T^{-1}(B) = \{(w, v) \in \Omega \times E\colon T(w, v) \in B\} \subset \mathcal{A} \times \beta_E\,,$$

where $\mathcal{A} \times \beta_E$ is the direct product of the $\sigma$-algebras $\mathcal{A}$ and $\beta_E$ defined in $\Omega$ and $E$, respectively.

**Lemma 10.2** ([136]). *Let $T\colon \Omega \times E \to E$ be a mapping such that $T(., v)$ is measurable for all $v \in E$, and $T(w, .)$ is continuous for all $w \in \Omega$. Then the map $(w, v) \mapsto T(w, v)$ is jointly measurable.*

**Definition 10.3** ([156]). A function $f\colon J \times E \times \Omega \to E$ is called random Carathéodory if the following conditions hold.
(i) The map $(x, y, w) \to f(x, y, u, w)$ is jointly measurable for all $u \in E$.
(ii) The map $u \to f(x, y, u, w)$ is continuous for almost all $(x, y) \in J$ and $w \in \Omega$.

Let $T\colon \Omega \times E \to E$ be a mapping. Then $T$ is called a random operator if $T(w, u)$ is measurable in $w$ for all $u \in E$ and it is expressed as $T(w)u = T(w, u)$. In this case we also say that $T(w)$ is a random operator on $E$. A random operator $T(w)$ on $E$ is called continuous (resp. compact, totally bounded, and completely continuous) if $T(w, u)$ is continuous (resp. compact, totally bounded, and completely continuous) in $u$ for all $w \in \Omega$. The details of completely continuous random operators in Banach spaces and their properties appear in Itoh [169].

**Definition 10.4** ([140]). Let $\mathcal{P}(Y)$ be the family of all nonempty subsets of $Y$ and $C$ a mapping from $\Omega$ to $\mathcal{P}(Y)$. A mapping $T\colon \{(w, y)\colon w \in \Omega, y \in C(w)\} \to Y$ is called a random operator with stochastic domain $C$ if $C$ is measurable (i.e., for all closed $A \subset Y$, $\{w \in \Omega, C(w) \cap A \neq \emptyset\}$ is measurable) and for all open $D \subset Y$ and all $y \in Y$, $\{w \in \Omega\colon y \in C(w), T(w, y) \in D\}$ is measurable. $T$ will be called continuous if every $T(w)$ is continuous. For a random operator $T$, a mapping $y\colon \Omega \to Y$ is called a random (stochastic) fixed point of $T$ if for $P$-almost all $w \in \Omega$, $y(w) \in C(w)$ and $T(w)y(w) = y(w)$ and for all open $D \subset Y$, $\{w \in \Omega\colon y(w) \in D\}$ is measurable.

https://doi.org/10.1515/9783110553819-010

Let $\emptyset \neq \Lambda \subset BC$, let $G: \Lambda \to \Lambda$, and consider the solutions of the random equation

$$G(w)u(t,x) = u(t,x,w); \quad w \in \Omega. \tag{10.1}$$

Inspired by the definition of the attractivity of solutions of integral equations (e.g., [36]), we introduce the following concept of attractivity of solutions for random equation (10.1).

**Definition 10.5.** Solutions of random equation (10.1) are locally attractive if there exists a ball $B(u_0, \eta)$ in the space $BC$ such that, for arbitrary random solutions $v = v(t,x,w)$ and $z = z(t,x,w)$ of equations (10.1) belonging to $B(u_0,\eta) \cap \Lambda$, we have that, for each $x \in [0,b]$ and $w \in \Omega$,

$$\lim_{t\to\infty} (v(t,x,w) - z(t,x,w)) = 0. \tag{10.2}$$

When the limit (10.2) is uniform with respect to $B(u_0,\eta) \cap \Lambda$, solutions of equation (10.1) are said to be uniformly locally attractive (or, equivalently, that solutions of (10.1) are locally asymptotically stable).

**Definition 10.6.** The solution $v = v(t,x,w)$ of random equation (10.1) is said to be globally attractive if (10.2) holds for each solution $z = z(t,x,w)$ of (10.1). If condition (10.2) is satisfied uniformly with respect to the set $\Lambda$, solutions of equation (10.1) are said to be globally asymptotically stable (or uniformly globally attractive).

In the sequel, we employ the following random fixed point theorem.

**Theorem 10.7** (Itoh [169]). *Let $X$ be a nonempty, closed, convex, bounded subset of a Banach space $E$, and let $N: \Omega \times X \to X$ be a compact and continuous random operator. Then the random equation $N(w)u = u$ has a random solution.*

## 10.2 Partial Hadamard Fractional Integral Equations with Random Effects

### 10.2.1 Introduction

This section deals with some existence results and Ulam stabilities for a class of random partial functional partial integral equations via Hadamard's fractional integral by applying random fixed point theorem with a stochastic domain.

This section deals with the existence of the Ulam stability of solutions to the Hadamard partial fractional integral equation of the form

$$\begin{aligned} u(x,y,w) &= \mu(x,y,w) \\ &\quad + \frac{1}{\Gamma(r_1)\Gamma(r_2)} \int_1^x \int_1^y \left(\log\frac{x}{s}\right)^{r_1-1} \left(\log\frac{y}{t}\right)^{r_2-1} \frac{f(s,t,u(s,t,w),w)}{st}\,dt\,ds\,; \\ &\quad \text{if } (x,y) \in J,\ w \in \Omega, \end{aligned} \tag{10.3}$$

where $J := [1, a] \times [1, b]$, $a, b > 1$, $r_1, r_2 > 0$, $(\Omega, \mathcal{A})$ is a measurable space, and $\mu: J \times \Omega \to \mathbb{R}$ and $f: J \times \mathbb{R} \times \Omega \to \mathbb{R}$ are given continuous functions.

### 10.2.2 Existence and Ulam Stabilities Results

In this section, we discuss the existence of solutions, and we present conditions for the Ulam stability for the Hadamard integral equation (10.3).

**Lemma 10.8** ([113]). *If $Y$ is a bounded subset of Banach space $X$, then for each $\epsilon > 0$ there is a sequence $\{y_k\}_{k=1}^{\infty} \subset Y$ such that*

$$\alpha(Y) \leq 2\alpha(\{y_k\}_{k=1}^{\infty}) + \epsilon .$$

**Lemma 10.9** ([202, 261]). *() If $\{u_k\}_{k=1}^{\infty} \subset L^1(J)$ is uniformly integrable, then $\alpha(\{u_k\}_{k=1}^{\infty})$ is measurable and for each $(x, y) \in J$*

$$\alpha\left(\left\{\int_0^x\int_0^y u_k(s,t)dtds\right\}_{k=1}^{\infty}\right) \leq 2\int_0^x\int_0^y \alpha(\{u_k(s,t)\}_{k=1}^{\infty})dtds .$$

**Lemma 10.10** ([195]). *Let $F$ be a closed and convex subset of a real Banach space, and let $G: F \to F$ be a continuous operator and $G(F)$ be bounded. If there exists a constant $k \in [0, 1)$ such that for each bounded subset $B \subset F$,*

$$\alpha(G(B)) \leq k\alpha(B) ,$$

*then $G$ has a fixed point in $F$.*

The following conditions will be used in the sequel.

(10.4.1) The function $w \mapsto \mu(x, y, w)$ is measurable and bounded for a.e. $(x, y) \in J$.

(10.4.2) The function $f$ is random Carathéodory on $J \times \mathbb{R} \times \Omega$.

(10.4.3) There exist functions $p_1, p_2: J \times \Omega \to [0, \infty)$ with $p_i(w) \in C(J, \mathbb{R}_+)$; $i = 1, 2$ such that for each $w \in \Omega$

$$|f(x, y, u, w)| \leq p_1(x, y, w) + \frac{p_2(x, y, w)}{1 + |u(x, y)|}|u(x, y, w)|$$

for all $u \in \mathbb{R}$ and a.e. $(x, y) \in J$.

(10.4.4) There exists a function $q: J \times \Omega \to [0, \infty)$ with $q(w) \in L^{\infty}(J, [0, \infty))$ for each $w \in \Omega$ such that for any bounded $B \subset \mathbb{R}$

$$\alpha(f(x, y, B, w)) \leq q(x, y, w)\alpha(B) , \quad \text{for a.e. } (x, y) \in J .$$

(10.4.5) There exists a random function $R: \Omega \to (0, \infty)$ such that

$$R(w) \geq \mu^*(w) + \frac{(p_1^*(w) + p_2^*(w))(\log a)^{r_1}(\log b)^{r_2}}{\Gamma(1 + r_1)\Gamma(1 + r_2)} ,$$

where

$$\mu^*(w) = \sup_{(x,y)\in J} |\mu(x,y,w)|\,, \quad p_i^*(w) = \sup_{(x,y)\in J}\text{ess}\, p_i(x,y,w); \quad i = 1,2\,.$$

(10.4.6) There exist $q_1, q_2 \colon J\times\Omega \to [0,\infty)$, with $q_i(.,w) \in L^\infty(J,[0,\infty))$, $i = 1,2$, such that for each $w \in \Omega$ and a.e. $(x,y) \in J$ we have

$$p_i(x,y,w) \le q_i(x,y,w,w)\Phi(x,y,w)\,.$$

(10.4.7) $\Phi(w) \in L^1(J,[0,\infty))$ for all $w \in \Omega$, and there exists $\lambda_\Phi > 0$ such that for each $(x,y) \in J$ we have

$$({}^H I_\sigma^r \Phi)(x,y,w) \le \lambda_\Phi \Phi(x,y,w)\,,$$

$$q^* = \sup_{(x,y,w)\in J\times\Omega}\text{ess}\; q(x,y,w)\,.$$

**Theorem 10.11.** *Assume* (10.4.1)–(10.4.5). *If*

$$\ell := \frac{4q^*(\log a)^{r_1}(\log b)^{r_2}}{\Gamma(1+r_1)\Gamma(1+r_2)} < 1\,,$$

*then integral equation* (10.3) *has a random solution defined on* $J$. *Furthermore, if conditions* (10.4.6) *and* (10.4.7) *hold, then the random equation* (10.3) *is generalized Ulam–Hyers–Rassias stable.*

*Proof.* From conditions (10.4.2) and (10.4.3), for each $w \in \Omega$ and almost all $(x,y) \in J$, we have that $f(x,y,u(x,y,w),w)$ is in $L^1$. Since the function $f$ is continuous, the indefinite integral is continuous for all $w \in \Omega$ and almost all $(x,y) \in J$. Again, as the map $\mu$ is continuous for all $w \in \Omega$ and the indefinite integral is continuous on $J$, $N(w)$ defines a mapping $N \colon \Omega\times C \to C$. Hence, $u$ is a solution for integral equation (10.3) if and only if $u = (N(w))u$.

We will show that the operator $N$ satisfies all conditions of Lemma 10.10. The proof will be given in several steps.

*Step 1: $N(w)$ is a random operator with a stochastic domain on $C$.* Since $f(x,y,u,w)$ is random Carathéodory, the map $w \to f(x,y,u,w)$ is measurable. Similarly, the product $(\log\frac{x}{s})^{r_1-1}(\log\frac{y}{t})^{r_2-1}\frac{f(s,t,u(s,t,w),w)}{st}$ of a continuous and measurable function is again measurable. Further, the integral is a limit of a finite sum of measurable functions; therefore, the map

$$w \mapsto \mu(x,y,w) + \int_0^x\int_0^y \left(\log\frac{x}{s}\right)^{r_1-1}\left(\log\frac{y}{t}\right)^{r_2-1}\frac{f(s,t,u(s,t,w),w)}{st\Gamma(r_1)\Gamma(r_2)}dtds$$

is measurable. As a result, $N$ is a random operator on $\Omega\times C$ to $C$.

Let $W \colon \Omega \to \mathcal{P}(C)$ be defined by

$$W(w) = \{u \in C \colon \|u\|_C \le R(w)\}\,,$$

with $W(w)$ bounded, closed, convex, and solid for all $w \in \Omega$. Then $W$ is measurable by Lemma [[140], Lemma 17]. Let $w \in \Omega$ be fixed; then from (10.4.4), for any $u \in w(w)$, we get

$$\begin{aligned}
&|(N(w)u)(x,y)| \\
&\le |\mu(x,y,w)| + \int_0^x\int_0^y \left|\log\frac{x}{s}\right|^{r_1-1}\left|\log\frac{y}{t}\right|^{r_2-1}\frac{|f(s,t,u(s,t,w),w)|}{st\Gamma(r_1)\Gamma(r_2)}dtds \\
&\le |\mu(x,y,w)| + \int_0^x\int_0^y \left|\log\frac{x}{s}\right|^{r_1-1}\left|\log\frac{y}{t}\right|^{r_2-1}\frac{|p_1(s,t,w)+p_2(s,t,w)|}{\Gamma(r_1)\Gamma(r_2)}dtds \\
&\le \mu^*(w) + \frac{(p_1^*(w)+p_2^*(w))(\log a)^{r_1}(\log b)^{r_2}}{\Gamma(1+r_1)\Gamma(1+r_2)} \\
&\le R(w)\,.
\end{aligned}$$

Therefore, $N$ is a random operator with stochastic domain $W$ and $N(w)\colon W(w) \to N(w)$. Furthermore, $N(w)$ maps bounded sets to bounded sets in $C$.

*Step 2: $N(w)$ is continuous.* Let $\{u_n\}$ be a sequence such that $u_n \to u$ in $\mathcal{C}$. Then, for each $(x,y) \in J$ and $w \in \Omega$, we have

$$\begin{aligned}
|(N(w)u_n)(x,y) - (N(w)u)(x,y)| &\le \int_0^x\int_0^y \left|\log\frac{x}{s}\right|^{r_1-1}\left|\log\frac{y}{t}\right|^{r_2-1} \\
&\quad\times \frac{|f(s,t,u_n(s,t,w),w) - f(s,t,u(s,t,w),w)|}{\Gamma(r_1)\Gamma(r_2)}dtds\,.
\end{aligned}$$

Using the Lebesgue dominated convergence theorem, we get

$$\|N(w)u_n - N(w)u\|_C \to 0 \ \text{ as } \ n \to \infty\,.$$

As a consequence of Steps 1 and 2, we can conclude that $N(w)\colon W(w) \to N(w)$ is a continuous random operator with stochastic domain $W$, and $N(w)(W(w))$ is bounded.

*Step 3: For each bounded subset $B$ of $W(w)$ we have $\alpha(N(w)B) \le \ell\alpha(B)$.* Let $w \in \Omega$ be fixed. From Lemmas 10.8 and 10.9, for any $B \subset W$ and any $\epsilon > 0$ there exists a sequence $\{u_n\}_{n=0}^{\infty} \subset B$, such that for all $(x,y) \in J$ we have

$$\begin{aligned}
&\alpha((N(w)B)(x,y)) \\
&= \alpha\left(\left\{\mu(x,y) + \int_1^x\int_1^y \left(\log\frac{x}{s}\right)^{r_1-1}\left(\log\frac{y}{t}\right)^{r_2-1}\frac{f(s,t,u(s,t,w),w)}{st\Gamma(r_1)\Gamma(r_2)}dtds;\ u\in B\right\}\right) \\
&\le 2\alpha\left(\left\{\int_1^x\int_1^y \left(\log\frac{x}{s}\right)^{r_1-1}\left(\log\frac{y}{t}\right)^{r_2-1}\frac{f(s,t,u_n(s,t,w),w)}{st\Gamma(r_1)\Gamma(r_2)}dtds\right\}_{n=1}^{\infty}\right) + \epsilon \\
&\le 4\int_1^x\int_1^y \alpha\left(\left\{\left(\log\frac{x}{s}\right)^{r_1-1}\left(\log\frac{y}{t}\right)^{r_2-1}\frac{f(s,t,u(s,t,w),w)}{st\Gamma(r_1)\Gamma(r_2)}dtds\right\}_{n=1}^{\infty}\right)dtds + \epsilon
\end{aligned}$$

$$\leq 4\int_1^x\int_1^y \left(\log\frac{x}{s}\right)^{r_1-1}\left(\log\frac{y}{t}\right)^{r_2-1}\frac{1}{\Gamma(r_1)\Gamma(r_2)}\alpha\left(\{f(s,t,u_n(s,t,w),w)\}_{n=1}^{\infty}\right)dtds+\epsilon$$

$$\leq 4\int_1^x\int_1^y \left(\log\frac{x}{s}\right)^{r_1-1}\left(\log\frac{y}{t}\right)^{r_2-1}\frac{1}{\Gamma(r_1)\Gamma(r_2)}q(s,t,w)\alpha\left(\{u_n(s,t,w)\}_{n=1}^{\infty}\right)dtds+\epsilon$$

$$\leq \left(4\int_1^x\int_1^y \left(\log\frac{x}{s}\right)^{r_1-1}\left(\log\frac{y}{t}\right)^{r_2-1}\frac{1}{\Gamma(r_1)\Gamma(r_2)}q(s,t,w)dsdt\right)\alpha\left(\{u_n\}_{n=1}^{\infty}\right)+\epsilon$$

$$\leq \left(4\int_1^x\int_1^y \left(\log\frac{x}{s}\right)^{r_1-1}\left(\log\frac{y}{t}\right)^{r_2-1}\frac{1}{\Gamma(r_1)\Gamma(r_2)}q(s,t,w)dtds\right)\alpha(B)+\epsilon$$

$$\leq \frac{4q^*(\log a)^{r_1}(\log b)^{r_2}}{\Gamma(1+r_1)\Gamma(1+r_2)}\alpha(B)+\epsilon$$

$$=\ell\alpha(B)+\epsilon\,.$$

Since $\epsilon>0$ is arbitrary,

$$\alpha(N(B))\leq \ell\alpha(B)\,.$$

Hence, from Lemma 10.10 it follows that for each $w\in\Omega$, $N$ has at least one fixed point in $W$. Since $\bigcap_{w\in\Omega} intW(w)\neq\emptyset$, the measurable selector of $intW$ exists. From Lemma 10.10, the operator $N$ has a stochastic fixed point, i.e., integral equation (10.3) has at least one random solution on $C$.

*Step 4: Generalized Ulam–Hyers–Rassias stability.* Set

$$q_i^*=\sup_{(x,y,w)\in J\times\Omega}\text{ess }q_i(x,y,w)\,;\quad i=1,2\,.$$

Let $u\colon\Omega\to C$ be a solution of inequality (9.8). By Theorem 10.11, there exists $v$, which is a solution of random equation (10.3). Hence,

$$v(x,y,w)=\mu(x,y,w)+\int_1^x\int_1^y \left(\log\frac{x}{s}\right)^{r_1-1}\left(\log\frac{y}{t}\right)^{r_2-1}\frac{f(s,t,v(s,t,w),w)}{st\Gamma(r_1)\Gamma(r_2)}dtds\,;\quad (x,y)\in J,\ w\in\Omega\,.$$

From conditions (10.4.6) and (10.4.7), for each $(x,y)\in J$ and $w\in\Omega$, we have

$$|u(x,y,w)-v(x,y,w)|\leq|u(x,y,w)-N(w)(u)|+|N(w)(u)-N(w)(v)|$$

$$\leq\Phi(x,y,w)+\int_1^x\int_1^y\left|\log\frac{x}{s}\right|^{r_1-1}\left|\log\frac{y}{t}\right|^{r_2-1}\frac{|f(s,t,u(s,t,w))-f(s,t,v(s,t,w))|}{\Gamma(r_1)\Gamma(r_2)}dtds$$

$$\leq\Phi(x,y,w)+\frac{1}{\Gamma(r_1)\Gamma(r_2)}\int_1^x\int_1^y\left|\log\frac{x}{s}\right|^{r_1-1}\left|\log\frac{y}{t}\right|^{r_2-1}$$

$$\times\left(2q_1^*+\frac{q_2^*|u(s,t,w)|}{1+|u|}+\frac{q_2^*|v(s,t,w)|}{1+|v|}\right)\frac{\Phi(s,t,w)}{st}dtds$$

$$\begin{aligned} &\leq \Phi(x,y,w) + 2(q_1^* + q_2^*)({}^H I_\sigma^r \Phi)(x,y,w) \\ &\leq [1 + 2(q_1^* + q_2^*)\lambda_\Phi]\Phi(x,y,w) \\ &:= c_{N,\Phi}\Phi(x,y,w)\,. \end{aligned}$$

Hence, random equation (10.3) is generalized Ulam–Hyers–Rassias stable. □

### 10.2.3 An Example

Let $E = \mathbb{R}$, $\Omega = (-\infty, 0)$ be equipped with the usual $\sigma$-algebra consisting of Lebesgue measurable subsets of $(-\infty, 0)$. Given a measurable function $u\colon \Omega \to C([1,e]\times[1,e])$, consider the partial random Hadamard integral equation

$$\begin{aligned} u(x,y,w) &= \mu(x,y,w) \\ &\quad + \int_1^x\int_1^y \left(\log\frac{x}{s}\right)^{r_1-1}\left(\log\frac{y}{t}\right)^{r_2-1}\frac{f(s,t,u(s,t,w),w)}{st\Gamma(r_1)\Gamma(r_2)}dtds \end{aligned} \tag{10.4}$$

for $(x,y)\in[1,e]\times[1,e]$, $w\in\Omega$, where

$$r_1, r_2 > 0\,,\quad \mu(x,y,w) = x\sin w + y^2\cos w;\ \ (x,y)\in[1,e]\times[1,e],$$

and

$$f(x,y,u(x,y)) = \frac{w^2xy^2}{(1+w^2+u(x,y,w)|)e^{x+y+3}},\quad (x,y)\in[1,e]\times[1,e],\ \ w\in\Omega\,.$$

The function $w \mapsto \mu(x,y,w) = x\sin w + y^2\cos w$ is measurable and bounded, with

$$|\mu(x,y,w)| \leq e + e^2\,;$$

hence, condition (10.4.1) is satisfied.

The map $(x,y,w) \mapsto f(x,y,u,w)$ is jointly continuous for all $u\in\mathbb{R}$, so jointly measurable for all $u\in\mathbb{R}$. Also, the map $u\mapsto f(x,y,u,w)$ is continuous for all $(x,y)\in[1,e]\times[1,e]$ and $w\in\Omega$. So the function $f$ is Carathéodory on $[1,e]\times[1,e]\times\mathbb{R}\times\Omega$.

For each $u\in\mathbb{R}$, $(x,y)\in[1,e]\times[1,e]$ and $w\in\Omega$ we have

$$|f(x,y,u,w)| \leq w^2xy^2\left(1+\frac{1}{e^3}|u|\right).$$

Hence, condition (10.4.3) is satisfied by $p_1^* = e^3$ and $p_1(x,y,w) = p_2^* = 1$. The condition $\ell < 1$ holds with $a = b = e$ and $q^* = \frac{1}{e^3}$. Indeed, for each $r_1, r_2 > 0$ we get

$$\begin{aligned} \ell &= \frac{4q^*(\log a)^{r_1}(\log b)^{r_2}}{\Gamma(1+r_1)\Gamma(1+r_2)} \\ &\leq \frac{4}{e^3\Gamma(1+r_1)\Gamma(1+r_2)} \\ &< 1\,. \end{aligned}$$

Condition (10.4.6) is satisfied by

$$\Phi(x, y, w) = w^2 w^2 xy^2, \quad \text{and } \lambda_\Phi = \frac{1}{\Gamma(1 + r_1)\Gamma(1 + r_2)} .$$

Indeed, for each $(x, y) \in [1, e] \times [1, e]$ we get

$$\begin{aligned}({}^H I_\sigma^r \Phi)(x, y, w) &\leq \frac{w^2 e^3}{\Gamma(1 + r_1)\Gamma(1 + r_2)} \\ &= \lambda_\Phi \Phi(x, y, w) .\end{aligned}$$

Finally, we can see that condition (10.4.7) is satisfied by $q_1(x, y, w) = 1$ and $q_2(x, y, w) = \frac{1}{e^3}$. Consequently, Theorem 10.11 implies that the Hadamard integral equation (10.4) has a solution defined on $[1, e] \times [1, e]$, and (10.4) is generalized Ulam–Hyers–Rassias stable.

## 10.3 Global Stability Results for Volterra–Hadamard Random Partial Fractional Integral Equations

### 10.3.1 Introduction

This section deals with the existence and stability of random solutions of a class of functional partial integral equations of Hadamard fractional order with random effects in Banach spaces.

The initial value problems of ordinary random differential equations have been studied in the literature on bounded as well as unbounded intervals. See, for example, Burton and Furumochi [114], Zielinski et al. [265], and the references therein.

In [8, 32], Abbas et al. studied existence and stability results for some classes of nonlinear differential and integral equations of fractional order. This section deals with the existence and the asymptotic behavior of random solutions to the nonlinear quadratic Volterra random partial integral equation of Hadamard fractional order

$$\begin{aligned}u(t, x, w) = f(t, x, u(t, x, w), w) + \frac{1}{\Gamma(r_1)\Gamma(r_2)} \int_1^t \int_1^x \left(\log \frac{t}{s}\right)^{r_1 - 1} \left(\log \frac{x}{\xi}\right)^{r_2 - 1} \\ \times g(t, x, s, \xi, u(s, \xi, w), w) \frac{d\xi ds}{s\xi} , \quad (t, x) \in J := [1, \infty) \times [1, b], \; w \in \Omega ,\end{aligned} \tag{10.5}$$

where $b > 1, r_1, r_2 \in (0, \infty)$, $\alpha, \beta, y \colon [1, \infty) \to [1, \infty)$, $(\Omega, \mathcal{A})$ is a measurable space, $f \colon J \times \mathbb{R} \times \Omega \to \mathbb{R}$ and $g \colon J_1 \times \mathbb{R} \times \Omega \to \mathbb{R}$ are given continuous functions, and $J_1 = \{(t, x, s, \xi) \colon 1 \leq s \leq t, 1 \leq \xi \leq s \leq b\}$. Our existence results are based on Itoh's random fixed point theorem. Also, we obtain some results about the global asymptotic stability of random solutions of the integral equation in question. Finally, we present an example illustrating the applicability of the imposed conditions.

### 10.3.2 Existence of Random Solutions and Global Stability Results

In this section, we are concerned with the existence and the asymptotic stability of random solutions for the Hadamard partial integral equation (10.5). The following conditions will be used in the sequel:

(10.5.1) The functions $f$ and $g$ are random Carathéeodory.

(10.5.2) There exist a constant $M, L > 0$ with $M < L$ and a nondecreasing function $\psi_1 : [0, \infty) \to (0, \infty)$ such that

$$|f(t, x, u, w) - f(t, x, v, w)| \leq \frac{M|u - v|}{(1 + t)(L + |u - v|)}$$

and

$$|f(t_1, x_1, u, w) - f(t_2, x_2, u, w)| \leq (|t_1 - t_2| + |x_1 - x_2|)\psi_1(|u|)$$

for each $(t, x), (t_1, x_1), (t_2, x_2) \in J$, $u, v \in \mathbb{R}$ and $w \in \Omega$.

(10.5.3) The function $t \to f(t, x, 0, 0, w)$ is bounded on $J \times \Omega$ with

$$f^* = \sup_{(t,x,w) \in J \times \Omega} f(t, x, 0, 0, w)$$

and

$$\lim_{t \to \infty} |f(t, x, 0, 0, w)| = 0 ; \quad x \in [1, b], \; w \in \Omega .$$

(10.5.4) There exist continuous measurable functions $\varphi : J \times \Omega \to \mathbb{R}_+$, $p : J_1 \times \Omega \to \mathbb{R}_+$ and a nondecreasing function $\psi_2 : [0, \infty) \to (0, \infty)$ such that

$$|g(t_1, x_1, s, \xi, u, w) - g(t_2, x_2, s, \xi, u, w)| \leq \varphi(s, \xi, w)(|x_1 - x_2| + |y_1 - y_2|)\psi_2(|u|)$$

and

$$|g(t, x, s, \xi, u, w)| \leq \frac{p(t, x, s, \xi, w)}{1 + t + |u|}$$

for each $(t, x), (s, t), (t_1, x_1), (t_2, x_2) \in J$, $u \in \mathbb{R}$, and $w \in \Omega$. Moreover, assume that

$$\lim_{t \to \infty} \int_1^t \int_1^x \left|\log \frac{t}{s}\right|^{r_1 - 1} \left|\log \frac{x}{\xi}\right|^{r_2 - 1} p(t, x, s, \xi, w) d\xi ds = 0 ; \quad x \in [1, b] .$$

**Theorem 10.12.** *Assume* (10.5.1)–(10.5.4), *then integral equation* (10.5) *has at least one random solution in the space BC. Moreover, the random solutions of* (10.5) *are globally asymptotically stable.*

*Proof.* Set $d^* := \sup_{(t,x,w) \in J \times \Omega} d(t, x, w)$, where

$$d(t, x, w) = \frac{1}{\Gamma(r_1)\Gamma(r_2)} \int_1^t \int_1^x \left|\log \frac{t}{s}\right|^{r_1 - 1} \left|\log \frac{x}{\xi}\right|^{r_2 - 1} p(t, x, s, \xi, w) d\xi ds .$$

From condition (10.5.4) we infer that $d^*$ is finite. Define a mapping $N\colon \Omega \times BC \to BC$ such that

$$N(w)u(t,x) = f(t,x,u(t,x,w),w) + \frac{1}{\Gamma(r_1)\Gamma(r_2)} \int_1^t \int_1^x \left(\log\frac{t}{s}\right)^{r_1-1} \left(\log\frac{x}{\xi}\right)^{r_2-1} \times g(t,x,s,\xi,u(s,\xi,w),w)\frac{d\xi ds}{s\xi}, \quad (t,x) \in J,\ w \in \Omega . \tag{10.6}$$

The maps $f$ and $g$ are continuous for all $w \in \Omega$. Again, as the indefinite integral is continuous on $J$, $N(w)$ defines a mapping $N\colon \Omega \times BC \to BC$. Then $u$ is a solution for integral equation (10.5) if and only if $u = N(w)u$.

Next we show that the function $N(w)u \in BC$ for any $u \in BC$ and each $w \in \Omega$. By considering the conditions of this theorem, for each $(t,x) \in J$ and $w \in \Omega$ we have

$$\begin{aligned}
|(Nw)u(t,x)| &\le |f(t,x,u(t,x,w),w) - f(t,x,0,w)| + |f(t,x,0,w)| \\
&\quad + \frac{1}{\Gamma(r_1)\Gamma(r_2)} \int_1^t \int_1^x \left|\log\frac{t}{s}\right|^{r_1-1} \left|\log\frac{x}{\xi}\right|^{r_2-1} \\
&\quad \times |g(t,x,s,\xi,u(s,\xi,w),w)|\frac{d\xi ds}{s\xi} \\
&\le \frac{M|u(t,x,w)|}{(1+t)(L+|u(t,x,w)|)} + |f(t,x,0,w)| \\
&\quad + \frac{1}{\Gamma(r_1)\Gamma(r_2)} \int_1^t \int_1^x \left|\log\frac{t}{s}\right|^{r_1-1} \left|\log\frac{x}{\xi}\right|^{r_2-1} \\
&\quad \times \frac{p(t,x,s,\xi)}{1+\alpha(t)+|u(s,\xi))|+|u(\gamma(s),\xi))|}\frac{d\xi ds}{s\xi} \\
&\le M + f^* + d^* .
\end{aligned}$$

Hence, $N(w)u \in BC$, and $N(w)$ transforms the ball $B_\eta := B(0,\eta)$ into itself, where $\eta = M + f^* + d^*$. We will show that $N\colon \Omega \times B_\eta \to B_\eta$ satisfies the assumptions of Theorem 10.7. The proof will be given in several steps.

*Step 1:* $N(w)$ is a random operator on $\Omega \times B_\eta$ into $B_\eta$. Since $f(t,x,u,w)$ is random Carathéodory, the map $w \to f(t,x,u,w)$ is measurable in view of Lemma 10.2. Similarly, the product $(\log\frac{t}{s})^{r_1-1}(\log\frac{x}{\xi})^{r_2-1}g(t,x,s,\xi,u(s,\xi,w),w)$ of a continuous and a measurable function is again measurable. Further, the integral is a limit of a finite sum of measurable functions; therefore, the map

$$w \mapsto N(w)u(t,x,w)$$

is measurable. As a result, $N(w)$ is a random operator on $\Omega \times B_\eta$ into $B_\eta$.

*Step 2: $N(w)$ is continuous.* Let $\{u_n\}_{n\in\mathbb{N}}$ be a sequence such that $u_n \to u$ in $B_\eta$. Then, for each $(t,x) \in J$ and $w \in \Omega$ we have

$$\begin{aligned}|N(w)u_n(t,x) - N(w)u(t,x)| &\le |f(t,x,u_n(t,x,w),w) - f(t,x,u(t,x,w),w)|\\&+ \frac{1}{\Gamma(r_1)\Gamma(r_2)}\int_1^t\int_1^x \left|\log\frac{t}{s}\right|^{r_1-1}\left|\log\frac{x}{\xi}\right|^{r_2-1}\\&\times \sup_{(s,\xi)\in J}|g(t,x,s,\xi,u_n(s,\xi,w),w) - g(t,x,s,\xi,u(s,\xi,w),w)|\frac{d\xi ds}{s\xi}\\&\le \frac{M}{L}\|u_n - u\|_{BC}\\&+ \frac{1}{\Gamma(r_1)\Gamma(r_2)}\int_1^t\int_1^x \left|\log\frac{t}{s}\right|^{r_1-1}\left|\log\frac{x}{\xi}\right|^{r_2-1}\\&\times \|g(t,x,.,.,u_n(.,.,w),w) - g(t,x,.,.,u(.,.,w),w)\|_{BC} d\xi ds\,.\end{aligned} \tag{10.7}$$

*Case 1.* If $(t,x) \in [1,T]\times[1,b]$, $T > 1$, then, since $u_n \to u$ as $n \to \infty$ and $f, g$ are continuous, (10.7) gives

$$\|N(w)u_n - N(w)u\|_{BC} \to 0 \text{ as } n \to \infty\,.$$

*Case 2.* If $(t,x) \in (T,\infty)\times[1,b]$, $T > 1$, then from (10.5.4) and (10.7) for each $(t,x) \in J$ we have

$$\begin{aligned}|N(w)u_n(t,x) - N(w)u(t,x)| &\le \frac{M}{L}\|u_n - u\|_{BC}\\&+ \frac{2}{\Gamma(r_1)\Gamma(r_2)}\int_1^t\int_1^x \left|\log\frac{\beta(t)}{s}\right|^{r_1-1}\left|\log\frac{x}{\xi}\right|^{r_2-1}\frac{p(t,x,s,\xi)}{s\xi}d\xi ds\\&\le \frac{M}{L}\|u_n - u\|_{BC} + 2d(t,x)\,.\end{aligned}$$

Thus, we get

$$|N(w)u_n(t,x) - N(w)u(t,x)| \le \frac{M}{L}\|u_n - u\|_{BC} + 2d(t,x,w)\,. \tag{10.8}$$

Since $u_n \to u$ as $n \to \infty$ and $t \to \infty$, (10.8) gives

$$\|N(w)u_n - N(w)u\|_{BC} \to 0 \text{ as } n \to \infty\,.$$

*Step 3: $N(w)(B_\eta)$ is uniformly bounded.* This is clear since $N(w)(B_\eta) \subset B_\eta$; $w \in \Omega$ and $B_\eta$ is bounded.

*Step 4: $N(B_\eta)$ is equicontinuous on every compact subset* $[1,a]\times[1,b]$ *of* $J$, $a > 1$. Let $w \in \Omega$, $(t_1,x_1), (t_2,x_2) \in [1,a]\times[1,b]$, $t_1 < t_2$, $x_1 < x_2$, and let $u \in B_\eta$. Then we have

$$
\begin{aligned}
&|N(w)u(t_2,x_2) - N(w)u(t_1,x_1)| \\
&\le |f(t_2,x_2,u(t_2,x_2,w),w) - f(t_2,x_2,u(t_1,x_1,w),w)| \\
&\quad + |f(t_2,x_2,u(t_1,x_1,w),w) - f(t_1,x_1,u(t_1,x_1,w),w)| \\
&\quad + \frac{1}{\Gamma(r_1)\Gamma(r_2)} \int_1^{t_2}\int_1^{x_2} \left|\log\frac{t_2}{s}\right|^{r_1-1} \left|\log\frac{x_2}{\xi}\right|^{r_2-1} \\
&\qquad \times |g(t_2,x_2,s,\xi,u(s,\xi,w),w) - g(t_1,x_1,s,\xi,u(s,\xi,w),w)|\,d\xi ds \\
&\quad + \left|\frac{1}{\Gamma(r_1)\Gamma(r_2)} \int_1^{t_2}\int_1^{x_2} \left(\log\frac{t_2}{s}\right)^{r_1-1} \left(\log\frac{x_2}{\xi}\right)^{r_2-1}\right. \\
&\qquad \times g(t_1,x_1,s,\xi,u(s,\xi,w),w)\,d\xi ds \\
&\quad - \frac{1}{\Gamma(r_1)\Gamma(r_2)} \int_1^{t_1}\int_1^{x_1} \left(\log\frac{t_2}{s}\right)^{r_1-1} \left(\log\frac{x_2}{\xi}\right)^{r_2-1} \\
&\qquad \left.\times g(t_1,x_1,s,\xi,u(s,\xi,w),w)\,d\xi ds\right| \\
&\quad + \frac{1}{\Gamma(r_1)\Gamma(r_2)} \int_1^{t_1}\int_1^{x_1} \left|\left(\log\frac{t_2}{s}\right)^{r_1-1} \left(\log\frac{x_2}{\xi}\right)^{r_2-1}\right. \\
&\quad \left.-\left(\log\frac{t_1}{s}\right)^{r_1-1} \left(\log\frac{x_1}{\xi}\right)^{r_2-1}\right| |g(t_1,x_1,s,\xi,u(s,\xi,w),w)|\,d\xi ds\,.
\end{aligned}
$$

Thus, we obtain

$$
\begin{aligned}
&|N(w)u(t_2,x_2) - N(w)u(t_1,x_1)| \\
&\le \frac{M}{L}(|u(t_2,x_2,w) - u(t_1,x_1,w)| + |u(t_2,x_2,w) - u(t_1,x_1,w)|) \\
&\quad + (|t_2 - t_1| + |x_2 - x_1|)\psi_1(\|u\|_{BC}) \\
&\quad + \frac{1}{\Gamma(r_1)\Gamma(r_2)} \int_1^{t_2}\int_1^{x_2} \left|\log\frac{t_2}{s}\right|^{r_1-1} \left|\log\frac{x_2}{\xi}\right|^{r_2-1} \\
&\qquad \times \varphi(t,x,s,\xi,w)(|t_2 - t_1| + |x_2 - x_1|)\psi_2(\|u\|_{BC})\,d\xi ds \\
&\quad + \frac{1}{\Gamma(r_1)\Gamma(r_2)} \int_{t_1}^{t_2}\int_1^{x_2} \left|\log\frac{t_2}{s}\right|^{r_1-1} \left|\log\frac{x_2}{\xi}\right|^{r_2-1} \\
&\qquad \times |g(t_1,x_1,s,\xi,u(s,\xi,w),w)|\,d\xi ds
\end{aligned}
$$

$$+\frac{1}{\Gamma(r_1)\Gamma(r_2)}\int_1^{t_2}\int_{x_1}^{x_2}\left|\log\frac{t_2}{s}\right|^{r_1-1}\left|\log\frac{x_2}{\xi}\right|^{r_2-1}$$

$$\times|g(t_1,x_1,s,\xi,u(s,\xi,w),w)|d\xi ds$$

$$+\frac{1}{\Gamma(r_1)\Gamma(r_2)}\int_{t_1}^{t_2}\int_{x_1}^{x_2}\left|\log\frac{t_2}{s}\right|^{r_1-1}\left|\log\frac{x_2}{\xi}\right|^{r_2-1}$$

$$\times|g(t_1,x_1,s,\xi,u(s,\xi,w),w)|d\xi ds$$

$$+\frac{1}{\Gamma(r_1)\Gamma(r_2)}\int_1^{t_1}\int_1^{x_1}\left|\left(\log\frac{t_2}{s}\right)^{r_1-1}\left(\log\frac{x_2}{\xi}\right)^{r_2-1}\right.$$

$$\left.-\left(\log\frac{\beta(t_1)}{s}\right)^{r_1-1}\left(\log\frac{x_1}{\xi}\right)^{r_2-1}\right||g(t_1,x_1,s,\xi,u(s,\xi,w),w)|d\xi ds\,.$$

Hence, we get

$$|N(w)u(t_2,x_2)-N(w)u(t_1,x_1)|$$

$$\leq\frac{M}{L}(|u(t_2,x_2,w)-u(t_1,x_1,w)|+|u(t_2,x_2,w)-u(t_1,x_1,w)|)$$

$$+(|t_2-t_1|+|x_2-x_1|)\psi_1(\eta)$$

$$+\frac{(|t_2-t_1|+|x_2-x_1|)\psi_2(\eta)}{\Gamma(r_1)\Gamma(r_2)}$$

$$\times\int_1^{t_2}\int_1^{x_2}\left|\log\frac{t_2}{s}\right|^{r_1-1}\left|\log\frac{x_2}{\xi}\right|^{r_2-1}\varphi(s,\xi)d\xi ds$$

$$+\frac{1}{\Gamma(r_1)\Gamma(r_2)}\int_{\beta(t_1)}^{\beta(t_2)}\int_1^{x_2}\left|\log\frac{\beta(t_2)}{s}\right|^{r_1-1}\left|\log\frac{x_2}{\xi}\right|^{r_2-1}p(t_1,x_1,s,\xi)|d\xi ds$$

$$+\frac{1}{\Gamma(r_1)\Gamma(r_2)}\int_1^{t_2}\int_{x_1}^{x_2}\left|\log\frac{t_2}{s}\right|^{r_1-1}\left|\log\frac{x_2}{\xi}\right|^{r_2-1}p(t_1,x_1,s,\xi)|d\xi ds$$

$$+\frac{1}{\Gamma(r_1)\Gamma(r_2)}\int_{t_1}^{t_2}\int_{x_1}^{x_2}\left|\log\frac{t_2}{s}\right|^{r_1-1}\left|\log\frac{x_2}{\xi}\right|^{r_2-1}p(t_1,x_1,s,\xi)|d\xi ds$$

$$+\frac{1}{\Gamma(r_1)\Gamma(r_2)}\int_1^{t_1}\int_1^{x_1}\left|\left(\log\frac{t_2}{s}\right)^{r_1-1}\left(\log\frac{x_2}{\xi}\right)^{r_2-1}\right.$$

$$\left.-\left(\log\frac{t_1}{s}\right)^{r_1-1}\left(\log\frac{x_1}{\xi}\right)^{r_2-1}\right|p(t_1,x_1,s,\xi,w)|d\xi ds\,.$$

From the continuity of $\varphi, p$ and as $t_1 \to t_2$ and $x_1 \to x_2$, the right-hand side of the preceding inequality tends to zero.

*Step 5: $N(w)(B_\eta)$ is equiconvergent.* Let $(t,x)\in J$, $w\in\Omega$ and $u\in B_\eta$. Then we have

$$\begin{aligned}
|N(w)u(t,x)| &\le |f(t,x,u(t,x,w),w)-f(t,x,0,w)+f(t,x,0,w)|\\
&\quad+\left|\frac{1}{\Gamma(r_1)\Gamma(r_2)}\int_1^t\int_1^x\left(\log\frac{t}{s}\right)^{r_1-1}\left(\log\frac{x}{\xi}\right)^{r_2-1}\right.\\
&\quad\left.\times\, g(t,x,s,\xi,u(s,\xi,w),w)\frac{d\xi ds}{s\xi}\right|\\
&\le \frac{M|u(t,x,w)|}{(1+t)(L+|u(t,x,w)|)}+|f(t,x,0,w)|\\
&\quad+\frac{1}{\Gamma(r_1)\Gamma(r_2)}\int_1^t\int_1^x\left(\log\frac{t}{s}\right)^{r_1-1}\left(\log\frac{x}{\xi}\right)^{r_2-1}\\
&\quad\times\frac{p(t,x,s,\xi,w)}{1+t+|u(s,\xi,w)|}d\xi ds\\
&\le \frac{M}{1+t}+|f(t,x,0,w)|\\
&\quad+\frac{1}{\Gamma(r_1)\Gamma(r_2)(1+t)}\int_1^t\int_1^x\left(\log\frac{\beta(t)}{s}\right)^{r_1-1}\left(\log\frac{x}{\xi}\right)^{r_2-1}p(t,x,s,\xi,w)d\xi ds\\
&\le \frac{M}{1+t}+|f(t,x,0,w)|+\frac{d^*}{1+t}\,.
\end{aligned}$$

Thus, for each $x\in[1,b]$ we get

$$|N(w)u(t,x)|\to 0,\quad \text{as}\ \ t\to+\infty\,.$$

Hence,

$$|N(w)u(t,x)-N(w)u(+\infty,x)|\to 0,\quad \text{as}\ \ t\to+\infty\,.$$

As a consequence of Steps 1–5, together with Lemma 1.57, we can conclude that $N\colon\Omega\times B_\eta\to B_\eta$ is continuous and compact. From an application of Theorem 10.7 we deduce that the operator equation $N(w)u=u$ has a random solution. This further implies that random integral equation (10.5) has a random solution.

*Step 6: The uniform global attractivity.* Let us assume that $u_0$ is a solution of integral equation (7.1) with the conditions of this theorem. Consider the ball $B(u_0,\eta^*)$ with $\eta^*=\frac{LM^*}{L-M}$, where

$$\begin{aligned}
M^* := \frac{1}{\Gamma(r_1)\Gamma(r_2)}\sup_{(t,x,w)\in J\times\Omega}&\left\{\int_1^t\int_1^x\left(\log\frac{t}{s}\right)^{r_1-1}\left(\log\frac{x}{\xi}\right)^{r_2-1}\right.\\
&\times|g(t,x,s,\xi,u(s,\xi,w),w)\\
&\left.-\,g(t,x,s,\xi,u_0(s,\xi,w),w)|d\xi ds;\ u\in BC\right\}.
\end{aligned}$$

Taking $w \in \Omega$ and $u \in B(u_0, \eta^*)$, we have

$$\begin{aligned}
&|N(w)u(t,x) - u_0(t,x,w)| = |N(w)u(t,x) - N(w)u_0(t,x)| \\
&\leq |f(t,x,u(t,x,w),w) - f(t,x,u_0(t,x,w),w)| \\
&\quad + \frac{1}{\Gamma(r_1)\Gamma(r_2)} \int_1^t \int_1^x \left(\log\frac{t}{s}\right)^{r_1-1} \left(\log\frac{x}{\xi}\right)^{r_2-1} \\
&\quad \times |g(t,x,s,\xi,u(s,\xi,w),w) - g(t,x,s,\xi,u_0(s,\xi,w),w)| \frac{d\xi ds}{s\xi} \\
&\leq \frac{M}{L}\|u - u_0\|_{BC} + M^* \\
&\leq \frac{M}{L}\eta^* + M^* = \eta^* .
\end{aligned}$$

Thus, we observe that $N(w)$ is a continuous function such that $N(w)(B(u_0, \eta^*)) \subset B(u_0, \eta^*)$. Moreover, if $u$ is a solution of integral equation (10.5), then

$$\begin{aligned}
&|u(t,x,w) - u_0(t,x,w)| = |N(w)u(t,x) - N(w)u_0(t,x)| \\
&\leq |f(t,x,u(t,x,w),w) - f(t,x,u_0(t,x,w),w)| \\
&\quad + \frac{1}{\Gamma(r_1)\Gamma(r_2)} \int_1^t \int_1^x \left(\log\frac{t}{s}\right)^{r_1-1} \left(\log\frac{x}{\xi}\right)^{r_2-1} \\
&\quad \times |g(t,x,s,\xi,u(s,\xi,w),w) - g(t,x,s,\xi,u_0(s,\xi,w),w)| d\xi ds .
\end{aligned}$$

Thus,

$$\begin{aligned}
|u(t,x,w) - u_0(t,x,w)| &\leq \frac{M}{L}|u(t,x,w) - u_0(t,x,w)| \\
&\quad + \frac{1}{\Gamma(r_1)\Gamma(r_2)} \int_1^t \int_1^x \left(\log\frac{t}{s}\right)^{r_1-1} \left(\log\frac{x}{\xi}\right)^{r_2-1} p(t,x,s,\xi,w) d\xi ds . \qquad (10.9)
\end{aligned}$$

Using (10.9), we get

$$\begin{aligned}
\lim_{t\to\infty} |u(t,x,w) - u_0(t,x,w)| &\leq \lim_{t\to\infty} \frac{L}{\Gamma(r_1)\Gamma(r_2)(L-M)} \int_1^t \int_1^x \left(\log\frac{t}{s}\right)^{r_1-1} \left(\log\frac{x}{\xi}\right)^{r_2-1} \\
&\quad \times p(t,x,s,\xi,w) d\xi ds = 0 .
\end{aligned}$$

Consequently, all random solutions of integral equation (7.1) are globally asymptotically stable. □

### 10.3.3 An Example

Let $\Omega = (-\infty, 0)$ be equipped with the usual $\sigma$-algebra consisting of Lebesgue measurable subsets of $(-\infty, 0)$. Given a measurable function $u : \Omega \to AC([1,\infty) \times [1,e])$, consider the partial Hadamard random fractional integral equation

$$u(t,x,w) = \frac{tx}{10(1+t+t^2+t^3+tw^2+w^2)}(1+\sin(u(t,x,w)))$$
$$+\frac{1}{\Gamma^2(q)}\int_1^t\int_1^x \left(\log\frac{t}{s}\right)^{q-1}\left(\log\frac{x}{\xi}\right)^{q-1}\frac{\ln(1+2x(s\xi)^{-1}|u(s,\xi)|)}{(1+t+|u(s,\xi)|)^2(1+x^2+t^4)}d\xi ds\,;$$
$$(t,x)\in[1,\infty)\times[1,e],\ w\in\Omega\,, \tag{10.10}$$

where $r_1 = r_2 = q > 0$,

$$f(t,x,u,w) = \frac{tx(1+\sin(u))}{10(1+t)(1+w^2+t^2)}$$

for $(t,x) \in J$ $w \in \Omega$ and $u \in \mathbb{R}$ and

$$g(t,x,s,\xi,u,w) = \frac{\ln(1+x(s\xi)^{-1}|u|)}{(1+t+|u|)^2(1+x^2+t^4)}$$

for $(t,x,s,\xi) \in J_1$ $w \in \Omega$ and $u \in \mathbb{R}$.

We can easily check that the assumptions of Theorem 10.12 are satisfied. In fact, clearly, the maps $(t,x,w) \mapsto f(t,x,u,w)$ and $(t,x,w) \mapsto g(t,x,s,\xi,u,w)$ are jointly continuous for all $u \in \mathbb{R}$ and, thus, jointly measurable for all $u \in \mathbb{R}$. Also, the maps $u \mapsto f(t,x,u,w)$ and $u \mapsto g(t,x,s,\xi,u,w)$ are continuous for all $(t,x) \in J$ and $w \in \Omega$. Thus, the functions $f$ and $g$ are Carathéodory; then condition (10.5.1) is satisfied. The function $f$ is continuous and satisfies (10.5.2), where $M = \frac{1}{10}$, $L = 1$. Also, $f$ satisfies (10.5.3), with $f^* = \frac{e}{10}$. Next, let us note that the function $g$ satisfies (10.5.4), where $p(t,x,s,\xi) = \frac{x(s\xi)^{-1}}{1+x^2+t^4}$. Also,

$$\lim_{t\to\infty} p(t,x)\int_1^t\int_1^x\left|\log\frac{t}{s}\right|^{q-1}\left|\log\frac{x}{\xi}\right|^{q-1} p(t,x,s,\xi)d\xi ds$$
$$= \lim_{t\to\infty}\frac{x}{1+x^2+t^4}\int_1^t\int_1^x\left|\log\frac{t}{s}\right|^{q-1}\left|\log\frac{x}{\xi}\right|^{q-1}\frac{d\xi ds}{s\xi}$$
$$= \lim_{t\to\infty}\frac{9x(\log t)^q}{1+x^2+t^4} = 0\,.$$

Hence, by Theorem 10.12, integral equation (10.10) has a random solution defined on $[1,\infty)\times[1,e]$, and the random solutions of this integral equation are globally asymptotically stable.

## 10.4 Multidelay Hadamard Fractional Integral Equations in Fréchet Spaces with Random Effects

### 10.4.1 Introduction

In this section, we present some results concerning the existence and Ulam stabilities of random solutions for some functional integral equations of Hadamard fractional order and random effects in Fréchet spaces.

Recently, some interesting results on the existence and Ulam stabilities of the solutions of some classes of differential equations were obtained by Abbas et al. [5, 24, 25, 28]. This section deals with the existence and Ulam stabilities of random solutions of the problem of Hadamard fractional integral equations

$$\begin{aligned} u(t,x,w) &= \mu(t,x,w) + f(t,x,({}^{H}I_{\sigma}^{r}u)(t,x,w),u(t,x,w),w) \\ &\quad + \frac{1}{\Gamma(r_1)\Gamma(r_2)}\int_1^t\int_1^x \left(\log\frac{t}{s}\right)^{r_1-1}\left(\log\frac{x}{y}\right)^{r_2-1} \\ &\quad \times g(t,x,s,y,u(s-\tau_1,y-\xi_1,w),\dots,u(s-\tau_m,y-\xi_m,w),w)\frac{dyds}{sy}\,, \\ &\quad \text{if } (t,x)\in J:=[1,+\infty)\times[1,b],\ w\in\Omega\,, \end{aligned} \tag{10.11}$$

$$u(t,x,w)=\Phi(t,x,w)\,,\quad \text{if } (t,x)\in\tilde{J}:=[-T,\infty)\times[-\xi,b]\backslash(1,\infty)\times(1,b],\ w\in\Omega\,, \tag{10.12}$$

where $b>1$, $\sigma=(1,1)$, $r=(r_1,r_2)$, $r_1,r_2\in(0,\infty)$, ${}^{H}I_{\sigma}^{r}$ is the Hadamard integral of order $r$, $\tau_i,\xi_i\geq -1$; $i=1\dots,m$, $T=\max_{i=1\dots,m}\{\tau_i\}$, $\xi=\max_{i=1\dots,m}\{\xi_i\}$, $(\Omega,\mathcal{A})$ is a measurable space, $\mu\colon J\times\Omega\to\mathbb{R}$, $f\colon J\times\mathbb{R}\times\mathbb{R}\times\Omega\to\mathbb{R}$, $g\colon J'\times\mathbb{R}\times\Omega\to\mathbb{R}$ are given continuous functions, and $J'=\{(t,x,s,y)\colon 1\leq s\leq t, 1\leq y\leq x\leq b\}$.

Our investigations are conducted in Fréchet spaces with the application of a stochastic fixed point theorem of Goudarzi for the existence of solutions of problem(10.11)–(10.12), and we prove that all solutions are generalized Ulam–Hyers–Rassias stable.

### 10.4.2 Existence of Random Solutions and Ulam stabilities results

Let us start by defining what we mean by a random solution of problem (10.11)–(10.12).

**Definition 10.13.** A function $u\in C$ is said to be a random solution of (10.11)–(10.12) if $u$ satisfies equation (10.11) on $J$ and (10.12) in $\tilde{J}$.

Now we are concerned with the existence and uniform global attractivity of random solutions for problem (10.11)–(10.12). Set

$$J_p:=[1,p]\times[1,b]\,,\quad J_p'=\{(t,x,s,y)\colon 1\leq s\leq t\leq p, 1\leq y\leq x\leq b\};\quad p\in\mathbb{N}\backslash\{0,1\}\,.$$

The following conditions will be used in the sequel:

(10.7.1) The functions $w \mapsto \mu(t,x,w)$ and $w \mapsto \Phi(t,x,w)$ are measurable for a.e. $(t,x) \in J_p$ or $(t,x) \in \tilde{J}$, respectively, and the functions $f$ and $g$ are random Carathéeodory.

(10.7.2) There exist continuous measurable functions $l, k\colon J_p \times \Omega \to \mathbb{R}_+$ such that

$$|f(t,x,u_1,v_1,w) - f(t,x,u_2,v_2,w)| \le l(t,x,w)|u_1 - u_2| + k(t,x,w)|v_1 - v_2|$$

for each $(t,x) \in J_p$, $u_1,u_2,v_1,v_2 \in \mathbb{R}$, and $w \in \Omega$. Moreover, assume that the function $(u,v) \mapsto f(t,x,u,v,w)$ satisfies

$$f(t,x,\lambda u,\lambda v,w) = \lambda f(t,x,u,v,w)\,; \quad \text{for } \lambda \in (0,1),\ (t,x) \in J_p, \text{ and } w \in \Omega\,.$$

(10.7.3) There exist continuous measurable functions $P_i\colon J'_p \times \Omega \to \mathbb{R}_+$, $i = 1,\dots,m$, such that

$$|g(t,x,s,y,u_1,\dots,u_m,w)| \le \sum_{i=1}^{m} P_i(t,x,s,y,w)|u_i|$$

for $(t,x,s,y) \in J'_p$, $u_i \in \mathbb{R}$, and $w \in \Omega$. Moreover, assume that the function $(u_1,\dots,u_m) \mapsto g(t,x,u_1,\dots,u_m,w)$ satisfies

$$g(t,x,\lambda u_1,\dots,\lambda u_m,w) = \lambda g(t,x,u_1,\dots,u_m,w)\,; \quad \text{for } \lambda \in (0,1),\ (t,x) \in J_p, \text{ and } w \in \Omega\,.$$

(10.7.4) There exist $Q_i\colon J_p \times \Omega \to [0,\infty)$, $i = 1,\dots,m$, with $Q_i(.,w) \in L^\infty(J_p,[0,\infty))$, $i = 1,\dots,m$, such that for each $w \in \Omega$ and a.e. $(t,x) \in J_p$ we have

$$P_i(t,x,s,y,w) \le \varphi(t,x,w)Q_i(s,y,w)\,, \quad i = 1\dots,m\,.$$

For any $p \in \mathbb{N}\backslash\{0,1\}$ set

$$\Phi^* = \sup_{(t,x,w)\in\tilde{J}\times\Omega} |\Phi(t,x,w)|\,, \quad \mu_p = \sup_{(t,x,w)\in J_p\times\Omega} |\mu(t,x,w)|\,,$$

$$f_p = \sup_{(t,x,w)\in J_p\times\Omega} |f(t,x,0,0,w)|\,,$$

$$k_p = \sup_{(t,x,w)\in J_p\times\Omega} k(t,x,w)\,, \quad l_p = \sup_{(t,x,w)\in J_p\times\Omega} l(t,x,w)\,,$$

$$P_{ip} = \sup_{(t,x,w)\in J_p\times\Omega} \int_1^t \int_1^x \left|\log\frac{t}{s}\right|^{r_1-1} \left|\log\frac{x}{y}\right|^{r_2-1} \frac{P_i(t,x,s,y,w)}{\Gamma(r_1)\Gamma(r_2)}\,dyds\,, \quad P_p = \sum_{i=1}^{m} P_{ip}\,.$$

**Theorem 10.14.** *Assume* (10.7.1)–(10.7.3). *If*

$$\ell_p := P_p + k_p + \frac{l_p(\log p)^{r_1}(\log b)^{r_2}}{\Gamma(1+r_1)\Gamma(1+r_2)} < 1\,, \tag{10.13}$$

*then problem* (10.11)–(10.12) *has at least one random solution in space C. Furthermore, if condition (10.7.4) holds, then problem* (10.11)–(10.12) *is generalized Ulam–Hyers–Rassias stable.*

*Proof.* Let $N\colon \Omega \times C \to C$ be the mapping defined by

$$N(w)u(t,x) = \begin{cases} \Phi(t,x,w), & (t,x) \in \tilde{J}\,, \\ \mu(t,x,w) + f(t,x,({}^{H}I_{\sigma}^{r}u)(t,x,w), u(t,x,w), w) & \\ + \frac{1}{\Gamma(r_1)\Gamma(r_2)} \int_1^t \int_1^x \left(\log\frac{t}{s}\right)^{r_1-1} \left(\log\frac{x}{y}\right)^{r_2-1} & w \in \Omega\,, \\ \times g(t,x,s,y,u(s-\tau_1, y-\xi_1, w), \dots, & \\ u(s-\tau_m, y-\xi_m, w), w)\frac{dyds}{sy}, & (t,x) \in J\,. \end{cases} \tag{10.14}$$

The maps $\Phi, \mu, f$, and $g$ are continuous for all $w \in \Omega$. Again, as the indefinite integral is continuous on $J$, $N(w)$ defines a mapping $N\colon \Omega \times C \to C$. Then $u$ is a random solution of problem (10.11)–(10.12) if and only if $u = N(w)u$.
For each $p \in \mathbb{N}\backslash\{0,1\}$ and any $w \in \Omega$ we can show that $N(w)$ transforms the ball $B_\eta := \{u \in C\colon \|u\|_p \le \eta_p\}$ into itself, where $\eta_p := \max\{\Phi^*, \eta_p'\}$, with

$$\eta_p' \ge \frac{\mu_p + f_p}{1 - \ell_p}\,.$$

Indeed, for any $w \in \Omega$ and each $u \in C$ and $(t,x) \in \tilde{J}$ we have

$$|N(w)u(t,x)| \le |\Phi(t,x,w)| \le \Phi^*\,,$$

and for any $w \in \Omega$ and each $u \in C$ and $(t,x) \in J_p$ we have

$$\begin{aligned} |N(w)u(t,x)| &\le |\mu(t,x,w)| + |f(t,x,({}^{H}I_{\sigma}^{r}u)(t,x,w), u(t,x,w), w)| \\ &\quad + \frac{1}{\Gamma(r_1)\Gamma(r_2)} \int_1^t \int_1^x \left|\log\frac{t}{s}\right|^{r_1-1} \left|\log\frac{x}{y}\right|^{r_2-1} \\ &\quad \times |g(t,x,s,y,u(s-\tau_1,y-\xi_1,w), \dots, u(s-\tau_m, y-\xi_m, w), w)|dyds \\ &\le |\mu(t,x,w)| + |f(t,x,0,0,w)| \\ &\quad + l(t,x,w)|({}^{H}I_{\sigma}^{r}u)(t,x,w)| + k(t,x,w)|u(t,x,w)| \\ &\quad + \frac{1}{\Gamma(r_1)\Gamma(r_2)} \int_1^t \int_1^x \left|\log\frac{t}{s}\right|^{r_1-1} \left|\log\frac{x}{y}\right|^{r_2-1} \\ &\quad \times \sum_{i=1}^{m} P_i(t,x,s,y)|u(s-\tau_i, y-\xi_i)|dyds \\ &\le \mu(t,x,w)| + |f(t,x,0,0,w)| + \eta_p' l(t,x,w)|{}^{H}I_{\sigma}^{r}1| + \eta_p' k(t,x,w) \\ &\quad + \frac{\eta_p'}{\Gamma(r_1)\Gamma(r_2)} \sum_{i=1}^{m} \int_1^t \int_1^x \left|\log\frac{t}{s}\right|^{r_1-1} \left|\log\frac{x}{y}\right|^{r_2-1} P_i(t,x,s,y)dyds \end{aligned}$$

$$
\begin{aligned}
&\le \mu_p + f_p + \frac{\eta'_p l_p (\log p)^{r_1} (\log b)^{r_2}}{\Gamma(1+r_1)\Gamma(1+r_2)} + \eta'_p k_p + \eta'_p \sum_{i=1}^{m} P_{ip} \\
&\le \mu_p + f_p + \eta'_p \left( P_p + k_p + \frac{l_p (\log p)^{r_1} (\log b)^{r_2}}{\Gamma(1+r_1)\Gamma(1+r_2)} \right) \\
&= \mu_p + f_p + \eta'_p \ell_p \\
&\le \eta'_p .
\end{aligned}
$$

Thus,

$$
\|N(u)\|_p \le \eta_p .
$$

Hence, $N(w)$ transforms the ball $B_\eta$ into itself. We will show that $N\colon \Omega \times B_\eta \to B_\eta$ satisfies the assumptions of [147, Theorem 3.1]. The proof will be given in three steps.

*Step 1. $N(w)$ is a random operator on $\Omega \times B_\eta$ into $B_\eta$.* Since $f(t,x,u,v,w)$ is random Carathéodory, the map $w \to f(t,x,u,v,w)$ is measurable in view of Lemma 10.2. Similarly, the product $(\log \frac{t}{s})^{r_1-1}(\log \frac{x}{\xi})^{r_2-1} g(t,x,u_1,\dots,u_m,w)$ of a continuous and a measurable function is again measurable. Further, the integral is a limit of a finite sum of measurable functions; therefore, the map

$$
w \mapsto N(w)u(t,x,w)
$$

is measurable. As a result, $N(w)$ is a random operator on $\Omega \times B_\eta$ into $B_\eta$.

*Step 2. $N(w)$ is continuous.* Let $\{u_n\}$ be a sequence such that $u_n \to u$ in $B_\eta$. Then for each $(x,y) \in J_p$ and $w \in \Omega$ we have

$$
\begin{aligned}
&|(N(w)u_n)(x,y) - (N(w)u)(x,y)| \\
&\le |f(t,x,({}^H I^r_\sigma u)(t,x,w), u_n(t,x,w), w) - f(t,x,({}^H I^r_\sigma u)(t,x,w), u(t,x,w), w)| \\
&\quad + \frac{1}{\Gamma(r_1)\Gamma(r_2)} \int_1^t \int_1^x \left|\log \frac{t}{s}\right|^{r_1-1} \left|\log \frac{x}{y}\right|^{r_2-1} \\
&\qquad \times |g(t,x,s,y,u_n(s-\tau_1,y-\xi_1,w),\dots,u_n(s-\tau_m,y-\xi_m,w),w) \\
&\quad - g(t,x,s,y,u(s-\tau_1,y-\xi_1,w),\dots,u(s-\tau_m,y-\xi_m,w),w)| \frac{dyds}{sy} \\
&\le l(t,x,w)^H I^r_\sigma |u_n(t,x,w) - u(t,x,w)| + k(t,x,w)|u_n(t,x,w) - u(t,x,w)| \\
&\quad + \frac{1}{\Gamma(r_1)\Gamma(r_2)} \int_1^t \int_1^x \left|\log \frac{t}{s}\right|^{r_1-1} \left|\log \frac{x}{y}\right|^{r_2-1} \\
&\qquad \times |g(t,x,s,y,u_n(s-\tau_1,y-\xi_1,w),\dots,u_n(s-\tau_m,y-\xi_m,w),w) \\
&\quad - g(t,x,s,y,u(s-\tau_1,y-\xi_1,w),\dots,u(s-\tau_m,y-\xi_m,w),w)| dyds .
\end{aligned}
$$

From the continuity of $g$ and ${}^H I^r_\sigma$ and using the Lebesgue dominated convergence theorem, we get

$$
\|N(w)u_n - N(w)u\|_p \to 0 \text{ as } n \to \infty .
$$

*Step 3. $N(w)$ is affine.* For each $u, v \in B_\eta$, $(t,x) \in J_p^*$ and any $\lambda \in (0,1)$ and $w \in \Omega$ we have

$$\begin{aligned} N(w)(\lambda u + (1-\lambda)v) &= \mu(t,x,w) + \lambda f(t,x,({}^H I_\sigma^r u)(t,x,w), u(t,x,w), w) \\ &+ \frac{\lambda}{\Gamma(r_1)\Gamma(r_2)} \int_1^t \int_1^x \left(\log\frac{t}{s}\right)^{r_1-1} \left(\log\frac{x}{y}\right) r_2 - 1 \\ &\times g(t,x,s,y,u(s-\tau_1, y-\xi_1, w), \dots, u(s-\tau_m, y-\xi_m, w), w) dyds \\ &+ (1-\lambda) f(t,x,({}^H I_\sigma^r u)(t,x,w), u(t,x,w), w) \\ &+ \frac{1-\lambda}{\Gamma(r_1)\Gamma(r_2)} \int_1^t \int_1^x \left(\log\frac{t}{s}\right)^{r_1-1} \left(\log\frac{x}{y}\right) r_2 - 1 \\ &\times g(t,x,s,y,u(s-\tau_1, y-\xi_1, w), \dots, u(s-\tau_m, y-\xi_m, w), w) dyds \\ &= \lambda N(w)(u) + (1-\lambda) N(w)(v) \,. \end{aligned}$$

Hence, $N(w)$ is affine.

As a consequence of Steps 1–3, together with [147, Theorem 3.1], we deduce that $N$ has a fixed point $v$ that is a random solution of problem (10.11)–(10.12).

*Step 4. Generalized Ulam–Hyers–Rassias stability.* Set

$$Q_{ip} = \operatorname*{sup\,ess}_{(s,y,w) \in J_p \times \Omega} Q_i(s,y,w) \,, \quad Q_p = \sum_{i=1}^m Q_{ip} \,.$$

Let $u : \Omega \to B_\eta$ be a solution of the inequality

$$\|u(t,x,w) - (N(w)u)(t,x)\|_p \le \varphi(t,x,w) \,, \quad \text{for a.e. } (t,x) \in J_p^*, \ w \in \Omega \,, \tag{10.15}$$

and $v$ a random solution of problem (10.11)–(10.12). Then $\|u\|_p \le \eta$, $\|v\|_p \le \eta$, and

$$v(t,x,w) = \begin{cases} \Phi(t,x,w) \,, & (t,x) \in \tilde{J} \,, \\ \mu(t,x,w) + f(t,x,({}^H I_\sigma^r v)(t,x,w), v(t,x,w), w) & \\ + \frac{1}{\Gamma(r_1)\Gamma(r_2)} \int_1^t \int_1^x \left(\log\frac{t}{s}\right)^{r_1-1} \left(\log\frac{x}{y}\right)^{r_2-1} & w \in \Omega \,, \\ \times g(t,x,s,y,v(s-\tau_1, y-\xi_1, w), \dots, & \\ v(s-\tau_m, y-\xi_m, w), w) \frac{dyds}{sy} \,; & (t,x) \in J \,. \end{cases}$$

For each $(t,x) \in \tilde{J}$ and any $w \in \Omega$ we have

$$\begin{aligned} |u(t,x,w) - v(x,y,w)| &\le |u(t,x,w) - N(w)(u(t,x,w))| \\ &\quad + |N(w)(u(t,x,w)) - N(w)(v(t,x,w))| \\ &\le \varphi(x,y,w) \,. \end{aligned}$$

Next, from condition (10.7.4), for each $(t,x) \in J_p$ and any $w \in \Omega$ we have

$$
\begin{aligned}
&|u(t,x,w) - v(x,y,w)| \le |u(t,x,w) - N(w)(u(t,x,w))| \\
&\quad + |N(w)(u(t,x,w)) - N(w)(v(t,x,w))| \\
&\le \varphi(x,y,w) + |f(t,x,({}^H I_\sigma^r u)(t,x,w), u(t,x,w)) - f(t,x,({}^H I_\sigma^r v)(t,x,w), v(t,x,w))| \\
&\quad + \frac{1}{\Gamma(r_1)\Gamma(r_2)} \int_1^t \int_1^x \left|\log \frac{t}{s}\right|^{r_1-1} \left|\log \frac{x}{y}\right|^{r_2-1} \\
&\quad \times |g(t,x,s,y,u(s-\tau_1, y-\xi_1, w), \ldots, u(s-\tau_m, y-\xi_m, w), w) \\
&\quad - g(t,x,s,y,v(s-\tau_1, y-\xi_1, w), \ldots, v(s-\tau_m, y-\xi_m, w), w)| \frac{dyds}{sy} \\
&\le \varphi(x,y,w) + l(t,x,w)|({}^H I_\sigma^r u)(t,x,w) - ({}^H I_\sigma^r v)(t,x,w)| \\
&\quad + k(t,x,w)|u(t,x,w) - v(t,x,w)| \\
&\quad + \frac{1}{\Gamma(r_1)\Gamma(r_2)} \int_1^t \int_1^x \left|\log \frac{t}{s}\right|^{r_1-1} \left|\log \frac{x}{y}\right|^{r_2-1} \varphi(t,x,w) \\
&\quad \times \left( \sum_{i=1}^m Q_i(s,y)(|u(s-\tau_i, y-\xi_i, w)| + |v(s-\tau_i, y-\xi_i, w)|) \right) dyds \\
&\le \varphi(x,y,w) + \ell_p |u(t,x,w) - v(t,x,w)| \\
&\quad + \frac{2\eta\varphi(t,x,w)}{\Gamma(r_1)\Gamma(r_2)} \int_1^t \int_1^x \left|\log \frac{t}{s}\right|^{r_1-1} \left|\log \frac{x}{y}\right|^{r_2-1} \left( \sum_{i=1}^m Q_{ip} \right) dyds \\
&\le \varphi(t,x,w) + \ell_p |u(t,x,w) - v(t,x,w)| \\
&\quad + \frac{2\eta Q_p \varphi(t,x,w)}{\Gamma(r_1)\Gamma(r_2)} \int_1^t \int_1^x \left|\log \frac{t}{s}\right|^{r_1-1} \left|\log \frac{x}{y}\right|^{r_2-1} dyds\,.
\end{aligned}
$$

Thus, for each $(t,x) \in J_p$ and any $w \in \Omega$ we obtain

$$
\begin{aligned}
|u(t,x,w) - v(x,y,w)| &\le \frac{\varphi(t,x,w)}{1-\ell_p} \left( 1 + \frac{2\eta Q_p}{\Gamma(r_1)\Gamma(r_2)} \int_1^t \int_1^x \left|\log \frac{t}{s}\right|^{r_1-1} \left|\log \frac{x}{y}\right|^{r_2-1} dyds \right) \\
&\le \frac{1}{1-\ell_p} \left( 1 + \frac{2\eta Q_p (\log p)^{r_1} (\log b)^{r_2}}{\Gamma(1+r_1)\Gamma(1+r_2)} \right) \varphi(t,x,w) \\
&:= c'_{N,\varphi} \varphi(t,x,w)\,.
\end{aligned}
$$

Hence, for each $(t,x) \in J_p^*$ and any $w \in \Omega$ we get

$$
|u(t,x,w) - v(x,y,w)| \le c_{N,\varphi} \varphi(x,y,w)\,,
$$

where $c_{N,\varphi} := \max\{1, c'_{N,\varphi}\}$. Consequently, random problem (10.11)–(10.12) is generalized Ulam–Hyers–Rassias stable. □

### 10.4.3 An Example

Let $\Omega = (-\infty, 0)$ be equipped with the usual $\sigma$-algebra consisting of Lebesgue measurable subsets of $(-\infty, 0)$. Given a measurable function $u\colon \Omega \to C([-1,\infty) \times [-2,e])$, consider the problem of Hadamard fractional order integral equations

$$\begin{aligned} u(t,x,w) &= \frac{xe^{3-2t}}{(1+w^2)(1+t+x^2)} + \frac{xc_pe^{-t-2}}{1+w^2}(e^{2p}|({}^HI^r_\sigma u)(t,x,w)| + e^p|u(t,x,w)|) \\ &\quad + \frac{1}{\Gamma(r_1)\Gamma(r_2)}\int_1^t\int_1^x\left(\log\frac{t}{s}\right)^{r_1-1}\left(\log\frac{x}{y}\right)^{r_2-1} \\ &\quad \times g(t,x,s,y,u(s-1,y-2,w),u(s-\frac{1}{2},y-\frac{2}{5},w),w) \\ &\quad \times \frac{1}{sy}dyds\,, \quad \text{if } (t,x) \in J := [1,+\infty) \times [1,e],\ w \in \Omega\,, \end{aligned} \tag{10.16}$$

$$u(t,x,w) = \frac{2}{(1+w^2)(2+t^2)(2+x^2)}\,, \quad \text{if } (t,x) \in \tilde{J},\ w \in \Omega\,, \tag{10.17}$$

where

$$\tilde{J} := [-1,\infty) \times [-2,e] \backslash (1,\infty) \times (1,e]\,, \quad r = (r_1,r_2) \in (0,\infty) \times (0,\infty)\,,$$

$$c_p = \frac{e^{-2}}{p^{\frac{-3}{4}}e + e^{-2+p} + \frac{e^{-2+2p}p^{r_1}}{\Gamma(1+r_1)\Gamma(1+r_2)}}\,, \quad p \in \mathbb{N}\backslash\{0,1\}\,,$$

$$g(t,x,s,y,u_1,u_2,w) = \frac{xc_ps^{\frac{-3}{4}}(|u_1|+|u_2|)\sin\sqrt{t}\sin s}{(1+w^2)(1+x^2+t^2)}\,, \quad \text{if } (t,x,s,y) \in J'\,,$$

and $u_1, u_2 \in \mathbb{R}$,

and

$$J' = \{(t,x,s,y)\colon 1 \le s \le t \text{ and } 1 \le x \le y \le e\}\,.$$

Set

$$\mu(t,x,w) = \frac{xe^{3-2t}}{(1+w^2)(1+t+x^2)}\,,$$

$$f(t,x,u,v,w) = \frac{xc_pe^{-t-2}}{1+w^2}(e^{2p}|u| + e^p|v|)\,; \quad p \in \mathbb{N}\backslash\{0,1\}\,.$$

We have $\mu_p = e^2$. The function $f$ is continuous and satisfies (10.7.2), with

$$l(t,x,w) = \frac{xc_pe^{-t-2+2p}}{c}1 + w^2\,, \qquad k(t,x,w) = \frac{xc_pe^{-t-2+p}}{1+w^2}\,,$$

$$l_p = c_pe^{-2+2p}\,, \qquad l_p = c_pe^{-2+p}\,.$$

Also, the function $g$ is continuous and satisfies (10.7.3), with

$$P_1(t,x,s,y,w) = P_2(t,x,s,y,w) = \frac{xc_ps^{\frac{-3}{4}}\sin\sqrt{t}\sin s}{(1+w^2)(1+x^2+t^2)}\,; \quad (t,x,s,y) \in J'\,,$$

$$P_p = c_pp^{\frac{-3}{4}}e\,.$$

Condition (10.7.4) is satisfied by

$$Q_1(s,y,w) = Q_2(s,y,w) = \frac{c_p s^{\frac{-3}{4}} \sin s}{1+w^2}, \quad \text{and} \quad \varphi(t,x,w) = \frac{x \sin\sqrt{t}}{1+x^2+t^2}.$$

Condition (10.13) holds, with $b = e$. Indeed, for each $p \in \mathbb{N}\backslash\{0,1\}$ we get

$$P_p + k_p + \frac{l_p (\log p)^{r_1} (\log b)^{r_2}}{\Gamma(1+r_1)\Gamma(1+r_2)} = c_p \left( p^{\frac{-3}{4}} e + e^{-2+p} + \frac{e^{-2+2p} p^{r_1}}{\Gamma(1+r_1)\Gamma(1+r_2)} \right) = e^{-2} < 1\,.$$

Hence, by Theorem 10.14, problem (10.16)–(10.17) has a random solution defined on $[-1,+\infty) \times [-2,e]$ and is generalized Ulam–Hyers–Rassias stable.

## 10.5 Notes and Remarks

The results of Chapter 10 are taken from Abbas et al. [8, 7, 6, 27]. Other results may be found in [5, 7, 21, 36, 40].

# Bibliography

[1] S. Abbas, E. Alaidarous, W. Albarakati and M. Benchohra, Upper and lower solutions method for partial Hadamard fractional integral equations and inclusions, *Discus. Math. Diff. Incl., Contr. Optim.* **35** (2015), 105–122.
[2] S. Abbas, E. Alaidarous, W. Albarakati and M. Benchohra, Existence and global stability results for Volterra type fractional Hadamard–Stieltjes partial integral equations, (Submitted).
[3] S. Abbas, E. Alaidarous, M. Benchohra and J. J. Nieto, Existence and stability of solutions for Hadamard–Stieltjes fractional integral equations, *Discrete Dyn. Nature Soc.* **2015** (2015), Article ID 317094, 6 pages.
[4] S. Abbas, W. Albarakati and M. Benchohra, Existence and attractivity results for Volterra type nonlinear Multi–Delay Hadamard–Stieltjes fractional integral equations, *Pan American Math. J.* **16**(1) (2016), 1–17.
[5] S. Abbas, W. A. Albarakati, M. Benchohra, M. A. Darwish and E. M. Hilal, New existence and stability results for partial fractional differential inclusions with multiple delay, *Ann. Polon. Math.* **114** (2015), 81–100.
[6] S. Abbas, W. A. Albarakati, M. Benchohra and J. Henderson, Existence and Ulam stabilities for Hadamard fractional integral equations with random effects, *Electron. J. Diffeential Equations* **2016**(25) (2016), 1–12.
[7] S. Abbas, W. Albarakati, M. Benchohra and J. Henderson, Ulam–Hyers–Rassias stability for multi-delay fractional Hadamard integral equations in Fréchet spaces with random effects, (Submitted).
[8] S. Abbas, W. A. Albarakati, M. Benchohra and E. M. Hilal, Global existence and stability results for partial fractional random differential equations, *J. Appl. Anal.* **21**(2) (2015), 79–87.
[9] S. Abbas, W. Albarakati, M. Benchohra and G. M. N'Guérékata, Existence and Ulam stabilities for Hadamard fractional integral equations in Fréchet spaces, *J. Fract. Calc. Appl.* **7**(2) (2016), 1–12.
[10] S. Abbas, W. Albarakati, M. Benchohra and J. J. Nieto, Existence and global stability results for Volterra type fractional Hadamard partial integral equations, (Submitted).
[11] S. Abbas, W. Albarakati, M. Benchohra and A. Petrusel, Existence and Ulam stability results for Hadamard partial fractional integral inclusions via Picard operators, *Stud. Univ. Babes-Bolyai, Ser. Math.* **61**(4) (2016), 409–420.
[12] S. Abbas, W. Albarakati, M. Benchohra and S. Sivasundaram, Dynamics and stability of Fredholm type fractional order Hadamard integral equations, *J. Nonlinear Stud.* **22**(4) (2015), 673–686.
[13] S. Abbas, W. Albarakati, M. Benchohra and J. J. Trujillo, Ulam stabilities for partial Hadamard fractional integral equations, *Arab. J. Math.* **5**(1) (2016) 1–7.
[14] S. Abbas and M. Benchohra, Darboux problem for implicit impulsive partial hyperbolic differential equations, *Electron. J. Differential Equations* **2011** (2011), 15 pp.
[15] S. Abbas and M. Benchohra, A global uniqueness result for fractional order implicit differential equations, *Comment. Math. Univ. Carolin.* **53**(4) (2012), 605–614.
[16] S. Abbas and M. Benchohra, Global stability results for nonlinear partial fractional order Riemann–Liouville Volterra–Stieltjes functional integral equations, *Math. Sci. Res. J.* **16**(4) (2012), 82–92.
[17] S. Abbas and M. Benchohra, Upper and lower solutions method for Darboux problem for fractional order implicit impulsive partial hyperbolic differential equations, *Acta Univ. Palacki. Olomuc.* **51**(2) (2012), 5–18.

https://doi.org/10.1515/9783110553819-011

[18] S. Abbas and M. Benchohra, Existence and stability of nonlinear fractional order Riemann–Liouville–Volterra–Stieltjes multi-delay integral equations, *J. Integral Equations Appl.* **25**(2) (2013), 143–158.

[19] S. Abbas and M. Benchohra, Nonlinear fractional order Riemann–Liouville Volterra–Stieltjes partial integral equations on unbounded domains, *Commun. Math. Anal.* **14**(1) (2013), 104–117.

[20] S. Abbas and M. Benchohra, On the generalized Ulam–Hyers–Rassias stability for Darboux problem for partial fractional implicit differential equations, *Appl. Math. E-Notes* **14** (2014), 20–28.

[21] S. Abbas and M. Benchohra, Ulam stabilities for the Darboux problem for partial fractional differential inclusions, *Demonstratio Math.* **XLVII**(4) (2014), 826–838.

[22] S. Abbas and M. Benchohra, Ulam–Hyers stability for the Darboux problem for partial fractional differential and integro-differential equations via Picard operators, *Results. Math.* **65** (1–2) (2014), 67–79.

[23] S. Abbas and M. Benchohra, *Advanced Functional Evolution Equations and Inclusions*, Developments in Mathematics, Springer, Cham, 2015.

[24] S. Abbas and M. Benchohra, Some stability concepts for Darboux problem for partial fractional differential equations on unbounded domain, *Fixed Point Theory* **16**(1) (2015), 3–14.

[25] S. Abbas and M. Benchohra, Uniqueness and Ulam stabilities results for partial fractional differential equations with not instantaneous impulses, *Applied Math. Computation* **257** (2015), 190–198.

[26] S. Abbas and M. Benchohra, Global convergence of successive approximations for the Darboux problem for implicit partial differential equations (Submitted).

[27] S. Abbas and M. Benchohra, Global stability results for Volterra–Hadamard random partial fractional integral equations (Submitted).

[28] S. Abbas, M. Benchohra and M. A. Darwish. New stability results for partial fractional differential inclusions with not instantaneous impulses, *Frac. Calc. Appl. Anal.* **18**(1) (2015), 172–191.

[29] S. Abbas, M. Benchohra and A. Hammoudi, Upper, lower solutions method and extremal solutions for impulsive discontinuous partial fractional differential inclusions, *Pan American Math. J.* **24**(1) (2014), 31–52.

[30] S. Abbas, M. Benchohra and J. Henderson, Asymptotic behavior of solutions of nonlinear fractional order Riemann–Liouville Volterra–Stieltjes quadratic integral equations, *Int. E. J. Pure Appl. Math.* **4**(3) (2012), 195–209.

[31] S. Abbas, M. Benchohra and J. Henderson, Ulam stability for partial fractional integral inclusions via Picard operators, *J. Frac. Calc. Appl.* **5**(2) (2014), 133–144.

[32] S. Abbas, M. Benchohra and J. Henderson, Partial Hadamard fractional integral equations, *Adv. Dynamical Syst. Appl.* **10**(2) (2015), 97–107.

[33] S. Abbas, M. Benchohra and J. J. Nieto, Global attractivity of solutions for nonlinear fractional order Riemann–Liouville Volterra–Stieltjes partial integral equations, *Electron. J. Qual. Theory Differ. Equ.* **81** (2012), 1–15.

[34] S. Abbas, M. Benchohra and J. J. Nieto, Functional implicit hyperbolic fractional order differential equations with delay, *Afr. Diaspora J. Math.* **15**(1) (2013), 74–96.

[35] S. Abbas, M. Benchohra and G. M. N'Guérékata, *Topics in Fractional Differential Equations*, Developments in Mathematics, **27**, Springer, New York, 2012.

[36] S. Abbas, M. Benchohra and G. M. N'Guérékata, *Advanced Fractional Differential and Integral Equations*, Nova Science Publishers, New York, 2015.

[37] S. Abbas, M. Benchohra and A. Petruşel, Ulam stabilities for the Darboux problem for partial fractional differential inclusions via Picard operators, *Electron. J. Qual. Theory Differ. Equ.* **2014**(51), 13 pages.
[38] S. Abbas, M. Benchohra, M. Rivero and J. J. Trujillo, Existence and stability results for nonlinear fractional order Riemann–Liouville Volterra–Stieltjes quadratic integral equations, *Appl. Math. Comput.* **247** (2014), 319–328.
[39] S. Abbas, M. Benchohra and S. Sivasundaram, Ulam stability for partial fractional differential inclusions with multiple delay and impulses via Picard operators, *J. Nonlinear Stud.* **20**(4) (2013), 623–641.
[40] S. Abbas, M. Benchohra and B. A. Slimani, Existence and Ulam stabilities for partial fractional random differential inclusions with nonconvex right hand side, *Pan American Math. J.* **25**(1) (2015), 95–110.
[41] S. Abbas, M. Benchohra and B. A. Slimani, Partial hyperbolic implicit differential equations with variable times impulses, *Stud. Univ. Babeş-Bolyai Math.* **60**(1) (2015), 61–73.
[42] S. Abbas, M. Benchohra and J. J. Trujillo, Upper and lower solutions method for partial fractional differential inclusions with not instantaneous impulses, *Prog. Frac. Diff. Appl.* **1**(1) (2015), 11–22.
[43] S. Abbas, M. Benchohra and A. N. Vityuk, On fractional order derivatives and Darboux problem for implicit differential equations, *Frac. Calc. Appl. Anal.* **15**(2) (2012), 168–182.
[44] S. A. Abd-Salam and A. M. A. El-Sayed, On the stability of a fractional-order differential equation with nonlocal initial condition, *Electron. J. Qual. Theory Differ. Equat.* **29** (2008), 1–8.
[45] R. P. Agarwal and B. Ahmad, Existence theory for anti-periodic boundary value problems of fractional differential equations and inclusions, *Comput. Math. Appl.* **62** (2011), 1200–1214.
[46] R. P. Agarwal, S. Arshad, D. O'Regan and V. Lupulescu, Fuzzy fractional integral equations under compactness type condition. *Fract. Calc. Appl. Anal.* **15** (2012), 572–590.
[47] R. P Agarwal, M. Benchohra and S. Hamani, Boundary value problems for fractional differential equations, *Adv. Stud. Contemp. Math.* **16**(2) (2008), 181–196.
[48] R. P. Agarwal, M. Benchohra and B. A. Slimani, Existence results for differential equations with fractional order impulses, *Mem. Differential Equations Math. Phys.* **44**(1) (2008), 1–21.
[49] R. P. Agarwal, M. Meehan and D. O'Regan, *Fixed Point Theory and Applications*, Cambridge University Press, Cambridge, 2001.
[50] B. Ahmad and J. R. Graef, Coupled systems of nonlinear fractional differential equations with nonlocal boundary conditions, *Panamer. Math. J.* **19** (2009), 29–39.
[51] B. Ahmad and J. J. Nieto, Existence of solutions for nonlocal boundary value problems of higher-order nonlinear fractional differential equations, *Abstrac. Appl. Anal.* **2009** (2009), Article ID 494720, 9 pages.
[52] B. Ahmad and J. J. Nieto, Anti-periodic fractional boundary value problems, *Comput. Math. Appl.* **62**(3) (2011), 1150–1156.
[53] B. Ahmad and J. J. Nieto, Existence of solutions for impulsive anti-periodic boundary value problems of fractional order, *Taiwaness J. Math.* **15**(3) (2011), 981–993.
[54] B. Ahmad and J. J. Nieto, Anti-periodic fractional boundary value problems with nonlinear term depending on lower order derivative, *Fract. Calc. Appl. Anal.* **15** (2012), 451–462.
[55] B. Ahmad, J. J. Nieto, A. Alsaedi and N. Mohamad, On a new class of anti-periodic fractional boundary value problems, *Abst. Appl. Anal.* (2013), 7 p.
[56] B. Ahmad and S. Sivasundaram, Theory of fractional differential equations with three-point boundary conditions, *Commun. Appl. Anal.* **12** (2008), 479–484.

[57] B. Ahmad and S. Sivasundaram, Existence results for nonlinear impulsive hybrid boundary value problems involving fractional differential equations. *Nonlinear Anal. Hybrid Syst.* **3**(3) (2009), 251–258.
[58] K. K. Akhmerov, M. I. Kamenskii, A. S. Potapov, A. E. Rodkina and B. N. Sadovskii, *Measures of noncompactness and condensing operators*, Birkhäuser Verlag, Basel, Boston, Berlin, 1992.
[59] R. Almeida, S. Pooseh and D. F. M. Torres, *Computational Methods in the Fractional Calculus of Variations*, World Scientific Publishing, USA, 2015.
[60] S. Almezel, Q. H. Ansari and M. A. Khamsi, *Topics in Fixed Point Theory*, Springer-Verlag, New York, 2014.
[61] C. Alsina and R. Ger, On some inequalities and stability results related to the exponential function. *J. Inequal. Appl.* **2** (1998), 373–380.
[62] J. C. Alvàrez, Measure of noncompactness and fixed points of nonexpansive condensing mappings in locally convex spaces, *Rev. Real. Acad. Cienc. Exact. Fis. Natur. Madrid* **79** (1985), 53–66.
[63] G. A. Anastassiou, *Advances on Fractional Inequalities*, Springer, New York Dordrecht Heidelberg London, 2011.
[64] J. Andres and L. Górniewicz, *Topological Fixed Point Principles for Boundary Value Problems*, Kluwer Academic Publishers, Dordrecht (2003).
[65] T. Aoki, On the stability of the linear transformation in Banach spaces, *J. Math. Soc. Japan* **2** (1950), 64–66.
[66] J. Appell, Implicit functions, nonlinear integral equations, and the measure of noncompactness of the superposition Operator. *J. Math. Anal. Appl.* **83** (1981), 251–263.
[67] J. Appell, J. Banaś and N. Merentes, *Bounded Variation and Around*, De Gruyter Studies in Nonlinear Analysis and Applications, **17**, Walter de Gruyter GmbH, Berlin/Boston, Germany, 2014.
[68] J.-P. Aubin and A. Cellina, *Differential Inclusions*, Springer-Verlag, Berlin, Heidelberg, New York, 1984.
[69] J.-P. Aubin and H. Frankowska, *Set-Valued Analysis*, Birkhauser, Basel, 1990.
[70] C. Avramescu, Some remarks on a fixed point theorem of Krasnoselskii, *Electr. J. of Qualitative Theory Differential Eqs.* **5** (2003), 1–15.
[71] J. M. Ayerbe Toledano, T. Dominguez Benavides and G. Lopez Acedo, *Measures of Noncompactness in Metric Fixed Point Theory*, Birkhauser, Basel, 1997.
[72] A. Babakhani and V. Daftardar-Gejji, Existence of positive solutions for $N$-term non-autonomous fractional differential equations, *Positivity* **9**(2) (2005), 193–206.
[73] A. Babakhani and V. Daftardar-Gejji, Existence of positive solutions for multi-term non-autonomous fractional differential equations with polynomial coefficients, *Electron. J. Differential Equations* **2006**(129), 12 pages.
[74] Z. Bai, On positive solutions of a nonlocal fractional boundary value problem, *Nonlinear Anal.* **72**(2) (2010), 916–924, 2010.
[75] D. D. Bainov and S. G. Hristova, Integral inequalities of Gronwall type for piecewise continuous functions, *J. Appl. Math. Stoc. Anal.* **10** (1997), 89–94.
[76] D. D. Bainov and P. S. Simeonov, *Systems with Impulsive Effect*, Horwood, Chichester, 1989.
[77] D. D. Bainov and P. S. Simeonov, *Impulsive Differential Equations: Periodic Solutions and Applications*, Pitman Monographs and Surveys in Pure and Applied Mathematics, **66**, Longman Scientific & Technical and John Wiley & Sons, Inc., New York, 1993.
[78] D. Baleanu, K. Diethelm, E. Scalas, and J. J. Trujillo, *Fractional Calculus Models and Numerical Methods*, World Scientific Publishing, New York, 2012.
[79] D. Baleanu, Z. B. Güvenç and J. A. T. Machado, *New Trends in Nanotechnology and Fractional Calculus Applications*, Springer, New York, 2010.

[80] D. Baleanu, J. A. T. Machado and A. C.-J. Luo *Fractional Dynamics and Control*, Springer, 2012.

[81] J. Banaś and K. Goebel, *Measures of Noncompactness in Banach Spaces*, Marcel Dekker, New York, 1980.

[82] J. Banaś and M. Mursaleen *Sequence Spaces and Measures of Noncompactness with Applications to Differential and Integral Equations*, Springer New Delhi Heidelberg New York, Dordrecht London, 2014.

[83] J. Banaś and L. Olszowy, Measures of noncompactness related to monotonicity, *Comment. Math. (Prace Mat.)* **41** (2001), 13–23.

[84] J. Banaś and T. Zając, A new approach to the theory of functional integral equations of fractional order. *J. Math. Anal. Appl.* **375** (2011), 375–387.

[85] M. Belmekki and M. Benchohra, Existence results for fractional order semilinear functional differential equations, *Proc. A. Razmadze Math. Inst.* **146** (2008), 9–20.

[86] M. Benchohra and F. Berhoun, Impulsive fractional differential equations with state dependent delay, *Commun. Appl. Anal.* **14**(2) (2010), 213–224.

[87] M. Benchohra and F. Berhoun, Impulsive fractional differential equations with variable times, *Comput. Math. Appl.* **59** (2010), 1245–1252.

[88] M. Benchohra, F. Berhoun, N. Hamidi and J. J. Nieto, Fractional differential inclusions with anti-periodic boundary conditions, *Nonlinear Anal. Forum* **19** (2014), 27–35.

[89] M. Benchohra, F. Berhoun and G. M. N'Guérékata, Bounded solutions for fractional order differential equations on the half-line, *Bull. Math. Anal. Appl.* **146**(4) (2012), 62–71.

[90] M. Benchohra and S. Bouriah, Existence and stability results for nonlinear implicit fractional differential equations with impulses, *Mem. Differ. Equ. Math. Phys.* **69** (2016), 15–31.

[91] M. Benchohra, S. Bouriah and M. Darwish, Nonlinear boundary value problem for implicit differential equations of fractional order in Banach spaces, *Fixed Point Theory*, **18** (2) (2017), 457–470.

[92] M. Benchohra, S. Bouriah and J. Henderson, Existence and stability results for nonlinear implicit neutral fractional differential equations with finite delay and impulses, *Comm. Appl. Nonlinear Anal.* **22** (2015), 46–67.

[93] M. Benchohra, A. Cabada and D. Seba, An existence result for non-linear fractional differential equations on Banach spaces, *Boundary Value Problems.* **2009** (2009), Article ID 628916, 11 pages.

[94] M. Benchohra, J. R. Graef and S. Hamani, Existence results for boundary value problems with nonlinear fractional differential equations, *Appl. Anal.* **87**(7) (2008), 851–863.

[95] M. Benchohra and S. Hamani, Boundary value problems for differential inclusions with fractional order, *Discus. Mathem. Diff. Incl., Contr. and Optim.* **28** (2008), 147–164.

[96] M. Benchohra, S. Hamani, J. J. Nieto and B. A. Slimani, Existence of solutions to differential inclusions with fractional order and impulses, *Electron. J. Differential Equations* **2010** (2010), No. 80, 1–18.

[97] M. Benchohra, S. Hamani and S. K. Ntouyas, Boundary value problems for differential equations with fractional order, *Surv. Math. Appl.* **3** (2008), 1–12.

[98] M. Benchohra, N. Hamidi and J. Henderson, Fractional differential equations with anti-periodic boundary conditions *Numer. Funct. Anal. Optim.* **34**(4) (2013), 404–414.

[99] M. Benchohra and B. Hedia, Positive solutions for boundary value problems with fractional order, *Int. J. Adv. Math. Sciences* **1**(1), (2013), 12–22.

[100] M. Benchohra, J. Henderson and S. K. Ntouyas, *Impulsive Differential Equations and Inclusions*, Hindawi Publishing Corporation, Vol. 2, New York, 2006.

[101] M. Benchohra, J. Henderson, S. K. Ntouyas and A. Ouahab, Existence results for fractional order functional differential equations with infinite delay, *J. Math. Anal. Appl.* **338**(2) (2008), 1340–1350.

[102] M. Benchohra and J. E. Lazreg, Nonlinear fractional implicit differential equations. *Commun. Appl. Anal.* **17** (2013), 471–482.

[103] M. Benchohra and J. E. Lazreg, Existence and uniqueness results for nonlinear implicit fractional differential equations with boundary conditions. *Romanian J. Math. Comput. Sc.* **4**(1) (2014), 60–72.

[104] M. Benchohra and J. E. Lazreg, Existence results for nonlinear implicit fractional differential equations. *Surv. Math. Appl.* **9** (2014), 79–92.

[105] M. Benchohra and D. Seba, Impulsive fractional differential equations in Banach spaces *Electron. J. Qual. Theory Differ. Equ.*. Spec. Edn. I (2009), No. 8, pp. 1–14.

[106] M. Benchohra and B. A. Slimani, Existence and uniqueness of solutions to impulsive fractional differential equations, *Electron. J. Differential Equations* **2009**(10) (2009), 1–11.

[107] M. Benchohra and M. S. Souid, Integrable solutions for implicit fractional order differential equations. *Transylvanian J. Math. Mechanics* **6** (2014), No. 2, 101–107.

[108] M. Benchohra and M. S. Souid, Integrable solutions for implicit fractional order functional differential equations with infinite delay, *Arch. Math. (Brno) Tomus* **51** (2015), 67–76.

[109] M. Benchohra and M. S. Souid, $L^1$-Solutions for implicit fractional order differential equations with nonlocal condition, *Filomat* **30** (6) (2016), 1485–1492.

[110] A. Bica, V. A. Caus and S. Muresan, Application of a trapezoid inequality to neutral Fredholm integro-differential equations in Banach spaces, *J. Inequal. Pure Appl. Math.* **7** (2006), Art. 173.

[111] H. F. Bohnenblust and S. Karlin, *On a theorem of ville. Contribution to the theory of games*. Annals of Mathematics Studies, no. 24. Princeton University Press, Princeton. N. G. 1950, 155–160.

[112] M. F. Bota-Boriceanu and A. Petrusel, Ulam–Hyers stability for operatorial equations and inclusions, *Analele Univ. I. Cuza Iasi* **57** (2011), 65–74.

[113] D. Bothe, Multivalued perturbation of m-accretive differential inclusions, *Isr. J. Math.* **108** (1998), 109–138.

[114] T. A. Burton and T. Furumochi, A note on stability by Schauder's theorem, *Funkcial. Ekvac.* **44** (2001), 73–82.

[115] T. A. Burton and C. Kirk, A fixed point theorem of Krasnoselskii–Schaefer type. *Math. Nachr.* **189** (1989), 23–31.

[116] L. Byszewski, Theorems about existence and uniqueness of solutions of a semilinear evolution nonlocal Cauchy problem, *J. Math. Anal. Appl.* **162** (1991), 494–505.

[117] L. Byszewski, Existence and uniqueness of mild and classical solutions of semilinear functional-differential evolution nonlocal Cauchy problem. *Selected problems of mathematics*, 25–33, 50th Anniv. Cracow Univ. Technol. Anniv. Issue, 6, Cracow Univ. Technol., Krakow, 1995

[118] L. Byszewski and V. Lakshmikantham, Theorem about the existence and uniqueness of a solution of a nonlocal abstract Cauchy problem in a Banach space, *Appl. Anal.* **40** (1991), 11–19.

[119] P. L. Butzer, A. A. Kilbas and J. J. Trujillo, Fractional calculus in the Mellin setting and Hadamard-type fractional integrals. *J. Math. Anal. Appl.* **269** (2002), 1–27.

[120] P. L. Butzer, A. A. Kilbas and J. J. Trujillo, Mellin transform analysis and integration by parts for Hadamard-type fractional integrals. *J. Math. Anal. Appl.* **270** (2002), 1–15.

[121] C. Castaing and M. Valadier, *Convex Analysis and Measurable Multifunctions*, Lecture Notes in Mathematics **580**, Springer-Verlag, Berlin-Heidelberg-New York, 1977.

[122] C. Cattani, H. M. Srivastava and X. J. Yang, *Fractional Dynamics*, de Gruyter, 2016.

[123] A. Cernea, Arcwise connectedness of the solution set of a nonclosed nonconvex integral inclusion, *Miskolc Math. Notes* **9**(1) (2008), 33–39.

[124] Y. Chang, A. Anguraj and P. Karthikeyan, Existence results for initial value problems with integral condition for impulsive fractional differential equations. *J. Fract. Calc. Appl.* **2**(7) (2012), 1–10.
[125] A. Chen and Y. Chen, Existence of Solutions to Anti-Periodic boundary value problem for nonlinear Fractional Differential Equations with Impulses. *Adv. Difference. Equat.* (2011), 17 pages.
[126] L. Chen and Z. Fan, On mild solutions to fractional differential equations with nonlocal conditions, *Electron. J. Qual. Theory Differ. Equ.* **53** (2011), 1–13.
[127] J. Chen, F. Liu, I. Turner and V. Anh, The fundamental and numerical solutions of the Riesz space-fractional reaction-dispersion equation, *The ANZIAM Journal* **50**(1) (2008), 45–57
[128] Y. J. Cho, Th. M. Rassias and R. Saadati, *Stability of Functional Equations in Random Normed Spaces*, Science-Business Media, Springer 52, 2013.
[129] C. Corduneanu, *Integral Equations and Applications*, Cambridge University Press, 1991.
[130] H. Covitz and S. B. Nadler Jr, Multivalued contraction mappings in generalized metric spaces, *Israel J. Math.* **8** (1970), 5–11.
[131] R. F. Curtain and A. J. Pritchard, *Functional Analysis in Modern Applied Mathematics* Academic Press, 1977.
[132] G. Darbo, Punti uniti in transformazioni a condominio non compatto, *Rend Sem. Mat. Univ. Padova* **24** (1955), 84–92.
[133] K. Deimling, *Nonlinear Functional Analysis*, Springer-Verlag, 1985.
[134] K. Deimling, *Multivalued Differential Equations*, Walter De Gruyter, Berlin, New York, 1992.
[135] K. Deng, Exponential decay of solutions of semilinear parabolic equations with nonlocal initial conditions, *J. Math. Anal. Appl.* **179** (1993), 630–637.
[136] B. C. Dhage, S. V. Badgire and S. K. Ntouyas, Periodic boundary value problems of second order random differential equations, *Electron. J. Qual. Theory Diff. Equ.* **21** (2009), 1–14.
[137] K. Diethelm, *The Analysis of Fractional Differential Equations*, Lecture Notes in Mathematics, 2010.
[138] A. El-Sayed and F. Gaafar, Stability of a nonlinear non-autonomous fractional order systems with different delays and non-local conditions. *Adv. Difference Equations* **47**(1) (2011), 12 pp.
[139] N. Engheta, Fractional curl operator in electromagnetics. *Microwave Opt. Tech. Lett.* **17**(1) (1998), 86–91.
[140] H. W. Engl, A general stochastic fixed-point theorem for continuous random operators on stochastic domains, *J. Math. Anal. Appl.* **66** (1978), 220–231.
[141] M. Faryad and Q. A. Naqvi, Fractional rectangular waveguide. *Progress In Electromagnetics Research* PIER **75** (2007), 383–396.
[142] P. Gavruta, A generalization of the Hyers–Ulam–Rassias stability of approximately additive mappings, *J. Math. Anal. Appl.* **184** (1994), 431–436.
[143] K. Goebel, *Concise Course on Fixed Point Theorems*, Yokohama Publishers, Japan, 2002.
[144] C. Goodrich and A. C. Peterson, *Discrete Fractional Calculus*, Springer International Publishing, Springer, New York, USA, 2016.
[145] L. Górniewicz, *Topological Fixed Point Theory of Multivalued Mappings*, Mathematics and Its Applications, 495, Kluwer Academic Publishers, Dordrecht, 1999.
[146] L. Górniewicz and T. Pruszko, On the set of solutions of the Darboux problem for some hyperbolic equations, *Bull. Acad. Polon. Sci. Math. Astronom. Phys.* **38** (1980), 279–285.
[147] H. R. Goudarzi, Random fixed point theorems in Fréchet spaces with their applications, *J. Math. Ext.* **8** (2) (2014), 71–81.
[148] J. R. Graef, J. Henderson and A. Ouahab, *Impulsive Differential Inclusions. A Fixed Point Approach*, de Gruyter, Berlin/Boston, 2013.
[149] A. Granas and J. Dugundji, *Fixed Point Theory*, Springer-Verlag, New York, 2003.

[150] A. Guezane-Lakoud and R. Khaldi, Solvability of a fractional boundary value problem with fractional integral condition, *Nonlinear Anal.* **75** (2012), 2692–2700.
[151] D. J. Guo, V. Lakshmikantham and X. Liu, *Nonlinear Integral Equations in Abstract Spaces*, Kluwer Academic Publishers, Dordrecht, 1996.
[152] Z. Guo and M. Liu, Existence and uniqueness of solutions for fractional order integrodifferential equations with nonlocal initial conditions. *Panamer. Math. J.* **21** (2011), 51–61.
[153] J. Hadamard, Essai sur l'étude des fonctions données par leur développment de Taylor, *J. Pure Appl. Math.* **4**(8) (1892), 101–186.
[154] J. Hale, J. Kato, Phase space for retarded equations with infinite delay, *Funkcial. Ekvac.* **21** (1978), 11–41.
[155] J. Hale and S. M. Verduyn Lunel, *Introduction to Functional Differential Equations*, Applied Mathematicals Sciences, 99, Springer-Verlag, New York, 1993.
[156] X. Han, X. Ma, G. Dai, Solutions to fourth-order random differential equations with periodic boundary conditions, *Electron. J. Differential Equations* **235** (2012) 1–9.
[157] J. Henderson and A. Ouahab, Impulsive differential inclusions with fractional order, *Comput. Math. Appl.* **59** (2010), 1191–1226.
[158] J. Henderson and A. Ouahab, A Filippov's theorem, some existence results and the compactness of solution sets of impulsive fractional order differential inclusions, *Mediterranean J. Math.* **9**(3), (2012) August 2012, 453–485.
[159] R. Hermann, *Fractional Calculus: An Introduction for Physicists*, World Scientific Publishing Co. Pte. Ltd. 2011.
[160] N. Heymans and I. Podlubny, Physical interpretation of initial conditions for fractional differential equations with Riemann–Liouville fractional derivatives, *Rheol. Acta* **45** (2006), 765–771.
[161] R. Hilfer, *Applications of Fractional Calculus in Physics*, World Scientific, Singapore, 2000.
[162] Y. Hino, S. Murakami and T. Naito, *Functional Differential Equations with Infinite Delay*, Springer-Verlag, Berlin, 1991.
[163] Ch. Horvath, Measure of non-compactness and multivalued mappings in complete metric topological spaces, *J. Math. Anal. Appl.* **108** (1985), 403–408.
[164] M. Hu and L. Wang, Existence of solutions for a nonlinear fractional differential equation with integral boundary condition, *Int. J. Math. Comp. Sc.* **7**(1) (2011).
[165] Sh. Hu and N. Papageorgiou, *Handbook of Multivalued Analysis*, Volume I: Theory, Kluwer Academic Publishers, Dordrecht, 1997.
[166] D. H. Hyers, On the stability of the linear functional equation, *Proc. Natl. Acad. Sci. U.S.A.* **27** (1941), 222–224.
[167] D. H. Hyers, G. Isac and Th. M. Rassias, *Stability of Functional Equations in Several Variables*, Birkhauser, 1998.
[168] R. W. Ibrahim, Stability for univalent solutions of complex fractional differential equations, *Proc. Pakistan Acad. Sci.* **49**(3) (2012), 227–232.
[169] S. Itoh, Random fixed point theorems with applications to random differential equations in Banach spaces, *J. Math. Anal. Appl.* **67** (1979), 261–273.
[170] K. W. Jun and H. M. Kim, On the stability of an n-dimensional quadratic and additive functional equation, *Math. Inequal. Appl.* **19**(9) (2006), 854–858.
[171] S.-M. Jung, On the Hyers–Ulam stability of the functional equations that have the quadratic property, *J. Math. Anal. Appl.* **222** (1998), 126–137.
[172] S.-M. Jung, *Hyers–Ulam–Rassias Stability of Functional Equations in Mathematical Analysis*, Hadronic Press, Palm Harbor, 2001.
[173] S.-M. Jung, Hyers–Ulam stability of linear differential equations of first order, II *Appl. Math. Lett.* **19** (2006), 854–858.

[174] S.-M. Jung, A fixed point approach to the stability of a Volterra integral equation, *Fixed Point Theory Appl.* **2007** (2007), Article ID 57064, 9 pages.
[175] S.-M. Jung, *Hyers–Ulam–Rassias Stability of Functional Equations in Nonlinear Analysis*, Springer, New York, 2011.
[176] S.-M. Jung and K. S. Lee, Hyers–Ulam stability of first order linear partial differential equations with constant coefficients *Math. Inequal. Appl.* **10** (2007), 261–266.
[177] K. Karthikeyan and J. J. Trujillo, Existence and uniqueness results for fractional integrodifferential equations with boundary value conditions, *Commun. Nonlinear Sci. Numer. Simulat.* **17** (2012) 4037–4043.
[178] A. A. Kilbas, Hadamard-type fractional calculus. *J. Korean Math. Soc.* **38**(6) (2001), 1191–1204.
[179] A. A. Kilbas, B. Bonilla and J. Trujillo, Nonlinear differential equations of fractional order in a space of integrable functions. *Dokl. Ross. Akad. Nauk* **374**(4), 445–449 (2000)
[180] A. A. Kilbas and S. A. Marzan, Nonlinear differential equations with the Caputo fractional derivative in the space of continuously differentiable functions, *Diff. Equat.* **41** (2005), 84–89.
[181] A. A. Kilbas, H. M. Srivastava and J. J. Trujillo, *Theory and Applications of Fractional Differential Equations*, North-Holland Mathematics Studies 204, Elsevier Science B.V., Amsterdam, 2006.
[182] G. H. Kim, On the stability of functional equations with square-symmetric operation, *Math. Inequal. Appl* **17**(4) (2001), 257–266.
[183] W. A. Kirk and B. Sims, *Handbook of Metric Fixed Point Theory*, Springer-Science + Business Media, B.V, Dordrecht, 2001.
[184] M. Kisielewicz, *Differential Inclusions and Optimal Control*, Kluwer, Dordrecht, The Netherlands, 1991.
[185] K. Kuratowski, Sur les espaces complets, *Fund. Math.* **15** (1930), 301–309.
[186] V. Lakshmikantham, D. D. Bainov and P. S. Simeonov; *Theory of Impulsive Differential Equations*, Worlds Scientific, Singapore, 1989.
[187] V. Lakshmikantham, S. Leela and J. Vasundhara, *Theory of Fractional Dynamic Systems*, Cambridge Academic Publishers, Cambridge, 2009.
[188] V. Lakshmikantham and J. Vasundhara Devi. Theory of fractional differential equations in a Banach space. *Eur. J. Pure Appl. Math.* **1** (2008), 38–45.
[189] V. Lakshmikantham and A. S. Vatsala, Theory of fractional differential inequalities and applications, *Commun. Appl. Anal.* **11**(3–4) (2007), 395–402.
[190] V. Lakshmikantham and A. S. Vatsala, Basic theory of fractional differential equations, *Nonlinear Anal.* **69** (2008), 2677–2682.
[191] V. Lakshmikantham and A. S. Vatsala, General uniqueness and monotone iterative technique for fractional differential equations *Appl. Math. Lett.* **21** (2008), 828–834.
[192] A. Lasota and Z. Opial, An application of the Kakutani–Ky Fan theorem in the theory of ordinary differential equations, *Bull. Acad. Pol. Sci. Ser. Sci. Math. Astronom. Phys.* **13** (1965), 781–786.
[193] V. L. Lazăr, Fixed point theory for multivalued $\varphi$-contractions, *Fixed Point Theory Appl.* **2011** 2011:50, 12 pages.
[194] C. Li, F. Zeng and F. Liu, Spectral approximations to the fractional integral and derivative, *Fract. Calc. Appl. Anal.* **15**(3) (2012), 383–406.
[195] L. Liu, F. Guo, C. Wu, and Y. Wu, Existence theorems of global solutions for nonlinear Volterra type integral equations in Banach spaces, *J. Math. Anal. Appl.* **309** (2005), 638–649.
[196] Q. Liu, F. Liu, I. Turner and V. Anh, Numerical simulation for the 3D seep age flow with fractional derivatives in porous media, *IMA Journ. of Appl. Math.* **74**(2) (2009), 201–229.
[197] A. J. Luo and V. Afraimovich, *Long-range Interactions, Stochasticity and Fractional Dynamics*, Springer, New York, Dordrecht, Heidelberg, London, 2010.

[198] F. Mainardi, *Fractional Calculus and Waves in Linear Viscoelasticity* An Introduction to Mathematical Models, World Scientific Publishing, USA, 2010.
[199] M. Martelli, A Rothe's type theorem for noncompact acyclic-valued map, *Boll. Un. Math. Ital.* **11** (1975), 70–76.
[200] K. S. Miller and B. Ross, *An Introduction to the Fractional Calculus and Differential Equations*, John Wiley, New York, 1993.
[201] V. D. Milman and A. A. Myshkis, On the stability of motion in the presence of impulses, *Sib. Math. J.* (in Russian), **1** (1960), 233–237.
[202] H. Mönch, Boundary value problems for nonlinear ordinary differential equations of second order in Banach spaces, *Nonlinear Anal.* **4** (1980), 985–999.
[203] S. A. Murad and S. Hadid, An existence and uniqueness theorem for fractional differential equation with integral boundary condition, *J. Frac. Calc. Appl.* **3**(6), (2012), 1–9.
[204] S. B. Nadler Jr., Multivalued contraction mappings, *Pacific J. Math.* **30** (1969), 475–488.
[205] Q. A. Naqvi and M. Abbas, Complex and higher order fractional curl operator in electromagnetics, *Opt. Commun.* **241** (2004), 349–355.
[206] G. M. N'Guérékata, A Cauchy problem for some fractional abstract differential equation with non local conditions, *Nonlinear Anal.* **70**(5) (2009), 1873–1876.
[207] G. M. N'Guérékata, Corrigendum: A Cauchy problem for some fractional differential equations, *Commun. Math. Anal.* **7** (2009), 11–11.
[208] M. Obloza, Hyers stability of the linear differential equation, *Rocznik Nauk-Dydakt. Prace Mat.* **13** (1993), 259–270.
[209] K. B. Oldham and J. Spanier, *The Fractional Calculus: Theory and Application of Differentiation and Integration to Arbitrary Order*, Academic Press, New York London, 1974.
[210] M. D. Ortigueira, *Fractional Calculus for Scientists and Engineers*, Lecture Notes in Electrical Engineering 84, Springer, Dordrecht, 2011.
[211] B. G. Pachpatte, On nonlinear integral and discrete inequalities in two independent variables, *Bul. Sti. Tech. Inst. Politehn. Timisoara* **40**(54) (1995), 29–38.
[212] B. G. Pachpatte, On Volterra–Fredholm integral equation in two variables, *Demonstratio Math.* **XL**(4) (2007), 839–852.
[213] B. G. Pachpatte, On Fredholm type integrodifferential equation, *Tamkang J. Math.* **39**(1) (2008), 85–94.
[214] B. G. Pachpatte, On Fredholm type integral equation in two variables *Differ. Equ. Appl.* **1** (2009), 27–39.
[215] N. A. Perestyuk, V. A. Plotnikov, A. M. Samoilenko and N. V. Skripnik, *Differential Equation with Impulse Effects, Multivalued Right-hand Sides with Discontinuities* Walter de Gruyter, Berlin/Boston, 2011.
[216] I. Petras, *Fractional-Order Nonlinear Systems: Modeling, Analysis and Simulation* Springer, Heidelberg Dordrecht London New York, 2011.
[217] T. P. Petru, A. Petrusel and J.-C. Yao, Ulam–Hyers stability for operatorial equations and inclusions via nonself operators, *Taiwanese J. Math.* **15** (2011), 2169–2193.
[218] A. Petruşel, Multivalued weakly Picard operators and applications, *Sci. Math. Japon.* **59** (2004), 167–202.
[219] I. Podlubny, *Fractional Differential Equations*, Academic Press, San Diego, 1999.
[220] S. Pooseh, R. Almeida, and D. Torres. Expansion formulas in terms of integer-order derivatives for the Hadamard fractional integral and derivative. *Numer. Funct. Anal. Optim.* **33**(3) (2012), 301–319.
[221] Y. Povstenko, *Fractional Thermoelasticity*, Solid Mechanics and Its Applications, **219**, Springer, New York, 2015.

[222] Y. Povstenko, *Linear Fractional Diffusion-Wave Equation for Scientists and Engineers*, Birkhäuser Mathematics, Springer, New York, 2015.

[223] J. M. Rassias, *Functional Equations, Difference Inequalities and Ulam Stability Notions (F.U.N.)*, Nova Science Publishers, Inc. New York, 2010.

[224] T. M. Rassias, On the stability of the linear mapping in Banach spaces, *Proc. Amer. Math. Soc.* **72** (1978), 297–300.

[225] T. M. Rassias and J. Brzdek, *Functional Equations in Mathematical Analysis*, Springer 86, New York Dordrecht Heidelberg London 2012.

[226] S. S. Ray, A new approach for the application of Adomian decomposition method for the solution of fractional space diffusion equation with insulated ends, *Applied Mathematics and Computation* **202**(2) (2008), 544–549.

[227] B. Ross, *Fractional Calculus and Its Applications*, Proceedings of the International Conference, New Haven, Springer-Verlag, New York, 1974.

[228] I. A. Rus, *Generalized Contractions and Applications*, Cluj University Press, Cluj-Napoca, 2001.

[229] I. A. Rus, Weakly Picard operators and applications, *Semin. Fixed Point Theory, Cluj-Napoca* **2** (2001), 41–57.

[230] I. A. Rus, Picard operators and applications, *Sci. Math. Jpn.* **58** (2003), 191–219.

[231] I. A. Rus, Fixed points, upper and lower fixed points: abstract Gronwall lemmas, *Carpathian J. Math.* **20** (2004), 125–134.

[232] I. A. Rus, Remarks on Ulam stability of the operatorial equations, *Fixed Point Theory* **10** (2009), 305–320.

[233] I. A. Rus, Ulam stability of ordinary differential equations, *Studia Univ. Babes-Bolyai, Math.* **LIV**(4)(2009), 125–133.

[234] I. A. Rus, Ulam stabilities of ordinary differential equations in a Banach space, *Carpathian J. Math.* **26** (2010), 103–107.

[235] I. A. Rus, A. Petruşel and A. Sîtămărian, Data dependence of the fixed points set of some multivalued weakly Picard operators, *Nonlinear Anal.* **52** (2003), 1947–1959.

[236] L. Rybinski, On Carathédory type selections, *Fund. Math.* **125** (1985), 187–193.

[237] J. Sabatier, P. Lanusse, P. Melchior and A. Oustaloup, *Fractional Order Differentiation and Robust Control Design, CRONE, H-infinity and Motion Control*, Intelligent Systems, Control and Automation: Science and Engineering, **77**, Springer, New York, 2015.

[238] P. Sahoo, T. Barman and J. P. Davim, *Fractal Analysis in Machining*, Springer, New York, Dordrecht, Heidelberg, London, 2011.

[239] S. G. Samko, A. A. Kilbas and O. I. Marichev, *Fractional Integrals and Derivatives. Theory and Applications*, Gordon and Breach, Yverdon, 1993.

[240] A. M. Samoilenko and N. A. Perestyuk, *Impulsive Differential Equations* World Scientific, Singapore, 1995.

[241] X. Su and L. Liu, Existence of solution for boundary value problem of nonlinear fractional differential equation, *Appl. Math.* **22**(3) (2007) 291–298.

[242] V. E. Tarasov, *Fractional Dynamics: Application of Fractional Calculus to Dynamics of Particles, Fields and Media*, Springer, Heidelberg; Higher Education Press, Beijing, 2010.

[243] V. Uchaikin and R. Sibatov, *Fractional Kinetics in Solids: Anomalous Charge Transport in Semiconductors, Dielectrics and Nanosystems*, World Scientific, 2013.

[244] S. M. Ulam, *Problems in Modern Mathematics*, Chapter 6, John Wiley and Sons, New York, USA, 1940.

[245] S. M. Ulam, *A Collection of Mathematical Problems*, Interscience Publishers, New York, 1968.

[246] S. Umarov, *Introduction to Fractional and Pseudo-Differential Equations with Singular Symbols*, Developments in Mathematics, **41**, Springer, New York, 2015.

[247] A. N. Vityuk and A. V. Golushkov, Existence of solutions of systems of partial differential equations of fractional order, *Nonlinear Oscil.* **7**(3) (2004), 318–325.

[248] F. Wang and Z. Liu, Anti-periodic fractional boundary value problems for nonlinear differential equations of fractional order, *Adv. in Difference Equat.* (2012), 12 pages.

[249] G. Wang, B. Ahmad and L. Zhang, Impulsive anti-periodic boundary value problem for nonlinear differential equations of fractional order. *Nonlinear Anal.* **74** (2011), 792–804.

[250] G. Wang, B. Ahmad and L. Zhang, Some existence results for impulsive nonlinear fractional differential equations with mixed boundary conditions, *Comput. Math. Appl.* **62** (2011), 1389–1397.

[251] G. Wang, B. Ahmad, L. Zhang and J. J Nieto, Comments on the concept of existence of solution for impulsive fractional differential equations, *Electron. Commun. Nonlinear Sci. Numer. Simulat.* **19** (2014), 401–403.

[252] J. Wang, M. Fečkan and Y. Zhou, Ulam's type stability of impulsive ordinary differential equations, *J. Math. Anal. Appl.* **395** (2012), 258–264.

[253] J. Wang, L. Lv and Y. Zhou, Ulam stability and data dependence for fractional differential equations with Caputo derivative, *Electron. J. Qual. Theory Differ. Equat.* **63** (2011), 1–10.

[254] J. Wang and Y. Zhang, Existence and stability of solutions to nonlinear impulsive differential equations in $\beta$-normed spaces, *Electron. J. Differential Equations* **83** (2014), 1–10.

[255] R. Węgrzyk, Fixed point theorems for multifunctions and their applications to functional equations, *Dissertationes Math. (Rozprawy Mat.)* **201** (1982), 28 pp.

[256] H. Ye, J. Gao and Y. Ding, A generalized Gronwall inequality and its application to a fractional differential equation, *J. Math. Anal. Appl.* **328** (2007), 1075–1081.

[257] K. Yosida, *Functional Analysis*, 6th edn. Springer-Verlag, Berlin, 1980.

[258] S. Zhai, X. Feng and Z. Weng, New high-order compact ADI algorithms for 3D nonlinear time-fractional convection-diffusion equation, *Math. Prob. Eng* **2013** (2013), 11 pages.

[259] L. Zhang and G. Wang, Existence of solutions for nonlinear fractional differential equations with impulses and anti-periodic boundary conditions, *Electron. J. Qual. Theory Differ. Equ.* **7** (2011), 1–11.

[260] S. Zhang, Positive solutions for boundary-value problems of nonlinear fractional differential equations, *Electron. J. Differential Equations* **36** (2006), 1–12.

[261] S. Zhang and J. Sun, Existence of mild solutions for the impulsive semilinear nonlocal problem with random effects, *Advances in Difference Equations* **19** (2014), 1–11.

[262] X. Zhang, J. Liu, L. Wei and C. Ma, Finite element method for Grünwald–Letnikov time-fractional partial differential equation, *Appl. Anal.* **92**(10) (2013), 2103–2114.

[263] Y. Zhou, *Basic Theory of Fractional Differential Equations*, World Scientific, Singapore, 2014.

[264] Y. Zhou, *Fractional Evolution Equations and Inclusions: Analysis and Control*, Elsevier Science, 2016.

[265] D. P. Zielinski and V. R. Voller, A random walk solution for fractional diffusion equations, *Inter. J. Numerical Meth. Heat Fluid Flow* **23** (2013), 7–22.

# Index



https://doi.org/10.1515/9783110553819-012

## De Gruyter Series in Nonlinear Analysis and Applications

**Volume 25**
Luboš Pick, Alois Kufner, Oldřich John, Svatopluk Fucík
Function Spaces. Volume 2, 2018
ISBN 978-3-11-027373-1, e-ISBN (PDF) 978-3-11-032747-2,
e-ISBN (EPUB) 978-3-11-038221-1, Set-ISBN 978-3-11-032748-9

**Volume 24**
Alexander A. Kovalevsky, Igor I. Skrypnik, Andrey E. Shishkov
Singular Solutions of Nonlinear Elliptic and Parabolic Equations, 2016
ISBN 978-3-11-031548-6, e-ISBN (PDF) 978-3-11-033224-7,
e-ISBN (EPUB) 978-3-11-039008-7, Set-ISBN 978-3-11-033225-4

**Volume 23/2**
Sergey G. Glebov, Oleg M. Kiselev, Nikolai N. Tarkhanov
Nonlinear Equations with Small Parameter.
Volume 2: Partial Differential Equations, 2018
ISBN 978-3-11-053383-5, e-ISBN (PDF) 978-3-11-053497-9,
e-ISBN (EPUB) 978-3-11-053390-3, Set-ISBN 978-3-11-053498-6

**Volume 23/1**
Sergey G. Glebov, Oleg M. Kiselev, Nikolai N. Tarkhanov
Nonlinear Equations with Small Parameter.
Volume 1: Ordinary Differential Equations, 2017
ISBN 978-3-11-033554-5, e-ISBN (PDF) 978-3-11-033568-2,
e-ISBN (EPUB) 978-3-11-038272-3, Set-ISBN 978-3-11-033569-9

**Volume 22**
Miroslav Bácak
Convex Analysis and Optimization in Hadamard Spaces, 2014
ISBN 978-3-11-036103-2, e-ISBN (PDF) 978-3-11-036162-9,
e-ISBN (EPUB) 978-3-11-039108-4, Set-ISBN 978-3-11-036163-6

**Volume 21**
Moshe Marcus, Laurent Véron
Nonlinear Second Order Elliptic Equations Involving Measures, 2013
ISBN 978-3-11-030515-9, e-ISBN (PDF) 978-3-11-030531-9, Set-ISBN 978-3-11-030532-6